NANO TECHNOLOGY FOR BATTERY RECYCLING, REMANUFACTURING, AND REUSING

NANO TECHNOLOGY FOR BATTERY RECYCLING, REMANUFACTURING, AND REUSING

Edited by

SIAMAK FARHAD
Associate Professor
Department of Mechanical Engineering
Interim Director of Center for Precision Manufacturing
The University of Akron
Akron, OH, United States

RAM K. GUPTA
Associate Professor
Department of Chemistry
Kansas Polymer Research Center
Pittsburg State University
Pittsburg, KS, United States

GHULAM YASIN
Senior Research Fellow
Institute for Advanced Study
College of Physics and Optoelectronic Engineering
Shenzhen University
Shenzhen, Guangdong, China

TUAN ANH NGUYEN
Principal Research Scientist
Institute for Tropical Technology
Vietnam Academy of Science and Technology
Hanoi, Vietnam

ELSEVIER

Elsevier
Radarweg 29, PO Box 211, 1000 AE Amsterdam, Netherlands
The Boulevard, Langford Lane, Kidlington, Oxford OX5 1GB, United Kingdom
50 Hampshire Street, 5th Floor, Cambridge, MA 02139, United States

Copyright © 2022 Elsevier Inc. All rights reserved.

No part of this publication may be reproduced or transmitted in any form or by any means, electronic or mechanical, including photocopying, recording, or any information storage and retrieval system, without permission in writing from the publisher. Details on how to seek permission, further information about the Publisher's permissions policies and our arrangements with organizations such as the Copyright Clearance Center and the Copyright Licensing Agency, can be found at our website: www.elsevier.com/permissions.

This book and the individual contributions contained in it are protected under copyright by the Publisher (other than as may be noted herein).

Notices

Knowledge and best practice in this field are constantly changing. As new research and experience broaden our understanding, changes in research methods, professional practices, or medical treatment may become necessary.

Practitioners and researchers must always rely on their own experience and knowledge in evaluating and using any information, methods, compounds, or experiments described herein. In using such information or methods they should be mindful of their own safety and the safety of others, including parties for whom they have a professional responsibility.

To the fullest extent of the law, neither the Publisher nor the authors, contributors, or editors, assume any liability for any injury and/or damage to persons or property as a matter of products liability, negligence or otherwise, or from any use or operation of any methods, products, instructions, or ideas contained in the material herein.

ISBN: 978-0-323-91134-4

For information on all Elsevier publications visit our website at
https://www.elsevier.com/books-and-journals

Publisher: Matthew Deans
Acquisitions Editor: Edward Payne
Editorial Project Manager: Mariana L. Kuhl
Production Project Manager: Punithavathy Govindaradjane
Cover Designer: Miles Hitchen

Typeset by TNQ Technologies

Contents

Contributors   xi
Preface   xv

SECTION 1   General and technical aspects

1. Market batteries and their characteristics   3
Shunli Wang, Yanxin Xie and Josep M. Guerrero

1. Chemical battery   3
2. Physical battery   26
3. Current status of battery technology at home and abroad   28
References   28

2. Available technologies for remanufacturing, repurposing, and recycling lithium-ion batteries: an introduction   33
Ashwani Pandey, Sarthak Patnaik and Soobhankar Pati

1. Introduction   33
2. State-of-health estimation   35
3. Remanufacturing   36
4. Repurposing   39
5. Recycling   40
6. Extraction of materials   45
7. Challenges and future trends   46
8. Conclusion   48
References   49

3. Nanotechnology and recycling, remanufacturing, and reusing battery   53
Giovani Pavoski, Amilton Barbosa Botelho Junior, Rebeca Mello Chaves, Thuany Maraschin, Leandro Rodrigues Oviedo, Thamiris Auxiliadora Gonçalves Martins, William Leonardo da Silva, Daniel Assumpção Bertuol and Denise Crocce Romano Espinosa

1. Introduction   53
2. Battery nanostructures   54
3. Processes for recovering and synthesizing nanomaterials   55
4. Nanotechnology for metal recovery from batteries   64
5. Synthesis of metallic nanoparticles and other materials   69
6. Future prospects   72

Acknowledgments	72
References	72

4. Promising technologies under development for recycling, remanufacturing, and reusing batteries: an introduction — 79
Amilton Barbosa Botelho Junior, Giovani Pavoski, Mauricio Dalla Costa Rodrigues da Silva, William Leonardo da Silva, Daniel Assumpção Bertuol and Denise Crocce Romano Espinosa

1. Introduction	79
2. Ionic liquids	82
3. Supercritical fluids	85
4. Nanohydrometallurgy	87
5. Organic acids	89
6. Deep-eutectic solvents	93
7. Biohydrometallurgy	94
8. Conclusion	96
Acknowledgments	96
References	97

5. Innovative strategies for recycling used batteries for brighter future — 105
Jonghyun Choi, Felipe M. de Souza and Ram K. Gupta

1. Introduction	105
2. Need for recycling batteries	106
3. Strategies for recycling battery components	108
4. Conclusion	118
References	119

6. Battery recycling for sustainable future: recent progress, challenges, and perspectives — 123
Felipe M. de Souza and Ram K. Gupta

1. Introduction	123
2. Battery market	125
3. Battery recycling	127
4. Challenges and perspectives	135
5. Conclusion	137
References	137

SECTION 2 Battery recycling and separation processes

7. Bio-inspired nanotechnology for easy-to-recycle lithium-ion batteries — 141
Congrui Jin and Jianlin Li

1. Introduction	141
2. Combating interfacial delamination	143

3.	Bio-inspired directional adhesives	146
4.	Interface for easy-to-recycle batteries	149
5.	Underlying mechanisms	153
6.	Concluding remarks and future perspectives	155
	References	155

8. Materials recycling using pH/thermal-responsive materials — 159
Muhammad Fahad Arain, Arsalan Ahmed and Muhammad Qamar Khan

1.	Introduction	159
2.	Recyclable pH-responsive nanomaterials	160
3.	Recyclable thermo-responsive nanomaterials	165
4.	Future trends	167
5.	Conclusion	167
	References	167

9. Pyrometallurgy-based applications in spent lithium-ion battery recycling — 171
Jiafeng Zhang

1.	Introduction	171
2.	Application of pyrometallurgy in pretreatment process	172
3.	Pyrometallurgy technology	173
4.	Summary and prospects	180
	References	182

10. Application of hydrometallurgy in spent lithium-ion battery recycling — 183
Jiafeng Zhang

1.	Introduction	183
2.	Leaching technology	192
3.	Separation and extraction technologies	207
	Further reading	215

11. Biohydrometallurgical recycling approaches for returning valuable metals to the battery production cycle — 217
Tannaz Naseri, Vahid Beigi, Ashkan Namdar, Arnavaz Keikavousi Behbahan and Seyyed Mohammad Mousavi

1.	Introduction	217
2.	Approaches for recycling spent lithium-ion batteries	219
3.	Types of biohydrometallurgical methods	219
4.	Microbes in bioleaching	222
5.	Bioleaching mechanism	224

    6. A brief summary of spent battery bioleaching   230
    7. Reuse of valuable metals in battery production   233
    8. Conclusion and perspectives   239
    Acknowledgments   240
    References   240

12. Technologies for separating nanomaterials from spent lithium-ion batteries   247

Jiafeng Zhang

    1. Introduction   247
    2. Pretreatment process   252
    3. Separation of nanomaterials from the collector fluid   254
    4. Separation of nanomaterials   258
    References   261

13. Separating battery nano/microelectrode active materials with the physical method   263

Hammad Al-Shammari and Siamak Farhad

    1. Introduction   263
    2. Materials   265
    3. Modeling   266
    4. Results and discussion   274
    5. Conclusions   282
    Nomenclature   283
    Abbreviations   283
    References   283

SECTION 3   Recycling battery materials

14. Recycling battery anode materials   289

Xing Ou and Long Ye

    1. Anode materials in lithium-ion batteries   289
    2. The imperative of anode materials recycling   289
    3. Graphite materials in lithium-ion batteries   291
    4. Recycling methods for anode materials   292
    5. Defects in industry recycling   299
    6. Prospects and challenges   300
    References   301

15. Recycling battery cathode materials — 303
Xing Ou and Wei Wang

1. Introduction — 303
2. Approaches for recycling battery cathode materials — 304
3. Challenges and perspectives — 316
References — 317

16. Recycling battery metallic materials — 321
Ziwei Zhao, Gurleen Kaur Walia, Ge Li and Tian Tang

1. Commercial batteries and recyclability of metallic materials — 321
2. Technologies for metal recycling — 330
3. Conclusion and future perspectives — 342
References — 343

17. Recycling battery casing materials — 349
Tony Lyon, Malena T.L. Staudacher, Thomas Mütze and Urs A. Peuker

1. Introduction — 349
2. Target — 349
3. Structure of cell housings — 350
4. Housing fraction and case disassembly — 351
5. Generating the housing fraction by mechanical processing — 352
6. Processing the housing fraction — 357
7. Process flowchart — 365
8. Summary — 368
References — 369

SECTION 4 Battery remanufacturing and reusing

18. Regeneration technologies for electrode nanomaterials of recycled batteries — 373
Xing Ou and Haiqiang Gong

1. Introduction — 373
2. Carbonaceous nanomaterials recycling — 376
3. Metal nanomaterials recycling — 380
4. Conclusions — 388
References — 388

19. LIB industry waste valorization for battery production — 391
Basudev Swain, Jae-chun Lee and Chan-Gi Lee

1. Introduction — 391
2. Waste characterization — 393
3. Cobalt extraction by acid leaching — 393
4. Selective separation of cobalt by solvent extraction from $LiCoO_2$ leach liquor — 401
5. Valorization through oxide synthesis — 416
6. Process development — 422
7. Conclusions — 424
References — 425

20. Recycling lithium, cobalt, and nickel for return to the battery production cycle — 427
Yue Yang, Shuya Lei and Rui Xu

1. Direct repair — 427
2. Extraction of metal followed by material regeneration — 429
References — 443

21. Effects of imperfect separation of recycled cathode active materials on remanufactured lithium-ion battery performance — 445
Hammad Al-Shammari and Siamak Farhad

1. Introduction — 445
2. Brief summary of the mathematical model — 446
3. Experiments — 447
4. Results and discussion — 447
5. Conclusions — 450
References — 450

22. Mechanical and physical processes of battery recycling — 455
Denis Manuel Werner, Thomas Mütze, Alexandra Kaas and Urs A. Peuker

1. Introduction — 455
2. Lithium-ion batteries — 456
3. Physical processing of end-of-life lithium-ion batteries — 457
4. Process groups — 471
5. Conclusion — 479
References — 480

Index — *487*

Contributors

Arsalan Ahmed
Department of Textile and Clothing, Faculty of Engineering and Technology, National Textile University Karachi Campus, Karachi, Sindh, Pakistan

Hammad Al-Shammari
Department of Mechanical Engineering, Jouf University, Sakaka, Saudi Arabia

Muhammad Fahad Arain
Department of Textile and Clothing, Faculty of Engineering and Technology, National Textile University Karachi Campus, Karachi, Sindh, Pakistan

Vahid Beigi
Biotechnology Group, Chemical Engineering Department, Tarbiat Modares University, Tehran, Iran

Daniel Assumpção Bertuol
Department of Chemical Engineering, Federal University of Santa Maria, Santa Maria, Brazil

Amilton Barbosa Botelho Junior
Department of Chemical Engineering, University of São Paulo, São Paulo, Brazil

Rebeca Mello Chaves
Department of Chemical Engineering, Federal University of Santa Maria, Santa Maria, Brazil

Jonghyun Choi
Department of Chemistry, Kansas Polymer Research Center, Pittsburg State University, Pittsburg, KS, United States

William Leonardo da Silva
Nanoscience Graduate Program, Franciscan University, Santa Maria, Brazil

Felipe M. de Souza
Department of Chemistry, Kansas Polymer Research Center, Pittsburg State University, Pittsburg, KS, United States

Denise Crocce Romano Espinosa
Department of Chemical Engineering, University of São Paulo, São Paulo, Brazil

Siamak Farhad
Advanced Energy & Manufacturing Laboratory, Mechanical Engineering Department, University of Akron, Akron, OH, United States

Haiqiang Gong
National Engineering Laboratory for High Efficiency Recovery of Refractory Nonferrous Metals, School of Metallurgy and Environment, Central South University, Changsha, Hunan, China

Josep M. Guerrero
Center for Research on Microgrids - CROM, Department of Energy Technology, Aalborg, Denmark

Ram K. Gupta
Department of Chemistry, Kansas Polymer Research Center, Pittsburg State University, Pittsburg, KS, United States

Congrui Jin
Department of Civil and Environmental Engineering, University of Nebraska—Lincoln, Lincoln, NE, United States

Alexandra Kaas
Institute of Mechanical Process Engineering and Mineral Processing, TU Bergakademie Freiberg, Freiberg, Germany

Arnavaz Keikavousi Behbahan
Department of Chemistry, Sharif University of Technology, Tehran, Iran

Muhammad Qamar Khan
Department of Textile and Clothing, Faculty of Engineering and Technology, National Textile University Karachi Campus, Karachi, Sindh, Pakistan

Jae-chun Lee
Mineral Resources Research Division, Korea Institute of Geoscience and Mineral Resources (KIGAM), Daejeon, Republic of Korea

Chan-Gi Lee
Advanced Materials & Processing Center, Institute for Advanced Engineering (IAE), Yongin, Republic of Korea

Shuya Lei
School of Minerals Processing and Bioengineering, Central South University, Changsha, Hunan, China

Jianlin Li
Electrification and Energy Infrastructures Division, Oak Ridge National Laboratory, Oak Ridge, TN, United States

Ge Li
Department of Mechanical Engineering, University of Alberta, Edmonton, AB, Canada

Tony Lyon
Institute of Mechanical Process Engineering and Mineral Processing, TU Bergakademie Freiberg, Freiberg, Germany

Thuany Maraschin
School of Technology, Pontifical Catholic University of Rio Grande do Sul, Porto Alegre, Brazil

Thamiris Auxiliadora Gonçalves Martins
Department of Chemical Engineering, University of São Paulo, São Paulo, Brazil

Seyyed Mohammad Mousavi
Biotechnology Group, Chemical Engineering Department, Tarbiat Modares University, Tehran, Iran; Modares Environmental Research Institute, Tarbiat Modares University, Tehran, Iran

Thomas Mütze
Helmholtz Institute Freiberg for Resource Technology, Helmholtz-Zentrum Dresden-Rossendorf, Freiberg, Germany

Ashkan Namdar
Faculty of Materials Science and Engineering, Khajeh Nasir Toosi University of Technology, Tehran, Iran

Tannaz Naseri
Biotechnology Group, Chemical Engineering Department, Tarbiat Modares University, Tehran, Iran

Xing Ou
National Engineering Laboratory for High Efficiency Recovery of Refractory Nonferrous Metals, School of Metallurgy and Environment, Central South University, Changsha, Hunan, China

Leandro Rodrigues Oviedo
Nanoscience Graduate Program, Franciscan University, Santa Maria, Brazil

Ashwani Pandey
School of Minerals, Metallurgical and Materials Science, IIT, Bhubaneswar, Odisha, India

Soobhankar Pati
School of Minerals, Metallurgical and Materials Science, IIT, Bhubaneswar, Odisha, India

Sarthak Patnaik
School of Minerals, Metallurgical and Materials Science, IIT, Bhubaneswar, Odisha, India

Giovani Pavoski
Department of Chemical Engineering, University of São Paulo, São Paulo, Brazil

Urs A. Peuker
Institute of Mechanical Process Engineering and Mineral Processing, TU Bergakademie Freiberg, Freiberg, Germany

Mauricio Dalla Costa Rodrigues da Silva
Department of Chemical Engineering, Federal University of Santa Maria, Santa Maria, Brazil

Malena T.L. Staudacher
Institute of Mechanical Process Engineering and Mineral Processing, TU Bergakademie Freiberg, Freiberg, Germany

Basudev Swain
Advanced Materials & Processing Center, Institute for Advanced Engineering (IAE), Yongin, Republic of Korea

Tian Tang
Department of Mechanical Engineering, University of Alberta, Edmonton, AB, Canada

Gurleen Kaur Walia
School of Electronics and Electrical Engineering, Lovely Professional University, Phagwara, Punjab, India

Shunli Wang
Southwest University of Science and Technology, Mianyang, Sichuan, China; Aalborg University, Aalborg, Denmark

Wei Wang
National Engineering Laboratory for High Efficiency Recovery of Refractory Nonferrous Metals, School of Metallurgy and Environment, Central South University, Changsha, Hunan, China

Denis Manuel Werner
LIBREC AG, Oensingen, Switzerland

Yanxin Xie
Southwest University of Science and Technology, Mianyang, Sichuan, China

Rui Xu
School of Minerals Processing and Bioengineering, Central South University, Changsha, Hunan, China

Yue Yang
School of Minerals Processing and Bioengineering, Central South University, Changsha, Hunan, China; Key Laboratory of Hunan Province for Clean and Efficient Utilization of Strategic Calcium-containing Mineral Resources, Central South University, Changsha, Hunan, China

Long Ye
National Engineering Laboratory for High Efficiency Recovery of Refractory Nonferrous Metals, School of Metallurgy and Environment, Central South University, Changsha, Hunan, China

Jiafeng Zhang
National Engineering Laboratory for High Efficiency Recovery of Refractory Nonferrous Metals, School of Metallurgy and Environment, Central South University, Changsha, Hunan, China

Ziwei Zhao
Department of Mechanical Engineering, University of Alberta, Edmonton, AB, Canada

Preface

The increased use of batteries in areas ranging from consumer electronics to automobiles, health care, and grid storage systems is expected to drive greater market demand. With the increased demand for batteries, the need for metals such as lithium, cobalt, and nickel is rising. About 50 thousand tons of lithium were used in 2018, and the amount is expected to increase every year. Industry analysts predict that by 2030, the worldwide volume of spent lithium-ion batteries will exceed two million metric tons per year. With the limited resource availability of these metals, it is crucial to recycle, remanufacture, and reuse batteries. In addition to the potential economic benefits, recycling batteries would substantially reduce the volumes of these materials going into landfills and being extracted from mines. Chemicals found in batteries can easily leak from the casing once the battery is in a landfill, contaminating the area and groundwater and creating a serious threat to the environment, ecosystems, and human health. Many recycling technologies have been developed in various research laboratories around the globe, and this work continues. This book aims to shed light on battery recycling, remanufacturing, and reuse technologies and highlights the role that nanotechnology can play to make these technologies economically and environmentally feasible. This book also explores how nanotechnology can enhance and improve battery recycling and remanufacturing technologies. This book has four sections. Section 1 provides a general and technical view to readers in the fields of battery recycling, remanufacturing, and reuse and discusses the role of nanotechnology in these areas. Section 2 focuses on the details of some battery recycling and separation processes. Section 3 emphasizes the recycling of various battery components and nanomaterials. Section 4 discusses battery remanufacturing and reuse. Overall, this book presents the current knowledge generated from recent research and work in battery recycling, remanufacturing, and reuse. Further, it underlines the role of nanotechnology to help researchers and engineers better understand the current border of knowledge in the field and provides a perspective on its prospects. This book is essential for academics and industry professionals working in batteries, energy materials, recycling, and nanotechnology.

Dr. Siamak Farhad
January 2022

SECTION 1

General and technical aspects

CHAPTER 1

Market batteries and their characteristics

Shunli Wang[1,2], Yanxin Xie[1] and Josep M. Guerrero[3]
[1]Southwest University of Science and Technology, Mianyang, Sichuan, China; [2]Aalborg University, Aalborg, Denmark; [3]Center for Research on Microgrids - CROM, Department of Energy Technology, Aalborg, Denmark

1. Chemical battery

A chemical battery refers to a device that can convert chemical energy into electrical energy. The main part includes the electrolyte solution and the positive and negative electrodes immersed in it. Chemical batteries can be divided into primary batteries, secondary batteries, and fuel cells.

1.1 Primary battery

1.1.1 Zinc-manganese battery

The zinc-manganese battery is a primary battery with manganese dioxide as the positive electrode, zinc as the negative electrode, and ammonium chloride aqueous solution as the main electrolyte. Zinc-manganese batteries are commonly known as dry batteries and as Le Clancy batteries in academia. They have the characteristics of heavy load, high current, strong continuous discharge ability, stable working voltage, excellent leak-proof performance, long storage time, and good low-temperature performance [1,2]. Zinc-manganese batteries are also known as alkaline dry batteries, alkaline zinc-manganese batteries, and alkaline-manganese batteries. They are the best-performing varieties in the zinc-manganese battery series and are suitable for large discharge capacity and long-term use. The internal resistance of the battery is lower, so the current generated is higher than that of an ordinary manganese battery. It is also environmentally friendly with a mercury content of only 0.025%, which does not need to be recycled [3,4]. Alkaline batteries are the most successful high-capacity dry batteries and one of the most cost-effective batteries overall [5]. The schematic diagram of the battery is shown in Fig. 1.1.

The expression of the electrochemical reaction of the zinc-manganese battery is shown in Eq. (1.1):

$$\begin{cases} \text{P: } MnO_2 + H^+ + e^- \rightarrow MnOOH \\ \text{N: } Zn + 2NH_4Cl \rightarrow Zn(NH_3)_2Cl_2 + 2H^+ + 2e^- \\ \text{T: } Zn + 2MnO_2 + 2NH_4Cl \rightarrow 2MnOOH + Zn(NH_3)_2Cl_2 \end{cases} \quad (1.1)$$

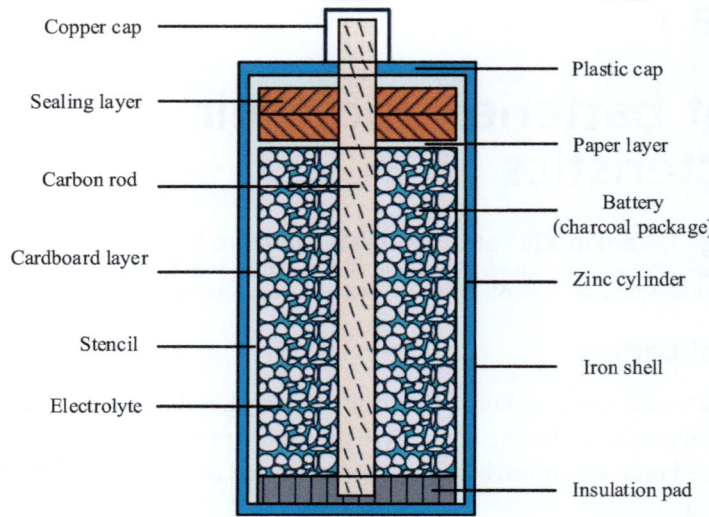

Figure 1.1 Schematic diagram of the structure of the zinc-manganese battery.

It can be seen from the reaction formula that the reduction reaction occurs during the discharge of the positive manganese dioxide, which reduces the H^+ concentration in the solution. Therefore, the pH value of the electrolyte and the alkalinity increases, causing the manganese dioxide electrode potential to move in a negative direction. When the negative electrode zinc discharges, an oxidation reaction occurs, and the concentration polarization of the zinc electrode moves the zinc electrode potential in a positive direction [6]. Therefore, when discharging, the battery voltage drops.

1.1.2 Alkaline battery

Alkaline batteries are also known as alkaline dry batteries, alkaline zinc-manganese batteries, and alkaline-manganese batteries. They are the best-performing varieties of the zinc-manganese battery series and are suitable for large discharge capacity and long-term use. The internal resistance of the battery is lower, so the current generated is larger than that of ordinary carbon batteries. Since this type of battery does not contain mercury, it can be disposed of with household garbage without deliberate recycling. Alkaline batteries are successful high-capacity dry batteries and one of the most cost-effective batteries [7]. Alkaline batteries use manganese dioxide as the positive electrode, zinc as the negative electrode, and potassium hydroxide as the electrolyte. Their characteristics are better than those of carbon batteries, and their electric capacity is large [8].

$$Zn + MnO_2 + H_2O \rightarrow Mn(OH)_2 + ZnO \qquad (1.2)$$

Alkaline batteries adopt the opposite electrode structure of ordinary batteries, increasing the relative area between the positive and negative electrodes. Moreover, the ammonium chloride and zinc chloride solutions are replaced by potassium hydroxide solution with high conductivity. The zinc of the negative electrode is changed from flake to granular, which increases the reaction area of the negative electrode. The electrical performance can be greatly improved by coupling with high-performance electrolytic manganese powder [9]. Generally, alkaline batteries of the same model have 3–7 times the capacity and discharge time of ordinary batteries, and the difference in low-temperature performance is even greater. Alkaline batteries are more suitable for high-current continuous discharge and high working voltage requirements. Occasionally, they are especially ideal for cameras, flashlights, razors, electric toys, CD players, high-power remote controls, wireless mice, keyboards, etc. Alkaline batteries can be roughly divided into the following three types.

1.1.2.1 Alkaline dry battery

The alkaline dry battery is a disposable battery with a relatively long life that was developed by improving on the common zinc-manganese acid battery. Also known as an alkaline manganese dry battery, its shape and size are the same as those of ordinary manganese batteries. Because of its improved performance, it is often used as a flash power source. The negative electrode material of the alkaline dry battery is zinc, the positive electrode is manganese dioxide, the electrolyte is sodium hydroxide or potassium hydroxide solution, and the positive electrode material is carbon rod. A positive electrode made of carbon rods can be found in the middle of its structure. It is wrapped with a fiber material impregnated with sodium hydroxide or potassium hydroxide solution, the outside is wrapped by manganese dioxide, and finally, a zinc cylinder constitutes the negative electrode. The outer shell of the battery is made of inert metal or plastic and is connected to the positive electrode, and the bottom of the battery is connected to the negative electrode through a collector pin.

The normal operating voltage of alkaline batteries is the same as that of common acid zinc-manganese batteries. Alkaline batteries have a long life, large capacity, low internal resistance, and will not leak after a long time like dry batteries, so they are the first choice to power personal audio devices, cameras, and other devices.

1.1.2.2 Alkaline button battery

Alkaline button batteries (LR batteries) are small primary batteries that use manganese dioxide for the positive electrode, zinc for the negative electrode, and alkaline aqueous solution for the electrolyte. The battery is cost-effective and widely used in toys and medical appliances. Murata's alkaline button batteries are all made in Japan. To protect the environment, the mercury usage rate is 0%. The product lineup of this series is shown in Table 1.1.

Fig. 1.2 is a cross-sectional view of the structure of the button battery. The yellow part is zinc or chromium positive. The blue part is the mercury oxide negative electrode.

Table 1.1 Series product lineup.

Type name	Electrical characteristics		Size		
	Nominal voltage (V)	Nominal capacitance (mAh)	Diameter (mm)	High (mm)	Weight (g)
LR41	1.5	45	7.9	3.60	0.6
LR1130	1.5	70	11.6	3.05	1.2
LR43	1.5	110	11.6	4.20	1.5
LR44	1.5	120	11.6	5.40	2.0

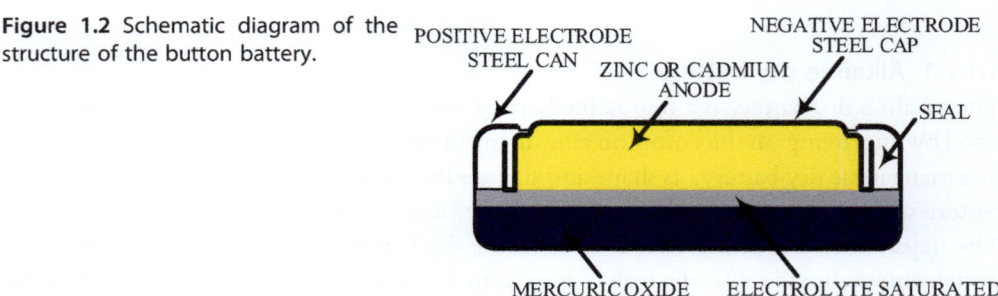

Figure 1.2 Schematic diagram of the structure of the button battery.

1.1.2.3 Silver oxide battery

The silver oxide battery, also known as the silver-zinc or zinc-silver oxide battery, uses silver oxide as the positive electrode, zinc as the negative electrode, and alkaline solution as the electrolyte. It has the advantages of stable discharge and higher energy-to-weight ratio. However, because of the higher cost of silver, it is generally used to make button batteries or for specific uses with higher requirements and that can withstand the high cost, such as certain torpedoes and submarines.

The voltage of the silver oxide battery is high relative to its counterparts. The nominal voltage of silver oxide primary batteries is marked as 1.55 V, higher than the 1.5 V for alkaline batteries and 1.35 V for mercury batteries. During the discharge process, the curve of voltage versus discharge time is flat; that is, it can maintain nearly the same voltage for a long time. It is different from the situation where the voltage of an alkaline battery decreases gradually as the power decreases. The chemical reaction is shown in Eq. (1.3):

$$Zn + Ag_2O \rightarrow ZnO + 2Ag \tag{1.3}$$

The reaction between the positive electrode and the negative electrode is shown in Eq. (1.4):

$$\begin{cases} Ag_2O + H_2O + 2e^- \rightarrow 2Ag + 2OH^- \\ Zn + 2OH^- \rightarrow ZnO + H_2O + 2e^- \end{cases} \quad (1.4)$$

Since the zinc in the negative electrode reacts with the components in the electrolyte, it is corroded and consumed. This reduces the useable capacity of the battery and generates hydrogen gas, causing the battery's internal pressure to increase and expand. A small amount of mercury is usually coated on the surface of zinc to suppress this effect, but it also causes mercury pollution to the environment from discarded batteries. The battery clip of the silver oxide battery can also be recycled. Silver oxide batteries should be recycled to avoid environmental pollution.

1.1.3 Organic electrolyte battery

The choice of electrolyte greatly influences lithium battery performance. The electrolyte must be chemically stable; especially, it should not easily decompose under higher potentials and higher-temperature environments and has higher ionic conductivity. Moreover, the anode and cathode materials must be inert and not corrode [10]. Due to the high charge and discharge potential of lithium-ion batteries and the chemically active lithium embedded in the anode material, the electrolyte must use organic compounds instead of water [11].

1.1.3.1 Lithium-manganese dioxide cell

The lithium-manganese dioxide battery uses lithium as the negative electrode and manganese dioxide as the positive electrode. The manganese dioxide battery has good low-rate and medium-rate discharge performance, a low price, and good safety performance. It is competitive with conventional batteries, so it is the first commercialized lithium battery.

Lithium-manganese dioxide batteries use metallic lithium as the negative electrode and appropriately heat-treated electrolytic manganese dioxide as the positive electrode. The electrolyte is lithium perchlorate dissolved in a mixed solvent of propylene carbonate/ethylene glycol dimethyl ether [12]. The discharge mechanism is different from the oxidation−reduction mechanism of general batteries. The positive electrode reaction is a typical embedded reaction. The chemical reaction formula of the battery is shown in Eq. (1.5):

$$\begin{cases} P: \ MnO_2 + xLi^+ + xe^- \rightarrow Li_xMnO_2 \\ \quad N: \ xLi^+ \rightarrow xLi^+ + xe^- \\ \quad T: \ MnO_2 + xLi^+ \rightarrow Li_xMnO_2 \end{cases} \quad (1.5)$$

As shown in Eq. (1.5), the negative lithium electrode undergoes an oxidation reaction when the battery is discharged. The formed lithium ions dissolve in the electrolyte solution, migrate to the manganese dioxide positive electrode, and are embedded in the manganese dioxide lattice. They promote the reduction of manganese in manganese dioxide from a four to three valence.

1.1.3.2 Lithium thionyl chloride battery

The chemical formula of the lithium thionyl chloride battery is $Li/SOCl_2$. It is the battery with the highest specific energy in the practical battery series, is nonrechargeable, and has a specific energy that can reach 590 W h/kg and 1100 W h/dm^3 [13,14]. This highest specific energy value is obtained by large-capacity, low-discharge rate, large-sized batteries.

$Li/SOCl_2$ batteries come in various sizes and structures. Their capacities range from cylindrical carbon-packed and wound electrode batteries as low as 400 mAh to square batteries up to 10,000 Ah. Many batteries have special sizes and structures that can meet special requirements [15]. The $Li/SOCl_2$ system originally had safety and voltage lag problems. Among them, safety problems are particularly likely to occur during high-discharge-rate discharge and overdischarge, and when the battery continues to discharge at low temperature after high-temperature storage, the voltage hysteresis phenomenon will obviously appear. The schematic diagram of the battery is shown in Fig. 1.3.

The $Li/SOCl_2$ battery consists of a negative lithium electrode, a positive carbon electrode, and a nonaqueous $SOCl_2:LiAlCl_4$ electrolyte. Thionyl chloride is both an electrolyte and a positive electrode active material. Other electrolyte salts, such as $LiAlCl_4$, have been used in specially designed batteries, but the electrolyte formulation is different, and the electrode performance is different. The composition of the negative electrode,

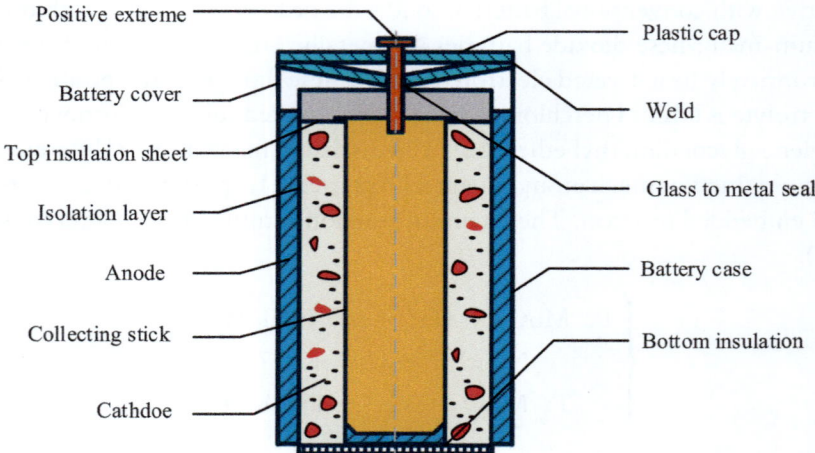

Figure 1.3 Schematic diagram of the Li/SOCl$_2$ battery structure.

positive electrode, and SOCl₂ should be selected by the manufacturer according to the expected battery performance [16–18]. The generally accepted overall reaction mechanism is shown in Eq. (1.6):

$$4Li + 2SOCl_2 \rightarrow 4LiCl + S + SO_2 \tag{1.6}$$

Sulfur and sulfur dioxide are dissolved in the excess thionyl chloride electrolyte. During discharge, due to the production of sulfur dioxide, a certain degree of pressure is generated. During storage, as soon as the negative lithium electrode comes into contact with the electrolyte, it reacts with the thionyl chloride electrolyte to form LiCl. The negative lithium electrode is protected by the LiCl film formed on it. This passivation film is beneficial to extend the storage life of the battery. But it will cause voltage hysteresis at the beginning of discharge. The voltage hysteresis is particularly obvious for batteries that have been stored for a long time at high temperatures and discharged at low temperatures [19].

The application of Li/SOCl₂ batteries is to take advantage of the series' high specific energy and long storage life. Cylindrical batteries discharged with low current can be used as power sources for CMOS memory, water, electricity, and other metering instruments and radio frequency identification (RFID) devices such as highway transit automatic electronic toll systems, program logic controllers, and wireless security alarm systems because the cost of these lithium batteries is higher [20]. At the same time, the safety of these batteries is still receiving special attention, and there are special requirements for their handling. Therefore, the application in the general consumer market is still restricted.

1.1.4 Air battery

Air battery is a kind of chemical battery. The construction principle is similar to that of a dry battery, except that its oxidant is taken from oxygen in the air. For example, there is an air battery that uses zinc as the anode, sodium hydroxide as the electrolyte, and porous activated carbon as the cathode. Therefore, oxygen in the air can be absorbed to replace the oxidant (manganese dioxide) in the general dry battery.

1.1.4.1 Zinc-air battery

The zinc-air battery is also known as a zinc-oxygen battery. Activated carbon adsorbs oxygen in the air or pure oxygen as the positive electrode active material, with zinc used as the negative electrode. Zinc-air batteries are primary batteries with ammonium chloride or caustic solution as the electrolyte [21]. Zinc-air batteries are divided into two systems, neutral and alkaline, represented by the letters A and P, respectively, with numbers indicating the battery models.

The charging process of the zinc-air battery is very slow. To solve this problem, the negative zinc plate or zinc particles of the zinc-air battery are usually oxidized to zinc oxide and become invalid [7,22]. Generally, the zinc plate or zinc particles and electrolyte are directly replaced so that the zinc-air is completely renewed. The chemical equations of the positive and negative electrodes and the overall reaction during discharge are shown in Eq. (1.7):

$$\begin{cases} \text{P:} \ 0.5O_2 + H_2O + 2e^- = 2OH^- \\ \text{N:} \ Zn + 2OH^- = ZnO + H_2O + 2e^- \\ \text{T:} \ 2Zn + O_2 = 2ZnO \end{cases} \quad (1.7)$$

Now zinc-manganese batteries use porous carbon electrodes containing platinum instead of manganese dioxide carbon packs, and the technology of zinc-air dry batteries has been developed.

1.1.4.2 Lithium-air battery
Lithium-air batteries are divided into two types: water-containing and nonaqueous. Among them, the discharge product of a lithium-air battery that does not contain water is Li_2O_2, which is insoluble in the electrolyte. Li_2O_2 has very poor conductivity, so lithium-air batteries that do not contain water usually have a larger overpotential [23–25].

In a water-containing lithium-air battery, what is produced due to the discharge reaction is not solid Li_2O_2 but LiOH (lithium hydroxide) that is easily dissolved in an aqueous electrolyte. After the lithium oxide accumulates on the air electrode, it will not cause the work to stop. Water, nitrogen, and other elements also do not pass through the partition wall of the solid electrolyte, so there is no danger of reacting with the lithium metal of the negative electrode [26]. Moreover, when charging, if a special positive electrode is configured for charging, it can also prevent the corrosion and aging of the air electrode caused by the charging.

1.1.4.3 Aluminum-air battery
The aluminum-air battery, as the name suggests, is a new type of battery that uses aluminum and air as battery materials. It is a pollution-free, long-lasting, stable, and reliable power supply, and it is very environmentally friendly. The battery structure and the raw materials used can be changed according to different practical environments and requirements with its great adaptability. It can be used both on land and in the deep sea. It can be used as a power battery and as a signal battery with long life and high specific energy. It is a very powerful battery with very broad application prospects.

The chemical reaction of the aluminum-air battery is similar to that of the zinc-air battery. The aluminum-air battery uses high-purity aluminum (containing 99.99% aluminum) as the negative electrode and oxygen as the positive electrode. Potassium hydroxide (KOH) and sodium hydroxide (NaOH) aqueous solutions are used as electrolytes. Aluminum ingests oxygen in the air and produces a chemical reaction when the battery is discharged. Aluminum and oxygen are converted into aluminum oxide [27,28]. The progress of the aluminum-air battery is very rapid, and its application in EVs has achieved good results. It is an air battery with great development prospects.

1.1.5 Storage battery

A storage battery is a chemical energy-storage device that is fully charged with direct current. At that time, electrical energy is stored in the battery as chemical energy [29]. When discharging, chemical energy is then converted into electrical energy. The storage battery is also known as "activated battery" The positive and negative active materials and the electrolyte do not directly contact, and the electrolyte is temporarily injected, or other methods are used to activate the battery before use.

During storage, the active material does not come into direct contact with the electrolyte. When in use, the electrolyte is injected or melted so that the battery is in a state of being discharged and has activity. This battery includes the seawater-activated Mg−AgCl battery, electrolyte-activated Zn−Ag_2O storage battery, thermally activated Ca/LiCl−KCl/WO_3, and Ca/LiCl−KCl/$CaCrO_4$ battery. This battery is used on specific occasions to meet the requirement for activation at any time.

1.1.6 Molten salt battery

With heat, the molten salt battery is also called a heat-activated reserve battery. The electrolyte is a nonconductive solid during storage. When in use, an electric igniter or striker mechanism is used to ignite the heating agent inside it to melt the electrolyte into an ion conductor to be activated as a reserve battery. It is a primary reserve battery that uses molten salt as the electrolyte and an automatic activation mechanism to ignite the heat source to melt and activate the electrolyte [30,31]. Thermal batteries include Mg/V_2O_5 thermal batteries, Ca/$PbSO_4$ thermal batteries, LAN sulfide thermal batteries, and lithium alloy iron sulfide thermal batteries.

A thermal battery is a high-temperature energy battery with a working temperature of 350−550°C. The positive and negative poles of the battery are often separated by an ion-conducting molten electrolyte. Fine metal oxides or ceramic powders added to the electrolyte as binders, such as MgO, SiO_2, and BN, play a role in fixing the electrolyte. The deformation and flow characteristics of the electrolyte layer are affected by factors such as temperature, pressure, composition, and binder content. Before the electrolyte melts, the battery is active and can be stored for a long time. Once the ignition is activated, it can be discharged at high power with a working current of several amperes per square centimeter for several seconds, or it can be discharged at low power for a long time for 1 h or more. Thermal batteries use some alkali metals or alkaline earth metals with small atomic weight, active chemical properties, and very negative electrode potential as anode materials (such as Li, Na, Ca, etc.), which have high specific power and specific energy [32]. Therefore, the thermal battery is ideal as a military power source.

Theoretically, FeS_2 has a medium potential that affects battery characteristics and results in poor thermal stability. Therefore, improved cathode materials for thermal batteries are needed. Among chromium compounds, Cr_2O_3 has a low potential, and CrO_3, Cr_2O_5, and CrO_2 lack thermal stability, so they are not suitable for thermal battery positive electrodes. $CaCrO_4$ has a long history as a positive electrode for thermal batteries. Other chromates such as $Li_2Cr_2O_7$ and $K_2Cr_2O_7$ are also commonly used as positive electrodes. In particular, $Li_2Cr_2O_7$ has a small molecular weight and the highest electrochemical capacity, which attracts people's attention. Because the melting point of lithium chromate is low at about 516°C, it is often necessary to add a binder to make the active material molten salt not flow when the electrode is working and keep the shape unchanged. The $Li_2Cr_2O_7$2LiAl pair has an open circuit voltage of 3.0 V and a 100 mA/cm^2 discharge cell voltage of 2.5–2.6 V, which is basically equivalent to lithium vanadium oxide (LVO). Among the manganese compounds, β_2MnO_2 has become quite common as a cathode material for other batteries, such as zinc manganese and lithium manganese batteries.

1.2 Secondary battery
1.2.1 Alkaline secondary battery
Alkaline storage battery is a general term for a storage battery that uses alkaline aqueous solution such as potassium hydroxide as the electrolyte [33]. Compared with lead-acid batteries, alkaline batteries are small in size, have high specific energy, high mechanical strength, stable working voltage, high specific power, and long service life.

1.2.1.1 Nickel-cadmium battery
The nickel–cadmium storage battery is an alkaline storage battery. The alkaline hydroxide in the battery is named after nickel and cadmium. Its positive electrode material is a mixture of nickel hydroxide and graphite powder, the negative electrode material is sponge mesh-like cadmium powder and cadmium oxide powder, and the electrolyte is usually potassium hydroxide and sodium hydroxide solution [34]. It has a long service life, low self-discharge, resistance to overcharge, overdischarge, and shock, stable and reliable performance, good fast charging performance, wide temperature range, and good safety performance [35].

However, the fatal disadvantage of nickel-cadmium batteries is that if they are not handled properly during the charging and discharging process, a serious "memory effect" will occur, which greatly reduces the service life. In addition, nickel-cadmium batteries contain metal cadmium that is harmful to the environment and the human body, so nickel-cadmium batteries are gradually withdrawing from the market [36].

The essence of the normal use of lithium batteries refers to the charging and discharging process, which is the basic principle of the battery. The basis of the battery's ability to charge and discharge is the mutual conversion of energy, as shown in Eq. (1.8).

$$\text{Electricity} \Leftrightarrow \text{Chemical energy} \tag{1.8}$$

When the battery is charged by the power source, electrical energy is converted into chemical energy, and the energy is stored inside the battery, that is, in the environment of the internal electrolyte, the cadmium (Cd) in the negative electrode loses electrons, and the hydrogen and oxygen in sodium hydroxide (NaOH). The root ion (OH^-) is chemically synthesized into cadmium hydroxide ($Cd(OH)_2$), which attaches to the anode and releases electrons at the same time [37]. The electrons travel along the wire to the cathode and react with the cathode's NiO(OH) and the water in the sodium hydroxide solution to form nickel hydroxide and hydroxide ions. The nickel hydroxide will adhere to the anode, and the hydroxide ions will return. In sodium hydroxide solution, the concentration of sodium hydroxide solution will not decrease with time. The electrochemical reaction of a nickel-cadmium battery is shown in Eq. (1.9).

$$\begin{cases} \text{P: } NiO(OH) + H_2O + e^- = Ni(OH)_2 + OH^- \\ \text{N: } Cd + 2OH^- + 2e^- = Cd(OH)_2 \\ \text{T: } Cd + 2NiO(OH) + 2H_2O = 2\,Ni(OH)_2 + Cd(OH)_2 \end{cases} \tag{1.9}$$

As shown in Eq. (1.9), the chemical reaction mechanism of the discharge process is that the lost Cd on the negative electrode becomes Cd^{2+} after removing two electrons, and then immediately combines with the two OH^- ions in the solution to form cadmium hydroxide $Cd(OH)_2$, which is deposited on the negative board. The active material on the positive plate is NiO(OH) crystal. Nickel is a positive trivalent ion (Ni^{3+}), and every two nickel ions in the crystal lattice can obtain two electrons transferred from the negative electrode from the external circuit to generate two divalent ions $2Ni^{2+}$. At the same time, the two hydrogen ions ionized by every two water molecules in the solution enter the positive plate and combine with the two oxygen anions on the crystal lattice to generate two hydroxide ions, and then combine with the original two oxygen ions on the crystal lattice. The hydroxide ions together with the two divalent nickel ions generate two nickel hydroxide crystals.

1.2.1.2 Ni—metal hydride battery

Ni-MH batteries have been developed based on Ni-Cd batteries. Compared with the nickel-cadmium battery, its biggest advantage is environmental friendliness, and there is no heavy metal pollution. The nickel-hydrogen battery is a positive electrode plate with nickel hydroxide as the main material. The negative electrode plate with hydrogen storage alloy as the main material has a protective ability. Diaphragm with good air

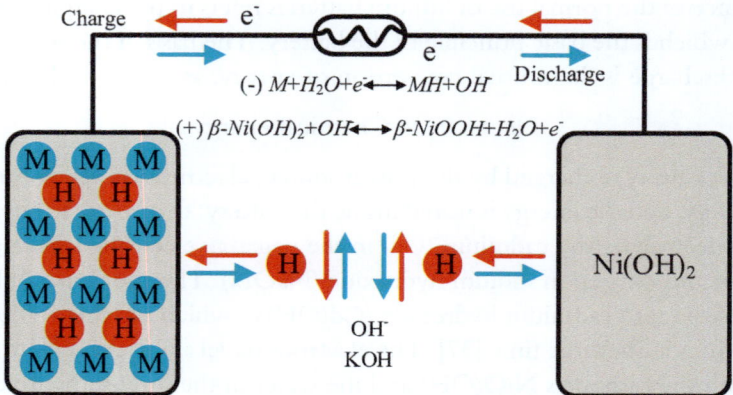

Figure 1.4 Electron transfer inside the battery.

permeability, alkaline electrolyte, metal shell, safety valve with automatic sealing, and other parts [38]. The internal electron transfer is shown in Fig. 1.4.

As the hydrogen storage alloy for the negative plate of the nickel–hydrogen battery, as the name implies, it is an alloy that can store hydrogen. Hydrogen is the smallest and most active element in the chemical periodic table. Different metal elements have different affinities with hydrogen. In the process of hydrogen absorption/desorption chemical reaction (reversible reaction), hydrogen storage alloy is also accompanied by exothermic/endothermic thermal reaction (reversible reaction). At the same time, an electrochemical reaction (reversible reaction) of charge/discharge is also generated [39].

The active materials of the positive electrode of the Ni-MH battery are NiO(OH) (when discharging) and $Ni(OH)_2$ (when charging). The electrolyte uses 30% potassium hydroxide solution, and the electrochemical reaction is shown in Eq. (1.10).

$$\begin{cases} \text{P:} \quad Ni(OH)_2 + OH^- \leftrightarrow NiO(OH) + H_2O + e^- \\ \text{N:} \quad xe^- + xH_2O + M \leftrightarrow xOH^- + MH_x \\ \text{T:} \quad xNi(OH)_2 + M \leftrightarrow xNiO(OH) + MH_x \end{cases} \quad (1.10)$$

As can be seen in Eq. (1.10), the reaction of the Ni-MH battery is similar to that of the Ni-Cd battery. The only difference is that the products produced during the charging and discharging of the negative electrode are different. In Eq. (1.10), during the charging and discharging process of the nickel-hydrogen battery, no intermediate soluble metal ions occur during the electrochemical reaction on the positive and negative electrodes, and no components in the electrolyte are consumed or generated. Therefore, the Ni-MH battery can be made into a sealed structure. The electrolyte of nickel-metal hydride batteries mostly uses KOH aqueous solution, and a small amount of LiOH is added. The diaphragm is made of porous vinyl on nonwoven fabric or nylon nonwoven fabric. To prevent the internal pressure of the battery from being too high at the end of the charging process, an explosion-proof device is installed in the battery. When the Ni-MH battery is overcharged, the gas pressure in the metal shell will gradually rise. When the pressure reaches a certain value, the pressure-limiting safety vent on the top cover opens so that the battery can be prevented from exploding due to excessive gas pressure.

When a Ni-MH battery is discharged, the NiO(OH) on the positive electrode is reduced to $Ni(OH)_2$ by electrons. The hydrogen atoms in the negative metal hydride (MHx) diffuse to the surface to form adsorbed hydrogen atoms. Then an electrochemical reaction takes place to produce water and hydrogen storage alloy. When the nickel-metal hydride battery is overdischarged, the NiO(OH) in the positive electrode active material has been consumed, and water molecules on the positive electrode will be reduced to hydrogen and OH^- ions. Due to the catalytic effect of the hydrogen storage alloy on the negative electrode, the OH^- ions react with hydrogen to generate water [40,41].

When overcharged, oxygen will be released on the positive electrode. Then it diffuses to the negative electrode to cause a depolarization reaction to generate OH^- ions. In the battery overcharge and overdischarge process, the reaction that occurs on the positive and negative electrodes is shown in Table 1.2.

It can be seen that the hydrogen storage alloy not only undertakes the role of hydrogen storage but also acts as a catalyst. When the battery is overcharged or overdischarged, the O_2 and H_2 produced by the positive electrode can be eliminated so that the battery has the ability to withstand overcharge and overdischarge [42]. However, as the charge and discharge cycle progresses, the catalytic ability of the hydrogen storage alloy gradually degrades, and the internal pressure of the battery will rise, which will eventually cause the battery to leak and fail. Compared with nickel-cadmium batteries, nickel-metal hydride batteries have the following significant advantages:

a. The energy density is high; the capacity of the same size battery is 1.5–2 times that of the nickel-cadmium battery.
b. It has good environmental compatibility, with no cadmium pollution.
c. It can be charged quickly with a large current, and the charge–discharge rate is high.
d. It has no obvious memory effect.

Table 1.2 Battery positive and negative chemical reaction.

Polarity	Reaction	Equation
Anode	Overcharge to release oxygen	$4OH^- \rightarrow O_2 + 2H_2O + 4e^-$
	Overdischarge oxygen evolution	$2H_2O + 2e^- \rightarrow 2OH^- + H_2$
Cathode	Overcharge consumes oxygen	$2H_2O + O_2 + 4e^- \rightarrow 4OH^-$
	Overdischarge oxygen consumption	$H_2 + 2OH^- \rightarrow 2H_2O + 2e^-$

e. It has good low-temperature performance and a strong ability to withstand overcharge and discharge.
f. The working voltage is the same as that of nickel-cadmium batteries, which is 1.32 V.

The Ni-MH battery is a replacement product of the Ni-Cd battery. The physical parameters of the battery, such as size, quality, and appearance, are completely interchangeable with nickel-cadmium batteries. The battery performance is basically the same, and the charge and discharge curves are similar. When the electricity is almost exhausted, the voltage drops suddenly. Therefore, it can completely replace nickel-cadmium batteries when in use without any modification of the equipment.

1.2.2 Plante battery
1.2.2.1 Lead-acid batteries

It is a storage battery whose electrodes are mainly made of lead and its oxides, and the electrolyte is a sulfuric acid solution. When a lead-acid battery is discharged, the main component of the positive electrode is lead dioxide, and the main component of the negative electrode is lead. In the charged state, the main components of the positive and negative electrodes are lead sulfate [43,44].

The nominal voltage of a single-cell lead-acid battery is 2.0 V, which can be discharged to 1.5 V and charged to 2.4 V. In applications, six single-cell lead-acid batteries are often connected in series to form a nominal 12 V lead-acid battery, as well as 24 V, 36 V, 48 V, etc. [45]. The battery structure is shown in Table 1.3.

Lead-acid battery charging is performed by connecting an external DC power supply to the battery for charging so that electrical energy is converted into chemical energy for storage. Discharge is the release of electrical energy from the battery to drive external devices [46].

When VRLA battery charging reaches its peak, the charging current is only used to decompose the water in the electrolyte. At this time, the positive electrode of the battery produces oxygen, and the negative electrode produces hydrogen. The gas will overflow from the battery, causing the electrolyte to decrease, and water needs to be added from time to time. On the other hand, at the end of charging or under overcharge conditions, the charging energy is used to split water. The oxygen generated in the positive electrode reacts with the spongy lead of the negative electrode, leaving a part of the negative electrode in an underfilled state and suppressing the generation of hydrogen gas in the negative electrode [47].

1.2.2.2 Small sealed lead battery

For various backup power sources, small portable equipment, and emergency lighting systems, small, sealed lead-acid batteries are the ideal power source because it has the advantages of being fully sealed, maintenance-free, high energy, long life, and so on. Therefore, this new type of battery has seen wide use in all aspects. The inside of the lead

Table 1.3 VRLA battery structure.

Component	Material	Function
Anode	The positive electrode is a Pb—Sb—Ca alloy fence that contains lead oxide as the active material.	Ensure sufficient capacity maintain battery capacity during long-term use and reduce self-discharge
Cathode	The negative electrode is a Pb—Sb—Ca alloy fence that contains a spongelike fiber active material.	Ensure sufficient capacity maintain battery capacity during long-term use and reduce self-discharge
Partition	The advanced microporous absorbent glass mat (AGM) separator keeps the electrolyte and prevents the positive electrode and the negative electrode from being short-circuited.	Prevent short circuit of positive and negative poles keep the electrolyte prevent active material from falling off the electrode surface
Electrolyte	In the electrochemical reaction of the battery, sulfuric acid acts as an electrolyte to conduct ions.	Enable electrons to transfer between the positive and negative active materials of the battery
Case and cover	Unless otherwise specified, the shell and cover are made of ABS resin.	Provide space for the battery positive and negative pole combination fence
Safety valve	The material is a synthetic rubber with excellent acid resistance and aging resistance.	When the internal pressure of the battery is higher than the normal voltage, gas is released to keep the pressure normal and prevent oxygen from entering.
Terminal	Depending on the battery, the positive terminal can be a connecting piece, rod, stud, or lead wire.	Sealed terminals help high current discharge and long service life.

storage battery is made up of a single cell battery connected in series, and the voltage of each cell is 2 V. Therefore, for a 6 V battery, there are three cells inside, and a 12 V battery has six cells. Each cell has the same structure. They are composed of positive and negative plates placed alternately and vertically and a separator placed in the center of the plates to absorb the electrolyte. Since the electrolyte is adsorbed on the diaphragm, and the gas generated inside during charging can be absorbed by the electrode plate and then reduced in the electrolyte, the battery can be completely sealed.

Rated voltage and rated capacity are the two basic parameters of lead storage batteries. Rated capacity is usually expressed in terms of 20-hour rate capacity. For example, 6 V

4.0 AH means to discharge at a current of four AH/20 H = 0.2 A, the average termination voltage per cell is 1.75 V, and the discharge can last for 20 h. Generally speaking, the larger the battery, the larger its capacity. The greater the weight, the greater the capacity. The capacity of the battery is directly proportional to the amount of metal lead used to make the battery; therefore, the larger and heavier the battery, the more lead inside, so the capacity is also larger.

1.2.3 Lithium secondary battery
1.2.3.1 Lithium-ion secondary battery
Lithium-ion secondary batteries generally include an electrode assembly, a container containing the electrode assembly, and an electrolyte. The electrode assembly includes two electrodes with opposite polarities and a separator. The separator includes a porous membrane containing clusters of ceramic particles. The porous membrane is formed by binding particle clusters with a binder. Each particle cluster is formed by sintering or by dissolving and recrystallizing all or part of the ceramic particles. The ceramic particles include a ceramic material having a bandgap. Each particle cluster may have the shape of a grape bunch or a thin layer, and may be formed by laminating flakes or flake-shaped ceramic particles [48–50].

1.2.3.2 Lithium polymer battery
Also known as a polymer lithium battery, it is a kind of chemical battery. Compared with the previous battery, it has the characteristics of high energy, miniaturization, and light weight [51]. Lithium polymer batteries have ultrathin characteristics and can be made into batteries of different shapes and capacities to meet the needs of various products. The theoretical minimum thickness is 0.5 mm.

The three elements of a general battery are a positive electrode, a negative electrode, and an electrolyte. The so-called lithium polymer battery refers to a battery system that uses polymer materials for at least one or more of the three elements. In the lithium polymer battery system, most of the polymer materials are used in the positive electrode and electrolyte. The positive electrode material uses conductive high molecular polymers or inorganic compounds commonly used in lithium-ion batteries. The negative electrode often uses lithium metal or lithium-carbon intercalation compound. The electrolyte is a solid or colloidal polymer electrolyte or an organic electrolyte [52,53]. Since there is no excess electrolyte in the lithium polymer, it is more reliable and stable.

Based on the advantages of lithium polymer batteries, they can be made into batteries of any shape and capacity to meet the needs of various products. They use aluminum-plastic packaging, and internal problems can be immediately manifested through the outer packaging. Even if there is a safety hazard, the lithium polymer battery will not explode; it will only swell. In polymer batteries, the electrolyte plays a dual function of separator and electrolyte. On the one hand, it separates the positive and negative materials like a separator to prevent self-discharge and short-circuit inside the battery. On the other hand, it conducts lithium ions between the positive and negative electrodes like an electrolyte. Polymer electrolyte has not only good electrical conductivity but also the characteristics of light weight, good elasticity, and easy film formation that are unique to polymer materials. It also conforms to the development trend of light weight, safety, high efficiency, and environmental protection of the chemical power supply.

There are two types of lithium-ion batteries: liquid lithium-ion batteries (LIBs) and lithium polymer batteries (PLIBs). A liquid lithium-ion battery refers to a secondary battery in which Li^+ intercalation compounds are positive and negative electrodes. The positive electrode uses the lithium compounds $LiCoO_2$, $LiNiO_2$, or $LiMn_2O_4$, and the negative electrode uses lithium-carbon intercalation compound $LixC_6$. The general chemical reaction formula is shown in Eq. (1.11):

$$\begin{cases} P: \ LiCoO_2 = Li_{1-x}CoO_2 + xLi^+ + xe^- \\ N: \ 6C + xLi^+ + xe^- = Li_xC_6 \\ T: \ LiCoO_2 + 6C = Li_{1-x}CoO_2 + Li_xC_6 \end{cases} \quad (1.11)$$

The main structure of the battery includes three elements: a positive electrode, a negative electrode, and an electrolyte. The principle of the lithium polymer battery is the same as that of liquid lithium, but the main difference is that the electrolyte is different from that of liquid lithium.

1.2.3.3 Lithium—iron phosphate secondary battery

Lithium iron phosphate is an electrode material for lithium-ion batteries with a chemical formula of $LiFePO_4$, which is mainly used in various lithium-ion batteries [54,55]. The battery performance analysis is as follows.

a. High energy density
   Its theoretical specific capacity is 170 mAh/g, and the actual specific capacity of the product can exceed 140 mAh/g (0.2 C, 25°C).
b. Safety
   It is the safest cathode material for lithium-ion batteries; it does not contain any heavy metal elements harmful to the human body.
c. Long life
   Under the condition of 100% DOD, it can charge and discharge more than 2000 times. (Reason: The lattice stability of lithium iron phosphate is good, and the insertion and extraction of lithium ions have little effect on the lattice. Therefore, it has good reversibility. The disadvantage is that the electrode ion conductivity is poor, it is not suitable for high current charge and discharge, and it is hindered in application. Solution: Cover the surface of the electrode with conductive materials and doping to modify the electrode.)

The service life of a lithium iron phosphate battery is closely related to its operating temperature. Using a temperature that is too low or too high will cause great hidden dangers in the charging and discharging process and use process.

d. Charging performance

The lithium battery with lithium iron phosphate cathode material can be charged at a high rate, and the battery can be fully charged within 1 h at the fastest.

1.2.3.4 Electronic storage battery

1.2.3.4.1 Sodium—sulfur battery Under normal circumstances, a sodium-sulfur battery consists of a positive electrode, a negative electrode, an electrolyte, a separator, and a casing. It is different from ordinary secondary batteries in that it is composed of molten electrodes and solid electrolytes. The active material of the negative electrode is molten sodium metal, and the active material of the positive electrode is liquid sulfur and sodium polysulfide molten salt [56,57]. The sodium-sulfur battery working mechanism is shown in Fig. 1.5.

Sodium and sulfur will store electrical energy through a chemical reaction. When the grid needs more electrical energy, it will convert chemical energy into electrical energy and release it [58]. The "flood storage" performance of the sodium-sulfur battery is very good. Even if the input current suddenly exceeds the rated power by 5—10 times, it can

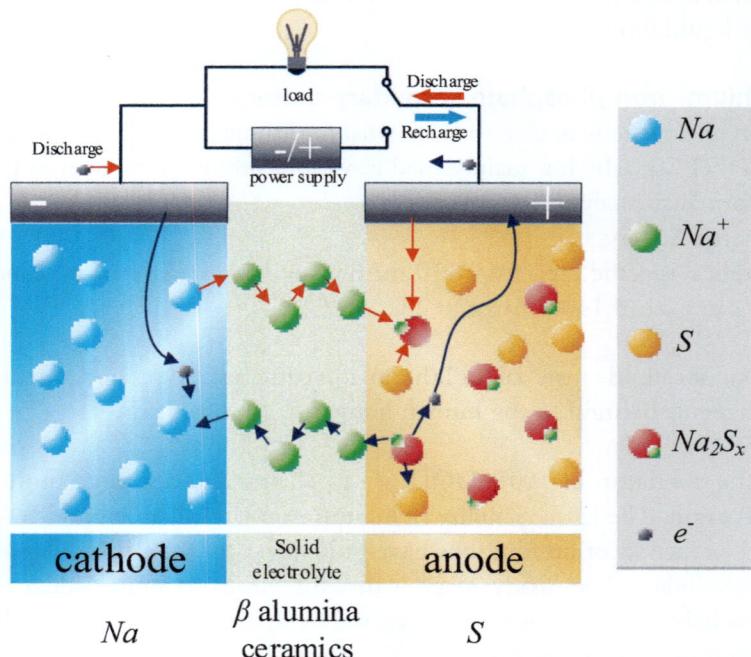

Figure 1.5 Sodium-sulfur battery working principle.

withstand it calmly and release it to the grid with stable power. This is especially useful for the smooth operation of large-scale urban power grids.

Although new energy sources such as solar energy and wind energy are clean, their power generation is very unstable. This will bring unexpected "peaks" to the entire power grid. The energy storage power station will collect all "green electricity" as per the order and then output it according to the demand of the grid.

The sodium-sulfur battery is a secondary battery with Na-beta-alumina (Al_2O_3) as the electrolyte and separator, and sodium metal and sodium polysulfide as the negative and positive electrodes, respectively. Sodium-sulfur batteries have unique advantages for energy storage, which are mainly reflected in the low raw materials and preparation costs, high energy and power density, high efficiency, freedom from site restrictions, and convenient maintenance [59].

1.2.3.4.2 Primary battery A device that generates electric current through an oxidation-reduction reaction is called a galvanic cell and can also be said to be a device that converts chemical energy into electrical energy [60]. Some primary batteries can constitute reversible batteries, and some primary batteries are not reversible batteries [14]. When the primary battery is discharged, the negative electrode undergoes an oxidation reaction, and the positive electrode undergoes a reduction reaction. For example, a copper-zinc primary battery is also called a Daniel battery. The positive electrode is a copper electrode and is immersed in a copper sulfate solution; the negative electrode is a zinc plate that is immersed in a zinc sulfate solution. The two electrolyte solutions are connected by a salt bridge, and the two poles are connected by wires to form a galvanic cell. The dry batteries used in daily life are made according to the principle of primary batteries [61]. The schematic diagram of the battery is shown in Fig. 1.6.

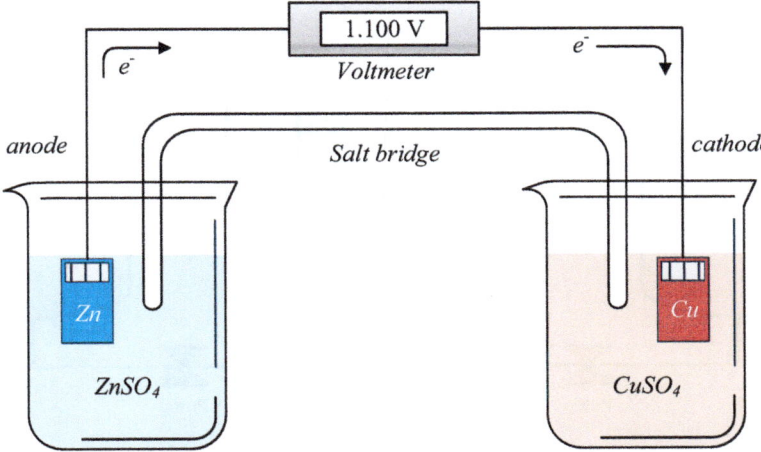

Figure 1.6 Schematic diagram of the primary battery structure.

Both the primary battery and the electrolytic cell are based on the oxidation-reduction reaction that occurs at the contact interface of an electronic conductor (such as a metal) and an ionic conductor (such as an electrolyte solution). The galvanic reaction is an exothermic reaction, generally an oxidation–reduction reaction. But it is different from the general redox reaction in that the electron transfer is not completed by the effective collision between the oxidant and the reducing agent, but the reducing agent loses electrons on the negative electrode to cause an oxidation reaction, and the electrons are transported to the positive electrode through an external circuit. The electrons obtained by the oxidant on the positive electrode undergo a reduction reaction, thereby completing the transfer of electrons between the reducing agent and the oxidant. The directional movement of the ions in the solution between the two electrodes and the directional movement of the electrons in the external wire forms a closed loop so that the two electrodes react continuously, an orderly electron transfer process occurs, an electric current is generated, and the conversion of chemical energy to electric energy is realized.

1.2.3.4.3 Zinc–bromine flow battery A zinc-bromine flow battery is a flow battery that belongs to energy-type energy storage. It can charge and discharge with a large capacity and a long life [62,63]. The overview diagram of the zinc-bromine flow battery is shown in Fig. 1.7.

Compared with other battery technologies, the zinc-bromine flow battery technology has the following characteristics:

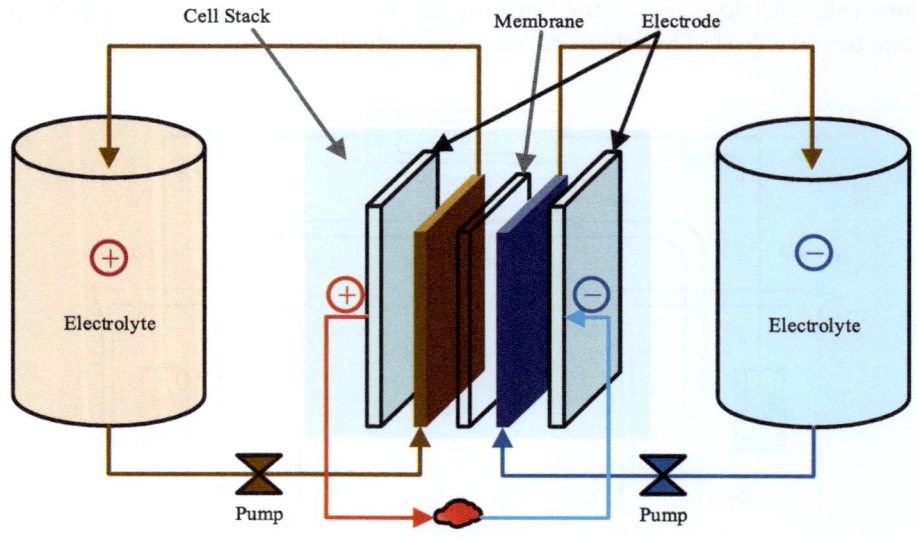

Figure 1.7 The overview diagram of the zinc-bromine flow battery.

(a) The zinc-bromine flow battery has a higher energy density. The theoretical energy density of the zinc-bromine flow battery can reach 435 W h/kg, and the actual energy density can reach 60 W h/kg;
(b) The electrolyte composition (except complex bromine) on both sides of the positive and negative electrodes is completely the same, there is no cross-contamination of the electrolyte, and the theoretical life of the electrolyte is unlimited;
(c) The flow of electrolyte is conducive to the thermal management of the battery system, which is difficult for traditional batteries;
(d) The discharge capacity of the battery is determined by the zinc loading on the electrode surface. The electrode itself does not participate in the charge and discharge reaction. The metallic zinc deposited on the surface during discharge can be completely dissolved in the electrolyte. Therefore, the zinc-bromine flow battery can frequently perform 100% deep discharge without affecting the performance and life of the battery;
(e) The electrolyte is an aqueous solution, and the main reactant is zinc bromide. It is often used as a completion fluid for drilling in oil fields. Therefore, the system is not suitable for fire, explosion, and other accidents, and its safety is high;
(f) The main components of the electrode and diaphragm materials used are plastic, free of heavy metals, recyclable, and environmentally friendly;
(g) The overall system cost is low, and it has commercial application prospects.

1.3 Fuel cell
1.3.1 Phosphoric acid fuel cell

Phosphoric acid fuel cell (PAFC) is currently the fastest commercialized fuel cell. As the name suggests, this battery uses liquid phosphoric acid as the electrolyte, usually in a silicon carbide matrix [64]. The operating temperature of phosphoric acid fuel cells is slightly higher than that of proton exchange membrane fuel cells and alkaline fuel cells, at about 150°C −200°C, but platinum catalysts on the electrodes are still needed to accelerate the reaction. The reaction on the anode and cathode is the same as that of the proton exchange membrane fuel cell, but due to its higher operating temperature, the reaction speed on the cathode is faster than that of the cathode of the proton exchange membrane fuel cell [65].

As shown in the figure "Basic Structure of PAFC" the battery uses 100% phosphoric acid electrolyte, which is solid at room temperature, and its phase transition temperature is 42°C. Hydrogen fuel is added to the anode, oxidized into protons under the action of a catalyst, and at the same time releases two free electrons. The hydrogen protons and phosphoric acid combine to form phosphoric acid protons and move to the positive electrode. The electrons move to the positive electrode, and the hydrated protons move to the cathode through the phosphoric acid electrolyte [66−68]. Therefore, on the positive electrode, electrons, hydrated protons, and oxygen generate water

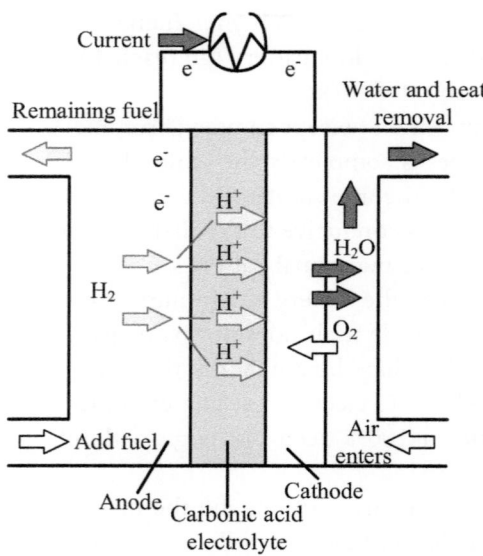

Figure 1.8 The specific electrode reaction schematic diagram.

molecules under the action of the catalyst. The specific electrode reaction schematic diagram is shown in Fig. 1.8.

Phosphoric acid fuel cells generally work at around 200°C, use platinum as a catalyst, and have an efficiency of more than 40%. Because it is not limited by carbon dioxide, the phosphoric acid fuel cell can use air as the cathode reaction gas and reformed gas as the fuel, which makes it very suitable for use as a stationary power station.

1.3.2 Molten carbonate fuel cell

Molten carbonate fuel cell is abbreviated as MCFC, which is a fuel cell composed of a porous ceramic cathode, porous ceramic electrolyte membrane, porous metal anode, and metal pole plate [69]. Its electrolyte is molten carbonate. MCFC has the advantages of higher working temperature and faster reaction speed; relatively low fuel purity requirements, which can be used for in-cell reforming of fuel; no noble metal catalysts and lower cost; liquid electrolytes are used, which is easier to operate [70]. The disadvantage is that it is difficult to manage liquid electrolytes under high-temperature conditions. During long-term operation, corrosion and leakage are serious, which reduces the battery life.

The electrolyte of MCFC is molten carbonate, which is generally a carbonate mixture of alkali metals Li, K, Na, and Cs. The separator material is $LiAiO_2$, and the positive and negative electrodes are lithium-added nickel oxide and porous nickel, respectively [71]. The working principle of MCFC is shown as in Fig. 1.9.

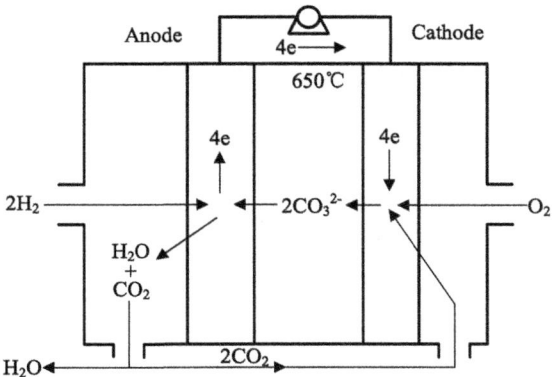

Figure 1.9 The working principle of a molten carbonate fuel cell.

The battery response of MCFC is as follows.

$$\begin{cases} \text{P:} \quad O_2 + 2CO_2 + 4e^- \rightarrow 2CO_3^{2-} \\ \text{N:} \quad 2H_2 + 2CO_3^{2-} \rightarrow 2CO_2 + 2H_2O + 4e^- \\ \text{T:} \quad O_2 + 2H_2 \rightarrow 2H_2O \end{cases} \quad (1.12)$$

It can be seen from the above reaction that the conductive ion of MCFC is CO2-3, CO_2 is the reactant at the cathode and the product at the anode. In fact, CO_2 is circulating during battery operation; that is, the CO_2 produced by the anode returns to the cathode to ensure continuous operation of the battery. The usual method is to burn the tail gas from the anode chamber to eliminate H_2 and CO, separate and remove water, and then return the CO_2 to the cathode for recycling [71,72].

1.3.3 Solid oxide fuel cell

The solid oxide fuel cell (SOFC) is a third-generation fuel cell. It is an all-solid chemical power-generation device that directly converts the chemical energy stored in fuel and oxidant into electrical energy at medium and high temperatures [73]. It is generally considered to be a fuel cell that will be widely used in the future, like the proton exchange membrane fuel cell (PEMFC) [74]. The schematic diagram of the battery is shown in Fig. 1.10.

The main components of the solid oxide fuel cell monomer are composed of electrolyte, anode, or fuel electrode, cathode, or air electrode, and connecting body or bipolar plate.

The working principle of the solid oxide fuel cell is the same as that of other fuel cells. In principle, it is equivalent to the "reverse" device of water electrolysis. The single cell is composed of an anode, a cathode, and a solid oxide electrolyte [75]. The anode is the place where the fuel is oxidized, and the cathode is the place where the oxidant is

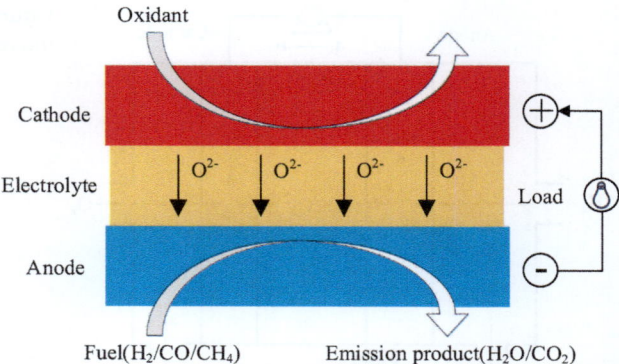

Figure 1.10 Schematic diagram of the structure of a solid electrolyte fuel cell.

reduced. Both electrodes contain catalysts that accelerate the electrochemical reaction of the electrodes [76]. When working, it is equivalent to a DC power supply, the anode is the negative pole of the power supply, and the cathode is the positive pole of the power supply.

The single battery can only generate a voltage of about 1 V, and the power is limited. To give the SOFC practical application possibilities, its power needs to be greatly increased. To this end, several single cells can be assembled into a battery pack in various ways (series, parallel, hybrid). The structure of the SOFC group is mainly tubular, planar, and three unique types, of which the planar type has become the development trend of SOFC due to its high power density and low production cost.

1.3.4 Polymer electrolyte fuel cell

The polymer electrolyte fuel cell, also known as the proton exchange membrane fuel cell (PEMFC), is a fuel cell that uses hydrogen-containing fuel and air to produce electricity and heat. The membrane electrode group and the collector plate are connected in series to form a fuel cell stack [77]. At present, especially hydrogen fuel cells have attracted the attention of power research and development personnel. Its compact structure, low working temperature (only 80°C), quick start-up, high power density, and long working life [78].

2. Physical battery

2.1 Solar battery

A solar cell is a kind of photoelectric semiconductor sheet that uses sunlight to generate electricity directly. It is also called "solar chip" or "photocell." As long as the illuminance meets a certain condition, it can instantly output voltage and generate current when there is a loop [79]. In physics, it is called solar photovoltaic (PV), or photovoltaic for short.

Solar cells are devices that directly convert light energy into electrical energy through the photoelectric or photochemical effect [80]. Crystalline silicon solar cells that work with the photovoltaic effect are the mainstream, while the implementation of solar cells with thin-film cells that work with the photochemical effect is still in its infancy.

Sunlight shines on the semiconductor p-n junction to form a new hole-electron pair. Under the action of the built-in electric field in the p—n junction, the light-generated holes flow to the p-area, the photogenerated electrons flow to the n-area, and a current is generated after the circuit is turned on. This is the working principle of photovoltaic effect solar cells. There are two methods for solar power generation, one is the light-heat-electric conversion method, and the other is the light-electric direct conversion method.

2.2 Thermoelectric battery

A thermoelectric battery uses temperature differences to directly convert heat energy into electrical energy. There are generally two materials used in thermoelectric batteries, metals and semiconductors [81]. The battery made of metal has a small Seebeck effect and is often used to measure temperature and radiation intensity. This kind of battery generally connects several thermocouples in series, exposing one of them to a heat source and fixing the other contact in a specific temperature environment [82]. The electromotive force generated in this way is equal to the sum of each electromotive force, which is then converted into temperature according to the measured electromotive force or intensity. For example, we often use it to measure the high temperature of smelting and heat treatment furnaces in our daily lives.

Since different metal materials have different free electron densities, when two different metal conductors are in contact, electron diffusion will occur on the contact surface. The diffusion rate of electrons is related to the electron density of the two conductors and is proportional to the temperature of the contact area.

2.3 Atomic energy battery

Atomic energy battery is also called radioisotope battery, also called radioisotope thermoelectric generator. This kind of thermoelectric generator is composed of some excellent semiconductor materials, such as bismuth telluride, lead telluride, germanium silicon alloy, and selenium compounds, which are connected in series with many materials. In addition, it is necessary to have a suitable heat source and energy converter to form a temperature difference between the heat source and the energy converter to generate electricity.

The heat source of a nuclear battery is a radioisotope. In the process of transformation, they will continuously emit energy much larger than ordinary matter in the form of rays with heat energy. This great energy has two beloved characteristics. One is that the amount and speed of energy released during transformation are not affected by temperature, chemical reactions, pressure, and electromagnetic fields in the external environment. Therefore, nuclear batteries are known for their strong anti-interference and accurate and reliable work. Another feature is that the transformation time is very long, which determines that the radioisotope battery can be used for a long time.

3. Current status of battery technology at home and abroad

Driven by the global new energy power generation, electric vehicles, and emerging energy-storage industries, many types of energy storage technologies have made considerable progress in recent years. In addition to the already commercialized pumped storage and cavern compressed air energy storage technologies, battery energy storage technologies, led by lithium-ion batteries, have begun to have commercial application potential on the source, network, and load side.

Battery energy storage technology uses the conversion between electrical energy and chemical energy to realize the storage and output of electrical energy. It has not only the technical characteristics of rapid response and two-way adjustment but also the technical advantages of strong environmental adaptability, small-scale decentralized configuration, and a short construction period. It breaks the traditional concept of source network load and breaks the inherent properties of all links of power generation, transmission, and distribution at the same time. It can assume and play different roles on the power supply side, grid side, and user side of the power system. As of the end of 2018, the global installed capacity of battery energy storage technology was 6058.9 MW, of which China's installed capacity was 1033.7 MW, with the United States, China, and South Korea ranking as the top three.

References

[1] Y. Liu, et al., Highly efficient dendrite suppressor and corrosion inhibitor based on gelatin/Mn^{2+} Co-additives for aqueous rechargeable zinc-manganese dioxide battery, Chem. Eng. J. (2021) 407.
[2] M.B. Lim, T.N. Lambert, B.R. Chalamala, Rechargeable alkaline zinc-manganese oxide batteries for grid storage: mechanisms, challenges and developments, Mater. Sci. Eng. R Rep. (2021) 143.
[3] Z.C. Li, et al., Batch-scale synthesis of porous zinc manganese oxide with large specific surface area for Li-ion battery anodes, Solid State Sci. 108 (2020).
[4] M.S. Zhang, et al., A high-energy-density aqueous zinc-manganese battery with a La-Ca co-doped epsilon-MnO(2)cathode, J. Mater. Chem. 8 (23) (2020) 11642−11648.
[5] X.Y. Li, et al., A Quasi-gel SiO_2/sodium alginate (SA) composite electrolyte for long-life zinc-manganese aqueous batteries, J. Inorg. Mater. 35 (8) (2020) 909.
[6] J.D. Huang, et al., High-performance aqueous zinc-manganese battery with reversible Mn^{2+}/Mn^{4+} double redox achieved by carbon coated MnO_x nanoparticles, Nano-Micro Lett. 12 (1) (2020).
[7] X.D. Shi, et al., Nitrogen and atomic Fe dual-doped porous carbon nanocubes as superior electrocatalysts for acidic H-2-O-2 PEMFC and alkaline Zn-air battery, J. Energy Chem. 59 (2021) 388−395.
[8] A. Semaan, et al., Severe vaginal burns in a 5-year-old girl due to an alkaline battery in the vagina, J. Pediatr. Adolesc. Gynecol. 28 (5) (2015) E147−E148.
[9] F. Yu, et al., Aqueous alkaline-acid hybrid electrolyte for zinc-bromine battery with 3V voltage window, Energy Storage Mater. 19 (2019) 56−61.

[10] J.E. Chen, et al., Tuning the solution structure of electrolyte for optimal solid-electrolyte-interphase formation in high-voltage lithium metal batteries, J. Energy Chem. 60 (2021) 178–185.

[11] Z. Chen, et al., Zinc/selenium conversion battery: a system highly compatible with both organic and aqueous electrolytes dagger, Energy Environ. Sci. 14 (4) (2021) 2441–2450.

[12] A. Baral, et al., Structure and activity of lysozyme on binding to lithium-manganese oxide nanocomposites prepared from seabed nodule, J. Phys. Chem. Solid. (2021) 151.

[13] H.Y. Song, M.H. Jung, S.K. Jeong, Electrochemical properties of acetylene black/multi-walled carbon nanotube cathodes for lithium thionyl chloride batteries at high discharge currents, J. Electrochem. Sci. Technol. 11 (4) (2020) 430–436.

[14] M.A. Zabara, B. Ulgut, Electrochemical Impedance Spectroscopy based voltage modeling of lithium Thionyl Chloride (Li\SOCl2) primary battery at arbitrary discharge, Electrochim. Acta (2020) 334.

[15] K. Li, et al., Space-confined construction of nitrogen-rich cobalt porphyrin-derived nanoparticulates anchored on activated carbon for high-current lithium thionyl chloride battery, Electrochim. Acta (2020) 353.

[16] D.H. Wang, et al., The effects of pore size on electrical performance in lithium-thionyl chloride batteries, Front. Mater. 6 (2019).

[17] Y. Zhang, et al., Improving electrocatalytic activity of fluorinated multi-walled carbon nanotubes modified with tetraaminophthalocyanines for lithium/thionyl chloride battery, Ionics 25 (4) (2019) 1459–1469.

[18] Y. Gao, et al., Amino-substituted binuclear phthalocyanines bonding with multi-wall carbon nanotube as efficient electrocatalysts for lithium-thionyl chloride battery, J. Mater. Res. 34 (6) (2019) 921–931.

[19] S. Li, et al., Designing Li-protective layer via $SOCl_2$ additive for stabilizing lithium-sulfur battery, Energy Storage Mater. 18 (2019) 222–228.

[20] E.A. Astafev, Wide frequency band electrochemical noise measurement and analysis of a $Li/SOCl_2$ primary battery, J. Solid State Electrochem. 23 (2) (2019) 389–396.

[21] Y.J. Chen, et al., Molten-salt-assisted synthesis of onion-like Co/CoO@FeNC materials with boosting reversible oxygen electrocatalysis for rechargeable Zn-air battery, J. Colloid Interface Sci. 596 (2021) 206–214.

[22] J.X. Gao, et al., Exploiting encapsulated FeCo alloy decorated N-doped hierarchically porous carbon electrocatalysts in rechargeable Zn-air batteries, J. Alloys Compd. (2021) 870.

[23] Z.Y. Guo, et al., A lithium air battery with a lithiated Al-carbon anode (vol 51, 676, 2015), Chem. Commun. 57 (30) (2021) 3724.

[24] N. Imanishi, O. Yamamoto, Perspectives and challenges of rechargeable lithium-air batteries, Mater. Today Adv. 4 (2019).

[25] Q.C. Liu, et al., Design and preparation of advanced materials for lithium-air batteries, Acta Chim. Sin. 75 (2) (2017) 137–146.

[26] L.L. Huang, et al., Carbon-based cathode materials for non-aqueous lithium-air batteries, Prog. Chem. 31 (10) (2019) 1406–1416.

[27] J.Y. Deng, et al., $NiMn_2O_4$-based Ni-Mn bimetallic oxides as electrocatalysts for the oxygen reduction reaction in Al?air batteries, Chem. Eng. J. (2021) 413.

[28] Z.J. Wang, et al., Ultrasonic-assisted hydrothermal synthesis of cobalt oxide/nitrogen-doped graphene oxide hybrid as oxygen reduction reaction catalyst for Al-air battery, Ultrason. Sonochem. (2021) 72.

[29] G.J. May, A. Davidson, B. Monahov, Lead batteries for utility energy storage: a review, J. Energy Storage 15 (2018) 145–157.

[30] Y.P. Gan, et al., Highly efficient synthesis of silicon nanowires from molten salt electrolysis cell with a ceramic diaphragm, J. Electron. Mater. 50 (9) (2021) 5021–5028.

[31] Z.J. Chen, et al., Closed-loop utilization of molten salts in layered material preparation for lithium-ion batteries, Front. Energy Res. 8 (2021).

[32] S.X. Wang, et al., An experimental and numerical examination on the thermal inertia of a cylindrical lithium-ion power battery, Appl. Therm. Eng. 154 (2019) 676–685.

[33] E. Shangguan, et al., Comparative structural and electrochemical study of spherical ZnO with different tap density and morphology as anode materials for Ni/Zn secondary batteries, J. Alloys Compd. (2021) 868.

[34] N.N. Yazvinskaya, et al., Probability investigation of thermal runaway in nickel-cadmium batteries with pocket electrodes, Int. J. Electrochem. Sci. 11 (7) (2016) 5850–5854.
[35] I.J. Park, et al., Recovery of cadmium in nickel-cadmium leaching solution by sulfide precipitation method, Kor. J. Metals Mater. 57 (11) (2019) 726–731.
[36] K. Pourabdollah, Development of electrolyte inhibitors in nickel cadmium batteries, Chem. Eng. Sci. 160 (2017) 304–312.
[37] Y.L. Gun'ko, et al., Digital simulation of discharge of nickel-cadmium batteries, Russ. J. Electrochem. 56 (12) (2020) 997–1010.
[38] H. Muhsen, A. Al-Muhtady, Optimized modeling of Ni-MH batteries primarily based on Taguchi approach and evaluation of used Ni-MH batteries, Turk. J. Electr. Eng. Comput. Sci. 27 (1) (2019) 197–212.
[39] S.P. Qiao, et al., Experimental study on storage and maintenance method of Ni-MH battery modules for hybrid electric vehicles, Appl. Sci. Basel 9 (9) (2019).
[40] M.I. Fedorova, et al., Extraction reprocessing of Fe,Ni-containing parts of Ni-MH batteries, Russ. J. Inorg. Chem. 66 (2) (2021) 266–272.
[41] A.J. Telmoudi, et al., Modeling and state of health estimation of nickel-metal hydride battery using an EPSO-based fuzzy c-regression model, Soft Comput. 24 (10) (2020) 7265–7279.
[42] L.Z. Ouyang, et al., Progress of hydrogen storage alloys for Ni-MH rechargeable power batteries in electric vehicles: a review (vol 200, pg 164, 2017), Mater. Chem. Phys. (2021) 258.
[43] P. Zhang, G.F. Liu, Data-driven recovery potential analysis and modeling for batteries recovery operations in electric bicycle industry, Discrete Dynam Nat. Soc. 2018 (2018).
[44] I.A. Azzollini, et al., Lead-acid battery modeling over full state of charge and discharge range, IEEE Trans. Power Syst. 33 (6) (2018) 6422–6429.
[45] S.J. Tang, et al., Non-contact detection of single-cell lead-acid battery electrodes' defects through conductivity reconstruction by magnetic induction tomography, Inverse Probl. Sci. Eng. 29 (13) (2021) 2470–2490.
[46] S. Bong, L. Azhari, Y. Wang, Laser ablation inductive coupled plasma mass spectroscopy (LA-ICP-MS) analysis on lead-acid battery system: development of evaluation method of sub-ppm metal impurity elements, J. Sustain. Metall. 7 (2) (2021) 610–619.
[47] S.Y. Tan, et al., Developments in electrochemical processes for recycling lead-acid batteries, Curr. Opini. Electrochem. 16 (2019) 83–89.
[48] Y. Yang, et al., Novel synthesis of porous Si-TiO_2 composite as a high-capacity anode material for Li secondary batteries, J. Alloys Compd. (2021) 872.
[49] H.Y. Li, S.H. Guo, H.S. Zhou, In-situ/operando characterization techniques in lithium-ion batteries and beyond, J. Energy Chem. 59 (2021) 191–211.
[50] K. Yoshida, et al., Sonochemical synthesis of Au/Pd nanoparticles on the surface of LiFePO/C cathode material for lithium-ion batteries, Jpn. J. Appl. Phys. 60 (Sd) (2021).
[51] M. Irfan, et al., Recent advances in high performance conducting solid polymer electrolytes for lithium-ion batteries, J. Power Sources (2021) 486.
[52] V.P.H. Huy, S. So, J. Hur, Inorganic fillers in composite gel polymer electrolytes for high-performance lithium and non-lithium polymer batteries, Nanomaterials 11 (3) (2021).
[53] Q. Zhou, J.J. Zhang, G.L. Cui, Rigid-flexible coupling polymer electrolytes toward high-energy lithium batteries, Macromol. Mater. Eng. 303 (11) (2018).
[54] M. Jin, et al., Dual-function $LiFePO_4$ modified separator for low-overpotential and stable Li-S battery, J. Alloys Compd. (2021) 873.
[55] W.G. Kidanu, et al., Enabling high-performance aqueous rechargeable Li-ion batteries through systematic optimization of TiS_2/$LiFePO_4$ full cell, Appl. Surf. Sci. (2021) 553.
[56] X. Ye, et al., Enabling a stable room-temperature sodium-sulfur battery cathode by building heterostructures in multichannel carbon fibers, ACS Nano 15 (3) (2021) 5639–5648.
[57] J.H. Zhou, et al., Sulfur in amorphous silica for an advanced room-temperature sodium-sulfur battery, Angew. Chem. Int. Ed. 60 (18) (2021) 10129–10136.
[58] S. Kandhasamy, et al., Operational strategies to improve the performance and long-term cyclability of intermediate temperature sodium-sulfur batteries, ChemElectroChem 8 (6) (2021) 1156–1166.

[59] Y.J. Wang, et al., Research progress toward room temperature sodium sulfur batteries: a review, Molecules 26 (6) (2021).
[60] M.Y. Chen, et al., A simplified analysis to predict the fire hazard of primary lithium battery, Appl. Sci. Basel 8 (11) (2018).
[61] E.A. Astafev, A.E. Ukshe, Y.A. Dobrovolsky, Measurement of electrochemical noise of a Li/MnO_2 primary lithium battery, J. Solid State Electrochem. 22 (11) (2018) 3597–3606.
[62] J.N. Lee, et al., Development of titanium 3D mesh interlayer for enhancing the electrochemical performance of zinc-bromine flow battery, Sci. Rep. 11 (1) (2021).
[63] J.H. Lee, et al., Dendrite-free Zn electrodeposition triggered by interatomic orbital hybridization of Zn and single vacancy carbon defects for aqueous Zn-based flow batteries, Energy Environ. Sci. 13 (9) (2020) 2839–2848.
[64] C. Park, et al., Analysis of a phosphoric acid fuel cell-based multi-energy hub system for heat, power, and hydrogen generation, Appl. Therm. Eng. (2021) 189.
[65] J.J. Zhang, et al., Advancement in distribution and control strategy of phosphoric acid in membrane electrode assembly of high-temperature polymer electrolyte membrane fuel cells, Acta Phys. Chim. Sin. 37 (9) (2021).
[66] Y. Cheng, et al., First demonstration of phosphate enhanced atomically dispersed bimetallic FeCu catalysts as Pt-free cathodes for high temperature phosphoric acid doped polybenzimidazole fuel cells, Appl. Catal. B Environ. (2021) 284.
[67] S. Wilailak, et al., Thermo-economic analysis of phosphoric acid fuel-cell (PAFC) integrated with organic ranking cycle (ORC), Energy (2021) 220.
[68] Z. Xie, et al., Enhanced low-humidity performance of proton exchange membrane fuel cell by incorporating phosphoric acid-loaded covalent organic framework in anode catalyst layer, Int. J. Hydrogen Energy 46 (18) (2021) 10903–10912.
[69] Y. Han, et al., An efficient hybrid system using a graphene-based cathode vacuum thermionic energy converter to harvest the waste heat from a molten hydroxide direct carbon fuel cell, Energy (2021) 223.
[70] M.X. Li, et al., Techno-economic and carbon footprint feasibility assessment for polygeneration process of carbon-capture coal-to-methanol/power and molten carbonate fuel cell, Energy Convers. Manag. (2021) 235.
[71] H. Meskine, et al., CO_2 electrolysis in a reversible molten carbonate fuel cell: online chromatographic detection of CO, Int. J. Hydrogen Energy 46 (28) (2021) 14913–14921.
[72] R. Cooper, et al., A feasibility assessment of a retrofit molten carbonate fuel cell coal-fired plant for flue gas CO_2 segregation, Int. J. Hydrogen Energy 46 (28) (2021) 15024–15031.
[73] B. Liu, et al., Rare-earth elements doped Nd_2CuO_4 as Cu-based cathode for intermediate-temperature solid oxide fuel cells, J. Alloys Compd. (2021) 870.
[74] X. Zhao, et al., Enhanced performance of $Gd_{0.6}Sr_{0.4}FeO_3$-delta electrode by Co_3O_4 incorporation for symmetrical solid oxide fuel cells, Mater. Lett. (2021) 295.
[75] S. Rauf, et al., Tailoring triple charge conduction in $BaCo_{0.2}Fe_{0.1}Ce_{0.2}Tm_{0.1}Zr_{0.3}Y_{0.1}O_3$-delta semiconductor electrolyte for boosting solid oxide fuel cell performance, Renew. Energy 172 (2021) 336–349.
[76] J. Shin, et al., Low-temperature processing technique of Ruddlesden-Popper cathode for high-performance solid oxide fuel cells, J. Alloys Compd. (2021) 868.
[77] L.L. Tian, et al., Enhanced performance and durability of high-temperature polymer electrolyte membrane fuel cell by incorporating covalent organic framework into catalyst layer, Acta Phys. Chim. Sin. 37 (9) (2021).
[78] Y.R. Xue, et al., Cost-effective hydrogen oxidation reaction catalysts for hydroxide exchange membrane fuel cells, Acta Phys. Chim. Sin. 37 (9) (2021).
[79] Q. Li, et al., Solar energy storage in the rechargeable batteries, Nano Today 16 (2017) 46–60.
[80] L.C. Kin, et al., Efficient area matched converter aided solar charging of lithium ion batteries using high voltage perovskite solar cells, ACS Appl. Energy Mater. 3 (1) (2020) 431–439.
[81] C. Wang, et al., Hybrid thermoelectric battery electrode FeS_2 study, Nano Energy 45 (2018) 432–438.
[82] Y. Lyu, et al., Electric vehicle battery thermal management system with thermoelectric cooling, Energy Rep. 5 (2019) 822–827.

CHAPTER 2

Available technologies for remanufacturing, repurposing, and recycling lithium-ion batteries: an introduction

Ashwani Pandey, Sarthak Patnaik and Soobhankar Pati
School of Minerals, Metallurgical and Materials Science, IIT, Bhubaneswar, Odisha, India

Remanufacturing: for reuse in EVs
   Repurposing/Reusing: reengineering for off-road, stationary storage applications
   Recycling: disassembling each cell in a battery and safely extracting the precious metals, chemicals, and by-products

1. Introduction

Currently, lithium-ion batteries (LIBs) cost around $150/kWh compared with $1100/kWh in 2011, and the cost is expected to decrease to $100/kWh within the next decade [1]. It is predicted that economies of scale will drive the cost down to $73/kWh by the end of 2030. It will be cost-competitive for mobility applications in such a scenario, leading to large-scale adaptation of LIBs in electric vehicles (EVs). This will introduce an enormous quantity of end-of-life LIBs and other components used to make battery packs. The LIB pack is a combination of cells connected in series and parallel with the help of electrical connectors to meet specific current and voltage requirements.

Fig. 2.1 shows a typical battery pack design assembled in our laboratory for an EV (three-wheeler) with a 7.2 kWh capacity. The battery pack design may further vary for different industry manufacturers according to the required space constraint and cell form factor as well as available manufacturing technologies [2]. Apart from these cell-level modules, the battery pack consists of a cooling system, battery management system, mechanical support and stack pressure system, outer casing, and various connectors [2].

Proper handling of end-of-life LIBs and other components will be beneficial for the environment and generate monetary value. As the material costs in an LIB pack are ~66% of overall pack costs, closed-loop manufacturing of LIB packs, including remanufacturing, reusing, and recycling, can reduce the cost substantially by reducing the requirement for pristine off-the-shelf materials and the burden of safe disposal. However, implementing closed-loop manufacturing is not very straightforward. As mentioned

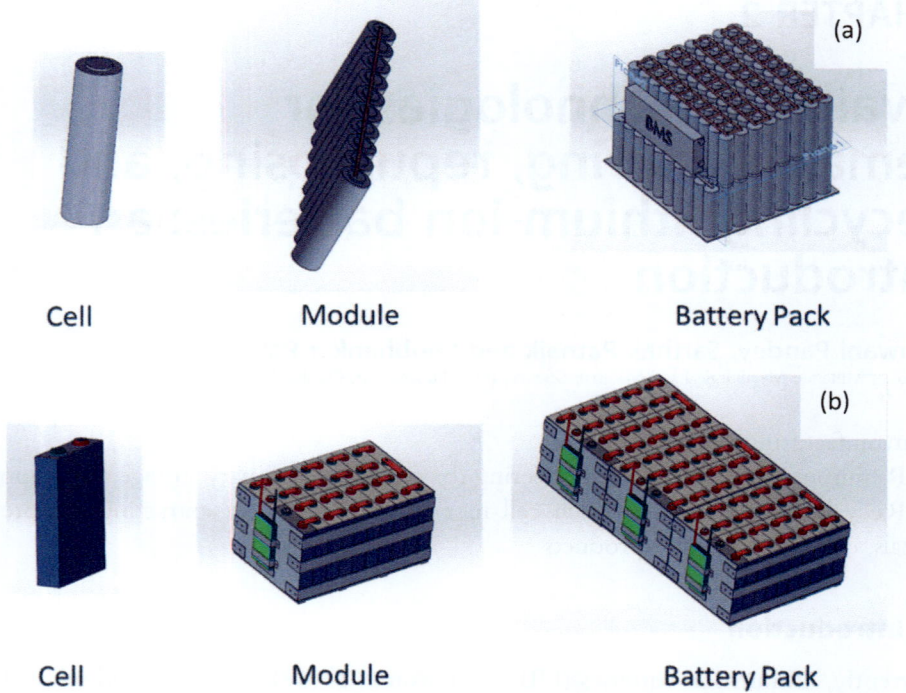

Figure 2.1 Assembly of (A) cylindrical cells and (B) prismatic cells along with battery management system.

above, the LIB pack consists of multiple cells and other components assembled in a specific way to achieve the desired performance. Due to this assembly, each cell in the LIB pack goes through a different charge—discharge cycle, leading to inherent cell-level internal resistance (IR) and temperature differences at different locations within the battery pack [3]. This further leads to the degradation and failure of certain cells in the battery pack at faster rates than other groups of cells. Therefore, it would not be wise and cost-effective to pick one particular process, i.e., remanufacturing, repurposing, or recycling, for the whole LIB pack. The state of degradation of individual cells in the LIB pack must be identified to pick any of the processes mentioned.

Fig. 2.2 shows the various steps to achieve the maximum value proposition of LIBs. The value chain of LIBs has a particular hierarchy in which remanufacturing is completed first, followed by repurposing and recycling. However, the end-of-life decision, whether a particular LIB module/submodule or component will be remanufactured, repurposed, or recycled, requires a criterion for benchmarking [4]. State of health (SoH) is the widely accepted criterion for predicting the remaining useful life and safety of LIB packs. In the

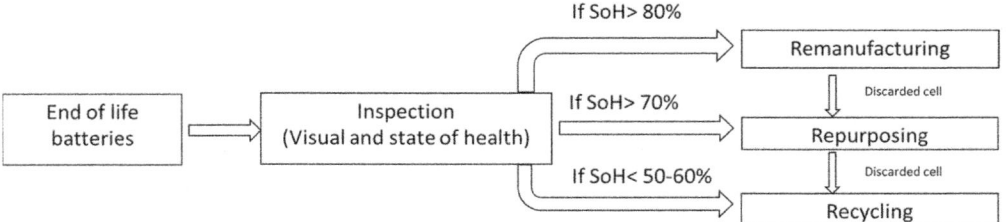

Figure 2.2 Processing end-of-life lithium-ion batteries.

next section, SoH and various estimation techniques are elaborated in detail so that a judicious end-of-life decision, viz. remanufacturing, repurposing, and recycling, can be made regarding LIBs. Also described are the various techniques used to remanufacture, reuse and recycle.

2. State-of-health estimation

SoH is a figure of merit that provides information regarding the condition of a battery pack or cell, compared with its performance at the time of manufacture, which is considered 100% in the ideal case. As expected, the SoH decreases with the application life of the battery until its eventual death [4]. The estimation of SoH can be divided into two broad methodological categories, experimental and model-based. In both methods, it is necessary to have the data for specific parameters, like the internal resistance, capacity, and impedance of the battery pack or individual modules. The experimental techniques to measure these parameters are direct current internal resistance, capacity, pulse discharged testing, and electrochemical impedance spectroscopy (EIS). Of these, EIS is more accurate in predicting the SoH of the battery pack since it provides information about internal resistance as well as system reactance [5]. However, EIS is time-consuming and requires stringent test conditions [6].

In model-based techniques, a mathematical model is used to estimate the SoH. The well-known Kalman filter-based algorithms are the most common model-based SoH estimation tool [7]. Least-squares-based algorithms, specifically recursive least-squares models, have also been used to estimate SoH. Recursive least-squares-based models are easy to implement and provide an accurate estimate of SoH [8]. Electrochemical model-based techniques consist of coupled partial differential equations which are used to describe the diffusion, migration, and reaction kinetics inside the battery. Solving electrochemical models is challenging. Doyle and Newman developed an electrochemical model that accounts for mass transfer, diffusion, migration, and reaction kinetics. This model is also called the "dualfoil" model in the literature. It is the most commonly used model for simulating the electrochemical process of LIBs and has been extensively validated by experiments [9].

Experimental results can also be used as input parameters in advanced equivalent-circuit models or machine learning (ML) algorithms to improve SoH estimation accuracy. ECM models use nonlinear differential equations to predict SoH. The challenge of the ECM model is using an appropriate equivalent circuit to represent the battery reaction mechanism [5]. Recently, with the availability of advanced computational facilities, machine learning-based SoH prediction models are gaining interest [10]. In this technique, the experimental data are first used to train the model to identify the most accurate relationship between the various parameters of the battery pack. This model is further used to predict the SoH of the batteries. Several ML techniques have been developed over the last few years [5]. Support vector regression model is one such ML technique that is nonparametric, accurate, and robust to predict SoH. Fuzzy logic-based ML algorithms have also been used for SoH estimation. This method can be used for nonlinear systems and provide accurate and robust results. Regression models and fuzzy logic-based models require quality and very diverse training data sets for accurate predictive results [11]. The need for a high-quality and diverse data set has been addressed to some extent in neural network-based ML models.

After benchmarking the battery pack using SoH as a parameter, the decision regarding the remaining life of the battery pack can be decided—whether to remanufacture, repurpose, or recycle. If the SoH is greater than 70%–80%, the battery pack is sent for remanufacturing; if it is less than 70%, it is sent for repurposing or recycling.

3. Remanufacturing

Remanufacturing is the process by which the damaged cells or modules in the battery pack are identified and replaced with new ones to bring the battery pack back to "life,"—i.e., the remanufactured pack performs at the same level as a new off-the-shelf battery pack [12]. Remanufacturing substantially reduces the material and energy needed in new battery pack manufacturing. It increases the average life of the battery pack, thus reducing costs and negative environmental impacts. The first step in remanufacturing is identifying the component, module, submodule, or cell that is no longer suitable for the current application. The next step is partial disassembly of the battery pack and removal of the damaged cell. After disassembly, only the damaged cells are replaced by new cells. The battery pack is then reassembled and tested to evaluate performance.

As mentioned, LIB packs are candidates for remanufacturing if the SoH is in the range of 70%–80%. However, the desired SoH may be beyond this range in many cases. For instance, battery packs used in EVs can function above an SoH level of 80%, while for stationary storage, the requirement may be just 50%–60% [13]. Generally, it is seen that SoH degradation of a battery pack is due to decreased performance of only 5%–30% of the cells rather than all the cells in the battery pack. Therefore, if these damaged cells are replaced with new cells at the right time, the SoH can be maintained at the desired level for quite some time. Mathew et al. reported the effect on SoH upon replacing damaged battery modules with new modules [14]. They showed that after 2000 cycles, the SoH falls below the desired SoH level of 80%, and with the replacement of damaged modules, the SoH of the overall pack is restored to the desired level, i.e., 80% SoH, repeatedly. The simulation shows that a battery pack can be used for more than 30,000 cycles with a carefully chosen cell replacement rate.

Fig. 2.3 shows the remanufacturing process flow. As mentioned, the LIB battery pack consists of different cells that can come in various form factors. Depending on the different form factors of the cells, the assembly and disassembly process varies. In the case of cylindrical cells, remanufacturing the packs is relatively difficult because many cells are required to assemble a useable battery module [15]. As a cylindrical cell has a maximum capacity of 10 Ah, several cells are required to make a pack, and monitoring cell-level SoH is tedious [16]. A smart battery management system can provide the SoH of battery modules connected in series; however, doing so at the cell level is impractical.

Modules and cells are connected via nickel or nickel-plated steel strips. These strips are welded to cells using spot, laser, bond wire, or ultrasonic welding [17]. Removing the weld joint from the cell generally damages cell integrity, and reusing it may not be a safe proposition. The connection between the busbar and connectors is glued with epoxy resin and difficult to remove during disassembly. Technologies to effectively remove these connections have yet to mature for remanufacturing at scale.

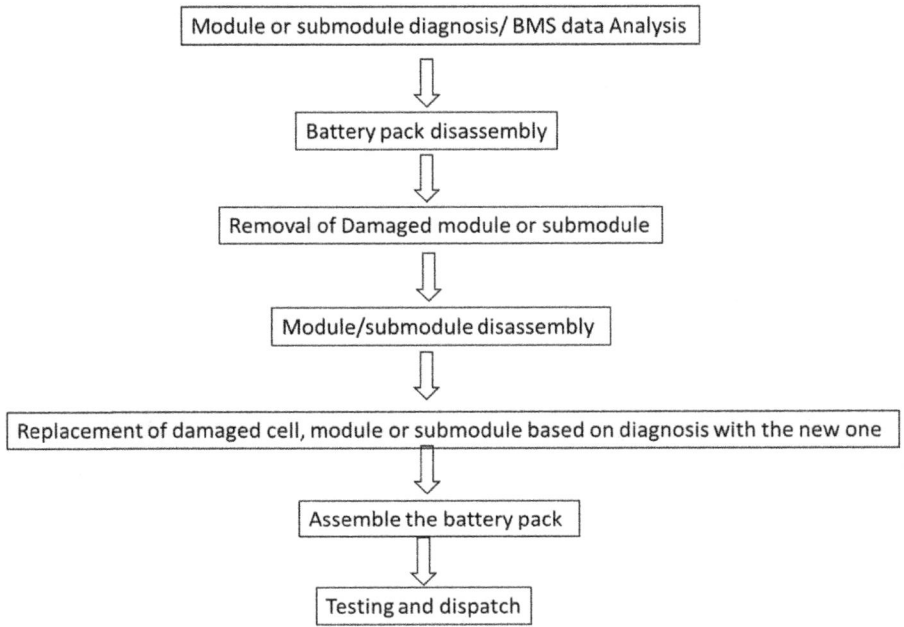

Figure 2.3 Battery remanufacturing process flowchart.

Another approach suggested by researchers is designing the battery pack to be easily remanufactured compared with current LIB packs [18]. For instance, Achim et al. [19] proposed welding one nickel plate to all parallel cells in the battery module so they can be removed easily compared with multiple strips. They suggested that if the welding is done at multiple locations on a cell at lower power instead of in one location with a larger weld area, the nickel plate can be easily dissembled. They also designed a special tool for removing the nickel plate, as shown in Fig. 2.4.

In the case of the prismatic cell LIB pack, the cell tabs are made of aluminum, and both tabs are placed on the same side separated by a vent plug. Aluminum busbars connect the cells, and the joining is done by laser welding. In the current prismatic cell pack design, double-sided adhesive pads join the stacks. Jens et al. [18] suggested an alternative design approach with nonwelded housing and tightening straps for pressure plates. They also studied different joining processes for connecting the prismatic cells like bond wire and conductive adhesive pastes. However, the aluminum connector has higher electrical resistance compared with laser-welded joints. Mechanically fastened joints with the help of screws also have higher electrical resistance, hence avoided. Automated disassembly is possible with milling or laser cutting of the connectors [20]. The heat generated during laser cutting must be controlled to reduce mechanical damage and thermal runaway. A special tool can be placed to collect the molten metal during laser cutting and protect the cell housing.

The third type of design is pouch cells, which are relatively delicate with less shape stability than cylindrical and prismatic form factors. To deal with this, pouch cells are glued to each other during the stack formation of the battery pack. The cells are often glued to aluminum sheets to allow thermal conductivity. Removing the cells from an Al sheet and from each other is very difficult and often results in a damaged cell. The cell tabs are welded with ultrasonic welding, and nondestructive separation is impossible because the tabs are about 0.20–0.3 mm thick. After separation, the surface is not suitable for further joining.

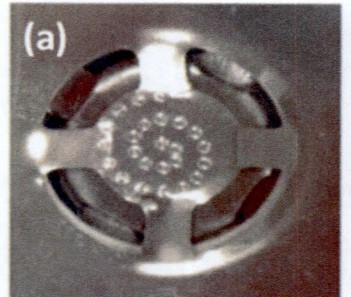
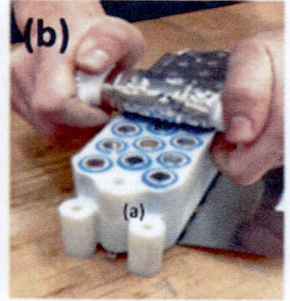
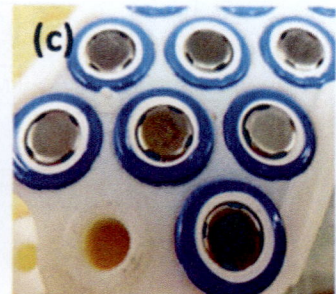

Figure 2.4 Design and tool for removing nickel strips (A) special joint shape with welding at multiple spots (B) removing one joint at a time with the help of special tool (C) Surface of cells after current collector removal [19].

Using cell frames to hold pouch cells can provide mechanical support as well as easy removal while remanufacturing pouch cell-based battery packs. The cell frame determines the stack height, and the cells are positioned on gap pads that compensate for the tolerances and adsorb the swelling due to the state of charge and aging of the cells.

Despite several suggested modifications based on laboratory research, there is no large-scale remanufacturing industry that can provide remanufactured LIBs for EVs. Most remanufacturing technologies are available for nonwelded cells; however, most battery pack manufacturers use welded architecture to manufacture battery packs, with rising demand and severe quality issues. Toyota and Johnson Controls recently registered process-specific patents for battery remanufacturing [21,22]. Tesla also announced its plan to provide remanufactured batteries for certain Tesla models. Eventually, most LIB manufacturers will have to employ remanufacturing to address environmental concerns and generate additional value. The remanufacturing technology used is expected to vary, but all manufacturers will have to ensure maximum safety during remanufacturing to avoid short-circuiting resulting in fire hazards [5]. Various safety protocols for remanufacturing, such as UL 1642, UL 2054, IEC 62,133, IEC 62,281, SAE J2464, and IEEE 1625 and 1725, are followed to evaluate the safety standard of the battery pack.

4. Repurposing

Repurposing is quite similar to remanufacturing since it also involves screening, evaluation, and disassembly, followed by modification of the pack assembly. However, in repurposing, the end application of the battery pack changes from its original application. Repurposing may require various thermal management systems, battery management systems, and high-voltage connectors. Repurposed batteries are generally used in less demanding applications such as energy-storage systems, uninterrupted power supply, electric forklifts, and low-speed EVs. The major concerns of repurposing are reliability and predicting the future useful life of the battery pack. In repurposing, the end-of-life battery pack is disassembled up to the cell level and again grouped after careful evaluation up to the cell level as required for the second-life application. The test protocols used to classify are similar to those used in remanufacturing. The individual cell performance parameters such as open-circuit voltage, internal resistance, lithium diffusion coefficient, and activation resistance are measured using the experimental and model-based techniques described in remanufacturing. The cell parameters vary within a wider range, and depending on those parameters, the cells are classified. The classification time spent for cell sorting affects the economics of repurposing. Manual classification is done based on comprehensive analysis and setting a threshold value for each cell performance parameter.

On the other hand, the ML technique facilitates the classification of cells with minimum human intelligence. However, the shortcoming of the ML algorithm is the need for a large, labeled data set. For LIBs, it is a tedious task compared with other fields, where data are images or text. This is because to generate labeled data set for LIBs, life cycle testing is needed, which takes ~1.5–2 years. Therefore, selecting ML models that effectively predict, even with a smaller number of labeled data, can benefit the process economics. For example, linear ML models like linear regression, logistic regression, Naïve–Bayes, K-nearest neighbors, etc. have been used to classify cells effectively to some degree [23]. More sophisticated nonlinear ML models like neural networks, convolutional neural networks, random forest, etc. can improve classification accuracy. To classify end-of-life LIBs, Lai et al. [23,24] and Jiang et al. [25] used genetic algorithm-based backpropagation piecewise linear fitting models and a K-means algorithm, respectively. Garg et al. [26] proposed experimental and ML self-organizing map clustering algorithm to sort and classify batteries. However, due to the absence of an aging model in the proposed classification method, its use in practical secondary-use applications is limited.

To illustrate repurposing, the classification of 15 Ah capacity cells discarded from an EV is used as a case study. The experimental data for OCV, capacity, pulsed discharge voltage, ohmic resistance (R_s), charge transfer resistance (R_{ct}), and the diffusion coefficient of lithium-ion (D_{Li}) are collected from the work of Lio et al. [27] and are provided in Table 2.1. The cells were classified into three groups based on the threshold value of these parameters. In the first group (Group 1), cells with a capacity in the range of 12–14 Ah, pulse discharge end voltage greater than 2.7 V, R_{ct} less than 7 miliohm, and D_{Li} greater than 6×10^{-14} cm^2/s have been placed. For selecting cells in less demanding stationary storage (Group 2), the thresholds for capacity, pulse discharge end voltage, Rct, and D_{Li} are kept in ranges of 10–14 Ah, 2.5 V, less than 10 miliohm, and 6×10^{-14} cm^2/s, respectively. The other cells that do not qualify for the thresholds set for Group 1 or Group 2 are sent for recycling.

It should be noted that the threshold values chosen for each parameter to classify the cells are based on previous scientific knowledge about cell parameters. These threshold cell parameters can vary depending on the required duty cycle of the end application.

5. Recycling

Recycling is the last link in the circular economy of LIBs. It includes the disassembly of the battery pack up to the cell level and extraction of the individual materials used to manufacture cells. The cells are around 80% of the total weight of the battery pack [28]. A cell consists of a lithium compound, graphite, and lithium salt that act as a cathode, an anode, and an electrolyte, respectively. Along with these, it contains organic binders, copper, and aluminum current collector [29]. The recycling process can be divided into physical and chemical processes. The major objectives of physical processes,

Table 2.1 Classification of retired cells for repurposed application.

Battery no.	Capacity (Ah)	Nominal voltage (V)	Capacity after test (Ah)	Pulse discharge voltage (5C)	Rs (mOhm)	Rct	D_{Li}	Group 1	Group 2
1	15	3.2	12.8	2.865	4.8	5.5	6.1	OK	OK
2	15	3.2	13.2	2.847	3.8	5	7	OK	OK
3	15	3.2	12.7	2.865	3.5	8	6.5	Not OK	OK
4	15	3.2	12.9	2.659	5	5.8	7.8	Not OK	OK
5	15	3.2	10.8	2.812	3.5	12	8	Not OK	Not OK
6	15	3.2	8.8	2.657	3.5	9	2	Not OK	Not OK
7	15	3.2	11.7	2.705	6.3	5.8	8.8	Not OK	OK
8	15	3.2	13.2	2.839	5.5	5.8	8	OK	OK
9	15	3.2	13.3	2.847	4.5	7	10.8	OK	OK
10	15	3.2	12.7	2.525	5	6	10.2	Not OK	OK
11	15	3.2	8	2.824	7	6.5	6	Not OK	Not OK
12	15	3.2	13.3	2.837	4	5	10.2	OK	OK
13	15	3.2	9.2	2.77	1.5	5.2	3.5	Not OK	Not OK
14	15	3.2	12	2.802	5	6	10	OK	OK
15	15	3.2	13.3	2.606	5	4.5	10	Not OK	OK
16	15	3.2	13.1	2.878	4	5	7	OK	OK
17	15	3.2	13.1	2.596	5.2	4.2	10.2	Not OK	OK
18	15	3.2	8.6	2.758	1	4.5	2	Not OK	Not OK
19	15	3.2	12.8	2.898	4	5	13	OK	OK
20	15	3.2	13.1	2.393	5.8	4.5	8.8	Not OK	Not OK

often called pretreatment processes, are separation of the cathode, anode, and separator and removal of the binder and other battery components.

Fig. 2.5 shows the schematic flowchart of various processes involved in preprocessing. The pretreatment regime plays a major role in improving the efficiency of subsequent recycling steps in LIBs and can be broadly classified into three major groups:

i. Discharging regime
ii. Liberation regime
iii. Separation regime

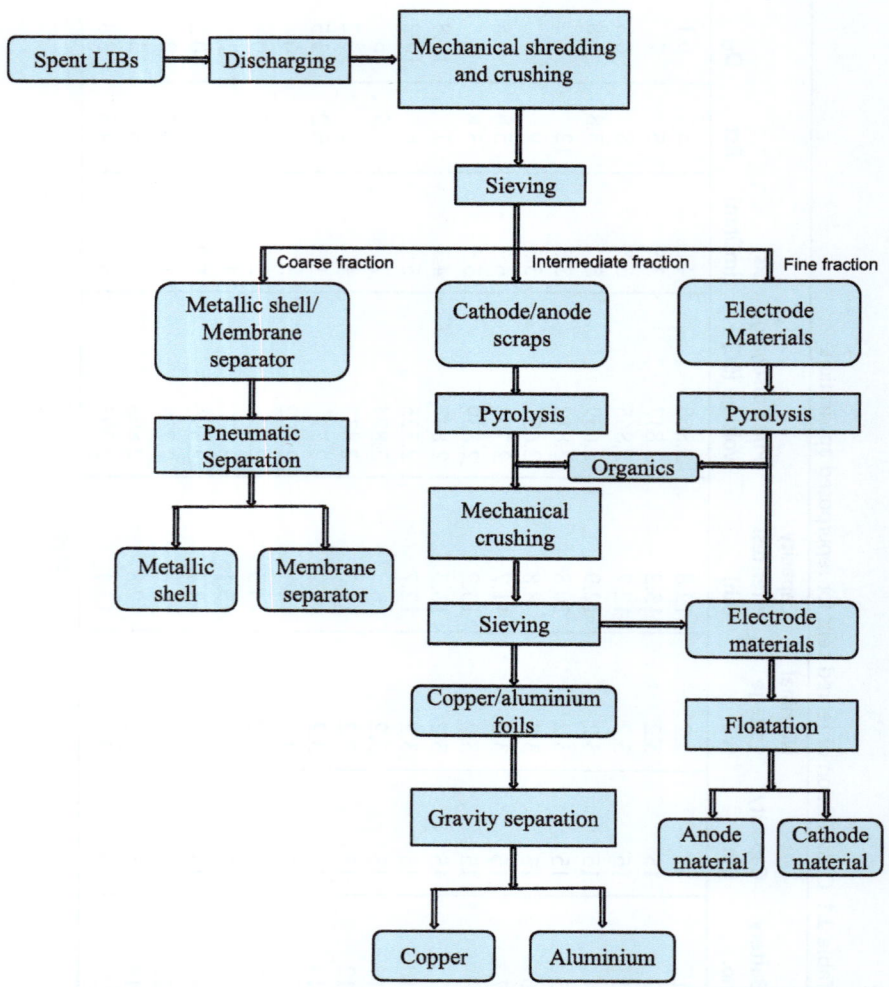

Figure 2.5 Steps involved in the pretreatment of lithium batteries for recycling.

5.1 Discharging regime

End-of-life lithium-ion cells, i.e., cells with an SoH of less than 40%, still have some remaining capacity. If the cells are not properly discharged, the charge can be suddenly released, causing a large amount of heat and thermal runaway, leading to catastrophe. Therefore, the first step in recycling is fully discharging the LIBs to their discharged state. Discharging can be done by either physical or chemical methods. In the physical discharge method, a load circuit is applied to LIBs to completely discharge the cells. Kruger et al. [30] used ohmic discharging to recycle NMC-based LIBs. Physical discharging makes LIBs electrochemically unstable. Yao et al. [31] report that physical discharging techniques are not ideally suited for large-scale industrial applications. On the other hand, with chemical methods, LIBs are immersed in chemical solutions with high electronic conductivity to short-circuit the electrodes. Generally, 0.8M NaCl, 0.8M $FeSO_4$, and 0.8M $MnSO_4$ solutions are used for the process. One downside of high-discharge-rate solutions is their need for proper disposal because they tend to corrode metallic shells by electrochemical reaction [32].

5.2 Liberation regime

In the liberation regime, individual components inside spent LIBs are liberated. It is a pretreatment process wherein batteries are first dismantled to isolate the major components such as plastics, metals, and current collectors with the anode and cathode. This is achieved by mechanically crushing the LIBs to reduce the granular size of the material for proper recycling. Preexisting crushing techniques include dry or wet impact crushing. Dry crushing is generally considered a better technique, as it favors subsequent sieving and separation. Various materials inside LIBs are present at different length scales, such as the metallic outer covering and battery separator measure about 0.25 mm, whereas current collector foils measure about 2 mm. Electrode materials in powder form in original LIBs are about 0.045 mm in size. As the length scales of individual components are quite different, the preliminary separation of different components can be done by combining the crushing and sieving processes based on the selective crushing property. DaCosta et al. [29] have proposed an alternate method where a direct mechanical crushing technique is used to crush spent LIBs without dismantling.

5.3 Separation regime

Separation is the process of separating different material components such as electrolyte, copper, aluminum, active anode material, casing material, separator, and active cathode material. Physical separation is performed according to differences in physical properties such as the density, size, ferromagnetism, conductivity, and hydrophobicity of the different materials. Various separation techniques are discussed below:

Eddy current separation: The eddy current separation technique is based on the significant conductivity difference between the various components of LIB electrodes. This process is employed after the liberation regime and membrane separator removal using air classification. This technique can be used to separate current collectors (Al, Cu) and the active cathode material. It has been successfully demonstrated to separate lithium iron phosphate, copper, and aluminum based on conductivity differences [33]. The presence of a rotating magnetic roller and uniform magnetic field generates eddy currents in nonferrous metals. Under the influence of these eddy currents and the magnetic field, the drop-down trajectory of the nonferrous metals is affected, leading to separation.

Electrostatic Separation: Eddy current separation is useful when the individual components are conductive. However, the spent LIBs contain a significant portion of the nonmetallic component. Electrostatic separation is used for separating the metallic and nonmetallic components. In this process, spent LIBs are milled and dried in the form of powders. Cathode and anode materials are separated from the remaining components by consecutive sieving. The separated cathode and milled anode powder are sent for electrostatic separation where components containing greater than 98% metal and the component containing greater than 99% polymers are separated. There is also the presence of an intermediate fraction which includes both metal and nonmetals, which is sent back to the previous step to repeat the process.

Gravity separation: Gravity separation is a technology that works on separation based on density differences. Spent LIBs are first crushed, and then the powder is taken for gravity separation. In the first step, polymer fraction along with smaller-sized crushed current collectors on which cathode and anode active material are present are collected. These are separated using a low volume of air. In the second stage, current collector foils, with the cathode and anode, of larger size are separated. In the third step, the metallic casing is recovered by applying a large air volume. In the case of poor prior liberation, anode or cathode materials adhere to foil material, which leads to difficulty in separating based on density difference.

Froth Floatation: Froth flotation is the technique used for the separation of fine particles based on the difference in their hydrophobicity values. Generally, when the anode and cathode active material particle sizes are very small, i.e., 0.045 mm, the effectiveness of gravity separation is reduced, and the separation of fine components is not possible. Froth floatation is an alternative method to separate these fine particles. The cathode material of a spent LIB is hydrophilic, while the anode material is hydrophobic. This property difference is exploited while applying froth flotation for their separation. Challenges in the process include organic components and binder residue, which decreases the hydrophobicity difference between cathode and anode materials [34]. It is advised that surface modification techniques like anaerobic pyrolysis [35], mechanical grinding [36], and cryogenic grinding [37] are applied to improve the efficiency of the froth floatation process.

After separating different components from spent LIBs, we have cathode and anode powders. As we know that anode and cathodes combine different elements such as Li, Co, Ni, Al, Li, and C, the individual elements require extraction to further reuse them in the cell manufacturing cycle. Various materials such as Ni, Co, Al, Cu, Li, and phosphorous can be extracted from obtained and separated materials. The material obtained after pretreatment will be sent for further refining to obtain the final secondary material. The final recycling product can be pure metal, alloy, compound, a solution containing metal ions, or slags.

6. Extraction of materials

The extraction of material can be broadly classified into three types:
  i. Hydrometallurgical processes
 ii. Pyrometallurgical processes
iii. Biohydrometallurgical processes

Generally, a combination of the above three processes is used to accomplish the final step in LIB recycling, i.e., the extraction of individual compounds or elements from the cathodic and anodic sides. In most processes, a combination of pyrometallurgical and hydrometallurgical steps is involved.

6.1 Hydrometallurgical processes

In this technique, aqueous solutions are used to leach heavy metals from the cathodic side. Reagents like H_2SO_4 and H_2O_2 are used to recover lithium in the form of its sulfate, as given in Reaction 1:

$$2LiCoO_2 \text{ (s)} + 3H_2SO_4 + H_2O_2 \rightarrow 2CoSO_4 \text{ (aq)} + Li_2SO_4 + 4H_2O + O_2 \text{ (Reaction } -1)$$

Lithium can be extracted through the precipitation reaction forming its corresponding carbonate or phosphate. (Li_2CO_3 or Li_3PO_4). The metals obtained through reagent recovery are then converted to compounds with other industrial applications—for example, $CoFe_2O_4$ and $MnCo_2O_4$. The type and concentration of leachant determine the product and its quality. Generally, metals are obtained as their carbonates, phosphates, or oxides in this process. These compounds can be further reduced to obtain metals in their elemental forms if required.

6.2 Pyrometallurgical processes

Pyrometallurgical processes are high-temperature processes where metal oxides are reduced using reductants, generally carbon, to alloys of heavy metals (metals like Co, Cu, Fe, Ni, etc.). This process is versatile, efficient, and continuous as metals are extracted in their molten states. The current collectors are also automatically smelted into the alloy obtained. The products of the pyrometallurgical processes are slag, alloy, and gases. Organic components used in the battery are burnt off. Generally, the pyrometallurgical process is coupled with the hydrometallurgical process, and the metals are reclaimed using alternate leaching steps. Plastics are burnt off along with the lithium salts that are present. However, this method has huge environmental drawbacks, such as producing toxic gases. Other demerits include high-energy costs and poor recovery of elements. This process is only used to recover costly heavy metals like Co and Ni.

6.3 Biohydrometallurgical processes

Biohydrometallurgical processes work on the principle of selectivity of microorganisms toward the concentration of metal ions present, enabling us to recover metals selectively. The process was inspired by early bioleaching processes employed in the mining industry. This is complementary to the hydrometallurgical and pyrometallurgical processes used to recycle LIBs. The field of application is relatively small and works only to extract metals such as iron and copper. This is not a standalone process at times and is used to complement certain preexisting recycling methods. This process remains a field of research, and widespread standalone application is still in its early days. The particle size on which the microorganisms act to recover metals is also a major factor. It has been reported that if the particle size is too small (less than 25 microns), cell activity is negatively affected.

7. Challenges and future trends

One of the major challenges of closed-loop manufacturing for remanufacturing, repurposing, and recycling is the lack of standardization in LIB pack design and battery chemistry. Large-scale remanufacturing and repurposing will be challenging unless efficient battery parameter monitoring tools are installed while designing the LIB pack. This will add to the up-front cost. Therefore, if it is not economically beneficial, it would be difficult to justify unless mandated by policy. The adaptation of LIB recycling at a commercial level is expected to be implemented within a short time frame. However, the existing technologies have many limitations.

Pyrometallurgical processes are generally faster than the other two processes since they can be applied directly to the physical form of battery scrap without the requirement of any chemical treatment. The disadvantage of pyrometallurgical processes is the emissions of gases such as oxides and furans [38]. They also release dust into the environment. Thus, recycling processes that involve pyrometallurgical steps must have dust collecting and gas cleaning mechanisms in place for the successful release of these gases into the atmosphere. This is an engineering challenge and requires capital for its implementation. One should note that the greatest disadvantage of the pyrometallurgical process is its inability to recover lithium.

On the other hand, hydrometallurgical processes eliminate the formation of gaseous elements and are less energy-intensive [39]. This makes them generally more economical. They are simple to operate and have low costs. The products obtained from hydrometallurgical processes provide high purity, as they are recovered by preferential leachants. The disadvantages of hydrometallurgy include undertaking certain required pretreatment steps. The pretreatment steps generally involve the physical separation of individual battery components, and this makes hydrometallurgy time-consuming. Further, the leachants involved are strong acids and bases, making handling difficult and posing environmental and personnel hazards. The effluents produced in the hydrometallurgical process also need subsequent treatment to be fit for discharge into the environment [40].

Biohydrometallurgical processes are relatively simpler and safer, require less energy and capital, enable easier management, and have a negligible environmental impact [41]. The leachants on which microbial growth is preferred for metal recovery are generally biological and hence pose no biological hazard. Biohydrometallurgical processes also have minimum effluent formation. The major disadvantage of the biohydrometallurgical process is its slow kinetics. As the process depends on microbial growth, external factors such as temperature, humidity, and pH play a major role in the yield obtained [42].

With the desire to increase energy density, LIBs have ever-changing chemistries. Currently, the major cathode compositions in the market are lithium cobalt oxides, lithium nickel manganese cobalt oxide (NMC), lithium iron phosphate (LFP), lithium nickel cobalt aluminum oxide, and lithium manganese oxide [43]. Fortunately, the anode is predominantly graphite, although many manufacturers incorporate some silicon in the anode. As the inflow of LIBs can come from any of the major prevalent chemistries, it complicates the recycling process, as there is no "one size fits all" recycling technique that can recycle all chemistries with the same kind of efficiency. With the new chemistries, the electrolyte composition has been modified to increase durability. These electrolyte additives can pollute the environment during recycling [44]. Solvents such as DMC, PC, and EC are used in LIBs, and they tend to react with carbon dioxide. These solvents have the potential to decompose with the application of heat and hence pose a risk for pyrometallurgical recycling techniques.

As discussed above, the conventional pyrometallurgical, hydrometallurgical, and biohydrometallurgical processes have certain shortcomings; hence, other recycling methods must be explored. Zhang et al. [45] developed a novel environmentally friendly process demonstrating the recyclability and resynthesizing ability of NMC cathode material. The process used trifluoroacetic acid as both the leaching and the separating agent. Recently, it was shown that it is possible to selectively recover lithium in spent LIBs by tweaking the age-old chlorination technology of metal extraction.

LIB electrode active materials, i.e., lithium compound cathode and graphite anode, have excellent microwave absorption properties. Hence, a recent attempt has been to use microwaves to reduce the electrodes after the physical separation step [46]. This process enables the extraction of the metal value in the spent LIBs. The lithium is obtained in the form of its carbonate and a magnetic fraction containing valuable metals such as cobalt, manganese, and nickel (a few of which are obtained in their oxide state).

Another method that can disrupt the current scheme of recycling is the direct recycling process, where battery constituents are directly recovered and used to make new LIBs with minimal processing. Complicated preprocessing processes used in the other methods, namely pyrometallurgy and hydrometallurgy, are targeted to be eliminated. Electrolyte extraction is done using supercritical CO_2, and this electrolyte can be recovered from CO_2 by reducing pressure and temperature [44]. Following quality assurance procedures, the recovered electrolyte can be directly used for battery manufacturing. After electrolyte recovery, the spent cells are dismantled and crushed. Then, physical techniques are used to separate individual components. Here, the major emphasis is on recovering the cathode material in its active form [47]. This cathode material can be used to make new batteries, provided it has sufficient lithium concentration. Possible relithiation steps are undertaken to increase the lithium concentration in case of irreversible lithium loss. Chen et al. [48] reported that LFP could be directly obtained from spent LIBs. The direct recycling process is energy-efficient, environmentally friendly, and has a high recovery rate of materials. However, the long-term working effects of these materials have not yet been evaluated.

8. Conclusion

With the rising EV demand and the need for a closed-loop circular economy, the concept of reusing lithium batteries is becoming popular. The closed-loop manufacturing of LIBs starting with remanufacturing, then repurposing, and finally recycling can benefit the LIB-based energy storage ecosystem. However, given the complexity of LIB manufacturing and use, it is not very straightforward. The first step would be to identify the optimal remanufacturing and repurposing points and their frequency during the overall working life of the LIB pack. Without this, it is not possible to exploit the full potential of remanufacturing or repurposing. State-of-the-art battery management systems and ML techniques are necessary to monitor SoH and battery parameters so that the right time for LIB remanufacturing and repurposing can be identified. The last link, recycling, has been extensively studied on a laboratory scale, and many techniques have been proposed. With different manufacturers using various chemistries, it will be difficult to standardize any particular recycling technique. However, some recycling techniques are being commercialized and may have significant impacts. In the future, direct recycling can provide a significant benefit by reducing the cost of making new materials for LIBs.

References

[1] Y. Zhao, et al., A review on battery market trends, second-life reuse, and recycling, Sustain. Chem. 2 (1) (2021) 167–205, https://doi.org/10.3390/suschem2010011.

[2] W. Cai, Lithium-Ion battery manufacturing for electric vehicles: a contemporary overview, in: J. Li, S. Zhou, Y. Han (Eds.), Advances in Battery Manufacturing, Service, and Management Systems, John Wiley & Sons, Inc., Hoboken, NJ, USA, 2016, pp. 1–28, https://doi.org/10.1002/9781119060741.ch1.

[3] Y. Hua, Sustainable value chain of retired lithium-ion batteries for electric vehicles, J. Power Sources (2020) 16.

[4] M. Galeotti, L. Cinà, C. Giammanco, S. Cordiner, A. Di Carlo, Performance analysis and SOH (state of health) evaluation of lithium polymer batteries through electrochemical impedance spectroscopy, Energy 89 (2015) 678–686, https://doi.org/10.1016/j.energy.2015.05.148.

[5] N. Noura, L. Boulon, S. Jemeï, A review of battery state of health estimation methods: hybrid electric vehicle challenges, World Electr. Veh. J. 11 (4) (2020) 66, https://doi.org/10.3390/wevj11040066.

[6] R. Xiong, Y. Zhang, J. Wang, H. He, S. Peng, M. Pecht, Lithium-ion battery health prognosis based on a real battery management system used in electric vehicles, IEEE Trans. Veh. Technol. 68 (5) (2019) 4110–4121, https://doi.org/10.1109/TVT.2018.2864688.

[7] Z.B. Omariba, L. Zhang, H. Kang, D. Sun, Parameter identification and state estimation of lithium-ion batteries for electric vehicles with vibration and temperature dynamics, World Electr. Veh. J. 11 (3) (2020) 50, https://doi.org/10.3390/wevj11030050.

[8] H. He, X. Zhang, R. Xiong, Y. Xu, H. Guo, Online model-based estimation of state-of-charge and open-circuit voltage of lithium-ion batteries in electric vehicles, Energy 39 (1) (2012) 310–318, https://doi.org/10.1016/j.energy.2012.01.009.

[9] R. Darling, J. Newman, Modeling side reactions in composite Li y Mn2 O 4 electrodes, J. Electrochem. Soc. 145 (3) (1998) 990–998, https://doi.org/10.1149/1.1838376.

[10] H. Pan, Z. Lü, H. Wang, H. Wei, L. Chen, Novel battery state-of-health online estimation method using multiple health indicators and an extreme learning machine, Energy 160 (2018) 466–477, https://doi.org/10.1016/j.energy.2018.06.220.

[11] A. Zenati, P. Desprez, H. Razik, Estimation of the SOC and the SOH of li-ion batteries, by combining impedance measurements with the fuzzy logic inference, in: IECON 2010 - 36th Annual Conference on IEEE Industrial Electronics Society, Glendale, AZ, USA, Nov. 2010, pp. 1773–1778, https://doi.org/10.1109/IECON.2010.5675408.

[12] E. Sundin, M. Lindahl, Rethinking product design for remanufacturing to facilitate integrated product service offerings, in: 2008 IEEE International Symposium on Electronics and the Environment, San Francisco, CA, USA, May 2008, pp. 1–6, https://doi.org/10.1109/ISEE.2008.4562901.

[13] L.C. Casals, Second life batteries lifespan_ rest of useful life and environmental analysis, J. Environ. Manag. (2019) 10.

[14] M. Mathew, Simulation of lithium ion battery replacement in a battery pack for application in electric vehicles, J. Power Sources (2017) 11.

[15] G. Harper, "Recycling Lithium-Ion Batteries from Electric Vehicles," p. 12.

[16] M. Galeotti, Performance Analysis and SOH (State of Health) Evaluation of Lithium Polymer Batteries through Electrochemical Impedance Spectroscopy, 2015, p. 9.

[17] M.J. Brand, P.A. Schmidt, M.F. Zaeh, A. Jossen, Welding techniques for battery cells and resulting electrical contact resistances, J. Energy Storage 1 (2015) 7–14, https://doi.org/10.1016/j.est.2015.04.001.

[18] J. Schafer, R. Singer, J. Hofmann, and J. Fleischer, "Challenges and Solutions of Automated Disassembly and Condition-Based Remanufacturing of Lithium-Ion Battery Modules for a Circular Economy," p. 6.

[19] A. Kampker, S. Wessel, F. Fiedler, F. Maltoni, Battery pack remanufacturing process up to cell level with sorting and repurposing of battery cells, J. Remanuf. 11 (1) (2021) 1–23, https://doi.org/10.1007/s13243-020-00088-6.

[20] G. Harper, et al., Recycling lithium-ion batteries from electric vehicles, Nature 575 (7781) (2019) 75–86, https://doi.org/10.1038/s41586-019-1682-5.
[21] "US2015093611A1".
[22] "WO2015016979A1".
[23] X. Lai, D. Qiao, Y. Zheng, W. Yi, A novel screening method based on a partially discharging curve using a genetic algorithm and back-propagation model for the cascade utilization of retired lithium-ion batteries, Electronics 7 (12) (2018) 399, https://doi.org/10.3390/electronics7120399.
[24] X. Lai, D. Qiao, Y. Zheng, M. Ouyang, X. Han, L. Zhou, A rapid screening and regrouping approach based on neural networks for large-scale retired lithium-ion cells in second-use applications, J. Clean. Prod. 213 (2019) 776–791, https://doi.org/10.1016/j.jclepro.2018.12.210.
[25] Y. Jiang, Y. Wang, C. Zhang, G. Su, J. Liu, "Research on group methods of second-use Li-ion batteries based on k-means clustering model, in: 2014 IEEE Conference and Expo Transportation Electrification Asia-Pacific (ITEC Asia-Pacific), Beijing, China, Aug. 2014, pp. 1–6, https://doi.org/10.1109/ITEC-AP.2014.6941098.
[26] A. Garg, Development of recycling strategy for large stacked systems: experimental and machine learning approach to form reuse battery packs for secondary applications, J. Clean. Prod. (2020) 17.
[27] Q. Liao, "Performance Assessment and Classification of Retired Lithium Ion Battery from Electric Vehicles for Energy Storage," p. 7.
[28] N. Vieceli, C.A. Nogueira, C. Guimarães, M.F.C. Pereira, F.O. Durão, F. Margarido, Hydrometallurgical recycling of lithium-ion batteries by reductive leaching with sodium metabisulphite, Waste Manag. 71 (2018) 350–361, https://doi.org/10.1016/j.wasman.2017.09.032.
[29] A.J. da Costa, J.F. Matos, A.M. Bernardes, I.L. Müller, Beneficiation of cobalt, copper and aluminum from wasted lithium-ion batteries by mechanical processing, Int. J. Miner. Process. 145 (2015) 77–82, https://doi.org/10.1016/j.minpro.2015.06.015.
[30] S. Krüger, C. Hanisch, A. Kwade, M. Winter, S. Nowak, Effect of impurities caused by a recycling process on the electrochemical performance of Li[Ni$_{0.33}$Co$_{0.33}$Mn$_{0.33}$]O$_2$, J. Electroanal. Chem. 726 (2014) 91–96, https://doi.org/10.1016/j.jelechem.2014.05.017.
[31] L.P. Yao, Q. Zeng, T. Qi, J. Li, An environmentally friendly discharge technology to pretreat spent lithium-ion batteries, J. Clean. Prod. 245 (2020) 118820, https://doi.org/10.1016/j.jclepro.2019.118820.
[32] J. Xiao, B. Niu, Q. Song, L. Zhan, Z. Xu, Novel targetedly extracting lithium: an environmental-friendly controlled chlorinating technology and mechanism of spent lithium ion batteries recovery, J. Hazard Mater. 404 (2021) 123947, https://doi.org/10.1016/j.jhazmat.2020.123947.
[33] H. Bi, H. Zhu, L. Zu, Y. Bai, S. Gao, Y. Gao, A new model of trajectory in eddy current separation for recovering spent lithium iron phosphate batteries, Waste Manag. 100 (2019) 1–9, https://doi.org/10.1016/j.wasman.2019.08.041.
[34] T. Zhang, Y. He, F. Wang, H. Li, C. Duan, C. Wu, Surface analysis of cobalt-enriched crushed products of spent lithium-ion batteries by X-ray photoelectron spectroscopy, Separ. Purif. Technol. 138 (2014) 21–27, https://doi.org/10.1016/j.seppur.2014.09.033.
[35] G. Zhang, Y. He, Y. Feng, H. Wang, X. Zhu, Pyrolysis-ultrasonic-assisted flotation technology for recovering graphite and LiCoO$_2$ from spent lithium-ion batteries, ACS Sustain. Chem. Eng. 6 (8) (2018) 10896–10904, https://doi.org/10.1021/acssuschemeng.8b02186.
[36] J. Yu, Y. He, Z. Ge, H. Li, W. Xie, S. Wang, A promising physical method for recovery of LiCoO$_2$ and graphite from spent lithium-ion batteries: grinding flotation, Separ. Purif. Technol. 190 (2018) 45–52, https://doi.org/10.1016/j.seppur.2017.08.049.
[37] J. Liu, et al., Recovery of LiCoO$_2$ and graphite from spent lithium-ion batteries by cryogenic grinding and froth flotation, Miner. Eng. 148 (2020) 106223, https://doi.org/10.1016/j.mineng.2020.106223.
[38] M. Zhou, B. Li, J. Li, Z. Xu, Pyrometallurgical technology in the recycling of a spent lithium ion battery: evolution and the challenge, ACS EST Eng. (2021), https://doi.org/10.1021/acsestengg.1c00067.
[39] A. Chagnes, B. Pospiech, A brief review on hydrometallurgical technologies for recycling spent lithium-ion batteries: technologies for recycling spent lithium-ion batteries, J. Chem. Technol. Biotechnol. 88 (7) (2013) 1191–1199, https://doi.org/10.1002/jctb.4053.

[40] P. Meshram, A. Mishra, Abhilash, R. Sahu, Environmental impact of spent lithium ion batteries and green recycling perspectives by organic acids — a review, Chemosphere 242 (2020) 125291, https://doi.org/10.1016/j.chemosphere.2019.125291.

[41] N. Bahaloo-Horeh, S.M. Mousavi, Enhanced recovery of valuable metals from spent lithium-ion batteries through optimization of organic acids produced by *Aspergillus niger*, Waste Manag. 60 (2017) 666—679, https://doi.org/10.1016/j.wasman.2016.10.034.

[42] N. Bahaloo-Horeh, F. Vakilchap, S.M. Mousavi, Bio-hydrometallurgical methods for recycling spent lithium-ion batteries, in: L. An (Ed.), Recycling of Spent Lithium-Ion Batteries, Cham: Springer International Publishing, 2019, pp. 161—197, https://doi.org/10.1007/978-3-030-31834-5_7.

[43] N. Mohamed, N.K. Allam, Recent advances in the design of cathode materials for Li-ion batteries, RSC Adv. 10 (37) (2020) 21662—21685, https://doi.org/10.1039/D0RA03314F.

[44] Y. Liu, D. Mu, R. Li, Q. Ma, R. Zheng, C. Dai, Purification and characterization of reclaimed electrolytes from spent lithium-ion batteries, J. Phys. Chem. C 121 (8) (Mar. 2017) 4181—4187, https://doi.org/10.1021/acs.jpcc.6b12970.

[45] X. Zhang, Y. Xie, H. Cao, F. Nawaz, Y. Zhang, A novel process for recycling and resynthesizing $LiNi_{1/3}Co_{1/3}Mn_{1/3}O_2$ from the cathode scraps intended for lithium-ion batteries, Waste Manag. 34 (9) (2014) 1715—1724, https://doi.org/10.1016/j.wasman.2014.05.023.

[46] S. Pindar, N. Dhawan, Rapid recycling of spent lithium-ion batteries using microwave route, Process Saf. Environ. Protect. 147 (2021) 226—233, https://doi.org/10.1016/j.psep.2020.09.012.

[47] B. Huang, Z. Pan, X. Su, L. An, Recycling of lithium-ion batteries: recent advances and perspectives, J. Power Sources 399 (2018) 274—286, https://doi.org/10.1016/j.jpowsour.2018.07.116.

[48] M. Chen, et al., Recycling end-of-life electric vehicle lithium-ion batteries, Joule 3 (11) (2019) 2622—2646, https://doi.org/10.1016/j.joule.2019.09.014.

CHAPTER 3

Nanotechnology and recycling, remanufacturing, and reusing battery

Giovani Pavoski[1], Amilton Barbosa Botelho Junior[1], Rebeca Mello Chaves[2], Thuany Maraschin[3], Leandro Rodrigues Oviedo[4], Thamiris Auxiliadora Gonçalves Martins[1], William Leonardo da Silva[4], Daniel Assumpção Bertuol[2] and Denise Crocce Romano Espinosa[1]

[1]Department of Chemical Engineering, University of São Paulo, São Paulo, Brazil; [2]Department of Chemical Engineering, Federal University of Santa Maria, Santa Maria, Brazil; [3]School of Technology, Pontifical Catholic University of Rio Grande do Sul, Porto Alegre, Brazil; [4]Nanoscience Graduate Program, Franciscan University, Santa Maria, Brazil

1. Introduction

With the advancement of technology, new materials are increasingly developed, and existing ones are improved. These are important steps in social and economic development. However, there is a certain concern with managing these materials when they are no longer being used—they become waste.

Batteries have experienced great technological development in recent years. Batteries have increased their efficiency by introducing nanomaterials to their composition, thus increasing their usage time, robustness, weight, and other characteristics. A battery recovery industry exists for batteries such as acid car batteries, but the methodologies for recycling newer battery forms are still being created. These factors are important because there is a concern with the actions of nanomaterials on the environment and human beings. From this, it is necessary to manage all materials, including those on the nanometric scale [1].

With this certainty of the need to recycle new materials, it is possible to point out the concept of a circular economy—that is, after recycling, the system of new materials from the handling of recycled matter. Some methodologies have consolidated and are used to recycle batteries and the nanomaterials within them. Methodologies such as hydrometallurgy, pyrometallurgy, and nanohydrometallurgy help in waste management and the separation of elements. In addition to these applied methodologies, research is ongoing to study nanomaterials using nanofibers, nanoparticles (NPs), and graphene, all of which are acquired from secondary sources, that is, residues [2].

In this chapter, some nanomaterials currently found in the composition of batteries will be presented. Also described are methodologies presently used to recycle batteries and nanomaterials, as well as the application of the circular economy and new materials synthesis from the recycling of these batteries.

2. Battery nanostructures

Batteries of various types and sizes are considered one of the most suitable approaches to store energy, and extensive research exists for various battery technologies and applications; however, the environmental impacts of large-scale battery use require further study [3]. Many kinds of batteries can be used to store energy, and there are many challenges in the electrode materials, electrolytes, and construction of these batteries. The research related to energy-storage systems is extremely active [4].

A substantial amount of research has been conducted on the fundamentals and practical applications of batteries. Motivated by resource exhaustion and environmental deterioration, people have made efforts to improve rechargeable battery performance and protect the environment by adopting a wide range of strategies. However, many challenges remain [5,6].

With the rapidly growing demand for higher energy density from industry, the fabrication of flexible and adaptable energy-storage materials combined with the development of high energy density, good mechanical properties, and long cycle life is imperative. In fact, it is a crucial need to maintain the living standards of modern society. As a result, unusual forms of electronics essentially require high-performance integrated flexible batteries for uninterrupted electrical power supply [7–9]. Therefore, nanomaterials are being applied to obtain advanced batteries built for reduced volume and weight [10].

Nanotechnology refers to techniques for using materials in nanoscale units ranging from 1 to 1000 nm [11]. Furthermore, control over the nanoscale assembly of various components of energy storage is being sought [12]. Consequently, emerging electrochemical energy-storage systems, including Na/K/Mg/Ca/Zn/Al-ion nanomaterials batteries, have been met with extensive interest in battery development [13].

Carbon-based materials with multiple functions represent the most standard materials for batteries due to their large surface area, good stability, and high conductivity [14]. In addition to the abovementioned carbon-based nanomaterial proprieties, carbon-based materials decorated with metal oxide NPs are considered promising and cost-effective in battery electrodes, which is ascribed to their low overpotential and notable electrocatalytic activity [15].

Metal oxide NPs with nonprecious elements (i.e., CeO_2, MnO_2, and ZrO_2) have been employed to modify carbon-based materials to obtain composite electrodes that improve reaction kinetics and reversibility at the battery anode due to low cost and high catalysis activity [15,16]. The deposition of metal oxides on graphite has been employed to enhance electrocatalytic activity [17]. Metal components, such as TiO_2 nanocoating [18], Nb-doped WO_3 [19], ZrO_2 [20], and $NiCoO_2$ [16], have been explored. In another study, graphite nanoflake decorated with a metal oxide (Li and Mn based) was acting anode and showed long-term cycling stability [9].

Graphene has been used as an electron conduction material for batteries because of its physical and electrochemical advantages, such as a large surface area and unique electron and photon transport mechanisms due to a two-dimensional monolayer carbon nanomaterial consisting of sp^2-hybridized carbon atoms arranged in a hexagonal crystalline structure [11,21]. Graphene has attracted considerable attention from battery researchers and presents applicability in various processes to fabricate battery components as in cathodes and anodes [5,8,22]. Furthermore, structures of graphene electrodes can be utilized for constructing flexible and stretchable battery electrodes [8]. Moreover, various sulfur and graphene nanocomposite designs have been synthesized for batteries [11]. Graphene-based nanomaterials such as reduced graphene oxide and graphene oxide have been widely applied in battery systems. In addition, due to enhanced conductivity and abundant oxygen functional groups, a significantly increased peak current density and improved reliability have been shown in batteries with composite electrodes [15].

Metal sulfides have attracted attention in the field of energy-storage systems due to their inherent electronic structures and physicochemical properties [15], and the interface of anodes and electrolytes/separators can be modified by nanosurface coating, nanocomposite metal lithium, and lithium nanoalloy, while the interface between cathodes and electrolytes/separators is designed with nanometal sulfide and is nanocarbon-based [23]. Table 3.1 provides a summary of the nanomaterials used in battery components.

Advances in nanotechnology represent a crucial element in the development of energy-storage systems. Nanomaterials are present in all the main battery components, such as materials based on carbon, metal oxides, metal sulfides, graphene and its derivatives (graphene oxide and reduced graphene oxide), Ni, Co, Fe, Al, and S. Hybrid nanomaterials, which act as multifunctional substitutes in traditional battery systems, have also been observed.

3. Processes for recovering and synthesizing nanomaterials

The technological world has entered a phase of mass consumption. New materials are being developed all the time. The manipulation of matter on a nanometric scale has followed this pattern of consumption and development. From this, a complete evaluation must be conducted of the issues of economic management, processing, recycling, and damage to the environment that any technological development can bring.

Some areas meet all the environmental requirements, but others still do not. There are still deficiencies in recycling. Thus, to combine economic and environmental interests, it is necessary to find ways to obtain raw materials (from natural or secondary sources), supply, recycle, and reuse. Only in this way will supply chains be sustainable.

Along these lines, it is possible to highlight batteries that are well evolved and have several components from valuable metals to nanometric materials and their composition [44]. Entering this field of nanotechnology requires a complete environmental assessment

Table 3.1 Summary of nanomaterials used in battery components.

Components	Nanomaterials	References
Anode	MnO_2	[5]
	TiO_2/Carbon nanofiber	[24]
	Zn/Graphene	[25]
	Zn/Carbon nanotubes	[26]
	SnS_2/Graphene	[27]
	MnO_2, Co_3O_4, and SnO_2, and graphene foam	[5,9].
	Metallic cadmium	[28]
	Mn_2O_3	[29]
	Based metal oxides and sulfides nanocomposites	[30]
	Titanium niobium oxide	[31]
	Co_3O_4/Graphene	[32]
	MoO_2/Graphene oxide	[33]
Cathode	MoS_2/Graphite	[34]
	Cr_2O_3/Graphite	[35]
	MnO_2/Graphene	[36]
	Graphene/Polyaniline	[37,38]
	SiO_2/LiNiCoMnO	[39]
	$Ni(OH)_2$	[28]
	Codoped NiS_2	[40]
	CoS_2	[41]
Separator	Al_2O_3 particles coated on polyethylene	[10]
	Cu-carboxylate/Graphene oxide	[42]
	Ni/S-doped graphene	[43]

of a circular economy. Recycling, recovering, and reusing materials with high added value enter this context. Only in this way is it possible to make a sustainable supply chain highlighted by nanomaterials [45].

Within this context, methodologies already developed to recover battery nanomaterials will be presented. Hydrometallurgy, pyrometallurgy, nanohydrometallurgy, thermal procedure-inert gas condensation, vacuum separation, oxidative polymerization, and the precipitation method are methodologies already tested on a bench scale with good efficiency. Table 3.2 shows some studies that have applied the process of recycling nanomaterials into batteries. The table shows the recycled nanomaterial, the method, and the percentage of recovery.

From the processes presented for recycling nanomaterials, a few stand out. Fig. 3.1, adapted from Dutta [2] shows the main steps in hydrometallurgy and pyrometallurgy processes to recover nanomaterials.

Table 3.2 Various recycled battery nanomaterials and their methodologies.

Nanomaterial of interest/ recycled battery	Method(s) used for recycling	Nanomaterial recovery percentage	References
ZnO nanoparticles/Zn alkaline batteries	Hydrometallurgy and liquid-liquid extraction	>98	[46]
Zn nanoparticles/Zn–Mn batteries	Inert gas condensation (thermal) and vacuum separation	>99	[47]
Co ferrites/Li-ion batteries	Sol-gel, hydrothermal method, coprecipitation	—	[48]
Graphite-polyaniline nanocomposites/ Zn–MnO_2 batteries	Oxidative polymerization	—	[49]
Graphene-MnO_2 nanocomposites/ Zn–MnO_2 batteries	Precipitation method	—	[50]

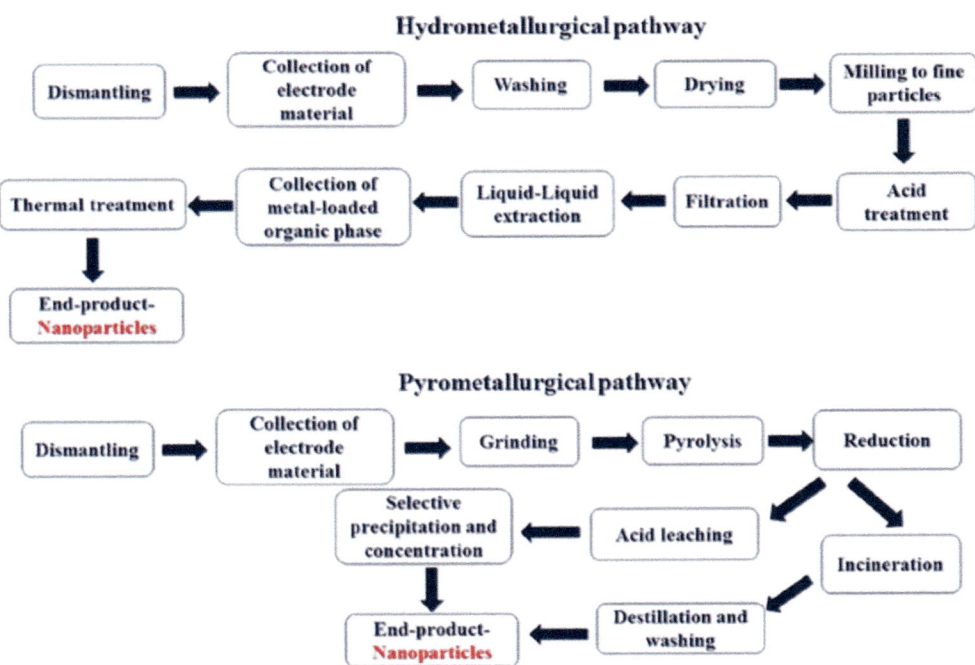

Figure 3.1 Steps in the hydrometallurgy and pyrometallurgy processes for recovering and synthesizing nanomaterials. *(From Dutta, T. et al. Recovery of nanomaterials from battery and electronic wastes: a new paradigm of environmental waste management. Renew. Sustain. Energy Rev. 82(2018), 3694–3704. https://doi.org/10.1016/j.rser.2017.10.094.)*

3.1 Hydrometallurgy

The hydrometallurgy process involves the extraction of metals and metallic compounds from materials such as batteries through leaching or dissolution in aqueous solutions (acidic or basic). Leached and in the liquid, metals can be separated (liquid–liquid), precipitated, and recovered through crystallization or reductive electrochemical processes [51].

Much research has already developed various hydrometallurgical processes, such as leaching, solvent extraction/ion exchange, agglomeration, and tailing/waste remediation [52–54].

As shown in Fig. 3.1, the steps of hydrometallurgy involve critical physical and chemical processes that must be studied and adjusted for each type of battery. Nowadays, there are many models and consequently many differences between the materials. Knowing the material (e.g., metal, oxide, ceramic, and plastic) and chemical composition of each part facilitates the steps "dismantling" and "collection of electrode material." In the initial stages, which are the physical ones, it is necessary to align process efficiency and energy cost. It is necessary to arrive at a methodology that aligns these sides, thus having the best result for "milling to fine particles." Theoretically, the smaller the particle size, the greater the surface area and better with liquid reagents. However, it is necessary to evaluate the cost of this process; the smaller the size, the more energy is needed.

In the "acid/basic treatment" step, the main parameters to be evaluated are reagent choice and leaching power, solution pH, solid:liquid ratio, reaction time, and temperature. These parameters can significantly influence the hydrometallurgy process and increase or decrease efficiency.

After leaching or "acid/basic treatment," the most common process for recovering nanomaterials from batteries is "liquid–liquid extraction." In this process, various solvents are used. Solvents can be partially or wholly immiscible. Usually, an aqueous phase and an organic phase are present. Metal ions that have been solubilized/dissolved (acid/basic treatment) by affinity tend to pass to the organic phase. This transfer process between phases has pH and solution concentrations as parameters. After the first selective separation, it is possible to reverse the process, transfer from the organic phase to the aqueous phase the metals of interest one at a time. In this step, it is also necessary to take care of the pH parameters and the concentrations of the solutions [55,56].

An excellent example of the application of hydrometallurgy and liquid-liquid separation is the research by Deep [46]. The author develops a technique for recovering ZnO (high-purity) NPs by recycling alkaline batteries in this research. These batteries are composed of $Zn-MnO_2$ structures and other metals such as Fe, Ni, Cd, and Cu. The hydrometallurgy process was carried out with the addition of 5M HCl together with a Zn extractant Cyanex 923. Parameters such as reaction time, reaction temperature, solid-liquid ratio (battery-HCl), and solid-extractor ratio (battery-Cyanex 923) were studied. The optimized reaction was carried out for 30 min at a temperature of 250°C. With the extraction of the Zn-Cyanex 923 complex, the Zn was quickly separated. The NPs obtained by recycling alkaline batteries were approximately 5 nm in diameter.

The efficiency of hydrometallurgy processes in batteries will depend on each case, the metal of interest, the reagent used. It was possible to observe that it can consist of simple processes, as in the previous research, or even more complicated processes, depending on the methodology. Thus, it is possible to state that hydrometallurgy is the initial process in this field of nanomaterials recycling. Further studies on more complex batteries and the method of separation of metals will still be necessary.

3.2 Pyrometallurgy

Pyrometallurgy is a process to make physical and chemical transformations in materials such as metals, minerals, and ores. Pyrometallurgy is also used to produce alloys and compounds or separate metals by density. This process involves high temperatures between 500°C and 2000°C [57].

Pyrometallurgy is one of the processes used in battery recycling. Pyrometallurgy recycling involves steps such as pyrolysis, reduction, incineration, and distillation [58]. Due to high temperatures, materials that decompose at lower temperatures are efficiently removed—for example, polymers. These decomposing materials can also supply energy and generate gases that help transport more volatile metals such as Hg and Cd. An advantage of the pyrometallurgy process is that it does not require pretreating of the materials to be recycled. A disadvantage is the possibility of the formation of toxic and harmful gases, so it is necessary to pay close attention to the vapors [59].

Among the steps of the pyrometallurgy process shown in Fig. 3.1, it is important to point out "pyrolysis" and "reduction." According to the methodology, these steps can facilitate the pyrometallurgy process by using less energy to strategically reduce gases or solvents. Reducing gases such as H_2, O_2, and the combination of CO/CO_2 lowers the temperature needed to reduce metal oxides for final metal separation. These gases are called reactive gases because they participate in chemical reactions. Another process is treating solvents such as organic acids that form volatile compounds, facilitating the pyrometallurgy process. It is essential to control the gases generated at this stage, especially organic solvents, because of their toxicity [60–62].

An example of battery recycling by pyrometallurgy in nanotechnology is the research of Marouf [63]. In this research, the authors recycle zinc-carbon batteries through homogeneous precipitation-calcination processes. Through this process, it is possible to obtain ZnO nanosheets. The authors first added acid to the zinc foil pieces to form an intermediate compound. Afterward, a heat treatment at 1000°C was carried out, and nanometric ZnO formation was proven. In this way, a simple, clean methodology is developed that can efficiently recycle batteries and use this recycling to generate a new nanomaterial.

In Fig. 3.2, the author Assefi [44] presented a flowchart of different methodologies for recycling different batteries by pyrometallurgy. Methodologies for recycling Li-ion batteries, Ni–MH (nickel-metal hydride battery), and Ni–Cd batteries are shown.

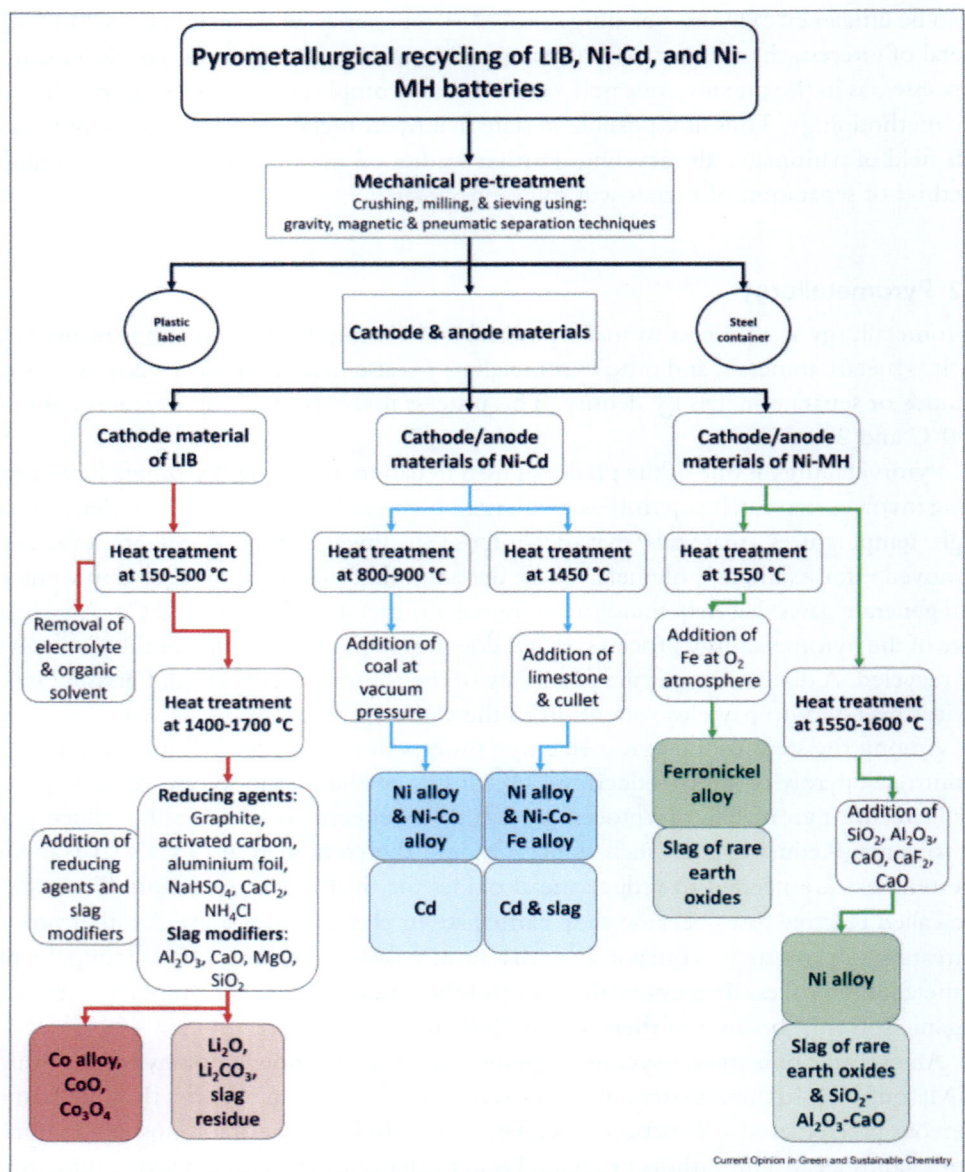

Figure 3.2 Methodologies for recycling Li-ion batteries, Ni–MH (nickel–metal hydride battery), and Ni–Cd batteries by pyrometallurgy. *(From Assefi, M. et al. Pyrometallurgical recycling of Li-ion, Ni–Cd and Ni–MH batteries: a minireview. Curr. Opin. Green Sustain. Chem. 24 (2020), 26–31. https://doi.org/10.1016/j.cogsc.2020.01.005.)*

3.3 Nanohydrometallurgy

Nanohydrometallurgy (or magnetic nanohydrometallurgy) consists of a new and sustainable approach to recover or extract noble metals from an aqueous solution containing electronic dispositive, such as batteries as substrate, as a hydrometallurgical alternative to conventional methods, applying nanotechnology to the mineral area [64]. The main advantages of nanohydrometallurgy are [65] (1) simple operation with the aid of magnets to separate metals with similar properties, (2) high efficiency, (3) the possibility to reuse metallic NPs used in the extraction process, and (4) an eco-friendly approach, one that does not use organic solvents or intense chemical processing. However, nanohydrometallurgy has some drawbacks, such as a partial loss of extraction capacity due to extremely acid or alkaline pH or high temperature, which can result in either dissolution of iron ions to solution or magnetization decay oxidation of iron(II) by saturated oxygen in the aqueous media [66]. Therefore, to protect the extract agent, several coatings, such as silica or diethylenetrietylaminepropylacetic have been developed by researchers and are used as complexing agents [67]. Fig. 3.3 shows a schematic representation of the process, indicating the main steps.

Nanomaterials used in nanohydrometallurgy are mainly maghemite (γ-Fe2O3), magnetite (Fe_3O_4) NPs, or nano zero-valent iron, once they show superparamagnetic properties and are responsive to an external magnetic force induced by a magnet, helping with the fast and efficient separation of noble metal ions from the aqueous solution [68]. These NPs are commonly labeled as support and have been functionalized with inert materials as complexing agents, such as amine and carboxylate ions [67]. Thus, the

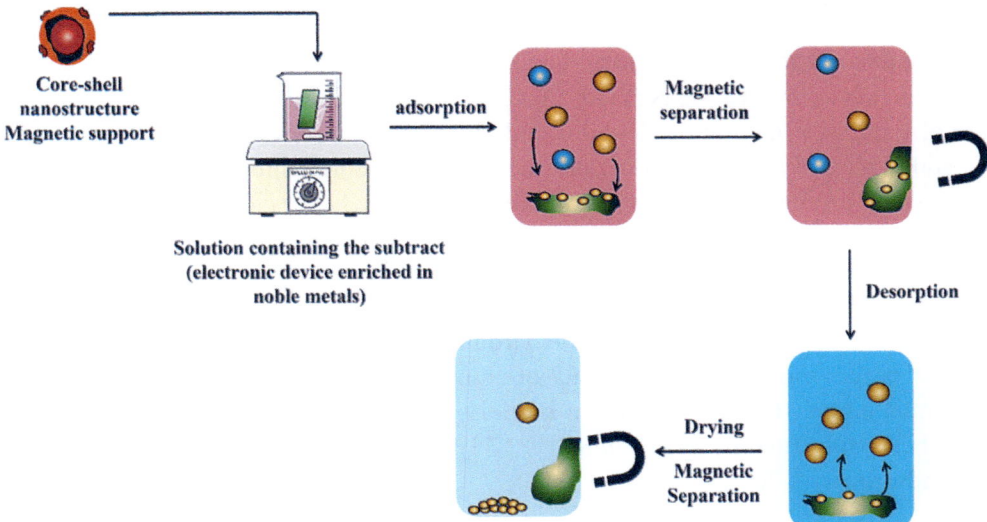

Figure 3.3 Nanohydrometallurgy scheme.

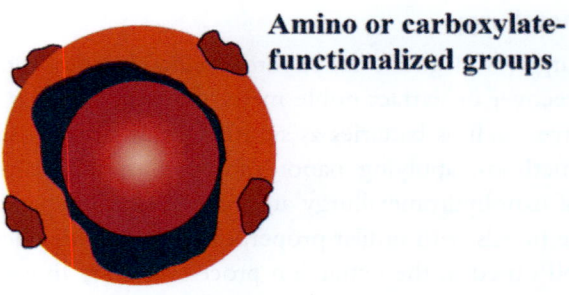

Figure 3.4 Scheme of core-shell support used in nanohydrometallurgy.

magnetization of support is essential to promote the oxidation-reduction reactions of Fe under ambient conditions and inert atmosphere [69], using various iron sources such as iron (II) chloride ($FeCl_2$), iron(II) sulfate ($FeSO_4$), and iron (II) phosphate ($Fe_3(PO_4)_2$) [70,71]. Fig. 3.4 shows a sketch of functionalized nanomaterial used in nanohydrometallurgy.

In this view, high recovery percentages of noble metals (e.g., La, Nd, Y, Ce, Pr, Sm, and Gd, to mention a few) are reported with this approach. To best our knowledge, the main section exploits the effect of rare metal concentration, solution pH, contact time, temperature, and the effect of the superparamagnetic nanostructure load incorporated to the extract agent.

3.3.1 Noble metal concentration

Noble metals such as La, Y, Nd, Gd, Dy, Eu, Tb are commonly found in the composition of batteries, cell phones, television, and other electronic devices [72]. When batteries are added to acid or alkaline solution, these metals dissolve into it as trivalent cations, X^{3+}, which can be easily captured by γ-Fe_2O_3 and Fe_3O_4 NPs [66]. Therefore, extraction of these noble metals is based on adsorption of X^{3+} ions onto amino- and carboxyl-functionalized γ-Fe_2O_3 and Fe_3O_4, which are nanomaterials with high surface area. Moreover, these NPs are surrounded by a mesoporous nanosilica coating to preserve the magnetic core from the chemical environment.

In this scenario, experimental data of adsorption of X^{3+} are usually fitted by the Langmuir model (Eq. 3.1), assuming monolayer coverage of the metal ions onto the core-shell nanostructure (MNPs@SiO_2−NH_2COO^-) after saturation of the support (nanoadsorbent) [67]:

$$q_{eq} = \frac{K_L q_{max} C_{eq}}{1 + K_L} \quad (3.1)$$

where q_{eq} is the noble metal concentration on the solid surface, in mg g^{-1}; C_{eq} is the noble metal concentration in the aqueous solution, in mg L^{-1}; K_L is the Langmuir constant related to the affinity between the metal ion and the core-shell nanostructure in L mg^{-1}; and q_{max} is the maximum amount of metal ion adsorbed per unit gram of support in mg g^{-1}.

Considering that the adsorption process is a surface phenomenon, the extraction is extremely dependent on the surface area of the MNPs@SiO$_2$—NH$_2$COO$^-$ and on a concentration gradient [73]. It relies on the fact that the greater the metal concentration in the solution, the greater the adsorption of metal ions, and consequently, higher extraction yields can be achieved [74].

3.3.2 Effect of solution pH and contact time

Adsorption equilibrium of X^{3+} ions can often be achieved after 100—150 min. Thus, the contact time seems to favor the extraction of noble metals from aqueous solutions under mild pH and temperature conditions [75]. Regarding pH, it exerts a strong effect on the extraction once it can protonate (turn the support to positive charge) or deprotonate (resulting in a negatively charged support) the functionalized core-shell nanostructure [76]. Higher noble metals amounts adsorbed are achieved at pH 4 or 8. However, at pH 9, 10, and 5, lower adsorption percentages of metals are reported due to the loss of the physicochemical stability of the core-shell nanostructure at these pH values [66]. Therefore, careful study of pH as a parameter should be taken into account for optimizing the nanohydrometallurgy process at either bench or full scale.

3.3.3 Effect of loading superparamagnetic nanostructures

The nanomagnetic core is responsible for the advantages of the nanohydrometallurgical process, such as ease of separation (with the aid of a magnet) and high reusability of the support [74]. Thus, increasing the volume of magnetic nanomaterials can improve noble metal extraction. However, some papers have reported that a high volume of magnetic NPs can lead to a loss in the stability of the support, resulting in degradation or even loss of magnetic properties [77].

Furthermore, a minimum or ideal volume of magnetic NPs is assumed to exist for use in the design and development of supports applied to nanohydrometallurgy for noble metal extraction. Some scientific works have reported iron-based magnetic loadings of 1%—10% of the total mass of the support, which resulted in 26%—99% for extraction of noble metals ions (X^{3+}) from aqueous solution even after five cycles using a tertiary mixture of lanthanide ions [65].

3.3.4 Effect of temperature

Although it does not affect the extraction process, the temperature has a tremendous effect on core-shell nanostructure stability and magnetization propriety. Nanohydrometallurgy is often carried out at low temperatures (near 20°C–25°C) once the magnetic properties of the core-shell nanostructure are preserved [67]. However, at high temperatures values, loss of efficiency of nanohydrometallurgy can occur due to magnetization decay of the support or the loss of physicochemical stability [78].

In addition, an increase in temperature can result in the dissolution of magnetic NPs in the aqueous solution, accompanied by a loss of extracting agent material during the process. However, the nanomagnetic core is then coated with supports materials, such as SiO_2-based materials, to enhance the temperature resistance of the support [79]. From this point of view, some literature reports high extraction percentages for Ln, Sm, and Nd after three cycles of magnetic nanohydrometallurgy extraction [66]. Thus, as a recommendation, the effect of temperature should be studied in the design stage of the core-shell nanostructure, as well as the indirect effect of temperature on the nanohydrometallurgy process.

3.3.5 Battery recycling, remanufacturing, and reuse

The noble metal content of batteries and electronic devices depends on the source of the material used in the extraction (residual or natural), the impurity content, and the age of the source (old or novel electronic device) [80]. Moreover, a difficulty commonly found in this extraction relies on the separation of lanthanide of similar physical and chemical properties [66]. Thus, more selective adsorption and extraction have been proposed in the literature [81].

After the extraction and magnetic separation of these lanthanides, they can be used as raw materials for remanufacturing batteries and electronic devices, once they show great chemical stability, as well as excellent optic-electronic properties, being used in medicine, energy storage, solar panels, and catalysis [82].

For the sake of clarity and to test our knowledge, Fig. 3.5 shows the number of scientific works developed in this view, with the application of noble metals for remanufacturing, reuse, and recycling of batteries and electronic devices.

According to Fig. 3.5, it can be noted that interest by the scientific community and industry has been increasing quickly due to the great versatility of nanohydrometallurgy, as well as its low cost and simple operation.

Therefore, nanohydrometallurgy has gained a special space in society and presents a sustainable and efficient alternative to conventional hydrometallurgical processes, with lower risk to the operator and the absence of intensive chemical processing and organic solvents.

4. Nanotechnology for metal recovery from batteries

The constant need for the implementation of new technologies in the electronics industry and the fact that batteries are utilized in many applications ranging from consumer

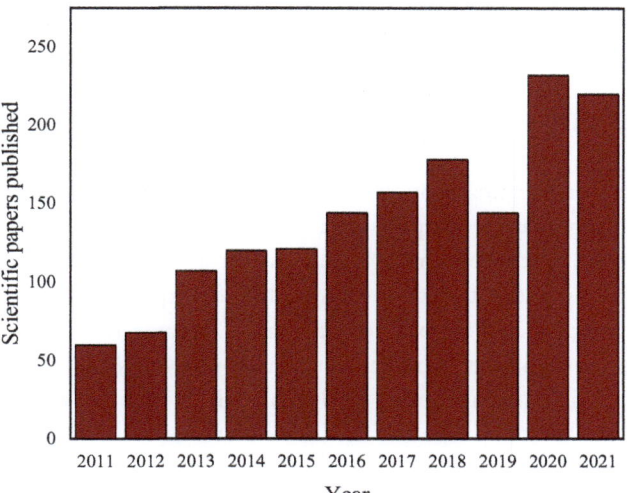

Figure 3.5 Scientific research published about recycling noble metals from January 2011 to July 2021.

electronics to hybrid electric vehicles justifies the predominance of batteries in the portable power industry. In addition to being limited resources, the primary sources of some metals with considerable content in batteries are unfeasible geopolitically, which makes their extraction impracticable [83].

One example is cobalt. Around 25% of the world's cobalt production is destined for battery manufacturing [84]. However, this element can cause problems of toxicity and environmental contamination, presenting a tremendous danger when improperly discarded [85].

Those are the main factors that drive the demand for recycling and recuperation of these materials. Battery recycling is garnering the interest of the public, industry, and academia because it comes across as a solution to eliminate the environmental risks arising from inappropriate disposal and help in the recovery of these high-value-added metals [83].

However, the importance of battery recycling goes beyond the consequences of inappropriate disposal and metal reuse; the approach used during the recycling process is also very important.

Considering that liquid-liquid extraction is extensively employed for the separation and extraction of metals and that the use of organic diluents can make this technique harmful to human health and the environment, nanotechnology has attracted attention in the solid phase extraction because in addition to the advantages of shorter processing time, low intrinsic costs and that there are no emissions of volatile organic compounds to the atmosphere, it still provides a greater contact area, as we shall see in more depth in the topics below [86,87].

Research, including the potential use of nanomaterials such as nanofibers, graphene, and graphene oxide, has been a highly effective strategy for the recovery of metals from LIBs.

4.1 Nanofibers

The solid-phase extraction of metal ions occurs through the immobilization of the organic extractor in a solid support, as the impregnation of a porous material by a liquid extractor or the polymerization of different resins in the presence of the extractor. The solid-phase extraction is recognized as a promising alternative to the classic liquid—liquid extraction, and as previously mentioned, overcomes many of the drawbacks of later techniques, providing shorter processing times, low solvent consumption, and lower costs [87].

However, the solid phase extraction method has a limitation when it comes to the contact area between the surface of the solid doped with the extractor and the solution containing the metals of interest. Solid-phase extraction has this disadvantage compared with liquid-liquid extraction—a relatively small contact area. The reduced contact area makes mass transport between the solid and the solution difficult, making the whole process less efficient. A solution to that issue is the use of nanomaterials doped with organic extractors [88,89].

For that very reason, nanofibers might be an effective and valued alternative to metal separation and recovery. Nanofibers doped with organic extractors offer unique behavior due to the large surface area offered, high porosity, flexibility as well as advantageous thermal, mechanical, and electronic properties [87].

Nanofibers are structures on the nanometer scale (10^{-9} m or 1 nm), usually produced by techniques such as self-assembly, template synthesis, phase separation, and electrospinning [90]. In addition to the techniques presented, Forcespinning is a way to produce nanofibers with diverse polymeric compositions and the potential for large-scale production [91].

In the example below (Fig. 3.6), we have images from a scanning electron microscope and energy dispersive spectroscopy elemental mapping, where we can observe nanofibers produced by the forcespinning technique after they were used in the recovery of metal. The nanofibers concerned were produced using a polymeric (Nylon 6) solution with the addition of an organic extractor, di-(2-ethylhexyl)phosphoric acid (DEHPA), and were placed in contact with a solution containing metals that are found in NiMH batteries (zinc and nickel) for recovery.

Fig. 3.6B—D shows that the chemical elements (carbon, oxygen, and phosphorus, respectively) compose the nylon 6 and DEHPA chains were homogeneously distributed in the nanofibers. Fig. 3.1E and 3.1F show the elements nickel and zinc, respectively, distributed on the surfaces of the nanofibers, confirming the efficient extraction of the metals. That satisfactory results were achieved by varying the following conditions: pH, the duration of contact and S:L ratio between the nanofiber and the aqueous solution [92].

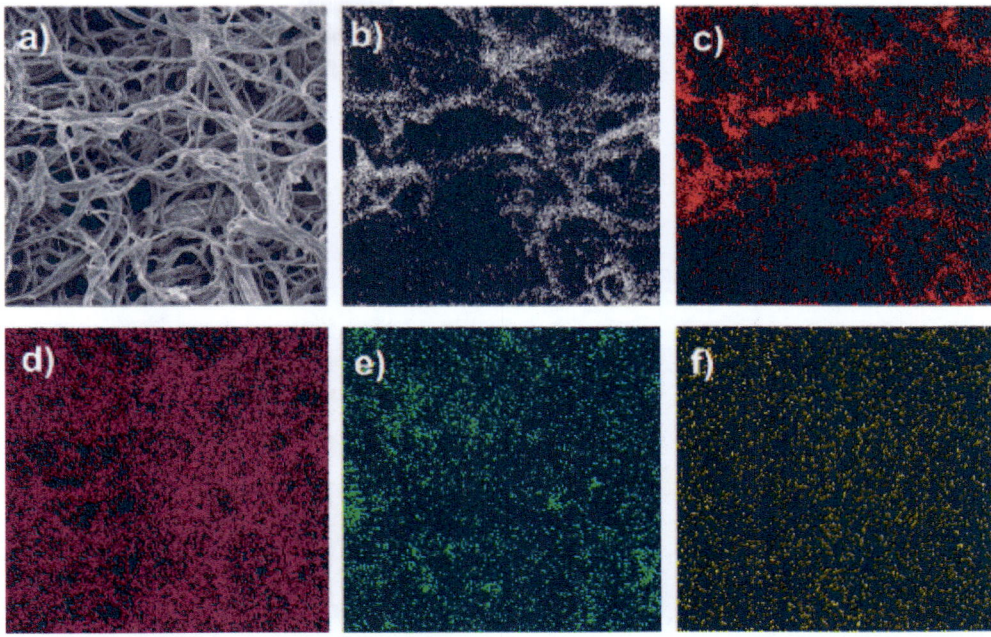

Figure 3.6 Scanning electron microscope images and energy dispersive spectroscopy elemental mapping: (a) Nylon 6-DEHPA nanofibers; (b) carbon; (c) oxygen; (d) phosphorus; (e) nickel; (f) zinc. *(From Nunes da Silva, F. et al. An eco-friendly approach for metals extraction using polymeric nanofibers modified with di-(2-ethylhexyl) phosphoric acid (DEHPA). J. Clean. Prod. 210(2019), 786–794. https://doi.org/10.1016/j.jclepro.2018.11.098.)*

4.2 Graphene and graphene oxide

Other nanostructured materials have been introduced for their use in solid-phase extraction. Due to its unique structures, carbon-based NPs have been among the most important trends in solid-phase extraction of organic compounds and metal ions. These materials include carbon nanotubes, carbon nanocones-disks, fullerenes, graphene, carbon nanofibers, as well as their functionalized forms [93]. Fig. 3.7 shows examples of carbon nanostructures.

Graphene is a carbonaceous material that may have a single-layer or few-layer thickness of sp2-hybridized atoms of carbon. The most popular method of graphene synthesis is based on the oxidation of graphite to graphene oxide, followed by the use of a reducing agent to conduct its reduction to graphene [94].

Graphene holds great promise for its applications, owing to its remarkable and unique electronic, thermal and mechanical properties, such as good thermal conductivity, fast mobility of charge carriers, high values for Young's modulus, and fracture strength [95].

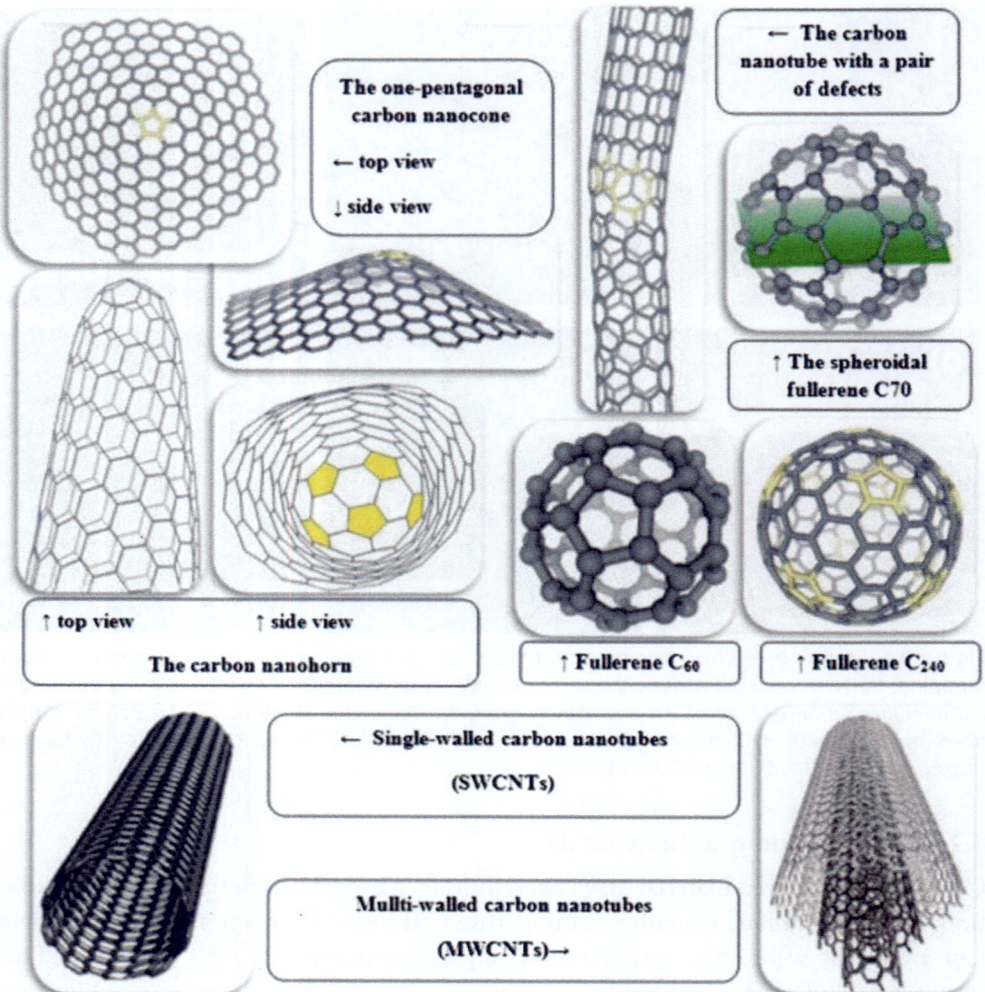

Figure 3.7 Examples of carbon nanomaterials. *(From Płotka-Wasylka, J. et al. Modern trends in solid phase extraction: new sorbent media. Trac. Trends Anal. Chem. (Regular ed.), 77(2016), 23–43. https://doi.org/10.1016/j.trac.2015.10.010.)*

However, there are some properties that make graphene a material with many advantages for solid-phase extraction, providing graphene with great potential as a good adsorbent. The most important are the following [87]:
- Its facility to be modified with functional groups, further increasing the selectivity of the solid phase extraction (especially the modification via graphene oxide that has many reactive groups);

- Ultrahigh specific surface area (theoretical value 2630 m^2/g), which enables high sorption capacity;
- Both sides of the planar sheets of graphene are available for molecule adsorption.

Finally, according to TIJING [96], the incorporation of graphene oxide in polymeric nanofiber membranes could further improve the selectivity of solid-phase extraction.

5. Synthesis of metallic nanoparticles and other materials

NPs have dimensions of less than 100 nm [97,98]. When NPs are compared with bulk materials, their study of synthesis, characterization, and application has attracted attention and relevance due to their high reactivity, the possibility of conjugating chemical systems linked to their surface or their functionalization, and their observation and monitoring [99,100].

This increase in reactivity is due to its increased total surface area after reducing its dimensions to the nanometer scale. When compared to their use in the same applications, materials close to this scale have their physical properties, such as optical, electrical, magnetic, thermal, and mechanical; chemical, such as catalytic; and biological, such as toxicity to bacteria and fungi, improved [101,102].

Records of the use of the optical properties of nanomaterials for the manufacture of glass and ceramics by Egyptians and Mesopotamians have existed since the 13th and 14th centuries BCE. In the fourth century CE, as an example of the application of NPs, the Lycurgus Cup is highlighted, which is exhibited at the British Museum in London, England; it is known for the dichroic aspect of the glass that makes it up. Its color changes as a result of light absorption by gold (AuNPs) and silver (AgNPs) NPs according to the angle of observation and illumination [103–105].

NPs can be classified in different ways, according to their physical and chemical characteristics, and can be nanomaterials based on carbon, metals, ceramics, semiconductors, polymers, and lipids [106–108].

The two main classes of carbon-based NPs are fullerenes and carbon nanotubes. Fullerenes or buck balls have pentagonal and hexagonal carbon units arranged as shapes C60 and C70. Carbon nanotubes are synthesized by deposition of vaporized carbon precursors from graphite by laser or electric arc onto metal particles, and they can be used in nanocomposites for gas adsorption applications for environmental remediation and as a support medium for different organic catalysts and inorganic [108,109].

Semiconductor NPs show their properties between metallic and nonmetallic materials. While some NPs are formed by lipid fractions, and others are composed of polymers, which are organic and functionalized, ceramic NPs are nonmetallic and inorganic solids and can be synthesized by heat and successive cooling [109].

Metallic NPs are synthesized from the use of metallic precursors, which are usually salts of this metal. Plasmonic NPs, such as CuNPs, AgNPs, and AuNPa, have a characteristic localized surface plasmon resonance, which is a clear spectral absorption band with specific peak scattering at wavelengths in the visible ultraviolet region [110,111].

According to the starting material for the production of NPs, two general methodologies are top-down and bottom-up techniques, and they can be divided into different processes according to dimensions, size distribution, shape, crystal structure, and composition (purity) expected from the NPs produced [103,106].

Top-down methods are physical approaches known as breaking, splitting, or deconstruction processes, in which bulk materials are reduced to nanometric dimensions. Mechanical milling, nanolithography, laser ablation, sputtering and thermal decomposition are some of the most widely used NP synthesis methods are included in this approach [107–109].

The bottom-up approach is also known as a constructive method that occurs through the biological and chemical processes of sedimentation and reduction, and it refers to the formation of NPs from atoms. As examples, the sol-gel, green synthesis, spinning, and biochemical synthesis techniques can be cited [101,108,112].

Table 3.3 shows a summary of the main synthesis techniques and a summary of their functioning related and described in the studies of [101,107,112,113].

Industrial waste such as batteries, tire rubber, plastic, cathode-ray-tube glass, aluminum foil, Cu waste generated from the printed circuit board manufacturing [114,115], biomass [116], and electronics scrap [117,118] can be used as a raw material for the synthesis of metal and carbon NPs as an opportunity to mitigate social, economic and environmental problems related to the inadequate disposal or treatment of these wastes [115].

Several ways to synthesize NPs from these wastes have been developed. Initially, the pretreatment of these materials must be carried out according to their composition and type. This step can encompass physical, chemical, and combined methods. Physical pretreatment is carried out by grinding, separating, and dismantling steps. Contaminants present in samples are solubilized by heating or through contact with chemical reagents such as acids, for example, HNO_3, HCl, H_2SO_4, in chemical treatment. Combined pretreatment involves the combination of the two processes (chemical and physical). After pretreatment, the production of these nanomaterials can be carried out using the methods of chemical/thermal activation, electric arc discharge, vacuum evaporation, and inert gas condensation, sodium borohydride reduction, and/or solvent thermal [115].

Hence, for the recovery of materials from these wastes through recycling, physical processing or sample preparation processes, recovery or concentration processes through pyrometallurgy, hydrometallurgy, biohydrometallurgy, or combined routes, followed by conventional synthetic methods can be carried out [117–119].

Table 3.3 Summary of the description of NP synthesis techniques included in the general top-down and bottom-up methodologies [101,107,112,113].

Top-down method	Technique summary
Mechanical milling	Mills are used to mechanically crush materials, and this method can be used to produce oxide- and carbide-strengthened aluminum alloys, wear-resistant spray coatings, Al/Ni/Mg/Cu-based nanoalloys, ball-milled carbon nanomaterials, and many other nanocomposite materials.
Nanolithography	This is a tool for the develop nanoarchitectures with the use of a focused beam of light or electrons, and it can be divided into two main types that are masked and maskless lithography. In masked lithography, nano–patterns are transferred over a surface area using a specific template, and it includes photolithography, nanoimprint, and soft lithography. The maskless lithography occurs without the involvement of a mask, and it can be achieved through a focused ion beam in combination with wet chemical etching.
Laser ablation	It involves using a laser beam that hits the target material. In the process, the source material or precursor vaporizes due to the high energy of the laser irradiation, resulting in the formation of nanoparticles (NPs). Metal NPs, carbon nanomaterials, oxide composites, and ceramics can be synthesized for this technique.
Sputtering	It is a process used to produce nanomaterials by bombarding solid surfaces with high-energy particles such as plasma or gas. Energetic gaseous ions bombard the target surface, causing the physical ejection of small atom clusters depending upon the incident gaseous-ion energy.

Bottom-up method	Technique summary
Sol-gel	This method is used for the development of metal-oxide–based nanomaterials, and it is so-called because during the synthesis, the liquid precursor is transformed to sol, and the sol is ultimately converted into a network structure that is called a gel. The conventional precursors for the generation of nanomaterials using the sol-gel method are metal alkoxides. In the first step, the hydrolysis of the metal oxide takes place in water or with the assistance of alcohol to form a sol. In the next step, condensation takes place, increasing the solvent viscosity to form porous structures that are left to age. After this stage, drying takes place, in which water and organic solvents are removed from the gel. Lastly, calcination is performed to achieve nanoparticles.
Biochemical	Extracts from plants and algae, yeasts, fungi, bacteria can be used together with chemical reagents for the production of nanoparticles (NPs).
Chemical vapor deposition	These methods have great significance in the generation of carbon-based nanomaterials. A thin film is formed on the substrate surface via the chemical reaction of vapor-phase precursors. In the generation of carbon nanotubes, for example, a substrate is placed in an oven and heated to high temperatures. A carbon-containing (such as hydrocarbon) gas is slowly introduced to the system as a precursor. At high temperatures, the decomposition of the gas releases carbon atoms, which recombine to form carbon nanotubes on the substrate.
Solvothermal and hydrothermal	These methods are generally carried out in closed systems such as in that a sealed vessel. Nanostructured materials are attained through a heterogeneous reaction carried out in an aqueous medium at high pressure and temperature around the critical point in the hydrothermal method. The difference among the techniques is that the solvothermal method is carried out in a nonaqueous medium.
Chemical reduction	In this synthesis, three main components are used, which are a metallic precursor (generally a salt), a reducing agent, and a stabilizer (which prevents the agglomeration and oxidation of these NPs).

6. Future prospects

Batteries go hand in hand with the development of nanomaterials today. With more applications and the need for a longer duration, the demand for raw material is increasing significantly. Natural sources, mainly metals, are scarce, and they also can contribute to the contamination of the environment and human beings. Thus, the application of the concepts of circular economy and obtaining materials from secondary sources is one of the best strategies for the future.

One of the concerns for the circular economy to be applied is the quality of nanomaterials synthesized from recycling. The quality must be very high, as new chemical alternatives, energy density, safety, and battery power costs are expected [120].

The methods discussed in this chapter are efficient to achieve good quality in synthesized nanomaterials. However, it is still necessary to invest more in research that combines recycling and separation of nanomaterials to reduce the cost and energy required. These processes need to address these areas. Methodologies such as ionic exchange, electrodialysis, nanohydrometallurgy are being adapted. Adapting these facts is a big step toward building a productive society with waste management and the synthesis of new high-quality materials.

Acknowledgments

Thanks to the University of Sao Paulo for supporting this project and to Fundação de Amparo à Pesquisa do Estado de São Paulo and Capes (FAPESP – grant 2012/51871-5, 2018/03483-6, 2018/11417-3, Sao Paulo Research Foundation) for financial support. This project was developed with the support of SemeAd (FEAUSP), FIA Fundação Instituto de Administração and Cactvs Instituto de Pagamento S.A. through the granting of assistance to a research project Bolsa SemeAd PQ Jr (Public Notice 2021.01).

References

[1] G.P. Nichols, Exploring the need for creating a standardized approach to managing nanowaste based on similar experiences from other wastes, Environ. Sci. Nano 3 (5) (2016) 946–952. https://doi.org/10.1039/c6en00214e.

[2] T. Dutta, et al., Recovery of nanomaterials from battery and electronic wastes: a new paradigm of environmental waste management, Renew. Sustain. Energy Rev. 82 (P3) (2018) 3694–3704. https://doi.org/10.1016/j.rser.2017.10.094.

[3] A.R. Dehghani-Sanij, et al., Study of energy storage systems and environmental challenges of batteries, Renew. Sustain. Energy Rev. 104 (2019) 192–208. https://doi.org/10.1016/j.rser.2019.01.023.

[4] J. Ma, et al., The 2021 battery technology roadmap, J. Phys. Appl. Phys. 54 (18) (2021) 183001. https://doi.org/10.1088/1361-6463/abd353.

[5] Y. Jiang, et al., Electrocatalytic activity of MnO2 nanosheet array-decorated carbon paper as superior negative electrode for vanadium redox flow batteries, Electrochim. Acta 322 (2019) 134754. https://doi.org/10.1016/j.electacta.2019.134754.

[6] M.M. Rahman, et al., Strategies, design and synthesis of advanced nanostructured electrodes for rechargeable batteries, Mater. Chem. Front. 5 (16) (2021) 5897–5931. https://doi.org/10.1039/D1QM00274K.

[7] Z.A. Alothman, et al., Innovative nanomaterials for energy storage: moving toward nature-inspired systems, Curr. Opin. Green Sustain. Chem. 32 (2021) 100520. https://doi.org/10.1016/j.cogsc.2021.100520.
[8] S.D. Kim, A. Sarkar, J. Ahn, Graphene-based nanomaterials for flexible and stretchable batteries, Small (2021) 2006262. https://doi.org/10.1002/smll.202006262.
[9] B. Yu, et al., Nanoflake arrays of lithiophilic metal oxides for the ultra-stable Anodes of lithium-metal batteries, Adv. Funct. Mater. 28 (36) (2018) 1803023. https://doi.org/10.1002/adfm.201803023.
[10] J.H. Ahn, et al., Nanostructured reactive alumina particles coated with water-soluble binder on the polyethylene separator for highly safe lithium-ion batteries, J. Power Sources 506 (2021) 230119. https://doi.org/10.1016/j.jpowsour.2021.230119.
[11] J.E. Knoop, S. Ahn, Recent advances in nanomaterials for high-performance Li–S batteries, J. Energy Chem. 47 (2020) 86–106. https://doi.org/10.1016/j.jechem.2019.11.018.
[12] X. Wang, et al., Recent progress in high-performance lithium sulfur batteries: the emerging strategies for advanced separators/electrolytes based on nanomaterials and corresponding interfaces, Chem. Asian J. 16 (2021) 2852–2870. https://doi.org/10.1002/asia.202100765.
[13] X. Xu, et al., Vanadium-based nanomaterials: a promising family for emerging metal-ion batteries, Adv. Funct. Mater. 30 (10) (2020) 1904398. https://doi.org/10.1002/adfm.201904398.
[14] J. Ding, et al., Review on nanomaterials for next-generation batteries with lithium metal anodes, Nano Select 1 (1) (2020) 94–110. https://doi.org/10.1002/nano.202000003.
[15] Y. Long, M. Ding, C. Jia, Application of nanomaterials in aqueous redox flow batteries, ChemNanoMat 7 (7) (2021) 699–712. https://doi.org/10.1002/cnma.202100124.
[16] Y. Xiang, W.A. Daoud, Binary $NiCoO_2$-modified graphite felt as an advanced positive electrode for vanadium redox flow batteries, J. Mater. Chem. 7 (10) (2019) 5589–5600. https://doi.org/10.1039/C8TA09650C.
[17] K. Amini, J. Gostick, M.D. Pritzker, Metal and metal oxide electrocatalysts for redox flow batteries, Adv. Funct. Mater. 30 (23) (2020) 1910564. https://doi.org/10.1002/adfm.201910564.
[18] W.-J. Lee, et al., Graphite felt modified by atomic layer deposition with TiO_2 nanocoating exhibits super-hydrophilicity, low charge-transform resistance, and high electrochemical activity, Nanomaterials 10 (9) (2020) 1710. https://doi.org/10.3390/nano10091710.
[19] D.M. Kabtamu, et al., Electrocatalytic activity of Nb-doped hexagonal WO_3 nanowire-modified graphite felt as a positive electrode for vanadium redox flow batteries, J. Mater. Chem. 4 (29) (2016) 11472–11480. https://doi.org/10.1039/C6TA03936G.
[20] H. Zhou, et al., ZrO_2-nanoparticle-modified graphite felt: bifunctional effects on vanadium flow batteries, ACS Appl. Mater. Interfaces 8 (24) (2016) 15369–15378. https://doi.org/10.1021/acsami.6b03761.
[21] N. Madima, et al., Carbon-based nanomaterials for remediation of organic and inorganic pollutants from wastewater. A review, Environ. Chem. Lett. 18 (4) (2020) 1169–1191, https://doi.org/10.1007/s10311-020-01001-0.
[22] L. Wu, Y. Dong, Recent progress of carbon nanomaterials for high-performance cathodes and anodes in aqueous zinc ion batteries, Energy Storage Mater. 41 (2021) 715–737, https://doi.org/10.1016/j.ensm.2021.07.004.
[23] J. Wang, et al., A highly flexible and lightweight MnO_2/graphene membrane for superior zinc-ion batteries, Adv. Funct. Mater. 31 (7) (2021) 2007397, https://doi.org/10.1002/adfm.202007397.
[24] Z. He, et al., Flexible electrospun carbon nanofiber embedded with TiO_2 as excellent negative electrode for vanadium redox flow battery, Electrochim. Acta 281 (2018) 601–610, https://doi.org/10.1016/j.electacta.2018.06.011.
[25] A. Xia, et al., Graphene oxide spontaneous reduction and self-assembly on the zinc metal surface enabling a dendrite-free anode for long-life zinc rechargeable aqueous batteries, Appl. Surf. Sci. 481 (2019) 852–859, https://doi.org/10.1016/j.apsusc.2019.03.197.
[26] M. Li, et al., A novel dendrite-free Mn^{2+}/Zn^{2+} hybrid battery with 2.3 V voltage window and 11000-cycle lifespan, Adv. Energy Mater. 9 (29) (2019) 1901469, https://doi.org/10.1002/aenm.201901469.

[27] V. Lakshmi, et al., Nanocrystalline SnS_2 coated onto reduced graphene oxide: demonstrating the feasibility of a non-graphitic anode with sulfide chemistry for potassium-ion batteries, Chem. Commun. 53 (59) (2017) 8272–8275, https://doi.org/10.1039/C7CC03998K.
[28] S. Rarotra, et al., Progress and challenges on battery waste management:A critical review, ChemistrySelect 5 (20) (2020) 6182–6193, https://doi.org/10.1002/slct.202000618.
[29] J. Yoon, et al., The effects of nanostructures on lithium storage behavior in Mn_2O_3 anodes for next-generation lithium-ion batteries, J. Power Sources 493 (2021) 229682, https://doi.org/10.1016/j.jpowsour.2021.229682.
[30] Y. Zhao, et al., A review on design strategies for carbon based metal oxides and sulfides nanocomposites for high performance Li and Na ion battery anodes, Adv. Energy Mater. 7 (9) (2017) 1601424, https://doi.org/10.1002/aenm.201601424.
[31] T. Yuan, et al., Recent advances in titanium niobium oxide anodes for high-power lithium-ion batteries, Energy Fuels 34 (11) (2020) 13321–13334, https://doi.org/10.1021/acs.energyfuels.0c02732.
[32] H. Kim, et al., Understanding origin of voltage hysteresis in conversion reaction for Na rechargeable batteries: the case of cobalt oxides, Adv. Funct. Mater. 26 (28) (2016) 5042–5050, https://doi.org/10.1002/adfm.201601357.
[33] J. Huang, et al., Tailoring MoO_2/graphene oxide nanostructures for stable, high-density sodium-ion battery anodes, Energy Technol. 3 (11) (2015) 1108–1114, https://doi.org/10.1002/ente.201500160.
[34] L. Wang, et al., 3D flower-like molybdenum disulfide modified graphite felt as a positive material for vanadium redox flow batteries, RSC Adv. 10 (29) (2020) 17235–17246, https://doi.org/10.1039/D0RA02541K.
[35] Y. Xiang, W.A. Daoud, Cr_2O_3-modified graphite felt as a novel positive electrode for vanadium redox flow battery, Electrochim. Acta 290 (2018) 176–184, https://doi.org/10.1016/j.electacta.2018.09.023.
[36] Z. Wang, et al., Layer-by-Layer self-assembled nanostructured electrodes for lithium-ion batteries, Small 17 (6) (2021) 2006434, https://doi.org/10.1002/smll.202006434.
[37] J. Han, et al., Rational design of nano-architecture composite hydrogel electrode towards high performance Zn-ion hybrid cell, Nanoscale 10 (27) (2018) 13083–13091, https://doi.org/10.1039/C8NR03889A.
[38] K. Li, et al., Integration of ultrathin graphene/polyaniline composite nanosheets with a robust 3D graphene framework for highly flexible all-solid-state supercapacitors with superior energy density and exceptional cycling stability, J. Mater. Chem. 5 (11) (2017) 5466–5474, https://doi.org/10.1039/C6TA11224B.
[39] C. Chen, et al., High-performance lithium ion batteries using SiO_2-coated $LiNi_{0.5}Co_{0.2}Mn_{0.3}O_2$ microspheres as cathodes, J. Alloys Compd. 709 (2017) 708–716, https://doi.org/10.1016/j.jallcom.2017.03.225.
[40] H. Guo, et al., Cobalt-doped NiS_2 micro/nanostructures with complete solid solubility as high-performance cathode materials for actual high-specific-energy thermal batteries, ACS Appl. Mater. Interfaces 12 (45) (2020) 50377–50387, https://doi.org/10.1021/acsami.0c13396.
[41] Z. Yuan, et al., Powering lithium–sulfur battery performance by propelling polysulfide redox at sulfiphilic hosts, Nano Lett. 16 (1) (2016) 519–527, https://doi.org/10.1021/acs.nanolett.5b04166.
[42] S. Bai, et al., Metal–organic framework-based separator for lithium–sulfur batteries, Nat. Energy 1 (7) (2016) 16094, https://doi.org/10.1038/nenergy.2016.94.
[43] G. Zhou, et al., Long-life Li/polysulphide batteries with high sulphur loading enabled by lightweight three-dimensional nitrogen/sulphur-codoped graphene sponge, Nat. Commun. 6 (1) (2015) 7760, https://doi.org/10.1038/ncomms8760.
[44] M. Assefi, et al., Pyrometallurgical recycling of Li-ion, Ni–Cd and Ni–MH batteries: a minireview, Curr. Opin. Green Sustain. Chem. 24 (2020) 26–31, https://doi.org/10.1016/j.cogsc.2020.01.005.
[45] P. Pati, S. Mcginnis, P.J. Vikesland, Waste not want not: life cycle implications of gold recovery and recycling from nanowaste, Environ. Sci. Nano 3 (5) (2016) 1133–1143, https://doi.org/10.1039/c6en00181e.

[46] A. Deep, et al., A facile chemical route for recovery of high quality zinc oxide nanoparticles from spent alkaline batteries, Waste Manag. 51 (2016) 190–195, https://doi.org/10.1016/j.wasman.2016.01.033.

[47] X. Xiang, et al., Preparation of zinc nano structured particles from spent zinc manganese batteries by vacuum separation and inert gas condensation, Separ. Purif. Technol. 142 (2015) 227–233, https://doi.org/10.1016/j.seppur.2015.01.014.

[48] L. Yao, et al., Synthesis of cobalt ferrite with enhanced magnetostriction properties by the sol–gel–hydrothermal route using spent Li-ion battery, J. Alloys Compd. 680 (2016) 73–79, https://doi.org/10.1016/j.jallcom.2016.04.092.

[49] X. Duan, et al., Manufacturing conductive polyaniline/graphite nanocomposites with spent battery powder (SBP) for energy storage: a potential approach for sustainable waste management, J. Hazard Mater. 312 (2016) 319–328, https://doi.org/10.1016/j.jhazmat.2016.03.009.

[50] J. Deng, et al., Facile preparation of MnO_2/graphene nanocomposites with spent battery powder for electrochemical energy storage, ACS Sustain. Chem. Eng. 3 (7) (2015) 1330–1338, https://doi.org/10.1021/acssuschemeng.5b00305.

[51] S.K. Bhargava, M.I. Pownceby, R. Ram, Hydrometallurgy, Metals 6 (5) (2016) 122, https://doi.org/10.3390/met6050122.

[52] Y. Fu, et al., Improved hydrometallurgical extraction of valuable metals from spent lithium-ion batteries via a closed-loop process, J. Alloys Compd. 847 (2020), https://doi.org/10.1016/j.jallcom.2020.156489.

[53] G. Gao, et al., A citric acid/$Na_2S_2O_3$ system for the efficient leaching of valuable metals from spent lithium-ion batteries, JOM 71 (10) (2019) 3673–3681, https://doi.org/10.1007/s11837-019-03629-y.

[54] C. Wei-Sheng, H. Hsing-Jung, Recovery of valuable metals from lithium-ion batteries NMC cathode waste materials by hydrometallurgical methods, Metals 8 (5) (2018) 321, https://doi.org/10.3390/met8050321.

[55] D.P. Mantuano, et al., Analysis of a hydrometallurgical route to recover base metals from spent rechargeable batteries by liquid–liquid extraction with Cyanex 272, J. Power Sources 159 (2) (2006) 1510–1518, https://doi.org/10.1016/j.jpowsour.2005.12.056.

[56] K. Provazi, et al., Metal separation from mixed types of batteries using selective precipitation and liquid–liquid extraction techniques, Waste management (Elmsford) 31 (1) (2011) 59–64, https://doi.org/10.1016/j.wasman.2010.08.021.

[57] N. Faris, et al., Application of ferrous pyrometallurgy to the beneficiation of rare earth bearing iron ores – a review, Miner. Eng. 110 (2017) 20–30, https://doi.org/10.1016/j.mineng.2017.04.005.

[58] J. Cui, L. Zhang, Metallurgical recovery of metals from electronic waste: a review, J. Hazard Mater. 158 (2) (2008) 228–256, https://doi.org/10.1016/j.jhazmat.2008.02.001.

[59] C.R. De Oliveira, A.M. Bernardes, A.E. Gerbase, Collection and recycling of electronic scrap: a worldwide overview and comparison with the Brazilian situation.(Report), Waste Manag. 32 (8) (2012) 1592.

[60] D. Gregurek, et al., Use of gases in pyrometallurgy, JOM 69 (6) (2017) 968–969, https://doi.org/10.1007/s11837-017-2306-x.

[61] M. Li, J. Liu, W. Han, Recycling and Management of Waste Lead-Acid Batteries: A Mini-Review, London, England, 2016, https://doi.org/10.1177/0734242X16633773.

[62] L. Zhu, M. Chen, Research on spent LiFePO 4 electric vehicle battery disposal and its life cycle inventory collection in China, Int. J. Environ. Res. Publ. Health 17 (23) (2020), https://doi.org/10.3390/ijerph17238828.

[63] S. Maroufi, et al., Waste-cleaning waste: Synthesis of ZnO porous nano-sheets from batteries for dye degradation, Environ. Sci. Pollut. Res. 25 (28) (2018) 28594–28600, https://doi.org/10.1007/s11356-018-2850-0.

[64] H.E. Toma, Magnetic nanohydrometallurgy: a nanotechnological approach to elemental sustainability, Green Chem. 1 (4) (2015) 2027–2041, https://doi.org/10.1039/c5gc00066a.

[65] U. Condomitti, et al., Magnetic nanohydrometallurgy: a promising nanotechnological approach for metal production and recovery using functionalized superparamagnetic nanoparticles, Hydrometallurgy 125 126 (2012) 148.

[66] F.M. de Melo, S.N. Almeida, H.E. Toma, Magnetic nanohydrometallurgy applied to lanthanide separation, Minerals 10 (530) (2020) 530, https://doi.org/10.3390/min10060530.

[67] S.D.N. Almeida, H.E. Toma, Neodymium(III) and lanthanum(III) separation by magnetic nanohydrometallurgy using DTPA functionalized magnetite nanoparticles, Hydrometallurgy 161 (2016) 22–28, https://doi.org/10.1016/j.hydromet.2016.01.009.

[68] T. Kegl, et al., Adsorption of rare earth metals from wastewater by nanomaterials: a review, J. Hazard Mater. 386 (2020), https://doi.org/10.1016/j.jhazmat.2019.121632.

[69] Y. Yuan, et al., Effect of surface modification on magnetization of iron oxide nanoparticle colloids, Langmuir 28 (36) (2012) 13051–13059, https://doi.org/10.1021/la3022479.

[70] M. Krajewski, et al., Structural and magnetic properties of iron nanowires and iron nanoparticles fabricated through a reduction reaction, Beilstein J. Nanotechnol. 6 (1) (2015) 1652–1660, https://doi.org/10.3762/bjnano.6.167.

[71] Y. Park, et al., One-step synthesis and functionalization of high-salinity-tolerant magnetite nanoparticles with sulfonated phenolic resin, Langmuir 35 (26) (2019) 8769–8775, https://doi.org/10.1021/acs.langmuir.9b00752.

[72] J.M.D. Coey, Hard magnetic materials: a perspective, IEEE Trans. Magn. 47 (12) (2011) 4671–4681, https://doi.org/10.1109/TMAG.2011.2166975.

[73] S.D.N. Almeida, H.E. Toma, Lanthanide Ion Processing from Monazite Based on Magnetic Nanohydrometallurgy vol. 189, Hydrometallurgy, 2019, https://doi.org/10.1016/j.hydromet.2019.105138.

[74] U. Condomitti, et al., Silver recovery using electrochemically active magnetite coated carbon particles, Hydrometallurgy 147 148 (2014) 241.

[75] F. Menegatti de Melo, et al., Magnetophoresis of superparamagnetic nanoparticles applied to the extraction of lanthanide ions in the presence of magnetic field, NanoWorld J. 03 (2017) 38–43, https://doi.org/10.17756/nwj.2017-044.

[76] T. Jesionowski, F. Ciesielczyk, A. Krysztafkiewicz, Influence of selected alkoxysilanes on dispersive properties and surface chemistry of spherical silica precipitated in emulsion media, Mater. Chem. Phys. 119 (1) (2010) 65–74, https://doi.org/10.1016/j.matchemphys.2009.07.034.

[77] W. Stöber, A. Fink, E. Bohn, Controlled growth of monodisperse silica spheres in the micron size range, J. Colloid Interface Sci. 26 (1) (1968) 62–69, https://doi.org/10.1016/0021-9797(68)90272-5.

[78] F.M. De Melo, et al., Superparamagnetic maghemite-based CdTe quantum dots as efficient hybrid nanoprobes for water-bath magnetic particle inspection, ACS Appl. Nano Mater. 1 (6) (2018) 2858–2868, https://doi.org/10.1021/acsanm.8b00502.

[79] D.G. da Silva, et al., Direct synthesis of magnetite nanoparticles from iron(II) carboxymethylcellulose and their performance as NMR contrast agents, J. Magn. Magn Mater. 397 (2016) 28–32, https://doi.org/10.1016/j.jmmm.2015.08.092.

[80] H. Zhang, et al., Selective extraction of heavy and light lanthanides from aqueous solution by advanced magnetic nanosorbents, ACS Appl. Mater. Interfaces 8 (14) (2016) 9523–9531, https://doi.org/10.1021/acsami.6b01550.

[81] Y. Cai, et al., Synthesis of core–shell structured Fe_3O_4 @carboxymethyl cellulose magnetic composite for highly efficient removal of Eu(III), Cellulose 24 (1) (2017) 175–190, https://doi.org/10.1007/s10570-016-1094-8.

[82] F.T. Edelmann, Lanthanides and actinides: annual survey of their organometallic chemistry covering the year 2007, Coord. Chem. Rev. 253 (21–22) (2009) 2515–2587, https://doi.org/10.1016/j.ccr.2009.06.019.

[83] R.G. Silva, J.C. Afonso, C.F. Mahler, Acidic leaching of li-ion batteries, Quim. Nova 41 (5) (2018) 581–586, https://doi.org/10.21577/0100-4042.20170207.

[84] M.K. Jha, et al., Recovery of lithium and cobalt from waste lithium ion batteries of mobile phone, Waste Manag. 33 (9) (2013) 1890–1897, https://doi.org/10.1016/j.wasman.2013.05.008.

[85] F. Wang, X. Lu, X.-Y. Li, Selective removals of heavy metals (Pb^{2+}, Cu^{2+}, and Cd^{2+}) from wastewater by gelation with alginate for effective metal recovery, J. Hazard Mater. 308 (2016) 75–83, https://doi.org/10.1016/j.jhazmat.2016.01.021.

[86] J.S. Cadore, D.A. Bertuol, E.H. Tanabe, Recovery of indium from LCD screens using solid-phase extraction onto nanofibers modified with Di-(2-ethylhexyl) phosphoric acid (DEHPA), Process Saf. Environ. Prot. 127 (2019) 141–150, https://doi.org/10.1016/j.psep.2019.05.011.

[87] J. Płotka-Wasylka, et al., Modern trends in solid phase extraction: new sorbent media, Regular ed. Trac. Trends Anal. Chem. 77 (2016) 23–43, https://doi.org/10.1016/j.trac.2015.10.010.

[88] J. Luo, et al., Removal of antimonite (Sb(III)) and antimonate (Sb(V)) from aqueous solution using carbon nanofibers that are decorated with Zirconium oxide (ZrO_2), Environ. Sci. Technol. 49 (18) (2015) 11115–11124, https://doi.org/10.1021/acs.est.5b02903.

[89] G.V.N. Rathna, et al., Studies on fabrication, characterization, and metal extraction using metal chelating nonwoven nanofiber mats of poly(vinyl alcohol) and sodium alginate blends, Polym. Eng. Sci. 53 (2) (2013) 321–333, https://doi.org/10.1002/pen.23267.

[90] Z.-M. Huang, et al., A review on polymer nanofibers by electrospinning and their applications in nanocomposites, Compos. Sci. Technol. 63 (15) (2003) 2223–2253, https://doi.org/10.1016/S0266-3538(03)00178-7.

[91] I. Wiraputra, et al., The design of mini-rotary forcespinning system for nanofiber synthesis, Procedia Eng. 170 (2017) 24–30, https://doi.org/10.1016/j.proeng.2017.03.005.

[92] F. Nunes da Silva, et al., An eco-friendly approach for metals extraction using polymeric nanofibers modified with di-(2-ethylhexyl) phosphoric acid (DEHPA), J. Clean. Prod. 210 (2019) 786–794, https://doi.org/10.1016/j.jclepro.2018.11.098.

[93] Y. Wen, et al., Recent advances in solid-phase sorbents for sample preparation prior to chromatographic analysis, Trends Anal. Chem. 59 (2014) 26.

[94] R. Sitko, B. Zawisza, E. Malicka, Graphene as a new sorbent in analytical chemistry, Trends Anal. Chem. 51 (2013) 33–43.

[95] Q. Liu, et al., Evaluation of graphene as an advantageous adsorbent for solid-phase extraction with chlorophenols as model analytes, J. Chromatogr. A 1218 (2) (2011) 197–204, https://doi.org/10.1016/j.chroma.2010.11.022.

[96] L.D. Tijing, et al., Nanofibers for water and wastewater treatment: recent advances and developments, in: X.-T. Bui, et al. (Eds.), Water and Wastewater Treatment Technologies. Energy, Environment, and Sustainability, Springer Singapore, Singapore, 2019, pp. 431–468, https://doi.org/10.1007/978-981-13-3259-3_20.

[97] I. Jahan, F. Erci, I. Isildak, Facile microwave-mediated green synthesis of non-toxic copper nanoparticles using *Citrus sinensis* aqueous fruit extract and their antibacterial potentials, J. Drug Deliv. Sci. Technol. 61 (2021) 102172, https://doi.org/10.1016/j.jddst.2020.102172.

[98] K. Kuroda, P. Keller, H. Kawasaki, Mild synthesis of single-nanosized plasmonic copper nanoparticles and their catalytic reduction of methylene blue, Colloid Interface Sci. Commun. 31 (2019) 100187, https://doi.org/10.1016/j.colcom.2019.100187.

[99] T.A. Saleh, Nanomaterials: classification, properties, and environmental toxicities, Environ. Technol. Innovat. 20 (2020) 101067, https://doi.org/10.1016/j.eti.2020.101067.

[100] H.E. Toma, Nanotecnologia Molecular – Materiais e Dispositivos, Blucher, São Paulo, 2016.

[101] N. Baig, I. Kammakakam, W. Falath, Nanomaterials: a review of synthesis methods, properties, recent progress, and challenges, Mater. Adv. 2 (2021) 1821–1871, https://doi.org/10.1039/D0MA00807A.

[102] M. Rafique, et al., History and fundamentals of nanoscience and nanotechnology, in: Nanotechnology and Photocatalysis for Environmental Applications, Elsevier Inc., 2020, pp. 1–25, https://doi.org/10.1016/b978-0-12-821192-2.00001-2.

[103] S. Bayda, et al., The history of nanoscience and nanotechnology: from chemical-physical applications to nanomedicine, Molecules 25 (2020) 1–15, https://doi.org/10.3390/molecules25010112.

[104] F.J. Heiligtag, M. Niederberger, The fascinating world of nanoparticle research, Mater. Today 16 (7–8) (2013) 262–271, https://doi.org/10.1016/j.mattod.2013.07.004.

[105] J. Jeevanandam, et al., Review on nanoparticles and nanostructured materials: history, sources, toxicity and regulations, Beilstein J. Nanotechnol. 9 (2018) 1050–1074, https://doi.org/10.3762/bjnano.9.98.

[106] M.F. Al-Hakkani, Biogenic copper nanoparticles and their applications: a review, SN Appl. Sci. 2 (3) (2020) 505, https://doi.org/10.1007/s42452-020-2279-1.

[107] S. Anu Mary Ealia, M.P. Saravanakumar, A review on the classification, characterisation, synthesis of nanoparticles and their application, IOP Conf. Ser. Mater. Sci. Eng. 263 (2017) 032019, https://doi.org/10.1088/1757-899X/263/3/032019.

[108] L.A. Kolahalam, et al., Review on nanomaterials: synthesis and applications, Mater. Today Proc. 18 (2019) 2182–2190, https://doi.org/10.1016/j.matpr.2019.07.371.

[109] I. Khan, K. Saeed, I. Khan, Nanoparticles: properties, applications and toxicities, Arab. J. Chem. 12 (2019) 908–931, https://doi.org/10.1016/j.arabjc.2017.05.011.

[110] P.N. Sudha, et al., Chapter 12 — nanomaterials history, classification, unique properties, production and market, in: Emerging Applications of Nanoparticles and Architecture Nanostructures, Elsevier Inc., 2018, p. 44, https://doi.org/10.1016/B978-0-323-51254-1/00012-9.

[111] S. Thota, Y. Wang, J. Zhao, Colloidal Au—Cu alloy nanoparticles: synthesis, optical properties and applications, Mater. Chem. Front. 2 (2018) 1074–1089, https://doi.org/10.1039/C7QM00538E.

[112] A. Biswas, et al., Advances in top-down and bottom-up surface nanofabrication: techniques, applications & future prospects, Adv. Colloid Interface Sci. 170 (2012) 2–27, https://doi.org/10.1016/j.cis.2011.11.001.

[113] E. Sánchez-López, et al., Metal-based nanoparticles as antimicrobial agents: an overview, Nanomaterials 10 (292) (2020) 1–39, https://doi.org/10.3390/nano10020292.

[114] N.V. Mdlovu, et al., Recycling copper nanoparticles from printed circuit board waste etchants via a microemulsion process, J. Clean. Prod. 185 (2018) 781–796, https://doi.org/10.1016/j.jclepro.2018.03.087.

[115] P. Samaddar, et al., Synthesis of nanomaterials from various wastes and their new age applications, J. Clean. Prod. 197 (2018) 1190–1209, https://doi.org/10.1016/j.jclepro.2018.06.262.

[116] I.A. Adelere, A. Lateef, A novel approach to the green synthesis of metallic nanoparticles: the use of agro-wastes, enzymes, and pigments, Nanotechnol. Rev. 5 (2016) 567–587, https://doi.org/10.1515/ntrev-2016-0024.

[117] R. Seif El-Nasr, et al., Environmentally friendly synthesis of copper nanoparticles from waste printed circuit boards, Separ. Purif. Technol. 230 (2020) 115860, https://doi.org/10.1016/j.seppur.2019.115860.

[118] M. Tatariants, et al., Antimicrobial copper nanoparticles synthesized from waste printed circuit boards using advanced chemical technology, Waste Manag. 78 (2018) 521–531, https://doi.org/10.1016/j.wasman.2018.06.016.

[119] S. Yousef, et al., A strategy for synthesis of copper nanoparticles from recovered metal of waste printed circuit boards, J. Clean. Prod. 185 (2018) 653–664, https://doi.org/10.1016/j.jclepro.2018.03.036.

[120] G. Karkera, M.A. Reddy, M. Fichtner, Recent developments and future perspectives of anionic batteries, J. Power Sources 481 (2021) 228877, https://doi.org/10.1016/j.jpowsour.2020.228877.

CHAPTER 4

Promising technologies under development for recycling, remanufacturing, and reusing batteries: an introduction

Amilton Barbosa Botelho Junior[1], Giovani Pavoski[1], Mauricio Dalla Costa Rodrigues da Silva[2], William Leonardo da Silva[3], Daniel Assumpção Bertuol[2] and Denise Crocce Romano Espinosa[1]

[1]Department of Chemical Engineering, University of São Paulo, São Paulo, Brazil; [2]Department of Chemical Engineering, Federal University of Santa Maria, Santa Maria, Brazil; [3]Nanoscience Graduate Program, Franciscan University, Santa Maria, Brazil

1. Introduction

The different types of batteries in the market can be highlighted as lead—acid (for vehicle use), alkaline, NiCd, NiMH, and Li-ion. However, the evolution of the domestic market has shown that the NiCd battery is being replaced by NiMH and mainly by lithium-ion batteries (LIBs). It occurs due to the environment's negative impact caused by cadmium in waste and the low specific energy and energy density. Moreover, LIBs have higher market growth and investment in improvements than NiCd and NiMH together, owing to the higher current density, use life, and nominal voltage [1–3].

The consumption of LIBs has been increasing to supply the electronic devices and electric and hybrid vehicles markets. Such batteries are composed, in general, of an anode, cathode, separator, aluminum, and copper foils as current collectors and electrolytes rolled in a metallic or plastic case. Commonly, the anode material is composed of graphite and a copper foil (current collector) glued with a polyvinylidene fluoride (PVDF) binder. The cathode material has different configurations: $LiCoO_2$ (LCO), $LiMn_2O_4$ (LMO), $LiFePO_4$ (LFP), $LiNiCoAlO_2$, and $LiNiMnCoO_2$ (NMC), for instance. The difference among their configurations is related to safety, energy and power density, lifespan, performance, and cost. The electrolyte is composed of $LiPF_6$ or other lithium salts dissolved in organic solvents (propylene carbonate, ethylene carbonate, or dimethyl sulfoxide). The separator is PP or PE [4–6].

The fast technological upgrades result in a growing generation of end-of-life LIBs. Furthermore, in 2020–2021, the consumption of electronic equipment for working or learning from home was also boosted by the COVID-19 pandemic, resulting in a rise in residues in further years [7].

Many approaches have been adopted to treat end-of-life LIBs, where the concepts of reduction, reuse, and recycling are essential to building up a circular economy [8]. The replacement of a few elements in battery manufacturing has been evaluated, as in cobalt. New batteries are produced with less cobalt (as NMC replacing NCO batteries) or without it (LMO and LFP) due to its high price and risk of short-term interruption [9].

The recovery of metals is convenient for economic and environmental reasons since the batteries could be a secondary raw material source and the treatment of the residue [2]. The main process of end-of-life batteries is recycling by pyrometallurgical and hydrometallurgical routes. Pyrometallurgy involves high temperature for metals evaporation and separation from the slag. Faster kinetic than hydrometallurgy is also a significant advance. The Umicore process treats different lithium-ion and NiMH batteries by thermal processing. At $700°C$ occurs the pyrolysis of plastic, and a nickel-cobalt-copper alloy and a slag rich in lithium, aluminum, silicon, and calcium are generated [10].

The Toxco process separates aluminum, copper, and the case by grinding and particle size separation. Then, the binder is removed at $400-700°C$, and the anode is separated by flotation, while the resulting slag is washed and heated at $400-850°C$ to obtain a new cathode remanufacturing of LIBs [11,12].

Regarding the NiCd recycling, the cadmium and nickel hydroxides are decomposed at $300°C$ and $230°C$, respectively. Cadmium vapor is obtained at $850-900°C$ and 10-4 bar (without reducing agent) followed by condensation for metallic cadmium (99.9% purity) [2].

On the other hand, hydrometallurgical processes are more accurate, easily controlled, less energy-intensive, and considered eco-friendly, and obtain high-pure products [13,14]. Bertuol et al. [15] extracted rare earth elements from NiMH batteries by sulfuric acid (H_2SO_4) 2 mol/L at $90°C$ for 30 min. The flowchart is depicted in Fig. 4.1. The batteries were milled, and Fe/Ni magnetic fraction was separated. After leaching, a mixture of rare earth elements is obtained at pH 1.2 with NaOH 5 mol/L. Recovery rate of cerium, lanthanum, neodymium and praseodymium are up to 97%. The rare-earth hydroxide may be calcinated for oxide obtaining [15].

Tanong et al. [16] studied the recycling of spent battery mixture (alkaline, Zn-C, NiMH, NiCd, and lithium-ion) by acid (inorganic and organic) and alkali leaching. The H_2SO_4 achieved the best results for extraction (>65%), while NaOH reached extraction efficiency lower than 10% (despite zinc — 21.5%). Electrolytic nickel is another product of the process. In addition, the recycling of spent battery mixture was studied considering the consumption behavior of the population [16].

Takahashi et al. [17] studied the recovery of cobalt from LIBs by acid leaching followed by solvent extraction. The inorganic acids evaluated were H_2SO_4, HCl (hydrochloric acid), and HNO_3 (nitric acid), as well as the use of hydrogen peroxide (H_2O_2) as a reducing agent. The combination $H_2SO_4 + H_2O_2$ achieved better results for cobalt leaching. According to the data presented by the authors, reducing leaching improves the leaching kinetics of cobalt—Eq. (4.1)—due to the conversion of Co(III) into Co(II) faster

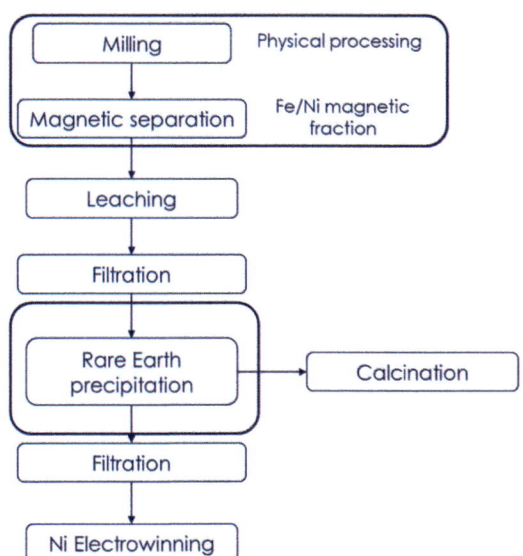

Figure 4.1 The flowchart of hydrometallurgical processing of NiMH for extraction of rare earth elements [15].

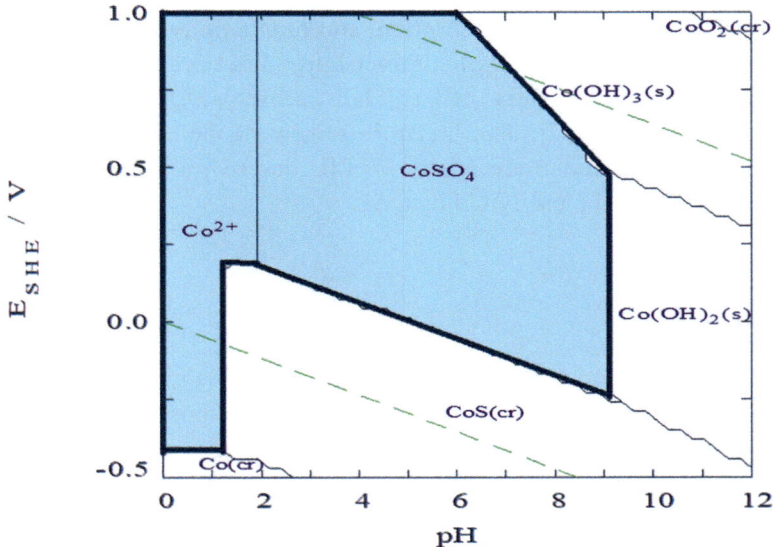

Figure 4.2 Pourbaix diagram of Co-S-H$_2$O system elaborated using Hydra and Medusa software [18,19].

in the presence of H$_2$O$_2$ [17]. Although the literature review, the Pourbaix Diagram of cobalt presented in Fig. 4.2 demonstrated that in leaching conditions (highlighted in green), the cobalt would be extracted without the need for H$_2$O$_2$ [17].

$$4LiCoO_{2(s)} + 6H_2SO_{4(aq)} + 2H_2O_2 \rightarrow 4CoSO_{4(aq)} + 2Li_2SO_{4(aq)} + 8H_2O + 2O_{2(g)} \quad (4.1)$$

OnTo recycling process is a hydrometallurgical route for remanufacturing of spent batteries. After shredding as comminution step, the sieve and magnetic separation remove the metallic case, aluminum, copper, and plastics. Then, cathode and anode are separated by dense media and leached for remanufacturing of batteries (new cathode and anode materials) [20].

Although the current processes for battery recycling, the search for green techniques toward sustainable development faces the need for new approaches for industrial processes [21]. Also, such techniques may achieve high extraction rates, not expensive, safe, and eco-friendly. Therefore, the ionic liquids, supercritical fluids, nanohydrometallurgy, organic acids, deep-eutectic solvents, and biohydrometallurgy are covered in this chapter.

Ionic liquids are considered green solvents with low solubility and non-flammability, used for both extraction and separation steps in the hydrometallurgical route [22,23]. On the other hand, deep eutectic solvents are a mixture of substances forming a eutectic mixture with a melting point lower than separated. Thus, they are similar to ionic liquids but cheaper, safer, and eco-friendly [24].

Supercritical fluids are also an important emerging technology owing to their high density, diffusivity, solubility, and reactivity alongside low viscosity [25]. In addition, nanotechnology has been used to separate metallic ions after leaching due to their large surface area-to-volume ratio, strong sorption, and high affinity for target metals [26].

The use of organic acids and biohydrometallurgy has been explored to replace the inorganic acids avoiding the release of Cl_2 (HCl leaching), SO_3 (H_2SO_4 leaching), and NO_x (HNO_3 leaching) [5,27]. The literature review of the emerging technologies is further discussed. Most studies are related to LIB due to the increasing consumption and replacement of NiMH and NiCd batteries.

2. Ionic liquids

Ionic liquids (ILs) are liquid salts at room temperature, with a melting bridge of less than 100°C, showing cations and anions in their chemical composition, with a delocalized charge [28,29]. Fig. 4.3 shows the main cations and anions present in the composition of ILs.

According to Fig. 4.3, it is possible to identify a series of cations and anions combinations for different ILs forms, directly affecting the physicochemical properties such as hydrophobicity and miscibility in organic solvents or water. For example, 1-n-butyl-3-methylimidazolium bis(trifluoromethylsulfonyl)imide ($C_4MImNTf_2$) and 1-n-butyl-3-methylimidazolium tetrafluoroborate (C_4MImBF_4) are characterized by being hydrophobic and hydrophilic, respectively, for the same cation (1-n-butyl-3-methylimidazolium), but with a variation in the anion (bis(trifluoromethyl-sulfonyl)imide or tetrafluoroborate) [31], according to Fig. 4.4.

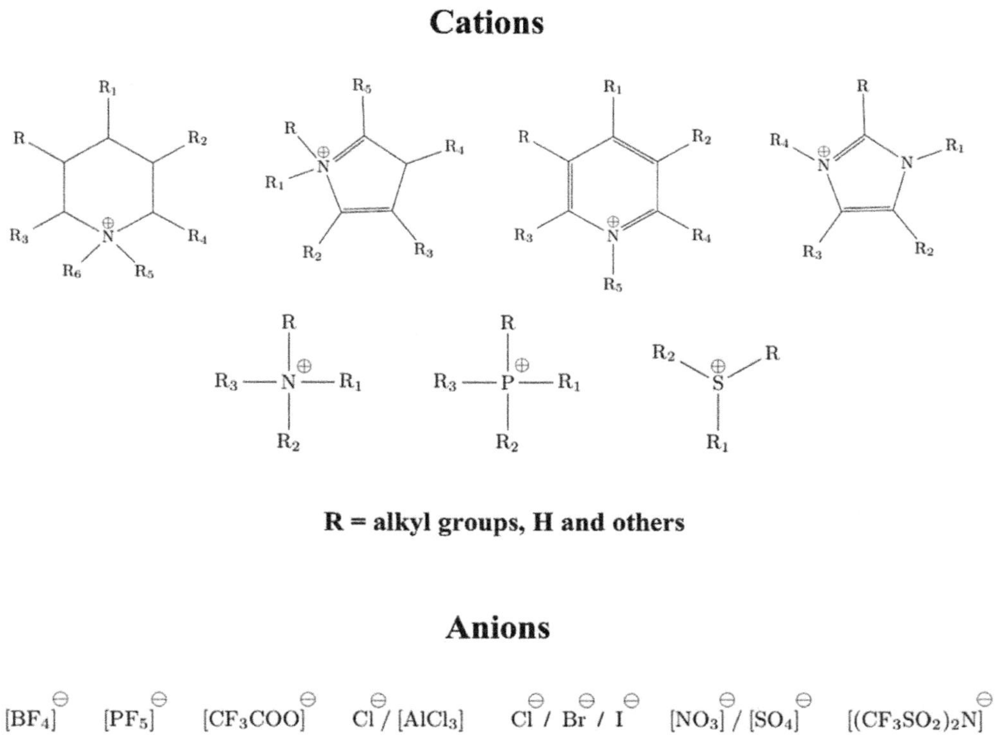

Cations

R = alkyl groups, H and others

Anions

$[BF_4]^-$ $[PF_5]^-$ $[CF_3COO]^-$ $Cl^-/[AlCl_3]^-$ $Cl^-/Br^-/I^-$ $[NO_3]^-/[SO_4]^-$ $[(CF_3SO_2)_2N]^-$

Figure 4.3 Main cations and anions present in the composition of ionic liquids [30].

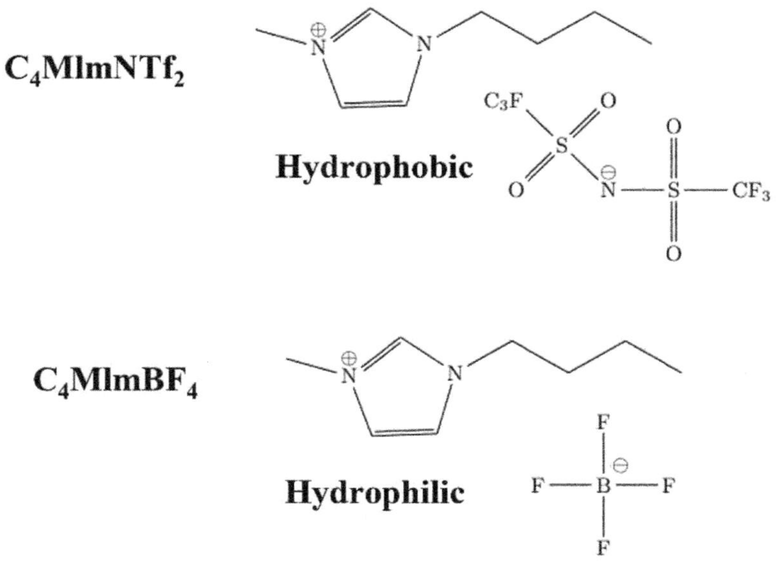

$C_4MlmNTf_2$

Hydrophobic

C_4MlmBF_4

Hydrophilic

Figure 4.4 Representation of ionic liquids with different anions and the same cation, with distinct hydrophobicity properties [32].

Moreover, classical structures of ILs can be modified by functionalization with chemical reactions, denominated task-ionic liquids (talk-ILs), in order to acquire specific characteristics, such as optimization in the properties of viscosity, melting point, solubility, and density, according to the purpose of the application of the ILs [33]. Thus, this flexibility means a wide range of applications of these characteristic compounds, according to Table 4.1.

Table 4.1 Summary of ionic liquids application in LIBs.

Application	Comments	Reference
Catalysis	Combining task-ionic liquids (ILs) and transition metals such as immobilized catalysts with two-phase increases the selectivity and textural properties (specific surface area and porosity), increasing photocatalytic activity and the possibility of catalyst reuse.	[34,35]
Gas capture	Task-ILs are used to increase the affinity with specific gases, such as CO_2 and SO_2, due to the functionalization with certain anions, mainly primary amine (NH_2) in the side chain, helping in the reversible capture of toxic gases, contributing to the environment and enabling the reuse of the task-IL.	[36]
Carbohydrate dissolution	ILs functionalized with ethers chains are characterized for affinity with sugars, dissolving specific carbohydrates such as β-D-glucose and α-cyclodextrin due to the low solubility in water, contributing to nutrient absorption.	[37]
Formation of monolayers	In order to increase hydrophobicity and biocompatibility, thiol-functionalized task-ILs are used to transfer properties of interest to the surface through the formation of organized monolayers (labeled as self-assembled monolayers - SAMs), contributing to the application as biomaterials and bone regeneration.	[38]
Metallic nanoparticles	Functionalization of ionic liquids with metallic nanoparticles promotes changes in the hydrophobicity properties due to the ion exchange between the ionic liquid counterion and the nanoparticle, promoting a change in the hydrophilic to the hydrophobic character.	[39]
Liquid-liquid extraction	Different functionalized ILs improve the extraction of different analytes, such as urea, thiourea, and thioether groups, increasing the extraction of metal ions from aqueous solutions.	[40]
New materials	Task-ILs with transition metals have been studied due to the acquisition of new properties, such as magnetic and electroactivity, providing a significant increase in the application in devices such as electrolytes for lithium batteries due to the low viscosity, high ionic conductivity, and good electrochemical stability.	[41]

According to Table 4.1, it is possible to verify a series of task-ILs applications, especially in new materials, as electrolytes for LIBs. Moreover, principles of sustainable development suggest the development of processes for the remanufacturing, repurposing, and recycling of LIBs applications. In this context, talk-ILs appear to replace conventional organic solvents as electrolytes for energy storage and conversion devices.

Thus, ILs have been applied as electrolytes in lithium and LIBs and supercapacitors due to the presence of a delocalized charge, ensuring anodic stability, flexible structure, and low viscosity [42].

Among the properties of the task-ILs such as excellent substitutes for conventional organic solvents for application in energy storage and conversion devices are [43]:
- neither flammable nor volatile;
- high chemical and electrochemical stability;
- stable liquids over a wide temperature range;
- high ionic conductivity.

Thus, these properties allow ILs to be used as electrolytes in more significant and safer devices, such as electric vehicles and industrial machinery. However, the numerous combinations of cations and anions for the formation of ionic liquids make them with a range of properties that are not entirely understood by the synergistic feat, requiring a better understanding between structure and property for the correct application. Regarding the application of ionic liquids as electrolytes in batteries and supercapacitors, aspects such as kinetic and thermodynamic stability of electrode materials and the effect of lithium salt need to be studied and evaluated.

Moreover, another aspect related to the use of ionic liquids as electrolytes corresponds to temperature range, since viscosity and conductivity are directly affected using slight variations concerning the temperature, compromising the electrochemical response. Thus, a decrease in the temperature promotes an increase in solution and charge transfer resistance. In contrast, at high temperatures, cycling is strongly affected by a decrease in charge overpotential, promoting electrolyte oxidation and favoring the formation of a layer resistive onto the surface [44].

Therefore, the main difference between conventional electrolytes and those based on ionic liquids is that ILs are made up exclusively of ions. Thus, this unique ionic nature helps in the charge compensation process, especially the anion (with less volume than the cations), resulting in high interface strength and excellent cycling.

3. Supercritical fluids

LIBs are constituted by many components (anode, cathode, current collectors, separator, and case, shown in Fig. 4.5), each of them with different compositions and requiring proper operations for recycling. In this process, pre-treatment is a key operation to guarantee the liberation of target elements (Co, Ni, and Li) from the other components,

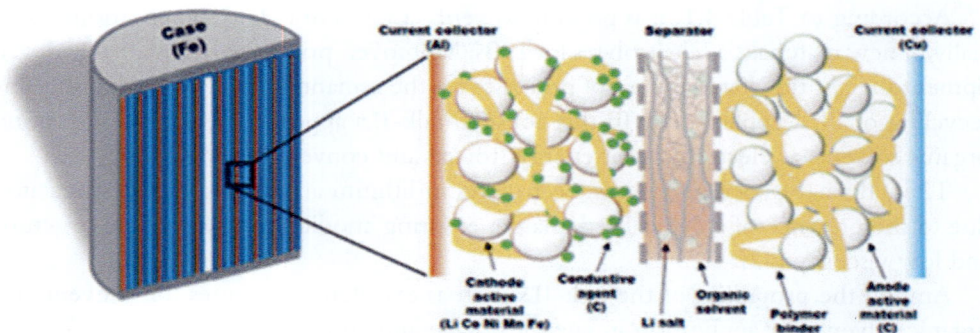

Figure 4.5 A schematic diagram of the LIB components and their constituent elements at issue [45].

especially polymer binders and electrolytes [45]. Supercritical fluid technologies have flourished in the past years as an alternative path due to their unique properties: high density, diffusivity, solubility, and reactivity alongside low viscosity make supercritical fluids favorable for extraction and oxidation processes when compared with other conventional processes. For example, carbon dioxide has been widely used as a supercritical fluid because of its low critical temperature (31.1°C) and critical pressure (79.8 bar) [46].

Electrolyte recovery using supercritical carbon dioxide (SC-CO_2) has been reported as an alternative process for reusing and recycling. Sloop and Parker [47] developed a system for electrolyte recycling from lithium batteries using SC-CO_2 as a solvent; however, information about this process in the literature is scarce [48]. The study of autoclave extraction with supercritical helium head pressure CO_2 method provided a reasonable recovery of organic carbonate solvents from electrolytes, but further purification is required to remove aging components from the final extract. Later, [49] developed a flow-through extraction method of electrolyte from LIBs using CO_2 in supercritical and liquid states combined with different co-solvents. However, the highest yield ($89.1 \pm 3.4\%$wt) of electrolyte extraction was achieved using liquid CO_2 at 25°C and 60 bar for 30 min with 0.5 mL/min acetonitrile/propylene carbonate (3:1) followed by an additional 20 min extraction with liquid CO_2.

Liu et al. [50] reported an 85% recovery of organic carbonate-based electrolytes from LIBs with SC-CO_2 at 23 MPa, 40°C, and 45 min of dynamic extraction under a constant flow rate of 4 L/min. The fractionation process demonstrated to be dependent on the medium polarity, which could be controlled by temperature, pressure, and the addition of cosolvents [51]. After purification, electrolyte reuse demonstrated satisfactory performances compared with commercial electrolytes [52].

Supercritical fluid extraction of the electrolyte from spent LIBs with CO_2 (40°C, 200 bar) and acetonitrile (5 mL/min) as cosolvent was performed to evaluate the formation of aging products and their influence on fading. This technique detected and identified 17

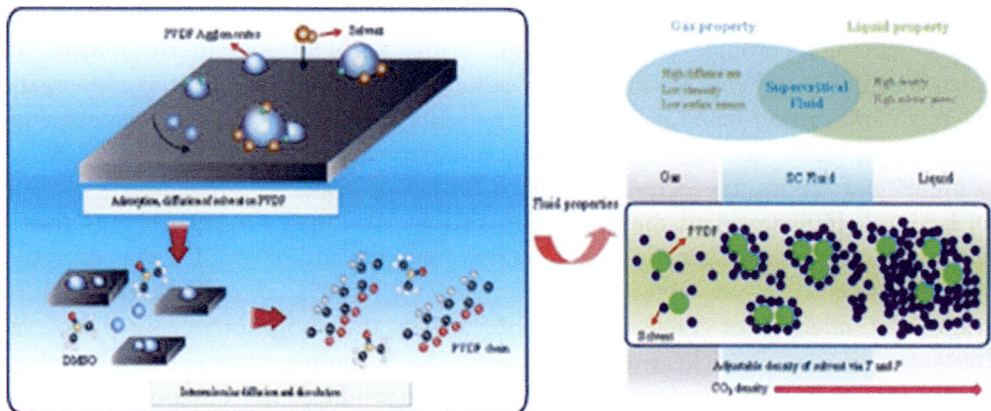

Figure 4.6 Schematic representation of PVDF dissolution in supercritical CO_2 process [54].

different aging products (7 of them not reported in the literature at that moment) using gas chromatography. Furthermore, the combination of supercritical extraction and gas chromatography provided a successful approach to investigate electrolyte composition with a substantial reduction of organic solvents on the graphite anode [53].

Regarding polymer recycling [54], reported the use of $SC-CO_2$ combined with dimethyl sulfoxide (DMSO) for extraction of PVDF, present in organic binders, from spent lithium-ion batteries, which is illustrated in Fig. 4.6. This method promoted 98.5%wt dissolution of PVDF in the $SC-CO_2$ − DMSO system at 70°C and 80 bar during 13 min of processing. Thus, the recovered PVDF preserved the same superficial chemical groups and content as the raw material, suggesting a sustainable approach to recycle organic binder and detach cathode material.

Beyond pre-treatment, supercritical fluids have also been applied in hydrometallurgical processes [55]: reported cobalt leaching from spent LIB using $SC-CO_2$ combined with sulfuric acid and hydrogen peroxide (cosolvents). More than 95% of cobalt was extracted in this process with reduced time and H_2O_2 concentration compared with conventional processes at atmospheric pressure.

4. Nanohydrometallurgy

Nanotechnology is the key technology of the 21st century, which has been widely used in recycling processes and waste treatment. However, as the technology is also applied to LIBs manufacturing, the recycling and management of their end-of-life products must be evaluated [56]. In this context, nanohydrometallurgy has been shown a promising approach for LIBs recycling and selective separation of valuable metals. Compared with traditional technologies, nanomaterials have a large surface area-to-volume ratio, strong sorption, high affinity toward the target element, and faster dissolution resulting in higher efficient rates [26].

There are several nanomaterials used for adsorption of metals functionalized with various chemical groups: polymer-derived nanoadsorbents; polymer-modified and metal-based nanosorbents; photocatalytic nanomaterials; biogenic, magnetic, superparamagnetic and carbon-based nanoparticles; nanofibers and nanocelluloses; nanozeolites; graphene and graphene oxide; and carbon and graphene quantum [26,57—59]. A few examples and application focused on LIBs is further presented.

Le et al. [60] described the adsorption of metallic ions by ZnO nanoadsorbents. Regarding the elements present in battery recycling, the authors studied the adsorption of copper, manganese, and nickel, and the efficiency of silver, lead, and cadmium was also evaluated. Under UV or visible light as photocatalysts, the adsorption of copper achieved 100% in 15 min, while nickel reached less than 10%. According to the authors, both light sources were insufficient for ZnO nanoadsorbents to produce electrons/holes for photo-reduction/photo-oxidation for nickel adsorption, indicating it would occur by physical mechanism [60].

Polymer-based nanosorbents exhibit a higher adsorption rate than normal-sized oxides due to the metal-ligand precipitation or formation of ternary ligands [26]. According to Ref. [61], polymer-derived ZnO nanoadsorbents may achieve up to 99% of metallic ions' removal efficiency in the LIB cathodes [61]. In this context, nanofibrillated celluloses would be used in the separation step of the hydrometallurgical route. The poly(methacrylic acid-co-maleic acid) grafted nanofibrillated cellulose demonstrated an extraction rate over 50% for a nickel at pH 5.0 [57].

In addition, biomaterials have been used as low-cost material, where it can be functionalized with amine, phosphate carboxyl, hydroxyl, phenol, phosphate, imidazole, thioether, or carbonyl groups. Furthermore, such materials may be used to create a bio-nanosorbent. For example, [62] developed a fungus extract-chitosan for nickel and copper adsorption, which reached adsorption capacity almost 1000 times higher than zeolites [62].

Among such techniques, magnetic nanoparticles have demonstrated results and recyclability close to industrial application. For example, [63] reported functionalized superparamagnetic nanoparticles to separate metals from the aqueous phase. The metallic ions captured by magnetite nanoparticles are then confined at the electrode surface using an external magnet. In this case, the nanoparticles are mixed with solution-containing metallic ions. Then, the solid-liquid separation would occur by applying an external magnetic field, where the superparamagnetic nanoparticles are attracted [63]. Finally, the elution process for ions release may be carried out by an acid solution, as the current technology for ion exchange resins [64—67].

Melo et al. [68] synthesized a superparamagnetic nanoparticle functionalized with diethylenetriaminepentaacetic acid using a crystalline Fe_3O_4 core with a SiO_2 protecting coating. According to the authors, the recovery of nickel, manganese, copper, and cobalt was similar to the commercial sorbents literature [68]. In the case of lanthanum, the results were inferior to those reported in the literature [69], which is essential for NiMH recycling (Fig. 4.1). According to Toma et al., the technique has excellent results for recovering copper, nickel, cobalt, and manganese compared with the literature meeting the principles of green chemistry [59,70].

The literature demonstrated that nanotechnology could be effective for the separation of metals through the recycling process. Furthermore, such technology is also used for LIB manufacturing. For instance, graphene may be used for anodes or as an additive for the cathode. Since it is made out of a single atomic layer of carbon, a better storage capacity is obtained when compared with graphite [71,72]. An improvement in energy storage was observed when graphene was added to the cathode [73].

Nanostructured materials as anode represent advantages in comparison with graphite. For example, titanate material produces faster charging times since lithium ions can be inserted into nanostructured than graphite. On the other hand, nanostructured separators made of silicon have been used to isolate the electrolyte from the electrodes, resulting in long inactive shelf life. The use of graphene or other nanostructures represents an additional issue for recycling or disposal as they move into the solution or waste stream, and new treatment approaches must be developed [56]. Despite that, nanotechnology has represented an outstanding tool for battery manufacturing and recycling.

5. Organic acids

Organic acids are outstanding on an industrial scale among the chemical products used for recycling, remanufacturing, and reusing batteries [5]. In hydrometallurgical processes of spent LIBs, Cl_2 (HCl leaching), SO_3 (H_2SO_4 leaching), and NO_x (HNO_3 leaching) gases may be released, causing several environmental and human damages. The costs of the processes rose for the treatment and disposal of hazardous gases and wastewaters to avoid environmental contamination. Moreover, the type of acid solution (H_2SO_4, HNO_3, or HCl, for instance) influences further separation and purification steps. If neutralization is required, the acid would not be recovered, affecting the economic feasibility [27].

Nonetheless, organic acids have advantages over inorganic acids, such as biodegradability, non-gases generation, easy management of wastewaters, and less environmental impact since microorganisms produce them. Acquisition costs for a small group of organic acids are higher than inorganic acids; however, their application is considered cost-effective due to environmental benefits and reduced process problems (corrosion caused by inorganic acids, for instance) [13,27].

The literature review of organic acids application in spent battery treatment is shown in Table 4.2. It is depicted different types of applications of LIBs (domestic or vehicles). Such batteries are widely used for computers, smartphones, electric cars, and several other electronic devices due to their low cost, sustainability, safety, and less toxicity in

Table 4.2 Summary of leaching conditions, efficiency rates, and batteries treated by organic acids gathered from the literature.

Li-ion battery type	Organic acids	Conditions	Leaching efficiency	References
LCO	Leaching agents: H_2SO_4, lactic, butyric, acetic and propionic; Reducing agents: H_2O_2 (6% v/v), glucose (0.09 mol/L) and lactose (0.09 mol/L)	S/L ratio = 20 g/L; H_2SO_4 1.25 mol/L + organic acids (from a fermentation effluent by an anaerobic microbial consortium) 0.75 mol/L; 86°C; 18.5 g/L, 300 rpm and 0.09 M lactose from MWP	Co = 93%; Li = 90%	[74]
NMC	Assisted by ultrasound: 37 kHz Leaching agent: Lemon juice (citric acid = 90 mg/g, malic acid = 0.86 mg/g, and ascorbic acid 1.24 mg/g); Reducing agent: H_2O_2	S/L ratio = 0.98% (w/v); lemon juice = 57.8% (v/v); H_2O_2 8.07% v/v	Li = 100%; Co = 96%; Ni = 96%	[75]
NMC	Leaching agents: Citric acid and DL-malic acid Reducing agents: H_2O_2	Citric acid: 1.5 mol/L, 30 min, 95°C, 2% v/v; DL-malic acid: 1.0 mol/L, 30 min, 95°C, 2% v/v;	Citric acid: Co = 95%, Li = 97%, Ni = 99%; DL-malic acid: Co = 98%, Li = 96%, Ni = 99%;	[76]
Li-Co battery (not specified)	Leaching agent: Acetic acid Reducing agent: H_2O_2	pH 1.69, 90°C, 4% v/v H_2O_2, S/L ratio = 10 g/L; 120 min	Li = 84%; Co = 54%	[77]
NMC (111)	Leaching agent: DL-malic Reducing agent: H_2O_2	S/L ratio = 40 g/L; 1.2 mol/L; H_2O_2 = 1.5% v/v; 80°C; 30 min	Li = 99%; Co = 94%; Ni = 95%; Mn = 96%	[78]

Table 4.2 Summary of leaching conditions, efficiency rates, and batteries treated by organic acids gathered from the literature.—cont'd

Li-ion battery type	Organic acids	Conditions	Leaching efficiency	References
LFP	Leaching agents: Oxalic acid, citric acid, acetic acid, HCl, H_2SO_4, HNO_3 Reducing agent: H_2O_2	S/L ratio = 120 g/L; acetic acid: 0.6 mol/L; H_2O_2: 6 % v/v; 50°C	Li = 95%; Fe =<1%; Al =<1%	[79]
NMC	Leaching agent: Citrus fruit juice	Citrus fruit juice = 50,000 mg/L; 90°C; 20 min	Li = ~100%; Mn = 99%; Ni = 98%; Co = 94%	[80]
LCO	Assisted by ultrasound Leaching agent: Acetic acid Reducing agent: H_2O_2	S/L ratio = 30 g/L, citric acid = 2 mol/L, H_2O_2 = 1.25 %v/v, 2 h, 60°C	Li = 92%; Co = 81%	[81]
LCO	Leaching agent: Glycine Reducing agent: Ascorbic acid	S/L ratio = 0.2 g $LiCoO_2$/100 mL; glycine = 0.5 mol/L; ascorbic acid = 0.02 M; 360 min; 80°C	Co = 95%	[82]
LCO	Leaching agent: Tartaric acid Reducing agent: Ascorbic acid	S/L ratio = 0.2 g $LiCoO_2$/100 mL; tartaric acid = 0.4 mol/L; ascorbic acid = 0.02 mol/L; 5 h; 80°C	Co =>95%	[83]
NMC (111)	Leaching agent: DL-malic acid Reducing agent: H_2O_2	S/L ratio = 3.0 g NCM/50 mL; 1.0 mol/L D,L-malic acid; 3 mL H_2O_2; 50°C; 30 min	~100%	[79]

comparison with the other types of batteries, resulting in different compositions and configurations [6,84]. The critical parameters evaluated were solid-liquid (S/L) ratio, temperature, time, acid concentration, and the use of reducing agent owing to increase the kinetics of leaching [85,86].

In Table 4.2, the data for different organic acids explored as leaching and reducing agents, such as lactic acid, butyric acid, acetic acid, propionic acid, citric acid, DL-malic acid, oxalic acid, glycine, and tartaric acid. The most common reducing agent is H_2O_2, but glucose, lactose, and ascorbic acid have also been studied as an alternative.

The extraction of metals is similar to inorganic acids. For instance, Ref. [17] achieved up to 99% cobalt extraction from LCO batteries using $H_2SO_4 + H_2O_2$ at pH 3, S/L ratio equals 1/5, at 50°C for 2h. Similar results were obtained using organic compounds, as [87] reported using 1 mol/L DL-malic acid + 3 mL H_2O_2, 3.0 g NCM battery/50 mL at 50°C for 30 min.

Eq. (4.1) shows the leaching of LCO batteries with H_2SO_4 under reducing conditions (H_2O_2), where lithium and cobalt sulfate are generated along with H_2O and O_2 gas. Acid leaching with citric acid (Eq. 4.2) and DL-malic acid (Eq. 4.3) is further presented under reducing agent (H_2O_2), where complexes of organic acids and metallic ions are formed [76].

$$4LiCoO_{2(s)} + 6C_6H_8O_{7(aq)} + 6H_2O_2 \rightarrow 2Co(C_6H_5O_7)_{2(aq)} + 2Li_3C_6H_5O_{7(aq)} + 15H_2O + \frac{9}{2}O_{2(g)}$$

(4.2)

$$4LiCoO_{2(s)} + 3C_4H_6O_{5(aq)} + H_2O_2 \rightarrow 2CoC_4H_4O_{5(aq)} + Li_2C_4H_4O_{5(aq)} + 4H_2O + O_{2(g)}$$

(4.3)

Organic compounds can also be applied as a reducing agent, as demonstrated by Ref. [74], where lactose achieved better results than H_2O_2 and glucose. In addition, [82,83] reported that ascorbic acid would be an excellent substitute for inorganic acids for sustainable goals [74,82,83].

The extraction of metals from LIBs by inorganic acids is improved under the ultrasound effect [17]. Also, the extraction by organic acids may be carried out using pure products and a mixture of organic compounds. Esmaeili et al. [75] evaluated the use of ultrasound to improve the leaching of lithium, cobalt, and nickel by lemon juice composed of citric acid (90 mg/g), malic acid (0.86 mg/g), and ascorbic acid (1.24 mg/g). The extraction achieved over 96% under the H_2O_2 effect [75]. Urias et al. [74] evaluated the mixture of H_2SO_4 and organic acids for acid leaching, reaching better results than H_2SO_4 alone. Among the organic acids, results demonstrated that the better leaching agents were citric, propionic, butyric, acetic, and lactic acid [74].

Yao et al. [87] recycled NMC-type batteries using DL-malic acid as a leaching agent and H_2O_2 as a reducing agent. During the reaction, Al foils were separated from the cathode with almost 0% of losses. After solid-liquid separation, the composition of metallic ions in the leaching solution was adjusted to obtain a new NMC cathode for remanufacturing. As clearly observed, organic acids can recycle spent batteries for different applications, such as metal salts with high purity and new cathode materials for reusing and remanufacturing batteries.

6. Deep-eutectic solvents

Deep eutectic solvents (DESs) are classified as a type of ionic fluid generally composed of two or three substances capable of self-association to form a eutectic mixture with a melting point lower than each substance (usually they are liquid at temperatures inferior to 150°C). DES presents similar physical and chemical properties to ionic liquids. However, the components are cheaper, safer, and eco-friendly. Additionally, DES is not flammable, not volatile, easy to synthesize and manipulate, and does not require purification. These advantages make it attractive from the point of view of green chemistry and large-scale industrial applications [24].

In the context of recycling, DES formed by choline chloride has been studied due to its ability to dissolve a range of metal oxides [88]. Foreman et al. [89] demonstrated the feasibility extraction of transition metals and lanthanides from a DES (composed of lactic acid and choline chloride) by 30% Aliquat 336 (v/v) in toluene and evaluated the effect of water content in the extraction of transition metals from the DES phase.

Albler et al. [90] presented a new approach to selectively recover nickel from the DES phase using DEHPA and a pyridyl pyrazole in Solvent 70, which is less harmful than the process using Aliquat 336/toluene and does not require pH adjustments between extractions. Amphlett et al. [91] described the speciation of Ni(II) and Co(II) in blends of choline chloride with glycerol and carboxylic acids (malonic acid, succinic acid, glutaric acid, lactic acid, and glycolic acid), highlighting the importance of process conditions for the selective separation, especially the presence of water in the DES. Moreover, selective fractionation of nickel and lithium using a DES-based on lidocaine and decanoic acid was also reported with a high extraction efficiency of nickel from the solution containing both metals [92].

A process for selective recovery of cobalt from spent LIB using choline chloride − ethylene glycol DES has been reported by Ref. [93]: 90% of cobalt was extracted from the end-of-life LIB during leaching with DES purification was performed through solvent extraction with DEHPA. After the purification, the cobalt was recovered as oxalate, which was later employed to produce lithium cobalt oxide, and the DES was recycled for another leaching, yielding cobalt extraction as the fresh DES mixture.

Using a DES composed of choline chloride and urea, [94] obtained 95% extraction efficiency of lithium and cobalt front spent LIB. These results were possible due to the strongly reduced power of the DES applied, reducing the time and temperature of processing. In addition, cubic cobalt oxide spinel was recovered from the DES liquor by dilution, precipitation, and calcination using oxalic acid and sodium hydroxide.

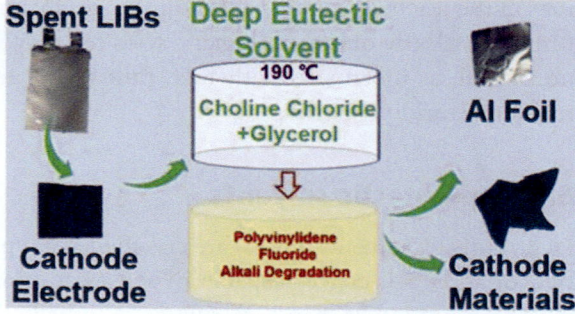

Figure 4.7 Separation of the aluminum foil from the cathode electrode using choline chloride + glycerol mixture as deep eutectic solvent [97].

Landa-Castro et al. [95] studied the leaching of NiMH batteries using a mixture of choline chloride and reline as DES to recover nickel (83.6%) and cobalt (53.2%) from the cathode. Thus, the obtained solution was used as an electrolytic bath for electrodeposition of particles of Ni-Co alloy (64.09% Ni and 35.90% Co) in a glassy carbon electrode.

Furthermore, deep eutectic solvents can be used for recycling other components from the LIB. For example, [96] reported a method of recovering different components of spent LIB using DES formed by combining choline chloride and ethylene glycol in a 1:2 M ratio. This process obtained efficiencies higher than 90% for lithium and cobalt leaching and separated the aluminum foil, PVDF binder, and conductive carbon by filtration. Cobalt was recovered from the DES through electrodeposition, and the remaining DES could be reused for another leaching.

Separation of the aluminum foil from spent LIB was also achieved using a choline chloride-glycerol DES and reaching 99.9% of peeling of the cathode material from the foil (Fig. 4.7). This result was credited to the deactivation of the PVDF binder attributed to alkali degradation of the acidic hydrogen atom in the matrix [97].

7. Biohydrometallurgy

Another way to avoid the landfill of batteries losing valuable metals is recycling by bioprocesses, known as biohydrometallurgy. The technology was previously developed to extract metals from primary sources (ores), but commercial plants for battery recycling are under development [98]. The main advantages are less energy consumption, low-toxic reagents, pollution reduction, and a simple, safe, and eco-friendly process. However, in the process, the microorganisms leach the metals from the spent batteries causing limitations in their growth solved by adaption before leaching reaction. Another limitation is the difficulty of maintaining the purity of the inoculated microorganism [99–101].

In the process, the microorganisms dissolve the metals from the battery by biooxidation, and they gain energy by rupturing the residue into their component metals. Thus, the reaction may aim different results: recycling of the entire battery to obtain different products (salts of the metals or metallic products), recovery of the main components into a new product, recycling to obtain new cathode and anode, or recycling to produce a new battery [98]. In the last case, the remanufacturing of batteries is beneficial in countries or regions that produce the batteries. In others, the production of new products from spent batteries or reuse may be advantageous [5,102,103].

Each microorganism has its own best conditions: pH, temperature, time, S/L ratio, and the composition of the culture medium. *Chemolithotrophic prokaryotes* and heterotrophic bacteria are the main microorganisms studied for bioleaching from secondary sources. Consortium (combination of different microorganisms) is studied for synergic bioleaching. In fungi, organic acids are produced and leach the metals from the scrap [7,98].

There are three mechanisms in bioleaching: redoxolysis, acidolysis, and complexolysis. The kinetic pathway relies on promoting redox potential, production of metabolites to complex metals after dissolution, and binding ability to metal substrates. As pointed by Ref. [98], the extraction yields also depend on more energy and carbon sources due to the complexity of the metals in the battery [98]. Nevertheless, the literature review has shown exciting data for recycling, remanufacturing, and reusing batteries using biohydrometallurgy.

The bioleaching of nickel, cobalt, lithium, manganese, aluminum, and copper was evaluated using adapted *Aspergillus niger* from mobile phone batteries. The use of non-adapted microorganisms was more advantageous, where organic acids were produced earlier than non-adapted resulting in a higher extraction rate. The acids generated by *Aspergillus niger* were gluconic, oxalic, malic, and citric. Eq. (4.4) shows the leaching using oxalic acid generated by the *Aspergillus niger*, the least efficient among them. At a pulp density of 1%w/v, the bioleaching rate was 100% for lithium, 94% for copper, 72% for manganese, 62% for aluminum, 45% for nickel, and 38% for cobalt [104].

$$4LiCoO_{2(s)} + 7C_2H_2O_{4(aq)} \rightarrow 2Co(C_4H_4O_4)_{2(aq)} + LiC_2HO_{4(aq)} + 4H_2O_{(l)} + 2CO_{2(g)} \qquad (4.4)$$

Bajestani et al. [105] studied the extraction of metals from spent household batteries using *Acidithiobacillus ferrooxidans*. It was evaluated the extraction of nickel, cadmium, and cobalt from NiCd and NiMH batteries. The microorganisms were adapted, and bioleaching achieved 85.6% for nickel, 66.1% for cadmium, and 90.6% for cobalt at pH 1 and an initial Fe(III) concentration of 9.7 g/L. In addition, it demonstrated that the *Acidithiobacillus ferrooxidans* could support the presence of the toxic metal cadmium [105].

Xin et al. [106] studied the bioleaching of LIBs from electric vehicles (LFP, LMO, and NMC) by autotrophic bioleaching. Despite the fast kinetic in the pyrometallurgical process and a few minutes in acid leaching of hydrometallurgical one, bioleaching takes a few days to achieve the maximum extraction. In this case, the extraction of metals from the NMC battery achieved 95% after 9 days. The process generated H_2SO_4, which reacted with metallic oxides and dissolved those [106].

The hybrid system of chemical and biological processes has also been advantageous for the treatment of spent batteries. For example, [107] used citric acid to remove the binder (PVDF), which leached lithium. In the combined system, the lithium extraction was higher than the chemical process using citric acid and *Lysinibacillus sp*, and the adsorption of cobalt by the biomass was enhanced in the system [107].

Biomass may also be used as a microorganism source for leaching as reducing agents [108]. *Phytolacca Americana* branch and tea waste powders combined with citric acid have been evaluated as leaching agents. Results compared with H_2O_2 demonstrated similar extraction efficiencies for cobalt and lithium. At 90°C, 1 mol/L of citric acid, 0.2 g/g of reducing agent for 120 min, and S/L ratio achieved up to 90% of cobalt efficiency using powders *Phytolacca Americana* branch, while the efficiency was 70% using tea waste and H_2O_2. In lithium, the extraction was close to 95% for all reducing agents [108].

Biohydrometallurgy has excellent potential for industrial purposes. The current lack in the literature is the selective separation of the metallic ions after bioleaching. Despite similar compositions, the presence of microorganisms may negatively impact the separation degree [5]. Biosorption or precipitation can be used for the separation of metallic ions. Live or dead biomass is used for the sorption of ions in batch or column scale. In precipitation, the microorganisms produce metabolites reacting with metallic ions, which results in precipitation [7,109,110].

8. Conclusion

Emerging technologies have been evaluated to achieve the goals for sustainable development. This chapter addressed the leading technologies for battery recycling, where LIBs get more attention due to their growing demand for electronic devices and electric/hybrid vehicles. The first part of this chapter focuses on presenting current technologies, where the hydrometallurgical route is highlighted for recycling, reuse, and remanufacturing. Then, the underdevelopment and promising technologies are further presented. The literature review has demonstrated that the extraction of metals from batteries is similar to the current hydrometallurgical process using inorganic acids. The main reagents and best extraction conditions were summarized for different applications. Finally, future trends are related to the pilot-scale of emerging technologies for economic evaluation.

Acknowledgments

Thanks to the University of Sao Paulo for supporting this project and to Fundação de Amparo à Pesquisa do Estado de São Paulo and Capes (FAPESP — grant 2012/51871-5, 2018/03483-6, 2018/11417-3, Sao Paulo Research Foundation) for financial support. This project was developed with the support of SemeAd (FEAUSP), FIA Fundação Instituto de Administração and Cactvs Instituto de Pagamento S.A. through the granting of assistance to a research project Bolsa SemeAd PQ Jr (Public Notice 2021.01).

References

[1] Christophe Pillot, The Rechargeable Battery Market and Main Trends 2011–2020. Avicenne Energy, 2019. URL, https://niobium.tech/en/pages/gateway-pages/pdf/technical-briefings/the-rechargeable-battery-market-and-main-trends-2018-2030 (Accessed 8 June 2021).

[2] D.C.R. Espinosa, M.B. Mansur, Recycling Batteries, Waste Electrical and Electronic Equipment (WEEE) Handbook, Elsevier Ltd, 2019, https://doi.org/10.1016/B978-0-08-102158-3.00014-8.

[3] M.A. Hannan, S.B. Wali, P.J. Ker, M.S.A. Rahman, M. Mansor, V.K. Ramachandaramurthy, K.M. Muttaqi, T.M.I. Mahlia, Z.Y. Dong, Battery energy-storage system: a review of technologies, optimization objectives, constraints, approaches, and outstanding issues, J. Energy Storage 42 (2021) 103023, https://doi.org/10.1016/j.est.2021.103023.

[4] G. Harper, R. Sommerville, E. Kendrick, L. Driscoll, P. Slater, R. Stolkin, A. Walton, P. Christensen, O. Heidrich, S. Lambert, A. Abbott, K. Ryder, L. Gaines, P. Anderson, Recycling lithium-ion batteries from electric vehicles, Nature 575 (2019) 75–86, https://doi.org/10.1038/s41586-019-1682-5.

[5] L.S. Martins, L.F. Guimarães, A.B. Botelho Junior, J.A.S. Tenório, D.C.R. Espinosa, Electric car battery: an overview on global demand, recycling and future approaches towards sustainability, J. Environ. Manag. 295 (2021) 113091, https://doi.org/10.1016/j.jenvman.2021.113091.

[6] J. Xie, Y.C. Lu, A retrospective on lithium-ion batteries, Nat. Commun. 11 (2020) 9–12, https://doi.org/10.1038/s41467-020-16259-9.

[7] E.R. Rene, M. Sethurajan, V. Kumar Ponnusamy, G. Kumar, T.N. Bao Dung, K. Brindhadevi, A. Pugazhendhi, Electronic waste generation, recycling and resource recovery: technological perspectives and trends, J. Hazard Mater. 416 (2021) 125664, https://doi.org/10.1016/j.jhazmat.2021.125664.

[8] T. Fujita, H. Chen, K. Wang, C. He, Y. Wang, G. Dodbiba, Y. Wei, Reduction, reuse and recycle of spent Li-ion batteries for automobiles: a review, Int. J. Miner. Metall. Mater. 28 (2021) 179–192, https://doi.org/10.1007/s12613-020-2127-8.

[9] M.B. Mansur, A.S. Guimarães, M. Petraniková, An overview on the recovery of cobalt from end-of-life lithium ion batteries, Miner. Process. Extr. Metall. Rev. 0 (2021) 1–21, https://doi.org/10.1080/08827508.2021.1883014.

[10] Umicore, Umicore Cobalt and Specialty Materials [WWW Document], 2020. URL, https://csm.umicore.com/en/about-us/ (Accessed 6 December 2020).

[11] Retriev Technologies, Recycling Technology, 2020. URL, https://www.retrievtech.com/recycling-technology (Accessed 6 December 2020).

[12] Retriev Technologies, Lithium Ion, 2020. URL, http://www.rathboneenergy.com/articles/sanyo_lionT_E.pdf (Accessed 6 December 2020).

[13] R. Golmohammadzadeh, F. Faraji, F. Rashchi, Recovery of lithium and cobalt from spent lithium-ion batteries (LIBs) using organic acids as leaching reagents: a review, Resour. Conserv. Recycl. 136 (2018) 418–435, https://doi.org/10.1016/j.resconrec.2018.04.024.

[14] A. Kumar, M. Holuszko, D.C.R. Espinosa, E-waste: an overview on generation, collection, legislation and recycling practices, Resour. Conserv. Recycl. 122 (2017) 32–42, https://doi.org/10.1016/j.resconrec.2017.01.018.

[15] D.A. Bertuol, A.M. Bernardes, J.A.S. Tenório, Spent NiMH batteries-the role of selective precipitation in the recovery of valuable metals, J. Power Sources 193 (2009) 914–923, https://doi.org/10.1016/j.jpowsour.2009.05.014.

[16] K. Tanong, L. Coudert, G. Mercier, J.F. Blais, Recovery of metals from a mixture of various spent batteries by a hydrometallurgical process, J. Environ. Manag. 181 (2016) 95–107, https://doi.org/10.1016/j.jenvman.2016.05.084.

[17] V.C.I. Takahashi, A.B. Botelho Junior, D.C.R. Espinosa, J.A.S. Tenório, Enhancing cobalt recovery from Li-ion batteries using grinding treatment prior to the leaching and solvent extraction process, J. Environ. Chem. Eng. 8 (2020) 103801, https://doi.org/10.1016/j.jece.2020.103801.

[18] A.B. Botelho Junior, D.B. Dreisinger, D.C.R. Espinosa, J.A.S. Tenório, Pre-reducing process kinetics to recover metals from nickel leach waste using chelating resins, Int. J. Chem. Eng. 1–7 (2018), https://doi.org/10.1155/2018/9161323, 2018.

[19] J.C. Izidoro, M.C. Kim, V.F. Bellelli, M.C. Pane, A.B. Botelho Junior, D.C.R. Espinosa, J.A.S. Tenório, Synthesis of zeolite a using the waste of iron mine tailings dam and its application for industrial effluent treatment, J. Sustain. Min. 18 (2019) 277–286, https://doi.org/10.1016/j.jsm.2019.11.001.

[20] O. Velázquez-Martínez, J. Valio, A. Santasalo-Aarnio, M. Reuter, R. Serna-Guerrero, A critical review of lithium-ion battery recycling processes from a circular economy perspective, Batteries 5 (2019) 5–7, https://doi.org/10.3390/batteries5040068.

[21] B. Soergel, E. Kriegler, I. Weindl, S. Rauner, A. Dirnaichner, C. Ruhe, M. Hofmann, N. Bauer, C. Bertram, B.L. Bodirsky, M. Leimbach, J. Leininger, A. Levesque, G. Luderer, M. Pehl, C. Wingens, L. Baumstark, F. Beier, J.P. Dietrich, F. Humpenöder, P. von Jeetze, D. Klein, J. Koch, R. Pietzcker, J. Strefler, H. Lotze-Campen, A. Popp, A sustainable development pathway for climate action within the UN 2030 Agenda, Nat. Clim. Change 11 (2021) 656–664, https://doi.org/10.1038/s41558-021-01098-3.

[22] I. Osada, H. de Vries, B. Scrosati, S. Passerini, Ionic-liquid-based polymer electrolytes for battery applications, Angew. Chem. Int. Ed. 55 (2016) 500–513, https://doi.org/10.1002/anie.201504971.

[23] L. Xu, C. Chen, M.L. Fu, Separation of cobalt and lithium from spent lithium-ion battery leach liquors by ionic liquid extraction using Cyphos IL-101, Hydrometallurgy 197 (2020) 105439, https://doi.org/10.1016/j.hydromet.2020.105439.

[24] Q. Zhang, K. De Oliveira Vigier, S. Royer, F. Jérôme, Deep eutectic solvents: syntheses, properties and applications, Chem. Soc. Rev. 41 (2012) 7108–7146, https://doi.org/10.1039/c2cs35178a.

[25] D.A. Bertuol, F.R. Amado, E.D. Cruz, E.H. Tanabe, Metal recovery using supercritical carbon dioxide, in: Green Sustainable Process for Chemical and Environmental Engineering and Science: Supercritical Carbon Dioxide as Green Solvent, 2019, pp. 85–103, https://doi.org/10.1016/B978-0-12-817388-6.00005-2.

[26] T. Mehrotra, S. Sinha, R. Singh, Application of nanotechnology in the remediation of heavy metal toxicity, in: New Trends in Removal of Heavy Metals from Industrial Wastewater, Elsevier Inc, 2021, https://doi.org/10.1016/b978-0-12-822965-1.00015-5.

[27] Y. Yao, M. Zhu, Z. Zhao, B. Tong, Y. Fan, Z. Hua, Hydrometallurgical processes for recycling spent lithium-ion batteries: a critical review, ACS Sustain. Chem. Eng. 6 (2018) 13611–13627, https://doi.org/10.1021/acssuschemeng.8b03545.

[28] J.P. Hallett, T. Welton, Room-temperature ionic liquids: solvents for synthesis and catalysis, Chem. Rev. 111 (2011) 3508–3576, https://doi.org/10.1021/cr1003248.

[29] T. Welton, Ionic liquids in catalysis, Coord. Chem. Rev. 248 (2004) 2459–2477, https://doi.org/10.1016/j.ccr.2004.04.015.

[30] J. Dupont, P.A.Z. Suarez, Physico-chemical processes im imidazolium ionic liquid, Phys. Chem. Chem. Phys. 8 (2006) 2441–2452, https://doi.org/10.1039/B602046A.

[31] E.I. Izgorodina, U.L. Bernard, P.M. Dean, J.M. Pringle, D.R. Macfarlane, The madelung constant of organic salts, Cryst. Growth Des. 9 (2009) 4834–4839, https://doi.org/10.1021/cg900656z.

[32] R.D. Rogers, K.R. Seddon, Ionic liquids: solvents of the future? Science 302 (2003) 792–793, https://doi.org/10.1126/science.1090313.

[33] Y. Dai, Y. Qu, S. Wang, J. Wang, Theoretical study on the interactions between ionic liquid and solute molecules for typical separation problems, Chem. Phys. Lett. 199 (2014) 366–372, https://doi.org/10.1016/j.cplett.2014.03.008.

[34] R. Giernoth, Task-specific ionic liquids, Angew. Chem. Int. Ed. 49 (2010) 2834–2839, https://doi.org/10.1002/anie.200905981.

[35] A.D. Sawant, D.G. Raut, N.B. Darvatkar, M.M. Salunkhe, Recent developments of task-specific ionic liquids in organic synthesis, Green Chem. Lett. Rev. 4 (2011) 41–54, https://doi.org/10.1080/17518253.2010.500622.

[36] E.D. Bates, R.D. Mayton, I. Ntai, J.H. Davis, CO2 capture by a task-specific ionic liquid, J. Am. Chem. Soc. 124 (2002) 926–927, https://doi.org/10.1021/ja017593d.

[37] S. Tang, G.A. Baker, H. Zhao, Ether-and alcohol-functionalized task-specific ionic liquid: attractive properties and applications, Chem. Soc. Rev. 41 (2012) 4030–4066, https://doi.org/10.1039/C2CS15362A.

[38] M. Koel, Ionic liquids in chemical analysis, Crit. Rev. Anal. Chem. 35 (2005) 177–192, https://doi.org/10.1080/10408340500304016.

[39] S. Lee, Functionalized imidazolium salts for task-specific ionic liquid and their applications, Chem. Commun. 1 (2006) 1049–1063, https://doi.org/10.1039/B514140K.

[40] A.E. Visser, R.P. Swatloski, W.M. Reichert, R. Mayton, S. Sheff, A. Wierzbicki, J.H. Davis, R.D. Rogers, Task-specific ionic liquids incorporating novel cations for the coordination and extraction of Hg^{2+} and Cd^{2+}: synthesis, characterization, and extraction studies, Environ. Sci. Technol. 36 (2002) 2523–2529, https://doi.org/10.1021/es0158004.

[41] W.L. Da Silva, B.C. Leal, A.L. Ziulkoski, P.W.N.M. Van Leeuwen, J.H.Z. Dos Santos, H.S. Schrekker, Petrochemical residue-derived silica-supported titania-magnesium catalysts for the photocatalytic degradation of imidazolium ionic liquids in water, Separ. Purif. Technol. 218 (2019) 191–199, https://doi.org/10.1016/j.seppur.2019.01.066.

[42] A. Fernicola, B. Scrosati, H. Ohno, Potentialities of ionic liquids as new electrolyte media in advanced electrochemical devices, Ionics 12 (2006) 95–102, https://doi.org/10.1007/s11581-006-0023-5.

[43] M. Armand, F. Endres, D.R. Macfarlane, H. Ohno, B. Scrosati, Ionic-liquid materials for the electrochemical challenges of the future, Nat. Mater. 8 (2009) 621–629, https://doi.org/10.1038/nmat2448.

[44] H. Zheng, H. Zhang, Y. Fu, T. Abe, Z. Ogumi, Temperature effects on the electrochemical behavior of spinel $LiMn_2O_4$ in quaternary ammonium-based ionic liquid electrolyte, J. Phys. Chem. B 109 (2005) 13676–13684, https://doi.org/10.1021/jp051238i.

[45] S. Kim, J. Bang, J. Yoo, Y. Shin, J. Bae, J. Jeong, K. Kim, P. Dong, K. Kwon, A comprehensive review on the pre-treatment process in lithium-ion battery recycling, J. Clean. Prod. 294 (2021) 126329, https://doi.org/10.1016/j.jclepro.2021.126329.

[46] D.A. Bertuol, F.R. Amado, E.D. Cruz, E.H. Tanabe, Metal recovery using supercritical carbon dioxide, in: Green Sustainable Process for Chemical and Environmental Engineering and Science, Elsevier, 2020, pp. 85–103, https://doi.org/10.1016/B978-0-12-817388-6.00005-2.

[47] S.E. Sloop, R. Parker, System and Method for Processing an End-of-Life or Reduced Performance Energy Storage and/or Conversion Device Using a Supercritical Fluid, Patent No. US 8,6067,107 B2, 2011.

[48] M. Grützke, V. Kraft, W. Weber, C. Wendt, A. Friesen, S. Klamor, M. Winter, S. Nowak, Supercritical carbon dioxide extraction of lithium-ion battery electrolytes, J. Supercrit. Fluids 94 (2014) 216–222, https://doi.org/10.1016/j.supflu.2014.07.014.

[49] M. Grützke, X. Mönnighoff, F. Horsthemke, V. Kraft, M. Winter, S. Nowak, Extraction of lithium-ion battery electrolytes with liquid and supercritical carbon dioxide and additional solvents, RSC Adv. 5 (2015) 43209–43217, https://doi.org/10.1039/C5RA04451K.

[50] J. Liu, D. Mu, R. Zheng, C. Dai, Supercritical CO_2 extraction of organic carbonatebased electrolytes of lithium-ion batteries, RSC Adv. 4 (2014) 54525–54531, https://doi.org/10.1039/C4RA10530C.

[51] M. Liu, D. Mu, Y. Dai, Q. Ma, R. Zheng, C. Dai, Analysis on extraction behaviour of lithium-ion battery electrolyte solvents in supercritical CO_2 by gas chromatography, Int. J. Electrochem. Sci. 11 (2016) 7594–7604, https://doi.org/10.20964/2016.09.03.

[52] Y. Liu, D. Mu, R.H. Li, Q. Ma, R. Zheng, C. Dai, Purification and characterization of reclaimed electrolytes from spent lithium-ion batteries, J. Phys. Chem. C 121 (8) (2017) 4181–4187, https://doi.org/10.1021/acs.jpcc.6b12970.

[53] X. Monnighoff, A. Friesen, B. Konersmann, F. Horsthemke, M. Grützke, M. Winter, S. Nowak, Supercritical carbon dioxide extraction of electrolyte from spent lithium-ion batteries and its characterization by gas chromatography with chemical ionization, J. Power Sources 352 (2017) 56–63, https://doi.org/10.1016/j.jpowsour.2017.03.114.

[54] Y. Fu, J. Schuster, M. Petranikova, B. Ebin, Innovative recycling of organic binders from electric vehicle lithium-ion batteries by supercritical carbon dioxide extraction, Resour. Conserv. Recycl. 172 (2021) 105666, https://doi.org/10.1016/j.resconrec.2021.105666.

[55] D.A. Bertuol, C.M. Machado, M.L. Silva, C.O. Calgaro, G.L. Dotto, E.H. Tanabe, Recovery of cobalt from spent lithium-ion batteries using supercritical carbon dioxide extraction, Waste Manag. 51 (2016) 245–251, https://doi.org/10.1016/j.wasman.2016.03.009.

[56] S. Olapiriyakul, R.J. Caudill, A framework for risk management and End-of-Life (EOL) analysis for nanotechnology products: a case study in lithium-ion batteries, IEEE Int. Symp. Electron. Environ. (2008), https://doi.org/10.1109/ISEE.2008.4562877.

[57] W. Maatar, S. Boufi, Poly(methacylic acid-co-maleic acid) grafted nanofibrillated cellulose as a reusable novel heavy metal ions adsorbent, Carbohydr. Polym. 126 (2015) 199–207, https://doi.org/10.1016/j.carbpol.2015.03.015.

[58] A.P. Tom, Nanotechnology for sustainable water treatment—a review, Mater. Today Proc. (2021), https://doi.org/10.1016/j.matpr.2021.05.629.

[59] H.E. Toma, Magnetic nanohydrometallurgy: a nanotechnological approach to elemental sustainability, Green Chem. 17 (2015) 2027–2041, https://doi.org/10.1039/c5gc00066a.

[60] A.T. Le, S.Y. Pung, S. Sreekantan, A. Matsuda, D.P. Huynh, Mechanisms of removal of heavy metal ions by ZnO particles, Heliyon 5 (2019) e01440, https://doi.org/10.1016/j.heliyon.2019.e01440.

[61] V. Dhiman, N. Kondal, ZnO Nanoadsorbents: a potent material for removal of heavy metal ions from wastewater, Colloids Interface Sci. Commun. 41 (2021) 100380, https://doi.org/10.1016/j.colcom.2021.100380.

[62] A. Yildirim, M.F. Baran, H. Acay, Kinetic and isotherm investigation into the removal of heavy metals using a fungal-extract-based bio-nanosorbent, Environ. Technol. Innovat. 20 (2020) 101076, https://doi.org/10.1016/j.eti.2020.101076.

[63] U. Condomitti, A. Zuin, A.T. Silveira, K. Araki, H.E. Toma, Magnetic nanohydrometallurgy: a promising nanotechnological approach for metal production and recovery using functionalized superparamagnetic nanoparticles, Hydrometallurgy 125–126 (2012) 148–151, https://doi.org/10.1016/j.hydromet.2012.06.005.

[64] A.B. Botelho Junior, M.M. Jiménez Correa, D.C.R. Espinosa, J.A.S. Tenório, Study of the reduction process of iron in leachate from nickel mining waste, Braz. J. Chem. Eng. 35 (2018) 1241–1248, https://doi.org/10.1590/0104-6632.20180354s20170323.

[65] A.B. Botelho Junior, A. de A. Vicente, D.C.R. Espinosa, J.A.S. Tenório, Recovery of metals by ion exchange process using chelating resin and sodium dithionite, J. Mater. Res. Technol. 8 (2019) 4464–4469, https://doi.org/10.1016/j.jmrt.2019.07.059.

[66] A.B. Botelho Junior, A.D.A. Vicente, D.C.R. Espinosa, J.A.S. Tenório, Effect of iron oxidation state for copper recovery from nickel laterite leach solution using chelating resin, Separ. Sci. Technol. 55 (2020) 788–798, https://doi.org/10.1080/01496395.2019.1574828.

[67] I.D. Perez, A.B. Botelho Junior, P. Aliprandini, D.C.R. Espinosa, Copper recovery from nickel laterite with high-iron content: a continuous process from mining waste, Can. J. Chem. Eng. 98 (2020) 957–968, https://doi.org/10.1002/cjce.23667.

[68] F.M. Melo, A.T. Silveira, L.F. Quartarolli, F.F. Kaid, D.R. Cornejo, H.E. Toma, Magnetic behavior of superparamagnetic nanoparticles containing chelated transition metal ions, J. Magn. Mater. 487 (2019) 165324, https://doi.org/10.1016/j.jmmm.2019.165324.

[69] A.B. Botelho Junior, É.F. Pinheiro, D.C.R. Espinosa, J.A.S. Tenório, M. Baltazar, P.G. dos, Adsorption of lanthanum and cerium on chelating ion exchange resins: kinetic and thermodynamic studies, Separ. Sci. Technol. 00 (2021) 1–10, https://doi.org/10.1080/01496395.2021.1884720.

[70] H.E. Toma, Developing nanotechnological strategies for green industrial processes, Pure Appl. Chem. 85 (2013) 1655–1669, https://doi.org/10.1351/PAC-CON-12-12-02.

[71] M.F. El-Kady, Y. Shao, R.B. Kaner, Graphene for batteries, supercapacitors and beyond, Nat. Rev. Mater. 1 (2016) 1–14, https://doi.org/10.1038/natrevmats.2016.33.

[72] K. He, Z.Y. Zhang, F.S. Zhang, Synthesis of graphene and recovery of lithium from lithiated graphite of spent Li-ion battery, Waste Manag. 124 (2021) 283–292, https://doi.org/10.1016/j.wasman.2021.01.017.

[73] S.I. AL-Saedi, A.J. Haider, A.N. Naje, N. Bassil, Improvement of Li-ion batteries energy storage by graphene additive, Energy Rep. 6 (2019) 64–71, https://doi.org/10.1016/j.egyr.2019.10.019.

[74] P.M. Urias, L.H. dos Reis Menêzes, V.L. Cardoso, M.M. de Resende, J. de Souza Ferreira, Leaching with mixed organic acids and sulfuric acid to recover cobalt and lithium from lithium ion batteries, Environ. Technol. (2020) 1–11, https://doi.org/10.1080/09593330.2020.1772372.

[75] M. Esmaeili, S.O. Rastegar, R. Beigzadeh, T. Gu, Ultrasound-assisted leaching of spent lithium ion batteries by natural organic acids and H_2O_2, Chemosphere 254 (2020) 126670, https://doi.org/10.1016/j.chemosphere.2020.126670.

[76] B. Musariri, G. Akdogan, C. Dorfling, S. Bradshaw, Evaluating organic acids as alternative leaching reagents for metal recovery from lithium ion batteries, Miner. Eng. 137 (2019) 108–117, https://doi.org/10.1016/j.mineng.2019.03.027.

[77] H. Setiawan, H.T.B.M. Petrus, I. Perdana, Reaction kinetics modeling for lithium and cobalt recovery from spent lithium-ion batteries using acetic acid, Int. J. Miner. Metall. Mater. (2019), https://doi.org/10.1007/s12613-019-1713-0.

[78] C. Sun, L. Xu, X. Chen, T. Qiu, T. Zhou, Sustainable recovery of valuable metals from spent lithium-ion batteries using DL-malic acid: leaching and kinetics aspect, Waste Manag. Res. 36 (2018) 113–120, https://doi.org/10.1177/0734242X17744273.

[79] Y. Yang, X. Meng, H. Cao, X. Lin, C. Liu, Y. Sun, Y. Zhang, Z. Sun, Selective recovery of lithium from spent lithium iron phosphate batteries: a sustainable process, Green Chem. 20 (2018) 3121–3133, https://doi.org/10.1039/c7gc03376a.

[80] D. Pant, T. Dolker, Green and facile method for the recovery of spent Lithium Nickel Manganese Cobalt Oxide (NMC) based Lithium ion batteries, Waste Manag. 60 (2017) 689–695, https://doi.org/10.1016/j.wasman.2016.09.039.

[81] R. Golmohammadzadeh, F. Rashchi, E. Vahidi, Recovery of lithium and cobalt from spent lithium-ion batteries using organic acids: process optimization and kinetic aspects, Waste Manag. 64 (2017) 244–254, https://doi.org/10.1016/j.wasman.2017.03.037.

[82] G.P. Nayaka, K.V. Pai, G. Santhosh, J. Manjanna, Recovery of cobalt as cobalt oxalate from spent lithium ion batteries by using glycine as leaching agent, J. Environ. Chem. Eng. 4 (2016) 2378–2383, https://doi.org/10.1016/j.jece.2016.04.016.

[83] G.P. Nayaka, K.V. Pai, G. Santhosh, J. Manjanna, Dissolution of cathode active material of spent Li-ion batteries using tartaric acid and ascorbic acid mixture to recover Co, Hydrometallurgy 161 (2016) 54–57, https://doi.org/10.1016/j.hydromet.2016.01.026.

[84] F. Wu, J. Maier, Y. Yu, Guidelines and trends for next-generation rechargeable lithium and lithium-ion batteries, Chem. Soc. Rev. 49 (2020) 1569–1614, https://doi.org/10.1039/c7cs00863e.

[85] H. Pinegar, Y.R. Smith, Recycling of end-of-life lithium-ion batteries, part II: laboratory-scale research developments in mechanical, thermal, and leaching treatments, J. Sustain. Metall. 6 (2020) 142–160, https://doi.org/10.1007/s40831-020-00265-8.

[86] L. Zhuang, C. Sun, T. Zhou, H. Li, A. Dai, Recovery of valuable metals from LiNi 0.5 Co 0.2 Mn 0.3 O 2 cathode materials of spent Li-ion batteries using mild mixed acid as leachant, Waste Manag. 85 (2019) 175–185, https://doi.org/10.1016/j.wasman.2018.12.034.

[87] L. Yao, H. Yao, G. Xi, Y. Feng, Recycling and synthesis of LiNi1/3Co1/3Mn1/3O2 from waste lithium ion batteries using d,l-malic acid, RSC Adv. 6 (2016) 17947–17954, https://doi.org/10.1039/c5ra25079j.

[88] A.P. Abbott, G. Capper, D.L. Davies, K.J. McKenzie, S.U. Obi, Solubility of metal oxides in deep eutectic solvents based on choline chloride, J. Chem. Eng. Data 51 (4) (2006), https://doi.org/10.1021/je060038c.

[89] M.R.S. Foreman, Progress towards a process for the recycling of nickel metal hydride electric cells using a deep eutectic solvent, Cogent Chemistry 2 (1) (2016) 1139289, https://doi.org/10.1080/23312009.2016.1139289.

[90] Albler, K. Bica, M.R.S. Foreman, S. Holgersson, M.S. Tyumentsev, A comparison of two methods of recovering cobalt from a deep eutectic solvent: implications for battery recycling, J. Clean. Prod. 167 (2017) 806–814, https://doi.org/10.1016/j.jclepro.2017.08.135.

[91] J.T.M. Amphlett, M.D. Ogden, W. Yang, S. Choi, Probing Ni2+ and Co2+ speciation in carboxylic acid based deep eutectic solvents using UV/Vis and FT-IR spectroscopy, J. Mol. Liq. 318 (2020) 114217, https://doi.org/10.1016/j.molliq.2020.114217.

[92] G. Zante, A. Braun, A. Masmoudi, R. Barillon, D. Trébouet, M. Boltoeva, Solvent extraction fractionation of manganese, cobalt, nickel and lithium using ionic liquids and deep eutectic solvents, Miner. Eng. 156 (2020) 106512, https://doi.org/10.1016/j.mineng.2020.106512.

[93] P.G. Schiavi, P. Altimari, M. Branchi, R. Zanoni, G. Simonetti, M.A. Navarra, F. Pagnanelli, Selective recovery of cobalt from mixed lithium ion battery wastes using deep eutectic solvent, Chem. Eng. J. 417 (2021) 129249, https://doi.org/10.1016/j.cej.2021.129249.

[94] S. Wang, Z. Zhang, Z. Lu, Z. Xu, A novel method for screening deep eutectic solvent to recycle the cathode of Li-ion batteries, Green Chem. 22 (2020) 4473, https://doi.org/10.1039/D0GC00701C.

[95] M. Landa-Castro, J. Aldana-Gonzáles, M.G. Montes de Oca-Yemha, M. Romero-Romo, E.M. Arce-Estrada, M. Palomar-Pardavé, NieCo alloy electrodeposition from the cathode powder of Ni-MH spent batteries leached with a deep eutectic solvent (reline), J. Alloys Compd. 830 (2020) 154650, https://doi.org/10.1016/j.jallcom.2020.154650.

[96] M.K. Tran, M.T.F. Rodrigues, K. Kato, G. Babu, P.M. Ajayan, Deep eutectic solvents for cathode recycling of Li-ion batteries, Nat. Energy 4 (2019) 339–345, https://doi.org/10.1038/s41560-019-0368-4.

[97] M. Wang, Q. Tan, L. Liu, J. Li, A low-toxicity and high-efficiency deep eutectic solvent for the separation of aluminum foil and cathode materials from spent lithium-ion batteries, J. Hazard Mater. 380 (2019) 120846, https://doi.org/10.1016/j.jhazmat.2019.120846.

[98] J.J. Roy, B. Cao, S. Madhavi, A review on the recycling of spent lithium-ion batteries (LIBs) by the bioleaching approach, Chemosphere 282 (2021) 130944, https://doi.org/10.1016/j.chemosphere.2021.130944.

[99] A.H. Kaksonen, N.J. Boxall, Y. Gumulya, H.N. Khaleque, C. Morris, T. Bohu, K.Y. Cheng, K.M. Usher, A.M. Lakaniemi, Recent progress in biohydrometallurgy and microbial characterisation, Hydrometallurgy 180 (2018) 7–25, https://doi.org/10.1016/j.hydromet.2018.06.018.

[100] A.H. Kaksonen, X. Deng, T. Bohu, L. Zea, H.N. Khaleque, Y. Gumulya, N.J. Boxall, C. Morris, K.Y. Cheng, Prospective directions for biohydrometallurgy, Hydrometallurgy 195 (2020) 105376, https://doi.org/10.1016/j.hydromet.2020.105376.

[101] K.A. Komnitsas, Recent Advances in Hydro-and Biohydrometallurgy, Recent Advances in Hydro- and Biohydrometallurgy, 2019, https://doi.org/10.3390/books978-3-03921-300-9.

[102] J. Baars, T. Domenech, R. Bleischwitz, H.E. Melin, O. Heidrich, Circular economy strategies for electric vehicle batteries reduce reliance on raw materials, Nat. Sustain. 4 (2021) 71–79, https://doi.org/10.1038/s41893-020-00607-0.

[103] H.E. Melin, State-of-the-art in reuse and recycling of lithium-ion batteries—a research review, Circ. Energy Storage 1 (2020) 1–57.

[104] N. Bahaloo-Horeh, S.M. Mousavi, M. Baniasadi, Use of adapted metal tolerant Aspergillus Niger to enhance bioleaching efficiency of valuable metals from spent lithium-ion mobile phone batteries, J. Clean. Prod. 197 (2018) 1546–1557, https://doi.org/10.1016/j.jclepro.2018.06.299.

[105] M.I. Bajestani, S.M. Mousavi, S.A. Shojaosadati, Bioleaching of heavy metals from spent household batteries using Acidithiobacillus ferrooxidans: statistical evaluation and optimization, Separ. Purif. Technol. 132 (2014) 309–316, https://doi.org/10.1016/j.seppur.2014.05.023.

[106] Y. Xin, X. Guo, S. Chen, J. Wang, F. Wu, B. Xin, Bioleaching of valuable metals Li, Co, Ni and Mn from spent electric vehicle Li-ion batteries for the purpose of recovery, J. Clean. Prod. 116 (2016) 249–258, https://doi.org/10.1016/j.jclepro.2016.01.001.

[107] T. Dolker, D. Pant, Chemical-biological hybrid systems for the metal recovery from waste lithium ion battery, J. Environ. Manag. 248 (2019) 109270, https://doi.org/10.1016/j.jenvman.2019.109270.

[108] X. Chen, C. Luo, J. Zhang, J. Kong, T. Zhou, Sustainable recovery of metals from spent lithium-ion batteries: a green process, ACS Sustain. Chem. Eng. 3 (2015) 3104–3113, https://doi.org/10.1021/acssuschemeng.5b01000.

[109] I.S. Bădescu, D. Bulgariu, I. Ahmad, L. Bulgariu, Valorisation possibilities of exhausted biosorbents loaded with metal ions—a review, J. Environ. Manag. 224 (2018) 288—297, https://doi.org/10.1016/j.jenvman.2018.07.066.

[110] A.T. Ubando, A.D.M. Africa, M.C. Maniquiz-Redillas, A.B. Culaba, W.H. Chen, J.S. Chang, Microalgal biosorption of heavy metals: a comprehensive bibliometric review, J. Hazard Mater. 402 (2021) 123431, https://doi.org/10.1016/j.jhazmat.2020.123431.

CHAPTER 5

Innovative strategies for recycling used batteries for brighter future

Jonghyun Choi, Felipe M. de Souza and Ram K. Gupta
Department of Chemistry, Kansas Polymer Research Center, Pittsburg State University, Pittsburg, KS, United States

1. Introduction

Concern over electronic device performance is receiving great attention from the industry and scientific community. It has become necessary for society to function properly since most electronic devices are required for communication, information, connection, and data transfer. These electronic devices are mostly powered by batteries, a relatively fast technology that has received great attention due to its important role. Such technological advances yielded several types of batteries developed to improve durability, shelf-life, and efficiency, based on the principle of converting chemical energy into electricity. There is a considerable number of batteries available. Primary batteries are composed of carbon and zinc that provide relatively lower power and are not rechargeable. Yet, they are cheaper and used in domestic portable devices such as remote controllers. Secondary batteries are rechargeable and include lead-acid, nickel-based, lithium-based, sodium-based, aluminum-based, and metal-air batteries, among others.

Lead-acid batteries have electrodes made of Pb (negative electrode), active materials composed of lead oxides such as PbO_2 (positive electrode), and diluted H_2SO_4 as electrolyte. In addition, Ca be added to increase mechanical strength. Compared to more recent batteries, lead-acid batteries have a lower energy density; however, they have relatively higher power density as they can provide high currents over a short period. This property allows them to be used as starter motors for vehicles. Because of that, a massive amount of lead was extracted and used as starting material for the fabrication of car batteries. However, this high production rate quickly raised concerns due to the high toxicity of Pb as it is harmful to the environment as well as humans by causing severe kidney and brain damage. This situation pushed the development of batteries that contained other types of metals that could be used as an alternative for lead-acid batteries. Because of that, nickel-cadmium batteries were introduced a while after lead-acid batteries. This type of battery has the advantage of rechargeability and higher capacity. On the other hand, it has a higher internal resistance and a limited number of charging cycles. Yet, Ni-Cd batteries display satisfactory performance for portable devices and electronic equipment.

Taking a step further, nickel-metal hydride (Ni-MH) batteries have been developed that were initially unattractive because of their instability. However, with successful research into incorporating new metal alloys, Ni-MH has become suitable for batteries, even surpassing the specific capacitance of Ni-Cd. The high power density of this type of battery found application in hybrid-electric and electric vehicles [1]. This was an important achievement for the technology of batteries because, likewise, Pb, Cd is also a highly toxic metal. Hence, developing batteries that contain nontoxic metals such as Ni-MH was a valuable technological achievement by the fact that heavy metals, i.e., Pb, Cd, Hg, Cr, among others, can cause severe contamination into the environment as well as health issues since once a biological system absorbs them, they take a long time to be expelled, which leads them to continuously damage the organism. This process is named bioaccumulation, and one of the reasons for that is its high density, which makes it difficult for it to be excreted through a natural process. However, Ni-Cd and Ni-MH suffer from memory effects that cause them to decrease their capacitance if they are recharged several times when the battery is partially discharged [2].

Further research led to commercial lithium-ion batteries (LIBs) around the 1990s. The attractiveness of LIBs was due to their high energy density, low self-discharge, and lack of memory effect. A more recent design of LIBs consists of a two-electrode system separated by an electrolyte. Generally, lithium cobalt oxide ($LiCoO_2$) is used as the cathode, and carbon-based materials (i.e., graphite) are used as the anode. The electrolytes currently used are a mixture of organic liquid solutions of ethylene, dimethyl, diethyl, or ethyl methyl carbonates, as well as $LiPF_6$. The charging process can occur through a conversion or chemical transformation process. For example, in the case of $LiCoO_2$, Li^+ moves from its lattice structure in the direction of the carbon anode to perform the lithiation process, yielding lithiated graphite (LiC_6). The discharging is the opposite process, at which Li^+ moves back to CoO_2, reincorporating into its framework and releasing electrons in the process. This phenomenon is the core of the charging-discharging process for LIBs [3]. Yet, despite their efficiency and applicability, conventional LIBs carry safety issues due to the flammability of the electrolyte if it enters in contact with oxygen or incorrect charging. This situation led to the buildup of solid-state LIBs, which consists of a solid electrolyte that performs ionic conduction of Li^+ through its framework, whereas it does not conduct electrons. Examples of solid electrolytes are polymers, metal oxides, and metal sulfides. The repercussion of LIBs provided a positive insight into the overall efficiency of energy storage applications, and it shows great potential to become one of the next energy grids.

2. Need for recycling batteries

The many advantages of LIBs have made them useful for various applications, and as a result, the demand for LIBs is increasing annually. However, this also means that LIBs will generate a lot of used batteries when batteries' life spans come to an end. Considering the current use and demand of LIBs, it is estimated that over 11 million tons of batteries will be abandoned by 2030 [4]. A big problem arises from this huge waste: environmental pollution. Some components of the battery, such as organic electrolytes and heavy metals (Ni, Co, Mn), can result in a negative impact on the environment and human health if improperly treated [5]. Therefore, properly recycling wasted batteries has become important to protect our Earth. In addition to protecting the environment, recycling LIBs can reap great economic benefits. For example, Li and Co, which are the representative cathode materials of LIBs, are in great demand worldwide but relatively less abundant than other metals. With the increase in the consumption of Li and Co globally, it is estimated that there will be a demand for 423 kilotons of Li in 2025 and a demand for 235—430 kilotons of Co in 2030, respectively [6,7]. The demand for these compounds has continually increased; however, the natural resources are limited. For this reason, recycling spent batteries can incur positive economic benefits since some of the metals and essential compounds for battery components can be reused and recovered.

Waste battery recycling methods can be broadly divided into reuse and material recovery. Reusing these materials involves utilizing useful compounds, such as graphite and metal ions from the waste batteries, as components of other energy devices such as supercapacitors, LIBs, Zn-air batteries, and catalysts, etc. Fig. 5.1 shows the process of making MnO_2/graphene nanocomposites using spent $Zn-MnO_2$ acidic, dry batteries containing graphite and MnO_2 nanoparticles [8]. The Deng group obtained black powder from spent batteries via the facile recovery process. The spent battery powder was mixed with graphene oxide (GO) or reduced graphene oxide with different weight ratios. The nanocomposite materials were used as electrodes for supercapacitors, and the black powder/GO (5:1) nanocomposite electrode displayed the specific capacitance of 150 F/g at the current density of 1 A/g, along with high cycling stability with only 6.3% loss of initial specific capacitance after 1000 cycles. The material recovery process involves leaching valuable constituents such as metals or compounds from the spent battery. The main focus of the recovery strategy is to make products with high purity via physical or chemical methods. For example, the Ma group recovered Pb with a high purity of 99.77% from lead storage batteries using hydrometallurgical desulfurization and vacuum thermal reduction methods [9]. Similarly, the Huang group recycled the Cd from Ni-Cd batteries using vacuum metallurgy separation methods. The metallic Cd had 99.99% purity [10].

While the technology to recycle certain batteries, such as lead-acid batteries, is well-developed, the technology to recycle LIBs is still insufficient and immature. For example, lead-acid batteries contain lead which is a toxic heavy metal. It can lead to severe environmental pollution if the material is not treated well. However, the recycling technology for this battery is well developed and currently has a recycling rate of 99% [11]. So there is not much concern about lead-acid batteries or the subsequent treatment processes. The successful recycling of lead-acid batteries can serve as a benchmark for other battery recycling methods. A closer look at the lead-acid battery recycling process reveals

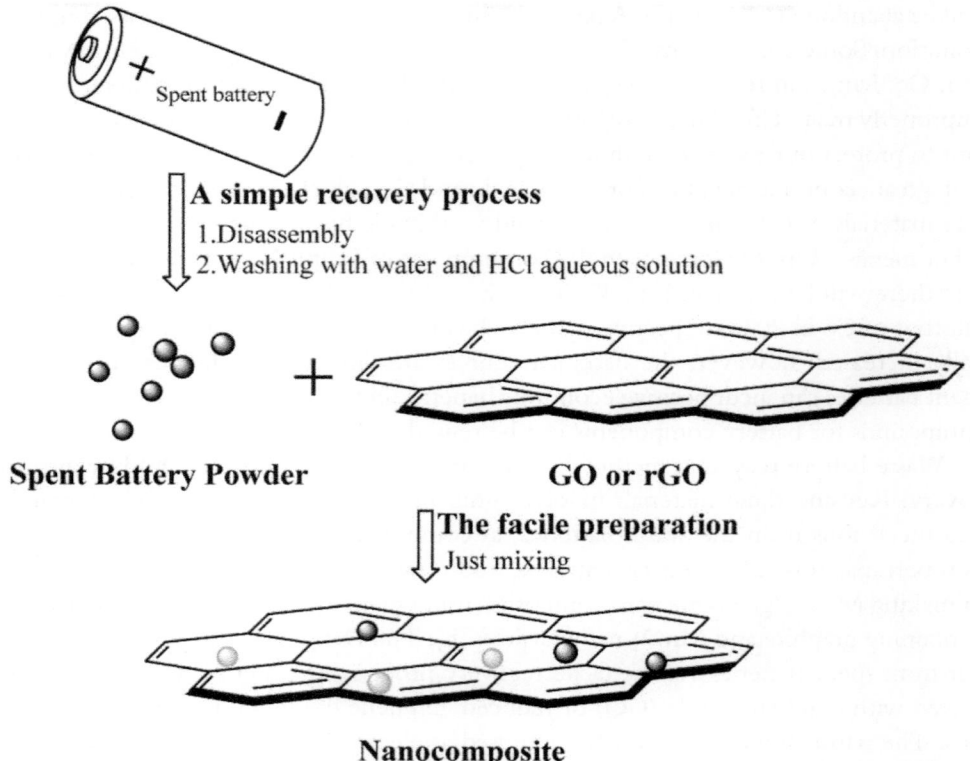

Figure 5.1 A spent battery powder containing graphite and MnO_2 nanoparticles was obtained from a Zn–MnO_2 spent battery via a facile recovery process. The powder was physically mixed with graphene oxide or sreduced graphene oxide, resulting in the synthesis of MnO_2/graphene nanocomposites. The material was used as an active material of a working electrode for a supercapacitor. *(Adapted with permission from J. Deng, X. Wang, X. Duan, P. Liu, Facile preparation of MnO_2/graphene nanocomposites with spent battery powder for electrochemical energy storage, ACS Sustain. Chem. Eng. 3 (2015), 1330–1338, Copyright (2015), American Chemical Society.)*

the use of effective techniques, such as pyrometallurgy and hydrometallurgical technologies [12]. Since these technologies have succeeded in almost completely recycling lead-acid batteries, these technologies are also being applied to spent LIBs. The following section will explain the innovative physical and chemical approaches for used battery recycling.

3. Strategies for recycling battery components

The increasing demand for electronic devices in nearly every area imposes the need to harvest the starting materials to manufacture electronic products. On top of that, the amount of raw materials required to sustain the demand tends to increase further due to the current investment in electric vehicles as an alternative to diminishing the dependence on nonrenewable sources. Yet, the materials used to produce batteries in cellphones, computers, smartwatches, electric engines, and other electronic devices are not renewable either. Because of that, it becomes necessary to develop technologies that allow the recovery of these starting materials. However, this process must be cost-effective to ensure a virtuous processing and recycling system. To address that, several strategies to recycle batteries are being proposed. The first step consists of separating the main components of an LIB, which, in simple terms, is composed of an anode, cathode, separator, electrolyte, and external shell. The anode comprises carbon-based materials, polyvinylidene fluoride (PVDF), a polymeric binder, a conductor material, and other additives coated over the Cu foil. The cathode is also composed of a carbon-based material and polymeric binder along with lithium transition metal oxides that include $LiCoO_2$, lithium nickel oxide ($LiNiO_2$), lithium manganese oxide ($LiMn_2O_4$), lithium iron phosphate ($LiFePO_4$), NCM ($LiNi_xCo_yMn_zO_2$, where $x + y + z = 1$) which are coated over an Al foil. The separator is a polymeric insulator material placed between the cathode and the anode [13]. Finally, the external shell can be either stainless steel or plastic. Fig. 5.2 shows the detailed composition of an LIB in a pie chart and the schematics for a general LIB. The recovery of raw materials from batteries is an inherently challenging process as it requires multiple steps to be a feasible and cost-effective approach. For that, a junction of both physical and chemical approaches is required.

3.1 Physical approach

The physical approach can be subdivided into two types which are manual and mechanical pretreatment. The physical separation takes place first by the following process of discharging, dismantling, and component separation, at which the most relevant part is to

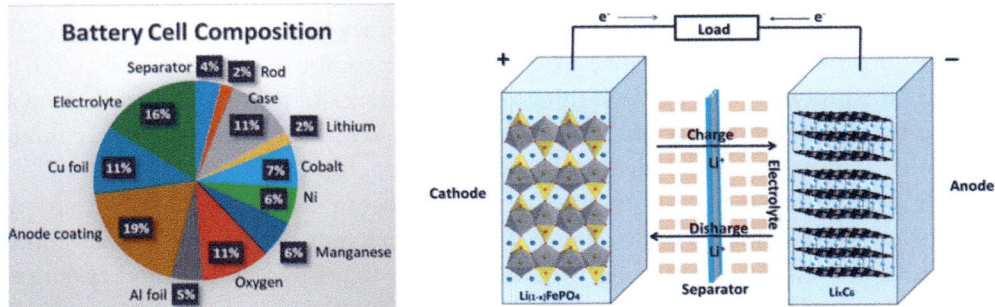

Figure 5.2 Detailed composition of a battery (left) and schematics for a general rechargeable lithium-ion battery (right). *(Adapted with permission from F. Arshad, L. Li, K. Amin, E. Fan, N. Manurkar, A. Ahmad, J. Yang, F. Wu, R. Chen, A comprehensive review of the advancement in recycling the anode and electrolyte from spent lithium ion batteries, ACS Sustain. Chem. Eng. 8 (2020), 13527–13554, Copyright (2020), American Chemical Society.)*

obtain the active materials from the cathode. Discharging is an important safety procedure to prevent a short circuit or an ignition of the battery. It can be performed by connecting the battery to a discharger or submerging the battery into a 5 to 10 wt.% solution of NaCl for complete discharge [14]. Optimal conditions for this process have been studied by Li et al. [14] that obtained the highest discharge efficiency of almost 72% in 358 min by using a 10 wt.% solution of NaCl. After that, the manual separation process can obtain the battery components: plastic or steel shells, organic separator, anode, and cathode. However, doing this process by using common tools is not feasible since it is a tedious and time-consuming task. Hence, it is more convenient to crush the batteries, followed by sieving and using a magnet to separate the components leading to particles with higher content of active materials from the cathode. The crushing process tends to obtain larger particle sizes for the plastic parts, Al, and Cu foils, whereas the active materials such as $LiCoO_2$ or carbon-based tend to become smaller. However, despite the robusticity of this method, it lacks selectivity as the inorganic and organic parts can blend, making it difficult to separate them [15]. Hence, proper control of parameters such as applied force, time, and sieving must be analyzed for further optimization.

The active materials that compose the electrode adhere to the current collectors through a binding polymer. The active materials that compose the cathode can be obtained by dissolving them from the Al foil's surface. Since PVDF is commonly used as the binder, the electrode can be submerged into an N-methyl pyrrolidone solution to separate the cathode from the Al due to their high polarity [16,17]. This step can be performed through a facile method of either submerging the electrode at 100 C for 1 hour or ultrasonication process at 80 C for 1 hour followed by an ultrasonic bath for around 20 min [18]. This is a convenient process, as the Al foil can be recovered, and the solvent can be reutilized. On top of that, other solvents such as dimethyl sulfoxide, N,N-dimethylacetamide, N,N-dimethylformamide, and ionic liquids can also be employed. However, like the crushing step, the dissolution of active material has drawbacks as some impurities are dragged with the active materials. Because of that, a calcining process may be required to remove PVDF. Furthermore, the solvents are relatively expensive, and a combination might be required to optimize the process. An example of this case is the presence of polytetrafluoroethylene in the battery, which is nonpolar and therefore not soluble in N-methylpyrrolidone; therefore, other solvents must be added to dissolve it [19].

A feasible option consists of thermal treatment to combust any impurity of carbon-based material or binder. Yet, it requires a filter to prevent the release of toxic fumes from the combustion process. Following that line, vacuum pyrolysis is another convenient approach. The vacuum and heat can weaken the adhesion of the active materials with the Al foil by decomposing them into smaller molecules. Yao and his team heated small fragments of a cathode and placed them into a furnace at 600 °C under a vacuum to remove the volatile components [18]. After that, the $LiCoO_2$ could be easily peeled from the Al foil. A summary of these pretreatment technologies is illustrated in Fig. 5.3.

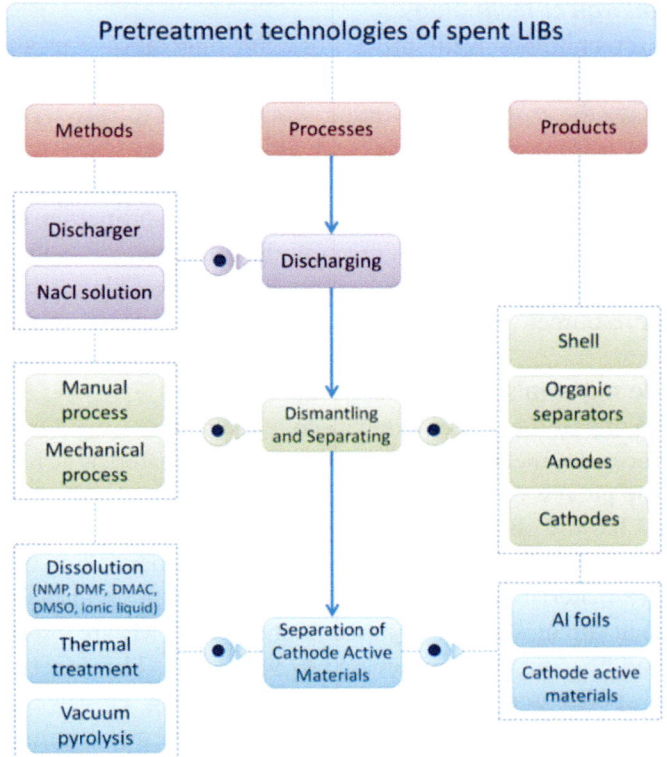

Figure 5.3 Pretreatment separation techniques for recycling batteries. *(Adapted with permission from Y. Yao, M. Zhu, Z. Zhao, B. Tong, Y. Fan, Z. Hua, Hydrometallurgical processes for recycling spent lithium-ion batteries: a critical review, ACS Sustain. Chem. Eng. 6 (2018), 13611−13627, Copyright (2018), American Chemical Society.)*

3.2 Chemical approach

Chemical approaches are effective methods for recycling batteries. Pyrometallurgical, hydrometallurgical, and biometallurgical technologies are mostly used to recover precious products from spent batteries. Most of the current research focuses on recovering cathode materials due to the substances being more valuable and high in demand. Thus, this section mainly explains the recycling of cathode materials using chemical approaches. Additionally, main anode materials (graphite) are crucial since they account for 10%−15% of the total battery price. Also, the materials constituting the anode of the spent LIBs include graphite, a current collector of Cu, and a fairly large amount of Li accumulated in the solid electrolyte interphase during the reversible reaction process. This fact promoted research on recycling the anode material of spent LIBs using hydrometallurgical technology. Thus, the recycling of anode materials will also be covered briefly.

3.2.1 Pyrometallurgical technology

Pyrometallurgical technology is the process of extracting valuable products by applying high temperatures to metal or compounds. Initially, this technology was used for mineral extraction, but later the scope of application was extended to recover materials, such as Zn, Ni, Cd, and Cu from spent batteries such as Zn—Mn dry, Ni—Cd, and Ni—H batteries. This technology has some advantages that can be applied to various raw materials, and waste liquid is not released during processing. These properties enable large-scale production beyond the lab [20]. For this reason, this technology is being applied to spent LIBs, which are increasingly being produced, making it possible to convert its components into useful resources.

3.2.1.1 Cathode materials

A typical method of pyrometallurgical technology is direct roasting. Direct roasting is the process that reduces the metal ions from active cathode materials at high temperatures. Fig. 5.4 shows the process of obtaining a metallic alloy using the direct roasting method [20]. First, spent LIBs and charcoal, a reductive agent, are put into a smelter and heated at a high temperature (over 1400 C). Then, slag-forming agents such as CaO and SiO_2 are added to separate materials with strong reducibilities, such as Li and Mg, and alloys composed of cobalt and nickel [19]. Materials such as Li and Mg are lost in slag as oxidized forms, and Co and Ni-containing alloys are recovered [21]. The direct roasting technology is simple to operate and can achieve high productivity. However, this technology requires high energy consumption as it necessitates high temperatures of over 1400 C. Also, during the roasting process, some components of the spent battery, such as plastic and organic electrolytes, burn off, producing toxic gases. In addition, Li-ion recovery through this technique is not well performed [5]. Although other types of pyrometallurgical technologies such as carbothermic reduction and thermite reduction have been used to increase the rate of Li-ion recovery, there are many necessary factors to use these technologies, which limit their mass production [22,23].

3.2.1.2 Anode materials

Since this process occurs at excessively high temperatures, recovering graphite on an industrial scale is nearly impossible. Electrolyte, steel, and aluminum, including the anode material, are all burned or go into slag. However, in laboratory settings, some methods, such as acid-treatments and carbothermal reduction reactions, are used to repair graphite [24]. For example, the Liu group conducted a study on graphite recycling from spent LIBs [25]. In the beginning, the graphite obtained from the spent battery had structural defects and many impurities. Thus, a chemical treatment was utilized sequentially with an acid-treatment and carbothermal reduction reaction process, which resulted in the successful reconstruction of the graphite. The reconstructed graphite was reused as an anode electrode in Li-ion batteries and showed high capacity and excellent rate capability.

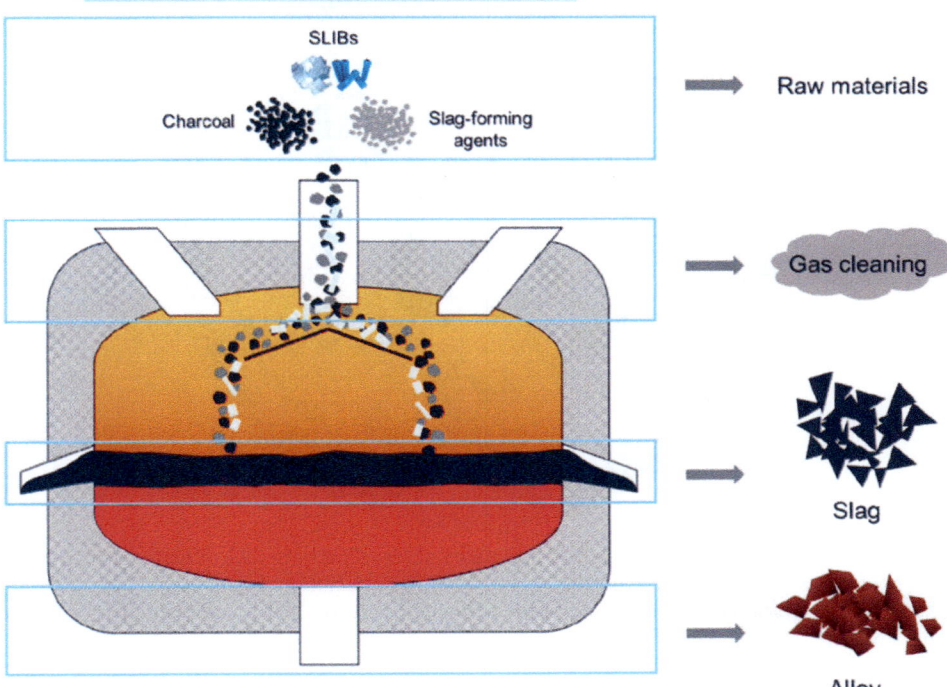

Figure 5.4 Schematic of pyrometallurgical technology: direct roasting process. Spent LIBs and charcoal are mixed into a smelter at over 1400 °C. Then slag-forming agents are added to separate cobalt or nickel-containing alloy compounds and irreducible materials (Li, Mg, etc.). Materials such as Li, Mg are lost in slag, and Co, Ni are recovered as alloy forms. *(Adapted with permission from Ref. M. Zhou, B. Li, J. Li, Z. Xu, Pyrometallurgical Technology in the Recycling of a Spent Lithium Ion Battery: Evolution and the Challenge, ACS ES&T Eng. (2021), Copyright (2021), American Chemical Society.)*

3.2.2 Hydrometallurgical technology

Hydrometallurgical technology is one of the most important recycling technologies to extract valuable materials from spent batteries. This technique includes pretreatment, leaching, separation, and recovery steps (Fig. 5.5). In the pretreatment process, each component is separated from used LIBs by sequentially going through a discharging, dismantling, and separation process. Whereas pyrometallurgical technology does not require pretreatment, the pretreatment for hydrometallurgical technology is crucial to improving valuable materials' recovery [26]. Then, the leaching process is used with a leaching agent to convert the components obtained from the pretreatment process into ions in a solution. When valuable materials are produced in solution form, the recovery of valuable compounds is completed through the separation processes, including techniques such as solvent extraction, chemical precipitation, and electrodeposition

Figure 5.5 Various recycling approaches and benefits of recycling batteries. *(Adapted with permission from A. Beaudet, F. Larouche, K. Amouzegar, P. Bouchard, K. Zaghib, Key challenges and opportunities for recycling electric vehicle battery materials, Sustain. Times. 12 (2020), 1−12, Copyright (2021), American Chemical Society.)*

[27,28]. This technology is effective for large-scale production owing to its many advantages of high recovery efficiency, high product purity, low energy consumption, and low hazardous gas emissions [29].

3.2.2.1 Cathode materials

After the pretreatment of the cathode materials, the leaching process is carried out to dissolve solid cathode materials into a solution using several leaching solutions such as inorganic acids, organic acids, and alkaline solutions. There are three main methods for leaching cathode materials: inorganic, organic, and alkaline leaching. It is worth noting that the leaching process is mainly used with reducing agents to increase leaching efficiency. Representative reducing agents include hydrogen peroxide (H_2O_2), sodium hydrogen sulfite ($NaHSO_3$), and glucose [30]. They play a role in reducing transition materials such as Co and Mn by participating in the redox reaction during the leaching process. For example, as shown in the Eh-pH diagrams in Fig. 5.6, Co^{2+} has a higher solubility than Co^{3+} at low temperatures [30]. During the leaching process, the presence of a reductant greatly influences the process by changing the chemical equilibrium. The leaching process was conducted using the glucose, reducing agent, in the phosphoric acid to dissolve Co and Li [31]. Glucose was oxidized to monocarboxylic acid during the leaching process, while it was reduced from Co(III) to Co(II). The existence of the reducing agent helped achieve a high recovery rate of 98% and near 100% for Co and Li, respectively.

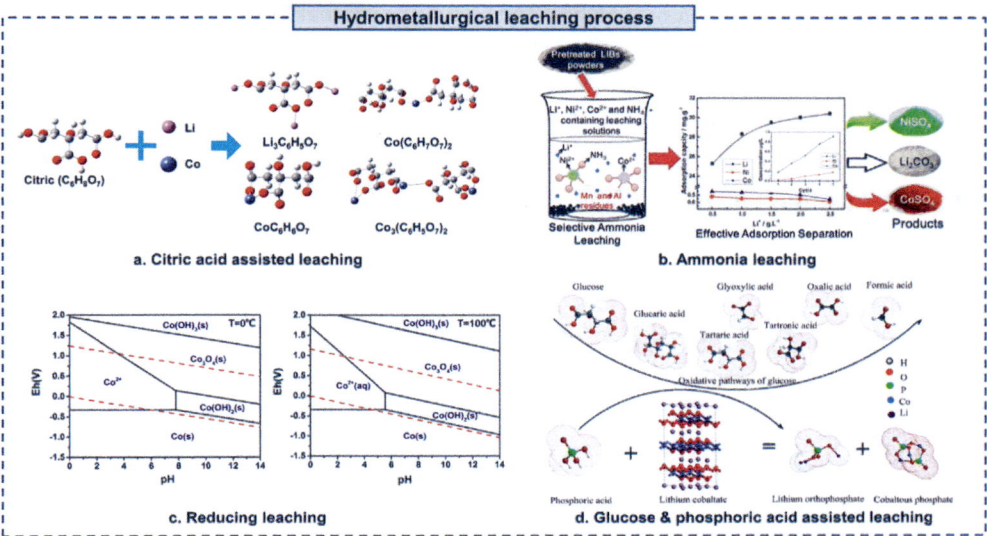

Figure 5.6 (A) Hydrometallurgical technology based on organic leaching is utilized to recover Co and Li from used LIBs in an acidic solution. Citric acid was used as an organic leaching agent, and lithium citrate and cobalt citrate are produced. (B) A selective ammonia leaching procedure is employed using an ammonia-based solution of the $NH_3 \cdot H_2O-NH_4HCO_3$. Highly efficient Li, Co, and Ni recovery is achieved from Li_2CO_3, $NiSO_4$, and $CoSO_4$. (C) The introduction of reductants during leaching shifts the chemical equilibrium according to the Eh-pH diagrams. (D) The reductant of glucose improves the efficiency of precious metal recovery from waste LIBs in phosphoric acid. While it oxidized to monocarboxylic acids, such as tartaric acid and formic acid, it reduced the metallic materials. *(Adapted with permission from A. Beaudet, F. Larouche, K. Amouzegar, P. Bouchard, K. Zaghib, Key challenges and opportunities for recycling electric vehicle battery materials, Sustain. Times. 12 (2020), 1–12, Copyright (2021), American Chemical Society.)*

3.2.2.1.1 Inorganic leaching The mechanism of inorganic leaching arises from the coordination of hydrogen ions and metal ions in acid media [19]. Representative leaching agents used for inorganic leaching include H_2SO_4, HCl, and HNO_3 [32]. These materials have very low pH values, which effectively leach metallic metals. However, it can lead to equipment corrosion during leaching since these materials are strong acids. Also, they can generate toxic substances such as Cl_2, SO_3, or NO_x by reacting with the cathode's active materials. Recently, phosphoric acid has been getting great attention because it is not only effective for the leaching of valuable metals from spent LIBs, but it also does not emit toxic gases [33].

3.2.2.1.2 Organic leaching Like inorganic leaching, the organic leaching process occurs during the interaction of hydrogen ions and metal ions in acid media. Organic leaching was designed to reduce toxic pollutants caused by inorganic leaching while upholding a high recovery efficiency. Citric acid, oxalic acid, malic acid, and acetic acid are mainly used as organic leaching agents [34,35]. Fig. 5.6A shows that citric acid and H_2O_2 were used to leach Co and Li from spent LIBs as leaching and reducing agents, respectively [36]. Citric acid is a strong acid that reacts with active cathode materials in acid media (which can be successfully leached from spent LIBs) and results in the high recovery of more than 90% of Co and almost 100% of Li.

3.2.2.1.3 Alkaline leaching Unlike inorganic and organic leaching, alkaline leaching occurs through the interaction between hydroxide ions and metal ions in alkaline solutions. Ammonia-based compounds such as $NH_3 \cdot H_2O$ and $(NH_4)_2CO_3$ are mostly used as alkaline leaching agents [30]. This leaching technique is attractive due to its efficient selective leaching of Li, Ni, and Co materials. Fig. 5.6B displays the selective leaching of valuable metals, such as Li, Co, and Ni from the used battery by using the $NH_3 \cdot H_2O-NH_4HCO_3$ solution and the reductant of H_2O_2 together [37]. Li_2CO_3, $NiSO_4$, and $CoSO_4$ were produced as products, and the recovery efficiencies of Li, Ni, and Co were 76.19%, 96.23%, and 94.57%, respectively.

The leachate obtained from the leaching process can be converted into valuable metals after the separation process. The separation process has been carried out using various strategies such as solvent extraction, chemical precipitation, and electrodeposition based on the difference in solubility and electrode potential [5]. These methods are efficient methods for separating metals from the leaching solution, but most solutions are too complex, so two or more separation processes are used together.

3.2.2.1.4 Solvent extraction Solvent extraction separates valuable metals from metal-containing solutions using extractants under optimum extraction conditions. The equation for metal extraction can be defined as below [38]:

$$M^2_{aq} + A^-_{org} + 2(HA)_{2,org} \rightarrow MA_2 \cdot 3HA_{org} + H^+_{aq} \ldots \quad (5.1)$$

Where $A^-_{org} + 2(HA)_{2,org}$ represents the solvent saponified by the reaction

$$Na^+_{aq} + (HA)_{2,org} \rightarrow NaA_{org} + H^+_{aq} \ldots \quad (5.2)$$

Representative materials of extractants include di(2-ethylhexyl) phosphoric acid (D2EHPA), diethylhexyl phosphoric acid, and 2-ethylhexylphosphonic acid mono-2-ethylhexyl ester [13]. The Chen group recycled spent LIBs using hydrometallurgical technology [39]. After leaching, they recovered more than 97% of Mn using cobalt-loaded D2EHPA as an extractant.

3.2.2.1.5 Chemical precipitation An insoluble precipitate is formed when metals in leachate are combined with anions such as OH^- and CO_3^{2-}. For example, the Shuguang group performed chemical precipitation using Na_2CO_3 [40]. The Na_2CO_3 compound was added to the Li ion-containing leaching solution, so the Li_2CO_3 precipitates were increased. The reaction can be described according to the equation:

$2Li + Na_2CO_3 \rightarrow Li_2CO_3 \downarrow + 2Na^+$. As a result, a high Li amount of 71% was achieved. This process is easy for separating simple metals, but there is a limit for compounds with complex structures, such as used LIBs. Therefore, this technique is mainly used with other separation techniques [19].

3.2.2.1.6 Electrodeposition The electrodeposition method can separate some metals by considering each material's voltage in the leaching solution. For example, the Barbieri group reused spent LIBs to make $Co(OH)_2$ and Co_3O_4 films @ ITO suitable as pseudocapacitor electrodes [41]. In detail, 0.85 V versus Ag/AgCl voltage was applied to the leaching solution. The $Co(OH)_2$ materials were electrodeposited on the ITO substrate, and a Co_3O_4 film was obtained by heat treatment of the $Co(OH)_2$ film at 450 C for 3 h.

3.2.2.2 Anode

It is difficult to reuse the material on an industrial scale using the hydrometallurgical method for recycling because the graphite's structure and chemical foam are not well retained during this process [13,29]. Therefore, this technique has difficulty in anode recovery. However, a few reports have been made on the recovery of anode materials using the water leaching process [42,43]. The binder mainly used for the anode is a styrene-butadiene copolymer and carboxyl methylcellulose. Since they are water-soluble, the water leaching process is often used to remove the binder [44]. For example, the Li group conducted a water leaching process using a DI water bath with a rotator, which exfoliated graphite from copper [43]. In addition, Li in the used graphite was water leached and precipitated using Na_2CO_3 compound.

3.2.3 Biometallurgical technology

Biometallurgical technology can extract valuable metals from batteries using microorganisms such as bacterial and fungi. As a type of bacteria, there are chemolithotrophic and acidophilic bacteria. The bacteria use $FeSO_4$ or elemental sulfur as an energy source to produce H_2SO_4 and Fe^{3+}, which can dissolve metals such as Li, Co, Ni, and Mn from the spent LIBs [45]. For example, the Xin group performed the bioleaching process using a mixed culture of acidophilic sulfur-oxidizing and iron-oxidizing bacteria [46]. The bioleaching mechanism of Li was most effective in the existence of the S in acidic media. However, the lowest Coleaching efficiency was observed when only elemental S was in the media. To simultaneously increase the recovery efficiency of Li and Co, elemental S and pyrite (FeS_2) were used together. Fe^{2+} served as a catalyze to help the leaching process of Li-ion and helped convert insoluble Co^{3+} into soluble Co^{2+}. Thus, the high leaching efficiency of Li and Co was achieved. In addition to bacteria, fungi are also considered effective microorganisms that can efficiently perform bioleaching because fungi can tolerate toxic substances well, have a rapid bioleaching rate, and grow well in alkaline and acid media [47]. The bioleaching process using *Aspergillus niger*, a type of fungus, to recover valuable metals from spent lithium-ion mobile phone batteries has been used [48]. Before bioleaching, the fungus was added into the sucrose medium containing spent LIBs to increase resistance to the toxicity of heavy metals. These adapted fungi were compared with unadapted fungi about the efficiency of bioleaching. The adapted fungi increased the rate of organic acid production and increased the leaching recovery of 100% Li and 94% Cu.

After the bioleaching process, a separation process must be performed using several methods such as solvent extraction, chemical precipitation, and electrodeposition, similar to hydrometallurgical methods. Compared with other recycling methods such as pyrometallurgical and hydrometallurgical methods, this process is more environmentally friendly, consumes less energy, and is an effective method with high recycling efficiency [49]. However, in the bioleaching process, microorganism cultivation takes a long time, and pulp density is low due to several reasons such as metal toxicity and reduced dissolved oxygen, so it is a challenge to be adopted for the recycling strategy in the industry [50]. The biometallurgical technology method is mainly used for the recovery process of the cathode material due to the difficulty in retaining the structure of the anode materials during the recycling process [13,29].

4. Conclusion

There are two types of methods used to recycle LIBs: physical and chemical approaches. Physical methods include discharging, dismantling, and component separation, etc. The physical method recovers the components by separating each component. However, since LIBs have a complex internal structure, there are limitations for precisely recycling valuable materials using only the physical method. The chemical approach can be used to overcome this problem, or it can be used in tandem with the physical method. Chemical approaches include strategies such as pyrometallurgical, hydrometallurgical, and biometallurgical technologies. Pyrometallurgical methods recycle some battery components by applying high temperatures to the metal that makes up the battery. Target materials, such as Co and Ni, are obtained as alloys, and the other materials are mostly accumulated in the slag. This technique can be easily used for large-scale production purposes, but it requires a lot of energy consumption and can lead to the emission of toxic substances. Hydrometallurgical techniques, combined with physical methods, include pretreatment, the leaching process, separation, and recovery steps to recover valuable materials. The leaching process includes inorganic, organic, and alkaline leaching, which dissolves the necessary substances into solution form. Most of this process is used in parallel with reducing agents to increase the efficiency of the leaching process. When valuable materials are leached, they are recovered through a separation process (like solvent extraction, chemical precipitation, and electrodeposition) based on each material's different solubility and electrode potentials. Biometallurgical technology is an efficient method of recycling.

The process includes reacting microorganisms, such as bacteria and fungi, with materials from spent LIBs. Although this technique has attracted a lot of attention due to its eco-friendly process and high efficiency of recovery, it has a low pulp density and takes a long time for microorganism cultivation, which is a big issue for mass production. It is also important that anode materials are recovered, but the research in this area is currently focused on the recycling of relatively more valuable cathode materials. Graphite was recycled using methods that included acid-treatment and carbon reduction reaction or water leaching, but these processes still pose many limitations for industrial production purposes.

References

[1] P. Gifford, J. Adams, D. Corrigan, S. Venkatesan, Development of advanced nickel/metal hydride batteries for electric and hybrid vehicles, J. Power Sources 80 (1999) 157–163.

[2] S.M.A.S. Bukhari, J. Maqsood, M.Q. Baig, S. Ashraf, T.A. Khan, Comparison of characteristics-lead acid, nickel based, lead crystal and lithium based batteries, in: 2015 17th UKSIM-AMSS International Conference on Modelling and Simulation, 2015, pp. 444–450.

[3] J. Xie, Y.-C. Lu, A retrospective on lithium-ion batteries, Nat. Commun. 11 (2020) 2499.

[4] D.J. Garole, R. Hossain, V.J. Garole, V. Sahajwalla, J. Nerkar, D.P. Dubal, Recycle, recover and repurpose strategy of spent Li-ion batteries and catalysts: current status and future opportunities, ChemSusChem. 13 (2020) 3079–3100.

[5] E. Fan, L. Li, Z. Wang, J. Lin, Y. Huang, Y. Yao, R. Chen, F. Wu, Sustainable recycling technology for Li-ion batteries and beyond: challenges and future prospects, Chem. Rev. 120 (2020) 7020–7063.

[6] F. Meng, J. McNeice, S.S. Zadeh, A. Ghahreman, Review of lithium production and recovery from minerals, brines, and lithium-ion batteries, Miner. Process. Extr. Metall. Rev. 42 (2021) 123–141.

[7] X. Fu, D.N. Beatty, G.G. Gaustad, G. Ceder, R. Roth, R.E. Kirchain, M. Bustamante, C. Babbitt, E.A. Olivetti, Perspectives on cobalt supply through 2030 in the face of changing demand, Environ. Sci. Technol. 54 (2020) 2985–2993.

[8] J. Deng, X. Wang, X. Duan, P. Liu, Facile preparation of MnO2/graphene nanocomposites with spent battery powder for electrochemical energy storage, ACS Sustain. Chem. Eng. 3 (2015) 1330–1338.

[9] Y. Ma, K. Qiu, Recovery of lead from lead paste in spent lead acid battery by hydrometallurgical desulfurization and vacuum thermal reduction, Waste Manag. 40 (2015) 151–156.

[10] K. Huang, J. Li, Z. Xu, Characterization and recycling of cadmium from waste nickel-cadmium batteries, Waste Manag. 30 (2010) 2292–2298.

[11] B.P.P. Lopes, V.R. Stamenkovic, Past, Present, and Future of Lead–Acid Batteries, (n.d.) 923–925.

[12] Z. Sun, H. Cao, X. Zhang, X. Lin, W. Zheng, G. Cao, Y. Sun, Y. Zhang, Spent lead-acid battery recycling in China—a review and sustainable analyses on mass flow of lead, Waste Manag. 64 (2017) 190–201.

[13] F. Arshad, L. Li, K. Amin, E. Fan, N. Manurkar, A. Ahmad, J. Yang, F. Wu, R. Chen, A comprehensive review of the advancement in recycling the anode and electrolyte from spent lithium ion batteries, ACS Sustain. Chem. Eng. 8 (2020) 13527–13554.

[14] J. Li, G. Wang, Z. Xu, Generation and detection of metal ions and volatile organic compounds (VOCs) emissions from the pretreatment processes for recycling spent lithium-ion batteries, Waste Manag. 52 (2016) 221–227.

[15] J. Xu, H.R. Thomas, R.W. Francis, K.R. Lum, J. Wang, B. Liang, A review of processes and technologies for the recycling of lithium-ion secondary batteries, J. Power Sources 177 (2008) 512–527.

[16] K. Liu, F.-S. Zhang, Innovative leaching of cobalt and lithium from spent lithium-ion batteries and simultaneous dechlorination of polyvinyl chloride in subcritical water, J. Hazard. Mater. 316 (2016) 19−25.

[17] L.-P. He, S.-Y. Sun, X.-F. Song, J.-G. Yu, Recovery of cathode materials and Al from spent lithium-ion batteries by ultrasonic cleaning, Waste Manag. 46 (2015) 523−528.

[18] L. Yao, Y. Feng, G. Xi, A new method for the synthesis of $LiNi_{1/3}Co_{1/3}Mn_{1/3}O_2$ from waste lithium ion batteries, RSC Adv. 5 (2015) 44107−44114.

[19] Y. Yao, M. Zhu, Z. Zhao, B. Tong, Y. Fan, Z. Hua, Hydrometallurgical processes for recycling spent lithium-ion batteries: a critical review, ACS Sustain. Chem. Eng. 6 (2018) 13611−13627.

[20] M. Zhou, B. Li, J. Li, Z. Xu, Pyrometallurgical Technology in the Recycling of a Spent Lithium Ion Battery: Evolution and the Challenge, ACS ES&T Eng, 2021.

[21] R. Ruismäki, A. Dańczak, L. Klemettinen, P. Taskinen, D. Lindberg, A. Jokilaakso, Integrated battery scrap recycling and nickel slag cleaning with methane reduction, Minerals 10 (2020).

[22] J. Li, G. Wang, Z. Xu, Environmentally-friendly oxygen-free roasting/wet magnetic separation technology for in situ recycling cobalt, lithium carbonate and graphite from spent $LiCoO_2$/graphite lithium batteries, J. Hazard. Mater. 302 (2016) 97−104.

[23] W. Wang, Y. Zhang, X. Liu, S. Xu, A simplified process for recovery of Li and Co from spent $LiCoO_2$ cathode using Al foil as the in situ reductant, ACS Sustain. Chem. Eng. 7 (2019) 12222−12230.

[24] A. Mayyas, D. Steward, M. Mann, The case for recycling: overview and challenges in the material supply chain for automotive li-ion batteries, Sustain. Mater. Technol. 19 (2019) e00087.

[25] K. Liu, S. Yang, L. Luo, Q. Pan, P. Zhang, Y. Huang, F. Zheng, H. Wang, Q. Li, From spent graphite to recycle graphite anode for high-performance lithium ion batteries and sodium ion batteries, Electrochim. Acta. 356 (2020) 136856.

[26] Z. Liang, C. Cai, G. Peng, J. Hu, H. Hou, B. Liu, S. Liang, K. Xiao, S. Yuan, J. Yang, Hydrometallurgical recovery of spent lithium ion batteries: environmental strategies and sustainability evaluation, ACS Sustain. Chem. Eng. 9 (2021) 5750−5767.

[27] S. Kim, J. Bang, J. Yoo, Y. Shin, J. Bae, J. Jeong, K. Kim, P. Dong, K. Kwon, A comprehensive review on the pretreatment process in lithium-ion battery recycling, J. Clean. Prod. 294 (2021) 126329.

[28] E.M.S. Barbieri, E.P.C. Lima, M.F.F. Lelis, M.B.J.G. Freitas, Recycling of cobalt from spent Li-ion batteries as β-Co(OH)2 and the application of Co_3O_4 as a pseudocapacitor, J. Power Sources 270 (2014) 158−165.

[29] V.T. Nguyen, J.C. Lee, J. Jeong, B.S. Kim, B.D. Pandey, Selective recovery of cobalt, nickel and lithium from sulfate leachate of cathode scrap of Li-ion batteries using liquid-liquid extraction, Met. Mater. Int. 20 (2014) 357−365.

[30] A. Beaudet, F. Larouche, K. Amouzegar, P. Bouchard, K. Zaghib, Key challenges and opportunities for recycling electric vehicle battery materials, Sustain. Times 12 (2020) 1−12.

[31] Q. Meng, Y. Zhang, P. Dong, Use of glucose as reductant to recover Co from spent lithium ions batteries, Waste Manag. 64 (2017) 214−218.

[32] W. Urbańska, Recovery of Co, Li, and Ni from spent li-ion batteries by the inorganic and/or organic reducer assisted leaching method, Minerals 10 (2020) 1−13.

[33] E.G. Pinna, M.C. Ruiz, M.W. Ojeda, M.H. Rodriguez, Cathodes of spent Li-ion batteries: dissolution with phosphoric acid and recovery of lithium and cobalt from leach liquors, Hydrometallurgy 167 (2017) 66−71.

[34] X. Zeng, J. Li, B. Shen, Novel approach to recover cobalt and lithium from spent lithium-ion battery using oxalic acid, J. Hazard. Mater. 295 (2015) 112−118.

[35] S. Natarajan, A.B. Boricha, H.C. Bajaj, Recovery of value-added products from cathode and anode material of spent lithium-ion batteries, Waste Manag 77 (2018) 455−465.

[36] L. Li, J. Ge, F. Wu, R. Chen, S. Chen, B. Wu, Recovery of cobalt and lithium from spent lithium ion batteries using organic citric acid as leachant, J. Hazard. Mater. 176 (2010) 288−293.

[37] H. Wang, K. Huang, Y. Zhang, X. Chen, W. Jin, S. Zheng, Y. Zhang, P. Li, Recovery of lithium, nickel, and cobalt from spent lithium-ion battery powders by selective ammonia leaching and an adsorption separation system, ACS Sustain. Chem. Eng. 5 (2017) 11489−11495.

[38] X. Zeng, J. Li, N. Singh, Recycling of spent lithium-ion battery: a critical review, Crit. Rev. Environ. Sci. Technol. 44 (2014) 1129–1165.
[39] X. Chen, Y. Chen, T. Zhou, D. Liu, H. Hu, S. Fan, Hydrometallurgical recovery of metal values from sulfuric acid leaching liquor of spent lithium-ion batteries, Waste Manag. 38 (2015) 349–356.
[40] S.G. Zhu, W.Z. He, G.M. Li, X. Zhou, X.J. Zhang, J.W. Huang, Recovery of Co and Li from spent lithium-ion batteries by combination method of acid leaching and chemical precipitation, Trans. Nonferrous Met. Soc. China (English Ed.) 22 (2012) 2274–2281.
[41] E.M.S. Barbieri, E.P.C. Lima, S.J. Cantarino, M.F.F. Lelis, M.B.J.G. Freitas, Recycling of spent ion-lithium batteries as cobalt hydroxide, and cobalt oxide films formed under a conductive glass substrate, and their electrochemical properties, J. Power Sources 269 (2014) 158–163.
[42] S. Saeki, J. Lee, Q. Zhang, F. Saito, Co-grinding LiCoO2 with PVC and water leaching of metal chlorides formed in ground product, Int. J. Miner. Process. 74 (2004) S373–S378.
[43] J. Li, Y. He, Y. Fu, W. Xie, Y. Feng, K. Alejandro, Hydrometallurgical enhanced liberation and recovery of anode material from spent lithium-ion batteries, Waste Manag. 126 (2021) 517–526.
[44] F.M. Courtel, S. Niketic, D. Duguay, Y. Abu-Lebdeh, I.J. Davidson, Water-soluble binders for MCMB carbon anodes for lithium-ion batteries, J. Power Sources 196 (2011) 2128–2134.
[45] J. Jegan Roy, M. Srinivasan, B. Cao, Bioleaching as an eco-friendly approach for metal recovery from spent NMC-based lithium-ion batteries at a high pulp density, ACS Sustain. Chem. Eng. 9 (2021) 3060–3069.
[46] B. Xin, D. Zhang, X. Zhang, Y. Xia, F. Wu, S. Chen, L. Li, Bioleaching mechanism of Co and Li from spent lithium-ion battery by the mixed culture of acidophilic sulfur-oxidizing and iron-oxidizing bacteria, Bioresour. Technol. 100 (2009) 6163–6169.
[47] N.B. Horeh, S.M. Mousavi, S.A. Shojaosadati, Bioleaching of valuable metals from spent lithium-ion mobile phone batteries using Aspergillus Niger, J. Power Sources 320 (2016) 257–266.
[48] N. Bahaloo-Horeh, S.M. Mousavi, M. Baniasadi, Use of adapted metal tolerant Aspergillus Niger to enhance bioleaching efficiency of valuable metals from spent lithium-ion mobile phone batteries, J. Clean. Prod. 197 (2018) 1546–1557.
[49] J. Zhao, B. Zhang, H. Xie, J. Qu, X. Qu, P. Xing, H. Yin, Hydrometallurgical recovery of spent cobalt-based lithium-ion battery cathodes using ethanol as the reducing agent, Environ. Res. 181 (2020) 108803.
[50] W. Lv, Z. Wang, H. Cao, Y. Sun, Y. Zhang, Z. Sun, A critical review and analysis on the recycling of spent lithium-ion batteries, ACS Sustain. Chem. Eng. 6 (2018) 1504–1521.

CHAPTER 6

Battery recycling for sustainable future: recent progress, challenges, and perspectives

Felipe M. de Souza and Ram K. Gupta
Department of Chemistry, Kansas Polymer Research Center, Pittsburg State University, Pittsburg, KS, United States

1. Introduction

Portable devices powered by batteries became major personal items due to their versatility in performing several tasks and by carrying important information and resources of individuals. Because of their convenience and fast information flux, they became a required tool for virtually every business and to connect people. Hence, this noticeable change in routine and culture after introducing portable devices made it unlikely for a society to function without them. In fact, from 2003 to 2017, there was an increase in the number of mobile phone shipments from 9.6 million to 1536 million, which made it the most consumed electronic segment worldwide [1]. Such a phenomenon can be attributed to two factors: (a) the necessity of fast information processing and flux and (b) the fast technological advances made with these devices. The evolution of these devices is illustrated in Fig. 6.1. However, one of the primary factors that allowed this technological surge was the introduction of Li–ion batteries (LIBs) into the market in 1991, which improved the charge-discharge cyclability, shelf-life, and performance of electronic devices. Further research on LIBs has provided overall optimization by decreasing battery size, weight, and toxicity while presenting a high capacity of 4200 mA h, compared with 500 mA h from Ni-Cd batteries, which were implemented before LIBs [1].

Because of this optimized performance, cellphones achieved a new level of importance as now they are used for several activities besides communication, such as multimedia, connection, enterprise features, and many more. On top of that, there was also an emergence of new products that became available to the public, such as Bluetooth speakers, drones, and flexible and wearable devices, among better versions of the current devices already established in the market. However, the overflow of new portable electronic devices has increased at a faster rate than the technology for improving battery performance. Because of that, the high technology products aside from cellphones, laptops, and tablets such as digital cameras, drones, smartwatches, smart glasses, and even some models of smart shoes demand battery systems that present higher energy density,

Figure 6.1 Evolution of portable electronic devices since 1983. *(Adapted with permission from Y. Liang, C.-Z. Zhao, H. Yuan, Y. Chen, W. Zhang, J.-Q. Huang, D. Yu, Y. Liu, M.-M. Titirici, Y.-L. Chueh, H. Yu, Q. Zhang, A review of rechargeable batteries for portable electronic devices, InfoMat. 1 (2019), 6–32, Copyright (2019) The Authors, some rights reserved; exclusive licensee John Wiley & Sons. Distributed under a Creative Commons Attribution License.)*

flexibility, safety, and environmental credentials along with lower cost. Despite demanding, these features can be incorporated into one set of products as it has been proved that it is an achievable goal. Hence, the development of new battery technologies such as lithium sulfide or solid-state batteries, for instance, is expected to provide suitable performance for these products. Among those two, solid-state batteries are receiving attention as they are less toxic and much safer than conventional LIBs. However, further research is still required to properly incorporate these novel batteries into mass production. It should be noted that the latest technologies can achieve satisfactory performance while promoting an eco-friendlier approach.

The concern with the environment is receiving more attention since the excessive use of nonrenewable sources increases the concentration of greenhouse gases. On top of that, most nations aim to decrease their dependence on nonrenewable sources to become more autonomous. Because of that, renewable energy derived from wind, tide, solar, biomass, and other types require a system to store and distribute the energy. Although batteries and supercapacitors are widely used to store and distribute energy, there is an issue regarding managing the resources used to build them. This is a relevant topic because vehicle and transportation companies are also investing in electrical engines to compose the future market, which will result in a strong impact on the distribution of resources used to manufacture active materials for batteries and supercapacitors. Despite the great demand, the investment in electrification of transport is a viable option, as it can decrease the demand for oil and move toward a more sustainable energy grid. Because of these factors, battery recycling technology can become a major process soon.

2. Battery market

Nonrenewable sources to obtain energy are dominant worldwide, accounting for almost 85% of energy consumption. In 2018 this percentage corresponded to 11,743.6 million tons of oil equivalent. Yet by 2040, one of the predictions is that this value will reach 17,651 million tons, whereas nonrenewable sources will still provide around 80% of the energy. This circumstance is concerning for the environment due to the large release of CO_2 and other greenhouse gases, which accounted for around 33 gigatons in 2018 [2]. Some efforts have been made to produce carbon-free electricity to move this scenario toward a more balanced situation. One of the promising technologies is founded on energy storage and electric vehicles (EVs). LIBs have great potential to serve as an alternative power source due to their high power density and energy efficiency. The convenience and relative cost efficiency of EVs compared with combustion vehicles has attracted more consumer demand and is considered a viable long-term investment [3]. This shift in fuel supply resulted in over 86% of Li-ion technology dominance among other electrochemical technologies such as sodium-based batteries, lead storage batteries, and flow batteries. Consequentially, there was an exponential increase in EV and plug-in hybrid electric vehicle sales for various countries. China, the USA, and Europe have the most sales at around 3.4 million units for battery-powered electric vehicles and around 22 million units for new EVs. Because of that, the LIB market is expected to reach nearly 100 billion dollars by 2025 [2]. Based on this data, the EV market is dealing with an ever-growing demand for manufacture to attend to the customers due to the transition from combustion to electric engines. This situation leads to the requirement to optimize batteries to take over for current LIBs. In that regard, progress in research has been made that has yielded batteries based on $LiCoO_2$, $LiFePO_4$, $LiNiO_2$, lithium nickel cobalt aluminum (NCA), and lithium nickel cobalt manganese (NCM). Therefore, it is necessary to develop efficient and sustainable methods for recycling Li and Co. Figure 6.2A demonstrates a plot of supply risk and circularity (C_m), and Fig. 6.2B demonstrates supply risk in the market value of active materials for cathodes. As noted, cobased cathodes such as NCMs and $LiCoO_2$ present higher risk and value than other metals, suggesting that recycling technologies should prioritize these resources, as they are also more profitable. Another viable strategy consists of employing high circularity and lower-cost materials to compose the mainstream of manufacture. Hence, Al, Ni, Mn, and Ni-based cathode

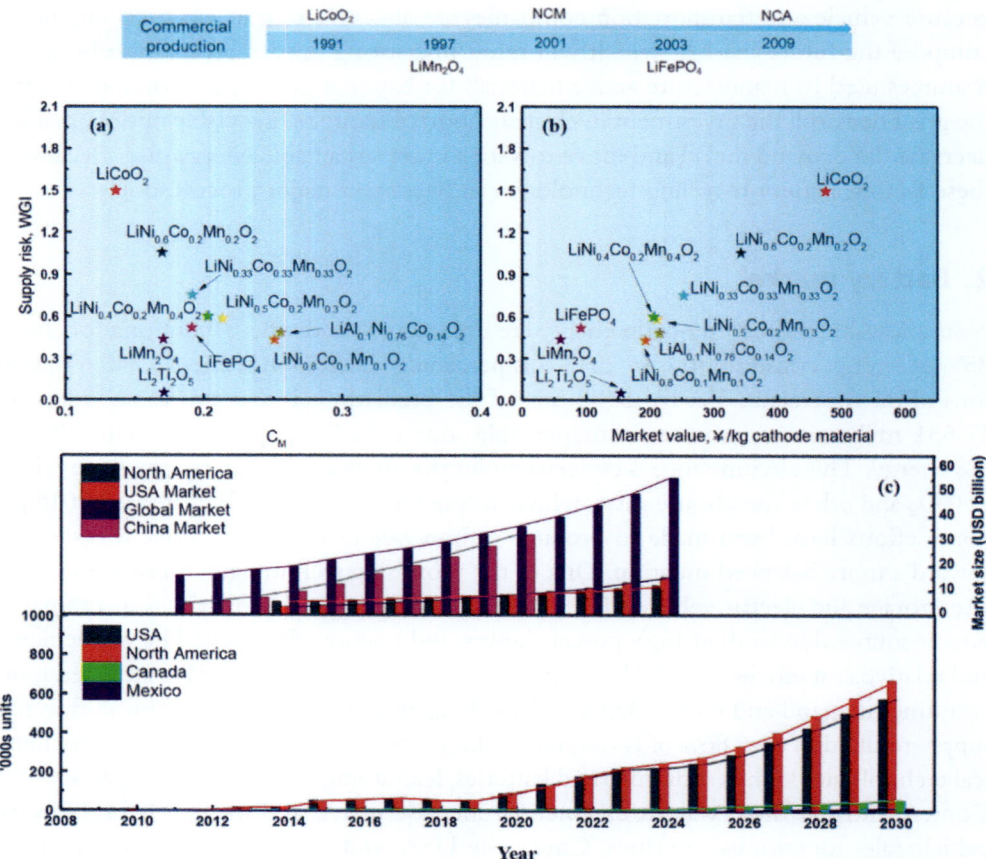

Figure 6.2 Supply risk based on World Governance Indicator in terms of (A) Circulability (C_m). (B) Value of cathode's active material. (C) Forecast from 2010 to 2030 of spent units of lithium-ion batteries in some countries. *(Adapted with permission from W. Lv, Z. Wang, H. Cao, Y. Sun, Y. Zhang, Z. Sun, A critical review and analysis on the recycling of spent lithium-ion batteries, ACS Sustain. Chem. Eng. 6 (2018), 1504–1521, Copyright (2017), American Chemical Society.)*

materials are more feasible to supply the demand. Thus, further research for performance optimization and establishing efficient recycling technologies should lead to a sustainable and profitable scenario. Accompanied to that, Fig. 6.2C shows the prediction of market size growth and an estimative for the amount of spent LIBs in different nations, indicating the necessity for recycling to reintroduce the metals used for cathodes into the manufacturing process.

The introduction of EV is still at its early stages, and it requires some precautions as it needs resources to be well managed to avoid a shortage or a crisis. For that, it's important to tackle the main points for this transition. The batteries in EVs generally last for 5—8 years when their performance (power and energy densities) decreases by 30% from their original value. Also, this process will occur simultaneously with the introduction and incentives for the purchase of EVs due to their green credentials [4,5]. However, a large volume of spent LIBs and low harvesting of raw materials for its manufacture can hinder sustainability. This issue can be prevented by reusing the energy storage systems for other applications and recycling processes to harvest and reprocess active materials [2]. This is a viable approach because EV batteries possess considerable capacitance suitable for secondary applications that require less energy. In addition, developing effective ways to recycle the materials that cannot be immediately used enables a way to close the production-consumption cycle.

3. Battery recycling

The recovery of valuable resources or the re-utilization of active materials is pivotal for the sustainable implementation of batteries in the coming years. Several companies have already foresighted this incoming scenario and are currently working on technologies and processes to address that. For example, BMW allied with Northvolt and Umicore to research a sustainable use for EV. Also, Audi sided with Umicore and is advancing in developing closed-cycle manufacture and reuse of active materials from the batteries. Currently, the main recycling technologies employed by companies are hydrometallurgy, pyrometallurgy, or a combination of both (Fig. 6.3). When these processes were first introduced to recover metals from batteries, only Co, Cu and Ni were recovered in the

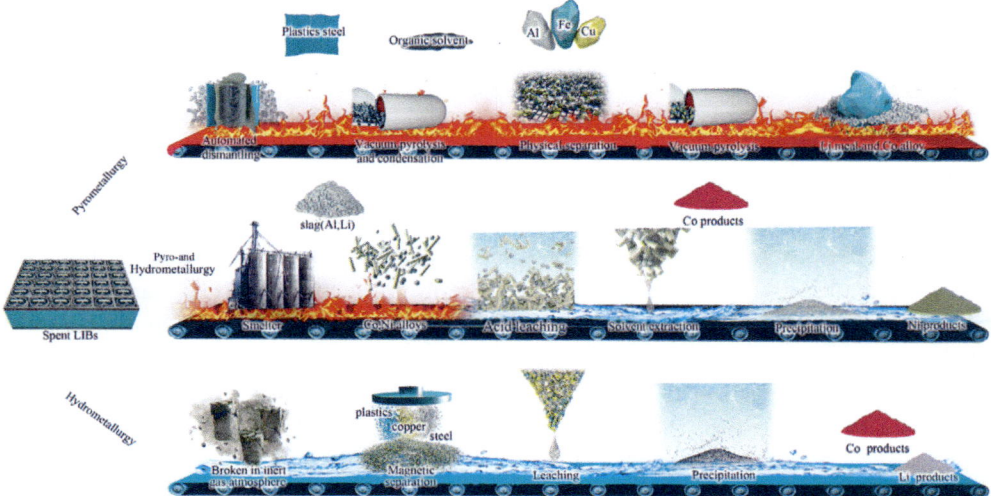

Figure 6.3 Schematic illustration for a general recycling process of a battery that includes the main three types performed in the industry: pyrometallurgy, pyro and hydrometallurgy, and hydrometallurgy. *(Adapted with permission from E. Fan, L. Li, Z. Wang, J. Lin, Y. Huang, Y. Yao, R. Chen, F. Wu, Sustainable recycling technology for Li-ion batteries and beyond: challenges and future prospects, Chem. Rev. 120 (2020), 7020–7063, Copyright (2020), American Chemical Society.)*

form of alloys, whereas Li was wasted. To address that, the hydrometallurgical process is used to recycle Li and Co with high purity. The usual recycling process consists of discharge or deactivation of the battery, mechanical dismantling, physical separation of magnetic parts, and pyro or hydrometallurgical process.

3.1 General recycling strategies

One of the main aspects of determining the future batteries' applications lies in finding efficient ways to recover their active materials in a facile, safe, and sustainable way. Since more than half of a battery is composed of metals such as Ag, Co, Cu, Zn, and other rare metals, it is important to recover these materials from an economic standpoint. However, from an environmental perspective, the recycling of batteries is perhaps more crucial as heavy metals such as Pb, Hg, and Cd can cause long-term contamination [6,7]. Hence, to avoid wastage and damage that these materials can cause is important to perform feasible recycling strategies. Based on that, Tanong et al. simulated a case in which there was a mixture of spent batteries such as Li-ion, lithium metallic (Li-M), Zn—C, Ni—Cd, and nickel-metal hydride [8]. The average value for the recovery of the metals Cd, Co, Mn, Ni, and Zn was 87.4% through a leaching step with H_2SO_4 and $Na_2S_2O_5$ followed by selective precipitation and dissolution or extraction. This procedure showed a universal approach to recover several types of metals. In addition, obtaining a solution of the cationic metals in high concentration to perform a separation step afterward seems to be more a more sustainable and economical approach for metal recovery [9]. Several strategies such as vacuum metallurgy separation, acid or alkaline leaching with simultaneous electrowinning, oxygen-free roasting with wet magnetic separation, reduction assisted pyrolysis, among others, can be used to recycle batteries [10]. All these strategies have demonstrated a yield of recovery over 95%. Hence, there are feasible and versatile approaches for recovering valuable metals with several methodologies, which opens several possibilities for the consolidation of battery recycling as a sustainable business.

Deng and his team showed an effective recycling approach by synthesizing MnO_2/graphene nanocomposites derived from spent battery powder (SBP) as starting material [11]. MnO_2 was obtained from a spent $Zn-MnO_2$ through simple manual milling followed by washing with an acidic solution to remove electrolyte NH_4Cl and Zn in the form of $ZnCl_2$. The nanocomposite synthesis based on MnO_2/reduced graphene oxide (rGO) or graphene oxide was obtained through a facile sonication at room temperature. This led to the formation of dispersion between the two components that entrapped MnO_2 into the porous of the carbon-based materials. The scheme for this approach is shown in Fig. 6.4. The SBP/rGO electrode displayed 150 F/g at 1 A/g and only 6.3% of capacitance loss after 1000 cycles. This improvement can be attributed to the synergy of rGO high surface area and conductivity (electric double layer capacitance) and the polarized structure and redox properties of MnO_2 (pseudocapacitance) that led to an overall improvement when these materials were combined.

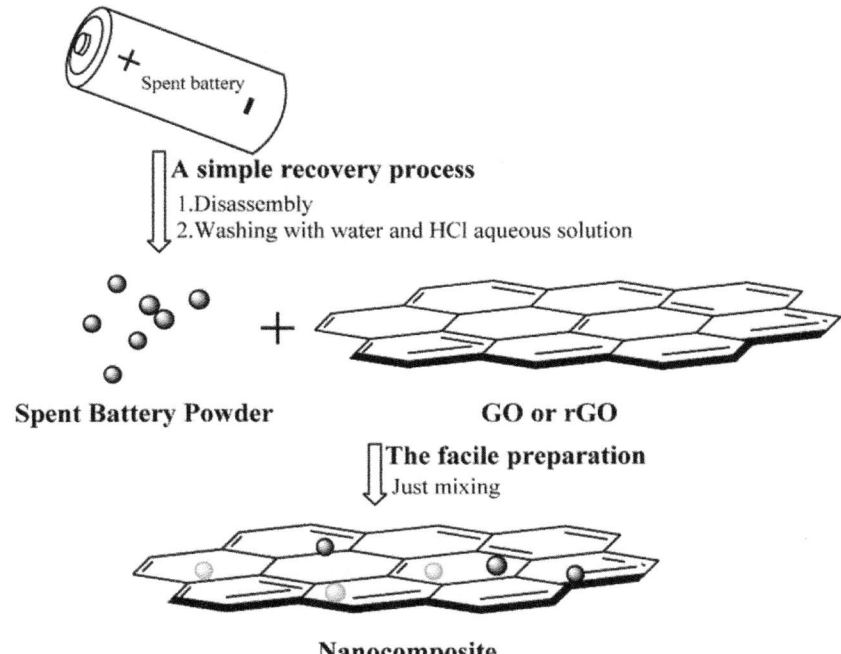

Figure 6.4 Facile process for the separation of spent battery powder and conversion into spent battery powder/reduced graphene oxide nanocomposite for energy storage application. *(Adapted with permission from J. Deng, X. Wang, X. Duan, P. Liu, Facile preparation of MnO₂/graphene nanocomposites with spent battery powder for electrochemical energy storage, ACS Sustain. Chem. Eng. 3 (2015), 1330–1338, Copyright (2015), American Chemical Society.)*

3.2 Recycling of electrode materials

Active materials used as cathodes in commercial batteries contain valuable lithium-based components that include lithium manganese oxide ($LiMn_2O_4$), lithium nickel oxide ($LiNiO_2$), ternary lithium oxides ($LiNi_xCo_yMn_zO_2$), lithium iron phosphate ($LiFePO_4$), and $LiCoO_2$. There is a great effort into the recycling approaches for cathode materials due to their relatively higher value in the market. For that, physical approaches are performed first to reclaim the electrode materials, including dismantling, scraping, grinding, magnetic separation, dissolution, pyrolysis, etc. Relatively high purity metals can be obtained through this method. However, it is a laborious process which makes it challenging for large-scale application. Hence, other technologies have gained force to extract active cathode materials such as pyrometallurgical, divided into four different processes: direct roasting, atmosphere or additive-assisted roasting, and cathode regeneration.

Direct roasting consists of high-temperature treatment ($\sim 1400°C$). The active cathode materials are burnt into a smelter, and the Al (current collector), electrolyte, and carbon-based anode function as fuel. Then, SiO_2 and CaO are introduced to separate thermally irreducible metals such as Li, Mg, and Al from valuable metals such as Co, Ni, or Mn. The alloys of these noble metals are obtained; however, the other materials are wasted in the slag, which makes this process nonsustainable [12]. Approaches to counter that drawback have been proposed, such as adding Na_2SO_4 or $CaCl_2$ into the smelter slag at 800°C for 1 h to turn Li into soluble salts. This additional step can lead to around 90% of leached Li [13,14]. Despite this extra step for the recovery of valuable metals, feasible strategies are used on a large scale that combines pyrometallurgical with hydrometallurgical. The latter consists of leaching with alkali or acid, chemical precipitation, and electrolysis process to effectively recover Li and Ni, which can be further resynthesized into $LiCoO_2$ and $Ni(OH)_2$ active materials. The active materials used as cathodes for batteries can be reintroduced into the manufacturing cycle. However, toxic fumes are released during the combustion of other battery components, leading to a gas treatment step to make it environmentally friendly.

Atmosphere-assisted roasting is a recycling process performed in an inert atmosphere (no oxygen). The initial advantage of this method is that it can be performed at lower temperatures compared with the direct roasting method since under vacuum or inert gas, there is a decrease in the partial pressure from gases such as O_2 or CO_2 formed during combustion. On top of that, Li is directly converted into Li_2O or Li_2CO_3. A strategy to recycle lithium batteries was performed by Li et al. where oxygen-free roasting accompanied with wet magnetic separation was performed with high recovery rates for Co (95.72%), Li_2CO_3 (98.93%), and graphite (91.05%) [15]. On top of satisfactory efficiency, the process was also solventless, saving on cost and making the process eco-friendlier. The schematic for the process is presented in Fig. 6.5. Other cathode active materials such as $LiMn_2O_4$ can also be recycled through a similar process yielding Li_2CO_3 and MnO as during heating the bonds Li—O and Mn—O are broken, allowing O atoms to be captured by carbon, which acts as a reducing agent. Through that, CO_2 is formed, leading to the eventual conversion of MnO and Li_2CO_3. The reduction process performed under the presence of carbon can be performed within the range of 650°C—750°C as Li_2CO_3 can volatilize in temperatures higher than 750°C.

Performing recycling with fewer steps makes it valuable for a scalable approach. Avoiding the formation of slags, solutions and diminishing the separation steps are also important to make the process sustainable and cost-effective. With that in mind, Wang and his team propose a recycling route in which Al was used as a reducing agent instead of C [16]. This approach leads to the formation of $LiAlO_2$ due to the reaction preference of $LiCoO_2$ with Al rather than C. First, CoO^- is tended to be formed due to the instability of Co(III), then it could reduce either to metallic Co when reacting with Al or it can be assimilated into $LiAlO_2$ after its reaction with Li_2O. After that, NaOH can be used to leach Al and Li, whereas Co can be recovered by acid treatment. Hence, this method is a step short as the active material from the cathode does not need to be separated from Al foil. The description of this approach is provided in Fig. 6.6.

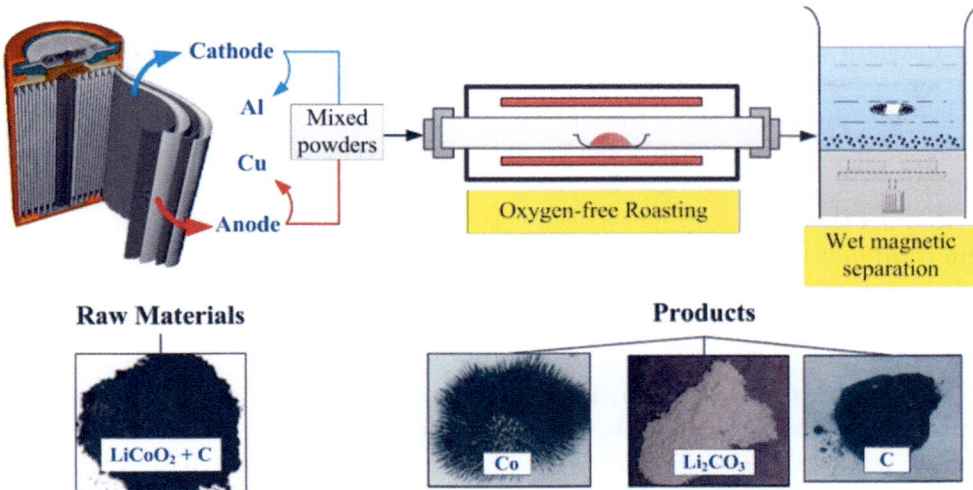

Figure 6.5 Scheme for the oxygen-free roasting followed by wet magnetic separation to recycle active cathode materials. *(Adapted with permission from J. Li, G. Wang, Z. Xu, Environmentally-friendly oxygen-free roasting/wet magnetic separation technology for in situ recycling cobalt, lithium carbonate and graphite from spent LiCoO₂/graphite lithium batteries, J. Hazard Mater. 302 (2016), 97–104, Copyright (2016), Elsevier.)*

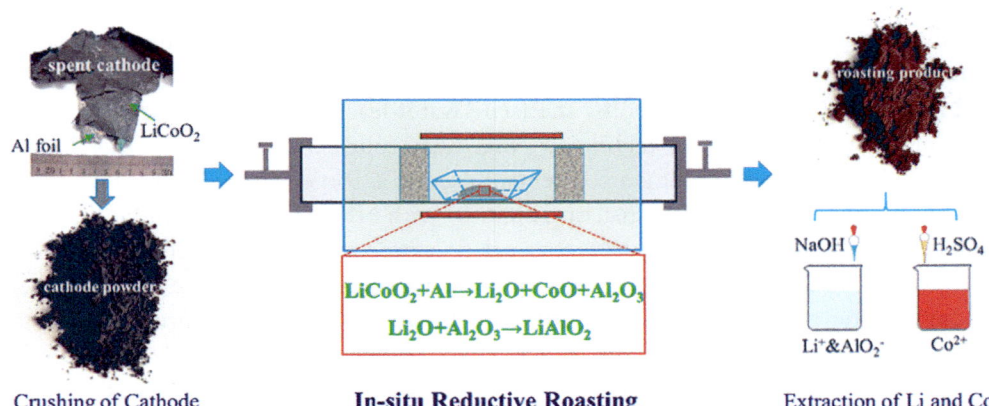

Figure 6.6 Scheme for the recycling of LiCoO$_2$ as active material blended with Al foil leading to the recovery of both Li and Co. *(Adapted with permission from W. Wang, Y. Zhang, X. Liu, S. Xu, A simplified process for recovery of Li and Co from spent LiCoO2 cathode using Al foil as the in situ reductant, ACS Sustain. Chem. Eng. 7 (2019), 12222–12230, Copyright (2019), American Chemical Society.)*

Ternary LIBs such as NCM and NCA can also be recycled through atmosphere-assisted roasting technology by heating the blended parts of the battery to 650°C for 1 h under the presence of argon, whereas water can be introduced as a bleaching agent for Li. The process yielded 99.1%, 97.7%, and 82.2% recovery of Co, Ni, and Li, respectively [17]. In another eco-friendly approach, CO_2 was used to carbonate NCA batteries at temperatures from 600°C to 800°C to obtain Li_2CO_3 along with CoO and NiO. Therefore, atmosphere-assisted technology is sustainable as it can be performed either through carbothermic or thermite reduction, as the second route can skip one separation process.

Another feasible strategy for recovering cathodes is additive-assisted roasting which is performed under the presence of acids such as HCl, H_2SO_4, or other inorganic salts. It can be summarized as a hydro leaching process performed in the dry phase [12]. Through that, stoichiometric amounts of reagents are required to ensure the highest efficiency. This process has inherently less waste than the others, which is one of the advantages of this method. On top of that, this approach allows the recovery of Li. The three main processes are roasting: sulfation, chlorination, and nitration. The sulfation method employs sulfur-based reagents to convert the cathode into the sulfonated form, which can be performed with sulfates, sulfuric acid, or sulfur gas. Shi and his colleagues used SO_2 under the presence of O_2 to form SO_3 along with argon to make the gaseous mixture [18]. Then, SO_3 reacted with $LiCoO_3$ at 700°C yielding CoO, $Li_2Co(SO_4)_2$, and Li_2SO_4. The process was deemed more efficient for the recovery of Li, reaching 99.5%. Another roasting approach at 600°C for 30 min on $LiCoO_2$ using $NaHSO_4 \cdot H_2O$ has been performed, which yielded Co_3O_4 and $LiNaSO_4$ [19]. However, the release of SO_2 and SO_3 deteriorates nearby instruments, leading to increased maintenance of the process. Hence, another path for sulfonation can be performed by roasting with H_2SO_4 with active cathode materials that yield sulfates of $MnSO_4$, $NiSO_4$, $CoSO_4$, and Li_2SO_4. Li can be recovered by the facile water leaching process.

Chlorination roasting is a technology that was applied in metallurgy earlier and has been adopted for recycling. It is a convenient process as some approaches achieve almost 100% recovery of Li and Co by reacting $LiCoO_2$ with NH_4Cl at 350°C for 20 min. This simpler process occurs through the thermal decomposition of NH_4Cl into HCl and NH_3. Then, HCl captures the O from $LiCoO_2$, whereas NH_3 acts as a reducing agent leading to the metallic Li and Co [20]. Along with lower temperature, this method can also be performed under air atmosphere, which when applied over $LiMn_2O_4$, tended to form Mn_3O_4 due to the higher oxidative potential of oxygen, whereas Li could be obtained with high purity.

Nitration roasting is another feasible technology as it can be performed at a lower temperature and reasonable time. Based on that, the active materials of a spent LIB were nitrated and exposed to 250°C for 1 h to calcinate the respective metallic nitrates [21]. Most of these metallic nitrates are converted into metal oxides, except $LiNO_3$ that has a decomposition temperature of around 600°C. Then, the water leaching process can be executed to recover the metals. The technology for the recovery of anode materials, mostly carbon-based, has received much attention compared with those of

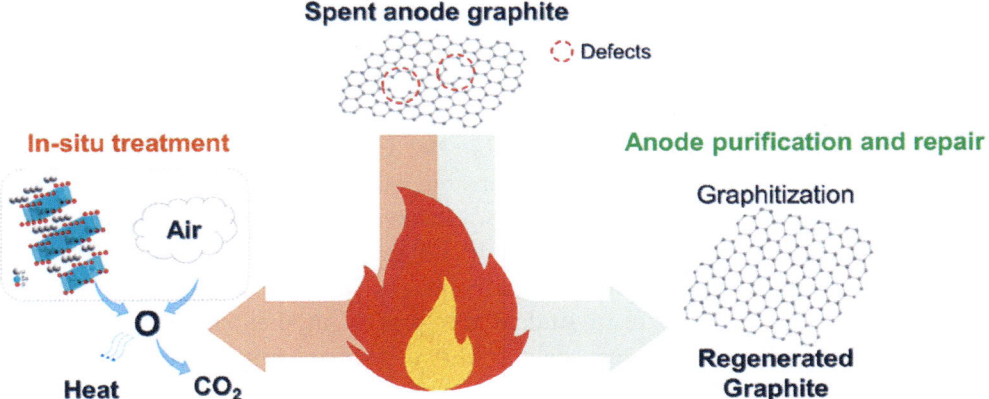

Figure 6.7 Conventional thermal treatment for the recovery of carbon-based anodic materials. *(Adapted with permission from M. Zhou, B. Li, J. Li, Z. Xu, Pyrometallurgical Technology in the Recycling of a Spent Lithium Ion Battery: Evolution and the Challenge, ACS ES&T Engineering (2021), Copyright (2021), American Chemical Society.)*

cathodes due to the lower cost of these materials. However, as the production of LIBs increases, this process may gain more force as graphite, which is the main component of commercial anodes, has its price from 8000 to 13,000 $/ton, which corresponds to around 10%−15% of the total cost of a battery. Most of the recovery for these materials is based on thermal treatments, as displayed in Fig. 6.7. However, it is noted that carbon tends to be used as a reducing agent to recover valuable metals in the cathode, which consumes it during the process.

3.3 Recycling of electrolyte and binders

An electrolyte is a medium responsible for the transport of ions within the battery, composed of Li-based salts such as LiPF6 and LiBF4 dissolved in organic solvents like ethylene, propylene dimethyl carbonates, and dimethoxymethane. The binders such as poly(vinylidene fluoride) (PVDF) are another important component used in commercial LIBs to increase the contact between the electrode and current collector. During the recycling process, the electrolyte should be removed first as it can decrease the efficiency of the recycling process and lead to more energy input and high impurity. On top of that, their toxicity can harm the human workforce accompanying the procedure. The first attempt at recycling electrolytes was using organic solvents such as N-methyl-2-pyrrolidone, dimethylformamide, and dimethylacetamide. However, their high cost and harm to the environment make this approach unsuitable. Hence, a simpler approach of thermal treatment for these components was adopted. For this process, the temperature range must be properly selected to separate the components based on their decomposition temperatures. Hence, organic matter should decompose without melting Al foil while the active materials from the cathode should degrade. Otherwise, the purity of the separated materials will decrease. Thus, for the proper execution of the pyrolytic process is important to know the decomposition temperatures for the components. For example, $LiPF_6$ decomposes in between 107 and 133°C, ethylene carbonate in around 326°C—350°C, PVDF in the range of 430°C—492°C and Al melts around 650°C—660 °C. Strategies to decompose the organic matter have been employed, such as using piranha solution over different temperatures. However, the process may become costly if the temperature exceeds 500°C, whereas PVDF may release F^-, which is harmful to the environment. An oxygen-free method is also an option for treating electrolytes and binders as Sun and his team studied this process in the temperature range of 450°C—700°C [22]. This process yielded low molecular weight organic molecules and metal oxides. A vacuum separated the former, whereas the latter were collected in a cold trap. In a follow-up study, the optimal temperature was 600°C executed for 30 min under the presence of oxalic acid [23]. Substant gases such as HF and COF_2 were generated despite the relative efficiency and lower energy input.

Microwave-assisted has also been proposed to further decrease energy input as this process can be performed around 360°C while providing better heat transfer [24]. This condition facilitates the breakdown of organic components into smaller molecules. However, if carried out for a long period, the Al foil can melt, damaging the relatively expensive instrument. These factors are a hinder to the scale-up of this process. Notably, oxygen-free pyrolysis offers the advantages of a close environment that prevents toxic gases from escaping. It also prevents the oxidation of Al foil and skips some of the separation process, which saves on cost. However, dealing with the release of fluorine or fluorine-based compounds that can corrode the instruments adds maintenance expenses. This challenge has led to the development of an attractive technology which is thermal defluorination. In this approach, the addition of CaO into the roasting process can promote PVDF decomposition under low temperatures. Also, CaO reacts with HF converting it into CaF_2, which prevents corrosion and is an extra step for F^- removal. This technique is used to specifically capture F^- to prevent the release of toxic fumes. Hence, it leads to an increase in cost and can decrease the final product's purity.

Non-destructive recovery is a valuable technique as it can be used to reutilize the electrolytes and binders, which are components with an inherent market value. Low-temperature volatilization can be used since the organic solvents in the electrolyte can be easily volatilized. For example, an industrial process was performed that had a productivity of 96.6 kg/h by performing the volatilization at 120°C for 150 min yielding more than 99% of the recovery for both $LiPF_6$ and organic electrolytes. Even though this method works separately from the other thermal process, it seems feasible due to its high yield and nondamaging process. The molten-salt approach is a robust green method for recovering active materials. Wang and his team used molten salt composed

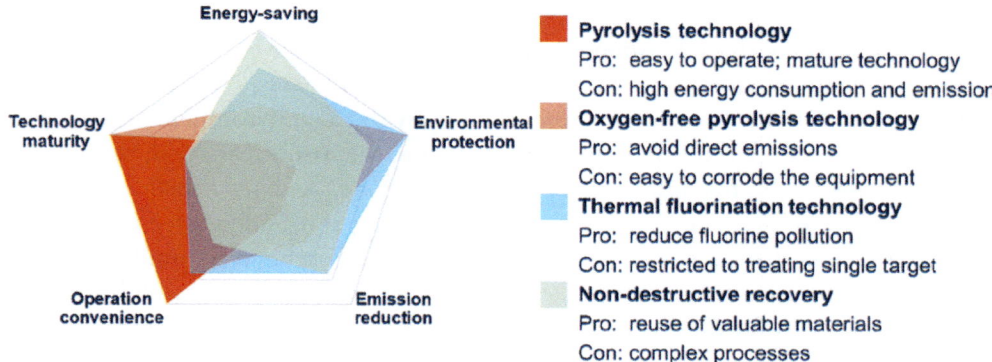

Figure 6.8 Comparison among the pyrometallurgical approaches for the recycling of binders and electrolytes. *(Adapted with permission from M. Zhou, B. Li, J. Li, Z. Xu, Pyrometallurgical Technology in the Recycling of a Spent Lithium Ion Battery: Evolution and the Challenge, ACS ES&T Engineering (2021), Copyright (2021), American Chemical Society.)*

of $AlCl_3-NaCl$ to melt PVDF [25]. This process can be performed at 160°C for 20 min and manages to separate 99.8 wt.% of active material from the current collector. On top of that, the molten $AlCl_3-NaCl$ can be reutilized up to four times without a considerable decrease in the recovery of cathodic material. PVDF can be recovered through low-pressure filtration by the end of the process, whereas the $AlCl_3-NaCl$ can be steamed dry to be reutilized. This process is relatively greener and requires only an absorbent for $AlCl_3 \cdot 6H_2O$ that can be formed during the process. Through the analysis of these approaches, it is notable that each one has its aspects, yet despite the inevitable adding cost, these processes make the recycling process greener. A summary of these strategies is provided in Fig. 6.8.

4. Challenges and perspectives

Managing resources has always been a challenge as energy consumption, transportation, and information technologies evolve, which demand active materials that can deliver higher efficiency, long-term use, eco-friendliness, and renewability. Since batteries are the powerhouse behind it, there is constant pressure to improve their performance. With the introduction of even higher performance devices and EVs, this type of technology is being considered a feasible option to provide energy and transportation as an alternative to petrochemicals. However, the overall process must be sustainable; otherwise, it can potentially culminate in another energetic crisis. The primordial aspect of implementing this technology is that batteries can be recycled, unlike petrochemicals. Hence, if a sustainable closed loop for the use of batteries and effective recycling of their components is achieved, then the use of batteries on a larger scale can become the new face of the future. For that, perhaps the most important factor lands of resource management. This factor should be depicted as a sustainable balance between harvesting resources and manufacturing from virgin raw materials, conscient use of devices and EVs as a society should be aware of the responsibility with the environment and effective recycling process to reimplement these materials into the mainstream with the same efficiency as there were first introduced even though there are several challenges for this process related to the decrease of cost during recycling and finding greener routes to avoid the release of toxic chemicals that would otherwise be as harmful. There has been a great effort from the scientific community and industries to establish recycling processes, especially for the more valuable components such as active materials for cathodes. On top of that, several technologies are being combined to improve the overall efficiency of the recycling process, mostly pyrometallurgical and hydrometallurgical. There is still plenty of room for optimizing these processes as the addition of several reagents and high-temperature steps is the main hurdle for this technology to become more accessible. Based on that, a summary of the technologies currently used to recycle batteries components such as cathodes, anodes, electrolytes, and binders is shown in Fig. 6.9.

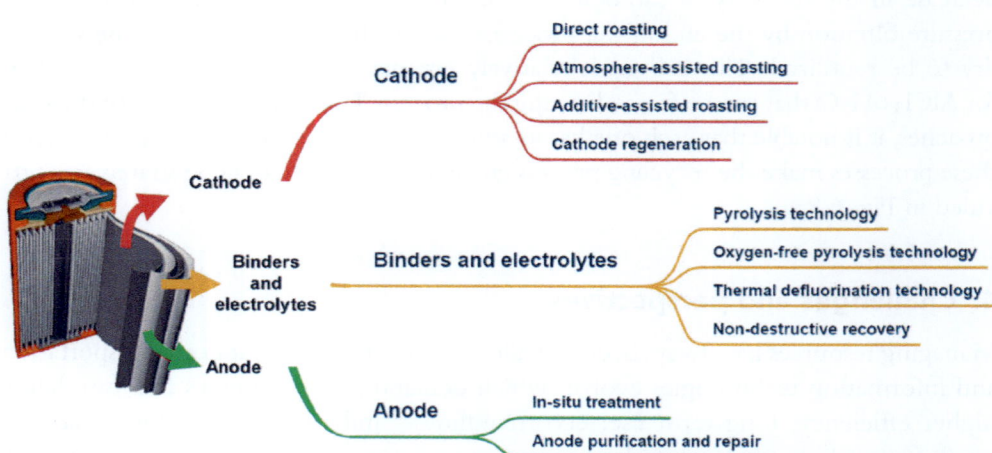

Figure 6.9 Schematics for the main technologies to recycle battery parts related to cathodes, anodes, binders, and electrolytes. *(Adapted with permission from M. Zhou, B. Li, J. Li, Z. Xu, Pyrometallurgical Technology in the Recycling of a Spent Lithium Ion Battery: Evolution and the Challenge, ACS ES&T Engineering (2021), Copyright (2021), American Chemical Society. The battery figure is reproduced with permission from J. Li, G. Wang, Z. Xu, Environmentally-friendly oxygen-free roasting/wet magnetic separation technology for in situ recycling cobalt, lithium carbonate and graphite from spent $LiCoO_2$/graphite lithium batteries, J. Hazard Mater. 302 (2016) 97–104, Copyright (2016), Elsevier.)*

5. Conclusion

In conclusion to this book chapter, it is noticeable that batteries recycling is a required step for the effective implementation of technologies that are likely to frame the next energy distribution around the globe. Scientists, as well as industries, are being able to develop practical approaches for the recycling of valuable active materials through feasible processes; however, it is necessary to improve recycling processes, so they require lower energy input and cost, as most processes require high thermal treatment (e.g., pyrometallurgical approach) or several reagents (e.g., hydrometallurgical approach). It deems necessary for the current and next generation of scientists to tackle these challenges to establish a sustainable cycle of resources that can be effectively reintroduced into the mainline of production. It is an important matter, since using alternative sources of energy and fuel is necessary to decrease the dependence on petrochemicals. Also, recycling batteries is a required step for the successful use of high-performance electronic devices, energy distribution, and the manufacture of EVs. More attention has been devoted to the recycling of valuable metals such as Li, Co, Ni, and Mn, at which the technologies applied for their recovery show relatively high efficiency by an order of 98%. However, even fewer valuable materials such as graphite that is used as an anode are becoming more relevant since technologies that can quickly replenish the demand in the manufacturing process can serve as a great help to maintain the flux of products. Therefore, the investment in recycling technologies has great potential to become a profitable and necessary sector for the future.

References

[1] Y. Liang, C.-Z. Zhao, H. Yuan, Y. Chen, W. Zhang, J.-Q. Huang, D. Yu, Y. Liu, M.-M. Titirici, Y.-L. Chueh, H. Yu, Q. Zhang, A review of rechargeable batteries for portable electronic devices, InfoMat. 1 (2019) 6—32.

[2] E. Fan, L. Li, Z. Wang, J. Lin, Y. Huang, Y. Yao, R. Chen, F. Wu, Sustainable recycling technology for Li-ion batteries and beyond: challenges and future prospects, Chem. Rev. 120 (2020) 7020—7063.

[3] P. Weldon, P. Morrissey, M. O'Mahony, Long-term cost of ownership comparative analysis between electric vehicles and internal combustion engine vehicles, Sustain. Cities Soc. 39 (2018) 578—591.

[4] W. Lv, Z. Wang, H. Cao, Y. Sun, Y. Zhang, Z. Sun, A critical review and analysis on the recycling of spent lithium-ion batteries, ACS Sustain. Chem. Eng. 6 (2018) 1504—1521.

[5] X. Zhang, L. Li, E. Fan, Q. Xue, Y. Bian, F. Wu, R. Chen, Toward sustainable and systematic recycling of spent rechargeable batteries, Chem. Soc. Rev. 47 (2018) 7239—7302.

[6] J. Patrício, Y. Kalmykova, P.E.O. Berg, L. Rosado, H. Åberg, Primary and secondary battery consumption trends in Sweden 1996—2013: method development and detailed accounting by battery type, Waste Manag. 39 (2015) 236—245.

[7] X. Wang, G. Gaustad, C.W. Babbitt, C. Bailey, M.J. Ganter, B.J. Landi, Economic and environmental characterization of an evolving Li-ion battery waste stream, J. Environ. Manage. 135 (2014) 126—134.

[8] K. Tanong, L. Coudert, M. Chartier, G. Mercier, J.-F. Blais, Study of the factors influencing the metals solubilisation from a mixture of waste batteries by response surface methodology, Environ. Technol. 38 (2017) 3167—3179.

[9] J. Li, X. Li, Q. Hu, Z. Wang, J. Zheng, L. Wu, L. Zhang, Study of extraction and purification of Ni, Co and Mn from spent battery material, Hydrometallurgy 99 (2009) 7—12.

[10] P. Liu, Recycling waste batteries: recovery of valuable resources or reutilization as functional materials, ACS Sustain. Chem. Eng. 6 (2018) 11176—11185.

[11] J. Deng, X. Wang, X. Duan, P. Liu, Facile preparation of MnO2/graphene nanocomposites with spent battery powder for electrochemical energy storage, ACS Sustain. Chem. Eng. 3 (2015) 1330—1338.

[12] M. Zhou, B. Li, J. Li, Z. Xu, Pyrometallurgical Technology in the Recycling of a Spent Lithium Ion Battery: Evolution and the Challenge, ACS ES&T Eng, 2021.

[13] N. Li, J. Guo, Z. Chang, H. Dang, X. Zhao, S. Ali, W. Li, H. Zhou, C. Sun, Aqueous leaching of lithium from simulated pyrometallurgical slag by sodium sulfate roasting, RSC Adv. 9 (2019) 23908—23915.

[14] H. Dang, N. Li, Z. Chang, B. Wang, Y. Zhan, X. Wu, W. Liu, S. Ali, H. Li, J. Guo, W. Li, H. Zhou, C. Sun, Lithium leaching via calcium chloride roasting from simulated pyrometallurgical slag of spent lithium ion battery, Separ. Purif. Technol. 233 (2020) 116025.

[15] J. Li, G. Wang, Z. Xu, Environmentally-friendly oxygen-free roasting/wet magnetic separation technology for in situ recycling cobalt, lithium carbonate and graphite from spent LiCoO2/graphite lithium batteries, J. Hazard Mater. 302 (2016) 97—104.

[16] W. Wang, Y. Zhang, X. Liu, S. Xu, A simplified process for recovery of Li and Co from spent LiCoO2 cathode using Al foil as the in situ reductant, ACS Sustain. Chem. Eng. 7 (2019) 12222—12230.

[17] Y. Ma, J. Tang, R. Wanaldi, X. Zhou, H. Wang, C. Zhou, J. Yang, A promising selective recovery process of valuable metals from spent lithium ion batteries via reduction roasting and ammonia leaching, J. Hazard Mater. 402 (2021) 123491.

[18] J. Shi, C. Peng, M. Chen, Y. Li, H. Eric, L. Klemettinen, M. Lundström, P. Taskinen, A. Jokilaakso, Sulfation roasting mechanism for spent lithium-ion battery metal oxides under SO2-O2-Ar atmosphere, J. Occup. Med. 71 (2019) 4473—4482.

[19] D. Wang, X. Zhang, H. Chen, J. Sun, Separation of Li and Co from the active mass of spent Li-ion batteries by selective sulfating roasting with sodium bisulfate and water leaching, Miner. Eng. 126 (2018) 28—35.

[20] J. Xiao, B. Niu, Z. Xu, Novel approach for metal separation from spent lithium ion batteries based on dry-phase conversion, J. Clean. Prod. 277 (2020) 122718.

[21] J. Mu, D.D. Perlmutter, Thermal decomposition of metal nitrates and their hydrates, Thermochim. Acta 56 (1982) 253—260.

[22] L. Sun, K. Qiu, Vacuum pyrolysis and hydrometallurgical process for the recovery of valuable metals from spent lithium-ion batteries, J. Hazard Mater. 194 (2011) 378—384.

[23] L. Sun, K. Qiu, Organic oxalate as leachant and precipitant for the recovery of valuable metals from spent lithium-ion batteries, Waste Manag. 32 (2012) 1575—1582.

[24] F. Diaz, Y. Wang, T. Moorthy, B. Friedrich, Degradation mechanism of nickel-cobalt-aluminum (NCA) cathode material from spent lithium-ion batteries in microwave-assisted pyrolysis, Met 8 (2018).

[25] M. Wang, Q. Tan, L. Liu, J. Li, Efficient separation of aluminum foil and cathode materials from spent lithium-ion batteries using a low-temperature molten salt, ACS Sustain. Chem. Eng. 7 (2019) 8287—8294.

SECTION 2

Battery recycling and separation processes

CHAPTER 7

Bio-inspired nanotechnology for easy-to-recycle lithium-ion batteries

Congrui Jin[1] and Jianlin Li[2]

[1]Department of Civil and Environmental Engineering, University of Nebraska—Lincoln, Lincoln, NE, United States; [2]Electrification and Energy Infrastructures Division, Oak Ridge National Laboratory, Oak Ridge, TN, United States

1. Introduction

Nowadays, lithium-ion batteries dominate the ever-growing consumer electronics and electric vehicles markets because of their high energy density, high power density, low self-discharge rate, zero memory effect, and quick charge acceptance [1–6]. However, battery lifetime, reliability, and manufacturing costs need to be improved for the technology to become more competitive. Lithium-ion batteries are now manufactured at a rate of several million units per month, and production capacity is expected to exceed 1 TWh by 2028 [7–9]. As a result, end-of-life batteries are driving increasing demand for spent battery collection systems [10].

Lithium-ion battery manufacturing consumes an enormous volume of scarce and expensive metallic materials. According to the existing literature, for the annual production of 20 million electronic vehicle batteries, the demand for cobalt equals current world mining production and will consume existing cobalt reserves in approximately 60 years. The nickel needed for this production rate would be larger by 200-fold than today's production capacity [11]. Although lithium makes up only a small portion of the total cost of battery manufacturing compared with other raw materials, lithium suppliers will experience serious pressure to cater to demand in the near future [12–15]. Therefore, if the recycling of spent lithium-ion batteries is not handled appropriately, it will waste immense amounts of valuable materials [16–20].

However, recycling technology is still in its nascent stage [21]. Recycling the entire lithium-ion battery is not yet economically profitable. Although diverse process chains have been applied or are under development to recycle batteries, a common, critical issue for battery recycling is separating the composite electrode film and metallic current collector [22,23]. A lithium-ion battery consists of a cathode, an anode, an organic electrolyte, and a separator. The cathode is typically made up of aluminum foil coated by a thin layer of lithium transition metal oxide; the anode consists of copper foil coated by a thin layer of graphite [5,9]. Each electrode also contains a polymeric binder that greatly facilitates adhesion between the electrode and current collector foil as well as cohesion among the electrode particles. A conductive additive is also included in each electrode to increase electronic conductivity. During recycling, an intensive milling step is usually applied to separate the composite electrode film from the metallic current collector. Although mechanical stressing loosens the composite film from the foil, it typically leads to smaller foil fragments that must be screened. Other methods have also been proposed, such as dissolving the electrode in a solvent and applying thermal decomposition to the electrode to weaken interfacial adhesion [24–28].

Hanisch et al. proposed a combined thermal and mechanical process to separate the metallic current collector from the composite electrode film [27]. First, thermal decomposition of the polymeric binder results in lower adhesion between the electrode and current collector as well as lower cohesion between the coated active material particles. Then, an air-jet separator separates electrode powder from the current collector foils with mesh sizes as small as 50 μm, guaranteeing high powder purities and preventing larger foil fractions from passing through the sieve. The separator is equipped with a higher airspeed and a lid to create impact stress on foils and agglomerates, as shown in Fig. 7.1. However, despite the feasibility and sophistication of the technique, implementation is complicated and still requires significant time and resources.

Alternatively, electrode films can be separated from current collectors using a solvent-based separation method. However, most solvents are highly toxic and harmful to operators. Some cause corrosion to the aluminum foil as well as lithium leaching. Recently, a green solvent, triethyl phosphate, has been used to separate the valuable cobalt-containing cathode from the aluminum current collector by dissolving the polymeric binder, as shown in Fig. 7.2 [28]. Electrochemically active materials are separated from

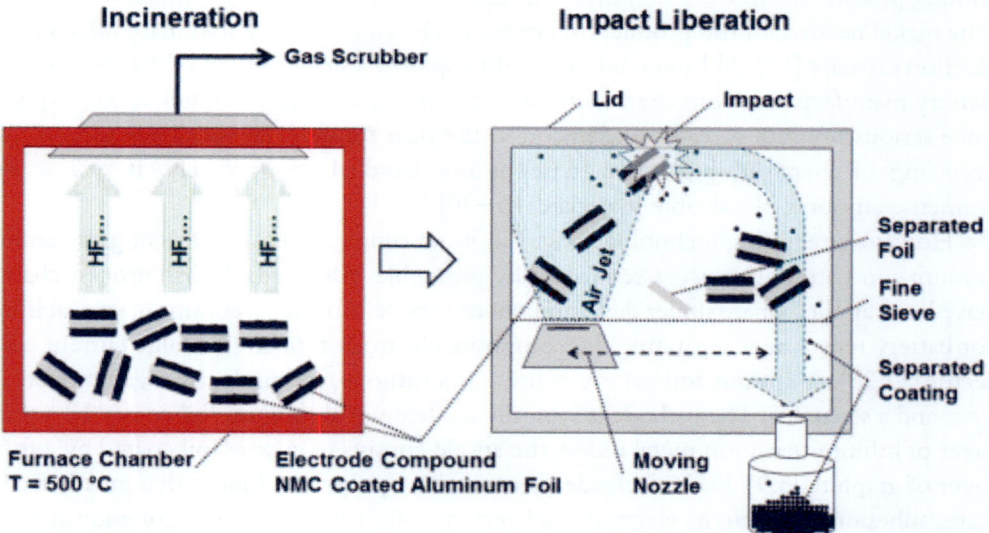

Figure 7.1 Schematic illustration of the experimental setup for the combined thermal and mechanical process. *(Adapted with permission from Ref. [27]. Copyright 2015, Elsevier B.V.)*

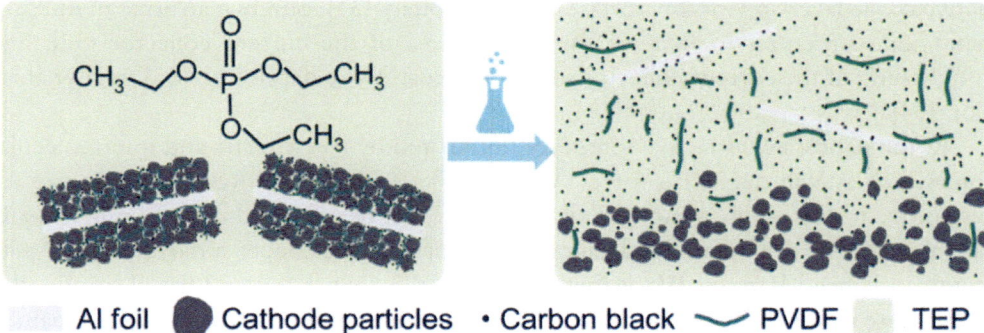

Figure 7.2 Schematic illustration of the molecular structure of the green solvent, i.e., triethyl phosphate (TEP), and the dissolution of the polymeric binder, i.e., poly(vinylidene fluoride) (PVDF) to release cathode particles from aluminum foil. *(Adapted with permission from Ref. [28]. Copyright 2021, American Chemical Society.)*

cathode scraps collected at the electrode manufacturing step without jeopardizing their physical characteristics, crystalline structure, and electrochemical performance. The recovered aluminum foils are clean without any sign of corrosion, and the polymeric binder can be recovered via a non-solvent-induced phase separation. This green solvent-based separation for cathode recovery is anticipated to attract significant interest from the lithium-ion battery manufacturing and recycling communities.

2. Combating interfacial delamination

On the other hand, delamination between the metallic current collector and composite film is an important cause of capacity decay and battery failure. When high-capacity electrode materials, such as germanium, silicon, or tin, are used as active materials to replace graphite, delamination becomes the key issue. Graphite material is known to have close to a 13% volume change during the lithation and delithation processes; if anode materials based on silicon, germanium, and tin are used, their volume change can be as high as 420%, 370%, and 260%, respectively, during cycling. Repeated volume changes significantly weaken the interfacial adhesion between the electrode and the current collector. Moreover, both copper and aluminum are susceptible to environmental degradation—aluminum to pitting corrosion and copper to environmentally assisted cracking—that can further weaken interfacial adhesion. When delamination occurs, the electrode loses contact with the current collector, resulting in higher interfacial resistance and even deactivating the electrode from redox reaction.

Therefore, various strategies have been developed in battery manufacturing to increase the adhesion between the current collector and the composite film, such as using a highly adhesive binder [29,30], carbon coating or surface treatment to remove the native oxide layer and modify surface hydrophobicity [31], building an array of micro-/nano-scale structures to increase the surface area of the current collector [32], and roughening of the current collector surface by electrolytic deposition [33] or laser ablation [34].

At the nanoscale, attractive forces are dominated by electrostatic and intermolecular forces such as Van der Waals forces. Van der Waals forces exist for all surfaces and are insensitive to surface chemistry. Their magnitude can be increased by enlarging the surface area. Zheng et al. developed nanorod structures on the copper surface, which significantly enhanced Van der Waals forces between the copper foil and the electrode film [32]. These nanorods, as shown in Fig. 7.3, are similar in physical structure to the spatulas in a gecko's foot and improve adhesion by increasing the surface contact area with the electrode film. The method involves simple chemical treatment of the copper to create structured nanorods of copper hydroxide that grow from the copper surface.

In a similar study, Su et al. developed nanoscale surface morphologies, as shown in Fig. 7.4, to improve the interfacial adhesion between solid electrolyte and vanadium oxide electrode film [35]. Patterned vanadium oxide films are produced via a standard lithographic process. Oxygen plasma and annealing treatments are then performed to

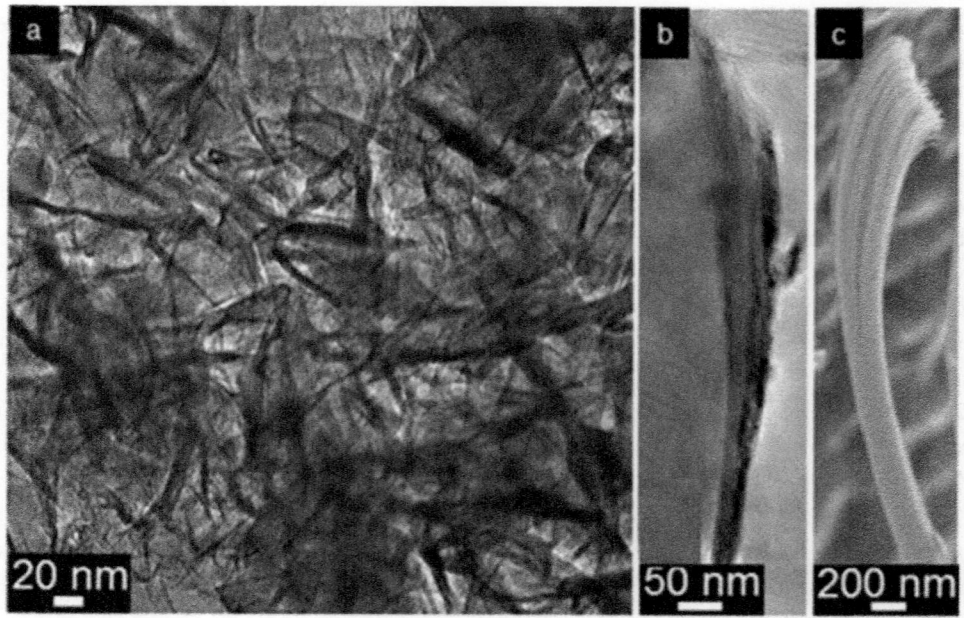

Figure 7.3 (A) Microscopic image of copper hydroxide nanofiber strands scratched off from the treated copper surface. (B) Microscopic image of a single copper hydroxide strand. (C) Microscopic image of a spatula stalk on a gecko's foot. *(Adapted with permission from Ref. [32]. Copyright 2014, Wiley-VCH.)*

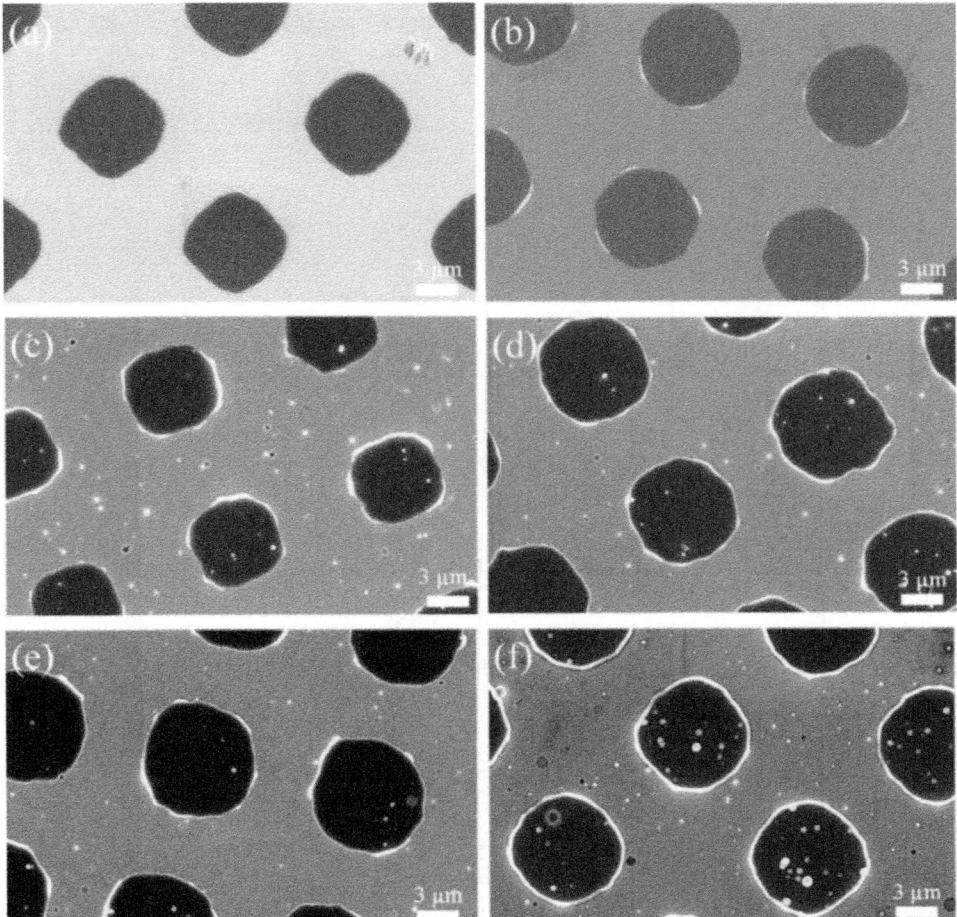

Figure 7.4 Patterned vanadium oxide films with island heights of (A) 28 nm, (B) 50 nm, (C) 100 nm, (D) 200 nm, (E) 400 nm, (F) 800 nm. *(Adapted with permission from Ref. [35]. Copyright 2015, Elsevier B.V.)*

increase surface roughness. The results demonstrate that surface modification, including roughness associated with the islands and island edges in the patterned films, has greatly enhanced the adhesion energy between the solid electrolyte and the electrode film, up to eight times higher than the increase in surface area.

Yang et al. electrochemically deposited tin–cobalt alloy films on nickel foam in an aqueous solution to achieve better battery performance [36]. Used as a current collector, nickel foam with a three-dimensional porous structure, as shown in Fig. 7.5, not only increases the electrical conductivity but also avoids the macrostructural deformation of tin-based electrodes and buffers the large volume changes that occur during lithium insertion and extraction.

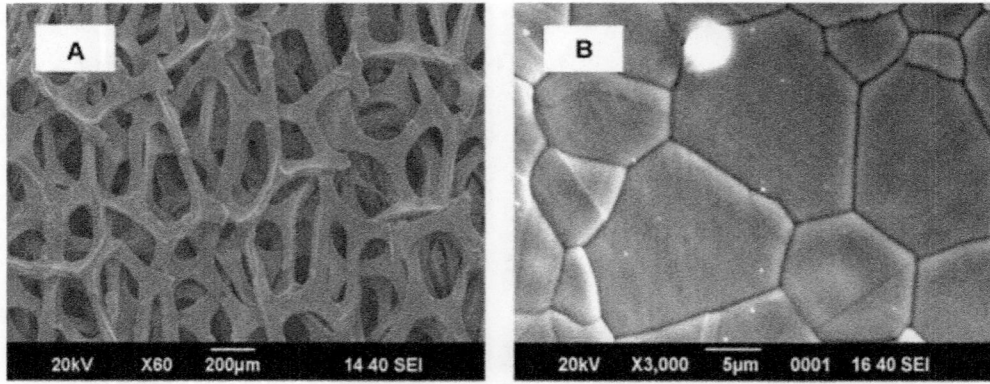

Figure 7.5 (A) Microscopic images of the nickel foam and (B) the nickel foam surface. The foamed nickel has a three-dimensional grid structure and with high porosity and a specific surface area with considerable, uniform strength. *(Adapted with permission from Ref. [36]. Copyright 2011, Elsevier B.V.)*

The recent trend to realize thick and three-dimensional electrodes for high-energy and high-power applications imply an urgent need for improved interfacial adhesion. To alleviate the delamination problem for thick electrodes, Hu et al. adapted a three-dimensional porous textile conductor, as shown in Fig. 7.6, as a replacement for metal current collectors [37]. The electrode materials were loaded into the three-dimensional pores of conductive textiles through a simple solution-based process. Compared with flat metal current collectors, this porous textile conductor greatly facilitates high active material mass loading on the battery electrode.

These studies show that the tuning of electrode topography plays a critical role in improving interfacial adhesion, which could indeed lead to enhanced battery lifetime. However, these processes can incur high production costs and cannot be easily implemented into roll-to-roll mass manufacturing. Most importantly, those technologies significantly increase the interfacial adhesion between the electrode and current collector and thus make the battery recycling process much more challenging.

3. Bio-inspired directional adhesives

The challenges of battery technologies have traditionally been understood in terms of electrochemistry. The mechanical aspect has been regarded as a secondary consideration within the design spectrum. However, as conventional methods are not satisfactory to assure the function of the interface between the metallic current collector and the composite electrode film—i.e., to display superior adhesion to effectively combat mechanical delamination during its lifetime while being easily separated at the end of its life during recycling—expert knowledge of contact mechanics, especially the adhesion and

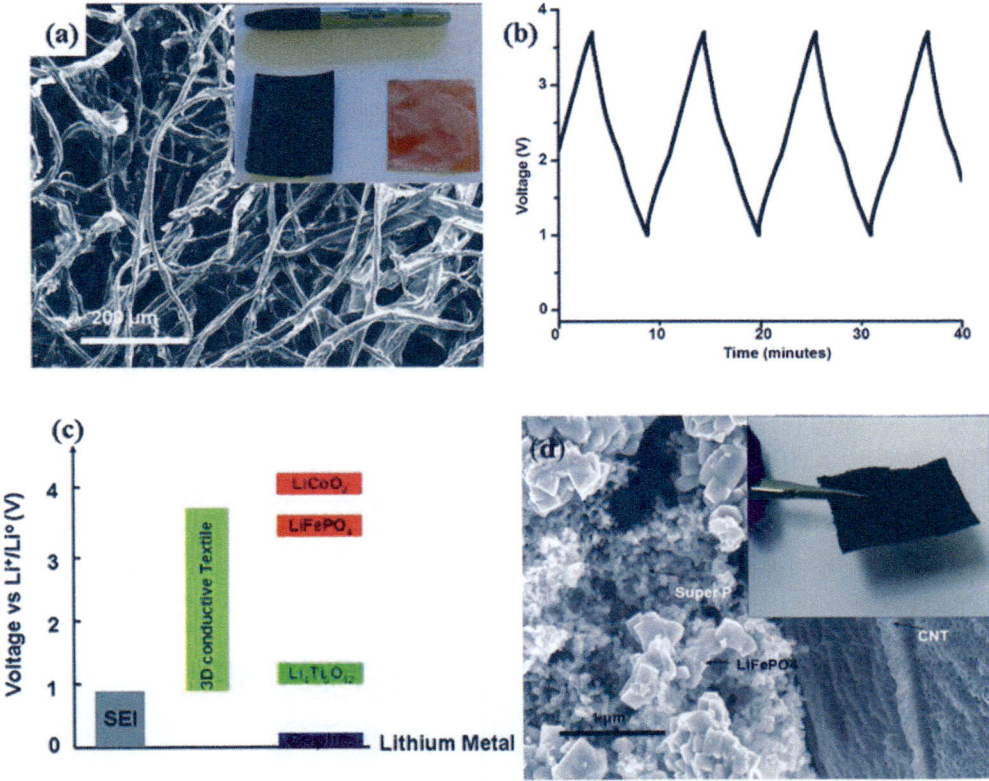

Figure 7.6 (A) Microscopic image of conductive textile after carbon nanotube coating with porous structure. (B) Charge–discharge profile of conductive textile shows good electrochemical stability in the window of 1–3.8 V. (C) Voltage stability window of 3D conductive textile current collector with widely used electrodes. (D) Microscopic image of interfaces among different components. *(Adapted with permission from Ref. [37]. Copyright 2011, Wiley-VCH.)*

delamination mechanisms of interfaces, is becoming essential for the development of next-generation easy-to-recycle batteries. Innovative research and revolutionary discoveries of new materials, structures, and interfacial designs are urgently needed.

In fact, by learning from nature, micro-/nanoscale near-surface architecture on the interface between the current collector and the composite electrode film can be designed for controllable and directional adhesion.

In nature, gecko foot-hairs demonstrate controllable and directional adhesion on a wide range of smooth and rough surfaces, mainly due to the angled and hierarchical micro-/nanoscale fibrillar structures on their feet [38–59]. The adhesion and shear strength of the gecko's angled fibrillar structures depends on mechanical deformations induced by vertical and lateral loading of its feet that actively control the contact area between the structures and the substrate. In past decades, various synthetic adhesives have been designed to mimic the directionality and controllability of these biological foot-hairs. Those adhesives are often referred to as bio-inspired structural adhesives.

Autumn et al. have shown that gecko foot-hairs possess a friction ratio of 5 to 1 for the "with" to "against" hair tilt directions [57]. The results of this study are being applied to the design of climbing robots, as shown in Fig. 7.7. The robot uses gecko-inspired structures that, while crude compared with those of a gecko, exhibit similar anisotropic frictional adhesion. Compared with isotropic adhesive materials, smoother and faster climbing is observed when utilizing the directional, microstructured, frictional adhesives rather than flat adhesive pads.

Murphy et al. developed adhesive structures that combine the high interfacial strength of mushroom tipped fibers with the directionality of fiber structures possessing both angled stalks and tip endings, as shown in Fig. 7.8 [59]. Angled base fiber arrays are fabricated from polyurethane. Directional shear and adhesion strength were observed with a with-to-against shear ratio of around 5 to 1 and an amazing adhesion ratio of 35 to 1. As a demonstration of the macroscale adhesion of the directional microfiber array, 1 centimeter square of a sample with a 14 degree tip angle was attached to a glass slide that supported a hanging weight of 1000 g in pure shear in the gripping direction, which equals interfacial shear strength of 100 kPa. When reversed to the releasing direction, the same sample

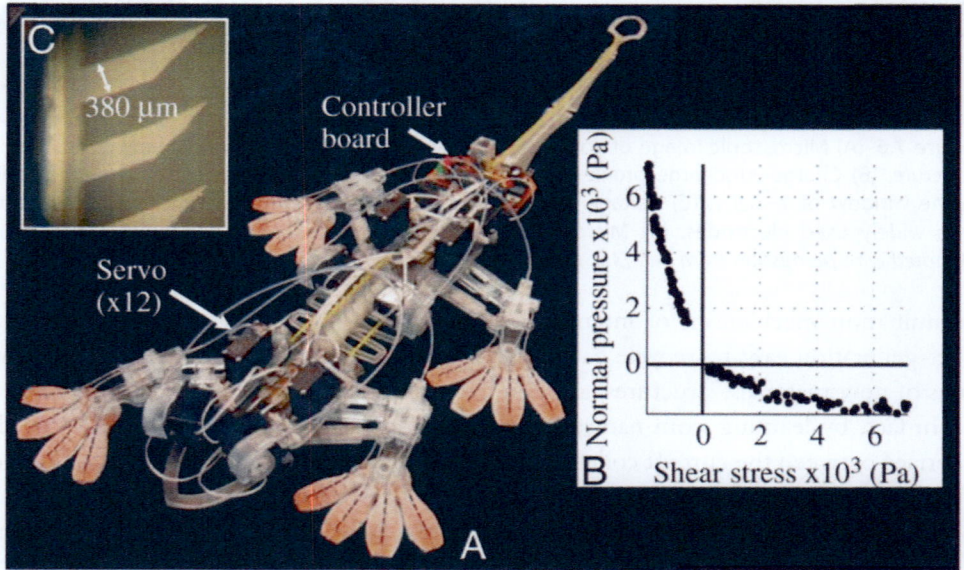

Figure 7.7 (A) Experimental climbing machine for testing anisotropic adhesive structures and force control strategies. (B) Experimental measurements of normal versus shear forces in an anisotropic frictional adhesive inspired by gecko setae. (C) Magnified view showing angled contact surface of frictional adhesive microarray. *(Adapted with permission from Ref. [57]. Copyright 2006, The Company of Biologists.)*

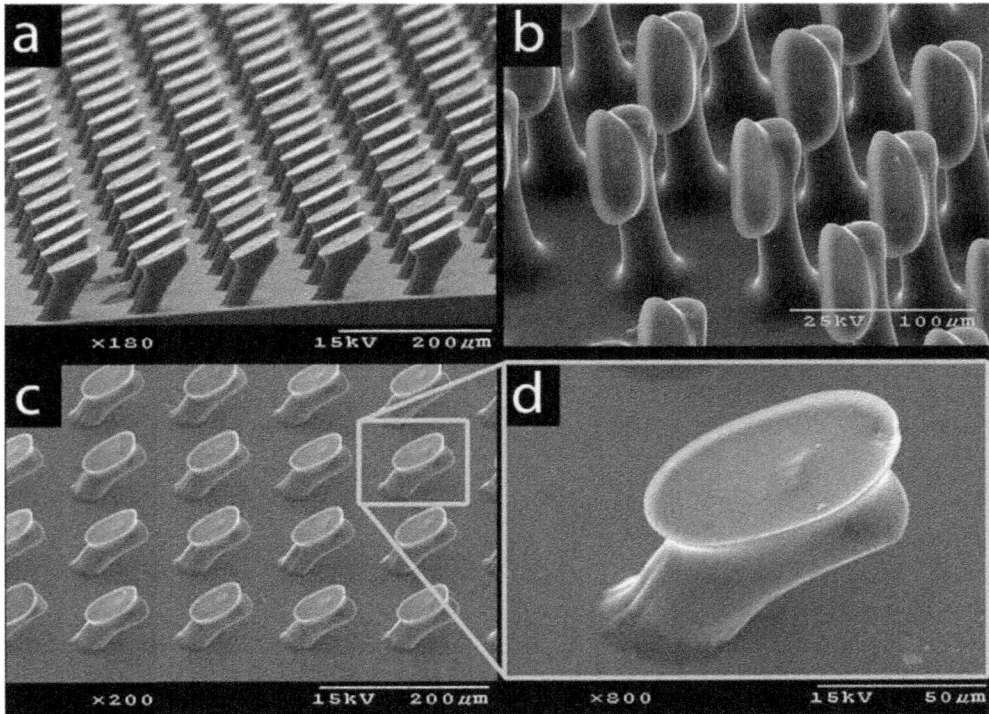

Figure 7.8 Microscopic image of arrays of 35-μm diameter angled polyurethane microfibers with angled mushroom tips. Tip orientation can be controlled to form tips with varying angles: (A) 34 degree; (B) 90 degree; and (C) 23 degree. Details of the tip can be seen in (D). These adhesives exhibit directional characteristics of gripping when loaded in one direction and self-releasing behavior when loaded in the opposite shear direction. *(Adapted with permission from Ref. [59]. Copyright 2009, Wiley-VCH.)*

supported only 200 g, as shown in Fig. 7.9. The fabrication methods described in this work can easily be extended to smaller scales and stiffer materials to more closely mimic the gecko's adhesive structures. The use of stiffer materials may improve durability and contamination resistance, which are common problems for softer materials. The high-magnitude anisotropic adhesion of these materials can be used to solve the interface issues associated with lithium-ion batteries.

4. Interface for easy-to-recycle batteries

To make easy-to-recycle batteries, directional adhesion is desired where interfacial separation along different directions requires different fracture energy. In a proof-of-concept investigation, Jin et al. tested the simplest triangular pattern [60], which is reviewed in detail in this section.

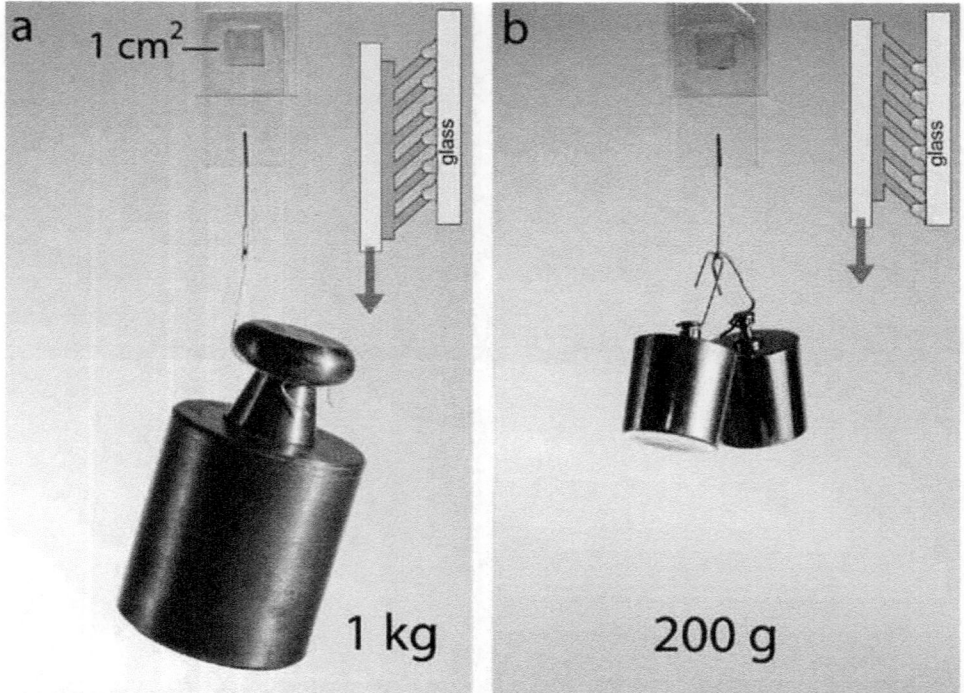

Figure 7.9 One centimeter square of directional polyurethane microfiber array with 14 degree angled tips adhering to smooth glass can support (A) 1000 g in shear in the gripping direction and (B) only 200 g in the releasing direction. The high magnitude anisotropic adhesion of these materials may enable efficient gripping and releasing of structures, such as in the area of climbing robots. *(Adapted with permission from Ref. [59]. Copyright 2009, Wiley-VCH.)*

In this study, two types of patterned surfaces, called Model A and Model B, were fabricated on the aluminum plates, both of which contained asymmetric right triangles separated by flat regions but with different geometrical parameters, as shown in Fig. 7.10.

A total of *20* aluminum plates with patterned surfaces were fabricated, with 10 for each pattern. Ten aluminum plates with flat surfaces were also fabricated to serve as a control. All the plates have a global dimension of 20 × 10 × 0.5 mm. The electrodes were coated onto the aluminum plates with patterned or flat surfaces. The composite cathode was composed of $LiNi_{0.5}Mn_{0.3}Co_{0.2}O_2$ (NMC532) particles, PVDF binders, and a porous carbon black matrix [61,62]. None of the electrodes were calendered, and the thickness of the composite film was approximately 50 μm.

To investigate the adhesion between the aluminum substrates and composite films, the composite films were peeled off using a universal testing machine according to the standard procedure ASTM D6862 [63]. The angle between the force direction and the interface was kept at 90 degree during the peeling tests. The initial part of the

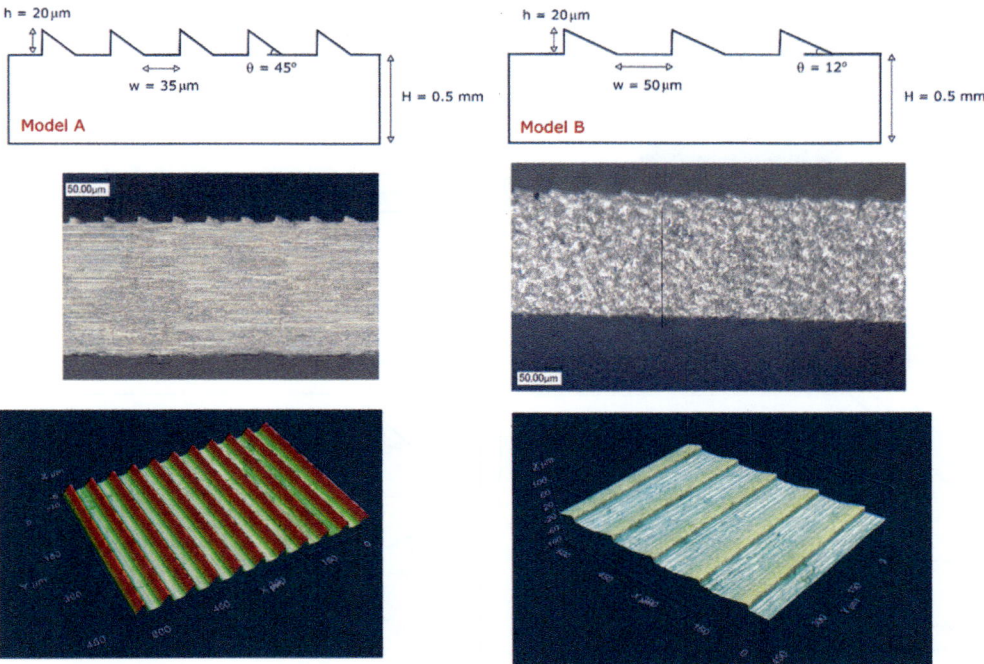

Figure 7.10 Two types of patterned surfaces were fabricated on the aluminum plates, both of which contain asymmetric right triangles separated by flat regions but with different geometrical parameters. Left panel: Model A; Right panel: Model B. For each model, a schematic drawing (not to scale) and a confocal microscope image are shown, respectively. *(Adapted with permission from Ref. [60]. Copyright 2020, Elsevier B.V.)*

composite film was manually peeled off and attached to an adhesive tape, and the remaining composite film was then peeled off at a constant rate of 0.7 mm/min. For each patterned interface, the composite film was peeled from the aluminum substrate in two different directions, as shown in Fig. 7.11. The force F needed to peel the composite films from the aluminum substrates was recorded. The peel energy can be deduced from the peel force F as F/M_t, with M_t being the width of the tape [64].

Since the triangular pattern is asymmetric, as the crack propagates from left to right, i.e., following peeling direction 1, it travels upward along the vertical side of the triangle and then downward along the hypotenuse. As the crack propagates from right to left, i.e., following peeling direction 2, it travels upward along the hypotenuse and then downward along the vertical side. In each direction, there is a flat region before the separation of the patterned region. The adhesion strength is expected to be different in different directions.

Fig. 7.12 shows the results of the peel energy for Models A and B, each with two peeling directions. Ten different measurements were conducted for each case. Error bars represent standard deviations based on 10 measurements. The results obtained

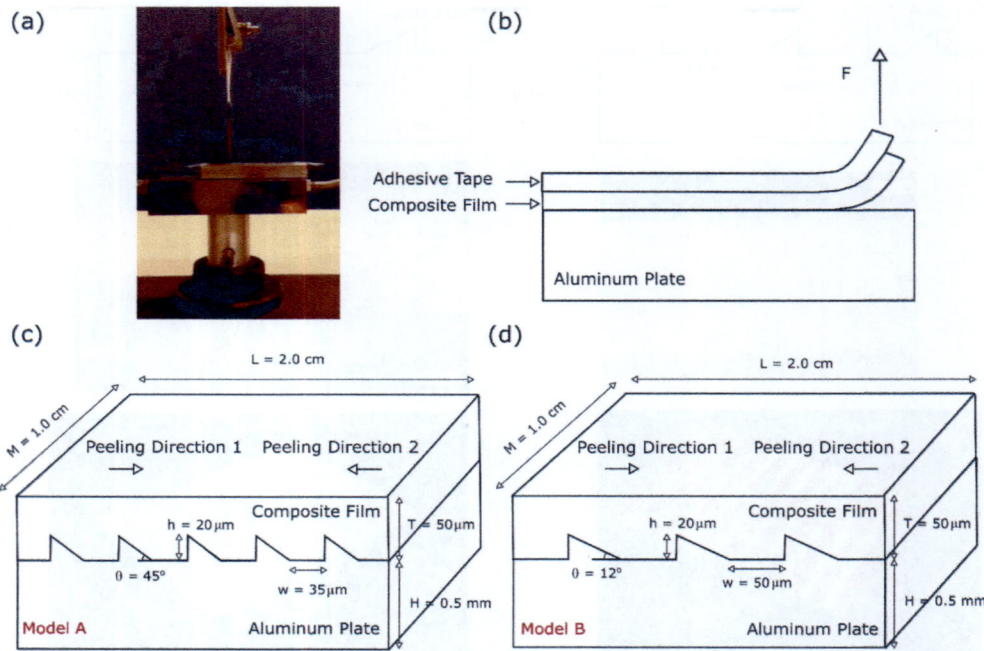

Figure 7.11 (A) To investigate the adhesion between the patterned current collectors and the composite films, the composite films were peeled off using a universal testing machine. An angle of 90 degree was maintained between the force direction and the interface during the peeling tests. (B) Schematics of the 90 degree peeling tests. For each patterned interface, (C) Model A and (D) Model B, the composite film was peeled off in two different directions. *(Adapted with permission from Ref. [60]. Copyright 2020, Elsevier B.V.)*

Figure 7.12 (A) The results of the peel energy are presented for Models A and B, each with two peeling directions. Error bars represent the standard deviations from 10 different tests. *MAD1*, Model A Direction 1; *MAD2*, Model A Direction 2; *MBD1*, Model B Direction 1; *MBD2*, Model B Direction 2. (B) The data were obtained from tests in which the composite film was completely peeled from the aluminum substrate. (C) If the composite film was not completely removed from the aluminum substrate, the data were considered not useful for analysis. *(Adapted with permission from Ref. [60]. Copyright 2020, Elsevier B.V.)*

from the flat control are presented as a reference. These data were obtained from tests in which the composite film was completely peeled from the aluminum substrate. If the composite film was not completely removed from the aluminum substrate, the data were considered not useful for analysis.

The peel energy for the flat control range from *22.8* N/m to *32.3* N/m. For Model A, the peel energy along peeling direction *1* varies in the range of *103.0* N/m to *117.0* N/m, while the peel energy along peeling direction *2* varies in the range of *116.5* N/m to *136.0* N/m. For Model B, the peel energy along peeling direction *1* varies in the range of *63.0* N/m to *77.0* N/m, whereas the peel energy along peeling direction *2* varies in the range of *80.5* N/m to *98.5* N/m. Enhanced adhesion of films deposited on patterned aluminum is observed compared with the flat control. Most importantly, directional adhesion is observed in both Model A and Model B. For both models, the peel energy along peeling direction *2* is larger than that along peeling direction *1*. The ratio of the peel energy along peeling direction *2* to that along the peeling direction z is approximately *1.15* for Model A and *1.28* for Model B, respectively, indicating that directional adhesion is slightly stronger in Model B, where $\theta = 12$ degree, than in Model A, where $\theta = 45$ degree.

5. Underlying mechanisms

The experimental results raised many interesting questions. Finite element analyses were thus performed to explain the adhesion measurement results and elucidate the underlying mechanisms. The commercial package Abaqus FEA was utilized for the finite element analysis [65]. Two patterned interface models, A and B, were simulated based on the actual geometries and materials used in the experiments.

For the two numerical models, the same loading condition is applied, where the first few nodes on the top edge are subjected to a fixed displacement boundary condition, and the bottom edge is fixed in a frictionless manner. Two opposite crack separation directions are examined numerically, as shown in Fig. 7.13.

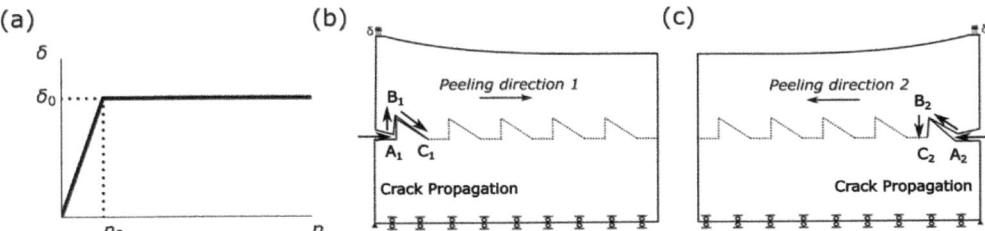

Figure 7.13 For the two numerical models, the same loading condition is applied, where the first few nodes on the top edge are subjected to a fixed displacement boundary condition, and the bottom edge is fixed in a frictionless manner. (A): The displacement δ increases linearly from 0 to δ_0 as numerical step n increases from 0 to n_0, and afterward, δ remains constant. (B) and (C): Two opposite crack separation directions are examined numerically. *(Adapted with permission from Ref. [60]. Copyright 2020, Elsevier B.V.)*

The energy release rates for patterned and flat interfaces are denoted as $G_{pattern}$ and G_{flat}, respectively. The energy release rate ratio R is defined as $G_{pattern}/G_{flat}$. Fig. 7.14 shows how R changes with the normalized apparent crack tip location x/λ_x for the two numerical models and each of the two respective peeling directions. The parameter x represents the projected length of the interface onto the horizontal direction, and λ_x represents the apparent or horizontally projected length of the interface in each period. The results demonstrate that both Model A and Model B show directional adhesion, and compared with Model B, adhesion is stronger but directional adhesion is weaker in Model A, which is consistent with the experimental results. The numerical simulation provides important insights into the mechanism of adhesion enhancement and directional adhesion achieved by the patterned interface.

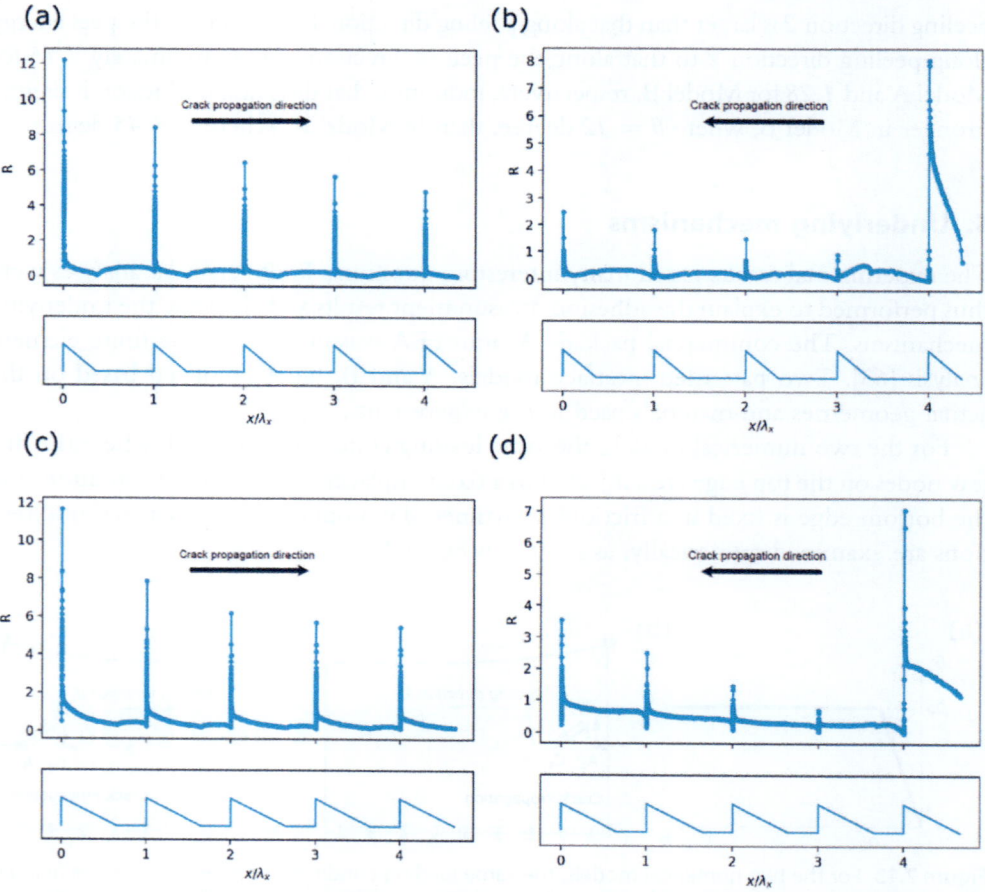

Figure 7.14 The energy release rate ratio R, defined as $G_{pattern}/G_{flat}$, plotted against the normalized apparent crack tip location x/λ_x in the patterned region. (A) Model A with peeling direction 1; (B) Model A with peeling direction 2; (C) Model B with peeling direction 1; and (D) Model B with peeling direction 2. *(Adapted with permission from Ref. [60]. Copyright 2020, Elsevier B.V.)*

6. Concluding remarks and future perspectives

Inspired by the controllable attachment and detachment ability of gecko foot-hairs, micro-/nanoscale near-surface architecture can be designed on the interface between the metallic current collector and the composite film of the electrode so that it displays controllable and directional adhesion—i.e., enhanced adhesion is obtained during its lifetime to cope with the substantial changes in volume of the composite film upon intercalation and deintercalation, whereas during recycling, the composite film can be easily peeled from the current collector in a certain direction.

In a proof-of-concept study, the simplest triangular pattern was tested, which successfully demonstrated the viability of this technology. Finite element simulation helps us gain a better understanding of the interfacial adhesion and delamination mechanisms, providing a scientific footing for realizing next-generation easy-to-recycle electronics.

In the future, it would be of great interest to explore more sophisticated patterns and the correlation between surface patterning and battery electrochemical performance. This technology will potentially make a significant contribution to increasing the recyclability of lithium-ion batteries. The application of this technology can also be extended to other electronic products to avoid an ever-growing volume of electronic waste.

References

[1] K. Ozawa, Lithium-ion rechargeable batteries with $LiCoO_2$ and carbon electrodes: the $LiCoO_2$/C system, Solid State Ionics 69 (1994) 212.
[2] Y.J. Nishi, Lithium ion secondary batteries; past 10 years and the future, J. Power Sources 100 (2001) 101.
[3] P.G. Bruce, B. Scrosati, J.M. Tarascon, Nanomaterials for rechargeable lithium batteries, Angew. Chem. Int. Ed. 47 (2008) 2930.
[4] M.S. Whittingham, Lithium batteries and cathode materials, Chem. Rev. 104 (2004) 4271.
[5] D.L. Wood, J. Li, C. Daniel, Prospects for reducing the processing cost of lithium ion batteries, J. Power Sources 275 (2015) 234.
[6] X. Su, Q. Wu, X. Zhan, J. Wu, S. Wei, Z. Guo, Advanced titania nanostructures and composites for lithium ion battery, J. Mater. Sci. 47 (2012) 2519.
[7] G.P. Hammond, T. Hazeldine, Indicative energy technology assessment of advanced rechargeable batteries, Appl. Energy 138 (2015) 559.
[8] B. Scrosati, J. Garche, Lithium batteries: status, prospects and future, J. Power Sources 195 (2010) 2419.
[9] J. Li, C. Daniel, D.L. Wood, Materials processing for lithium-ion batteries, J. Power Sources 196 (2011) 2452.
[10] M. Sun, X. Yang, D. Huisingh, R. Wang, Y. Wang, Consumer behavior and perspectives concerning spent household battery collection and recycling in China: a case study, J. Clean. Prod. 107 (2015) 775.
[11] M.A. Delucchi, C. Yang, A.F. Burke, J.M. Ogden, K. Kurani, J. Kessler, D. Sperling, An assessment of electric vehicles: technology, infrastructure requirements, greenhouse-gas emissions, petroleum use, material use, lifetime cost, consumer acceptance and policy initiatives, Philos. Trans. Ser. A 372 (2014) 20120325.
[12] J.B. Dunn, L. Gaines, J. Sullivan, M.Q. Wang, Impact of recycling on cradle-to-gate energy consumption and greenhouse gas emissions of automotive lithium-ion batteries, Environ. Sci. Technol. 46 (2012) 12704.

[13] S. Ziemann, M. Weil, L. Schebek, Tracing the fate of lithium: the development of a material flow model, Resour. Conserv. Recycl. 63 (2012) 26.

[14] D. Kushnir, B.A. Sanden, The time dimension and lithium resource constraints for electric vehicles, Resour. Policy 37 (2012) 93.

[15] P.W. Gruber, P.A. Medina, G.A. Keoleian, S.E. Kesler, M.P. Everson, T.J. Wallington, Global lithium availability, J. Ind. Ecol. 15 (2011) 760.

[16] J. Yan, R.F. Boehm, H.-X. Yang, Recycling of lithium-ion batteries, in: Handbook of Clean Energy Systems, John Wiley & Sons, Chichester, NY, USA, 2015.

[17] C. Hoyer, K. Kieckhafer, T.S. Spengler, Technology and capacity planning for the recycling of lithium-ion electric vehicle batteries in Germany, J. Bus. Econ. 85 (2014) 505.

[18] X. Zeng, J. Li, Spent rechargeable lithium batteries in e-waste: composition and its implications, Front. Environ. Sci. Eng. 8 (2014) 792.

[19] S. Ziemann, A. Grunwald, L. Schebek, D.B. Müller, M. Weil, The future of mobility and its critical raw materials, Rev. Métall. 110 (2013) 47.

[20] K. Richa, C.W. Babbitt, G. Gaustad, X. Wang, A future perspective on lithium-ion battery waste flows from electric vehicles, Resourc. Conserv. Recycl. 83 (2014) 63.

[21] L. Li, J.B. Dunn, X.X. Zhang, L. Gaines, R.J. Chen, F. Wu, K. Amine, Recovery of metals from spent lithium-ion batteries with organic acids as leaching reagents and environmental assessment, J. Power Sources 233 (2013) 180.

[22] C. Herrmann, A. Raatz, S. Andrew, J. Schmitt, Scenario-based development of disassembly systems for automotive lithium ion battery systems, Adv. Mater. Res. 907 (2014) 391.

[23] T. Zhang, Y. He, L. Ge, R. Fu, X. Zhang, Y. Huang, Characteristics of wet and dry crushing methods in the recycling process of spent lithium-ion batteries, J. Power Sources 240 (2013) 766.

[24] J. Li, S. Zhong, D. Xiong, H. Chen, Synthesis and electrochemical performances of $LiCoO_2$ recycled from the incisors bound of Li-ion batteries, Rare Met. 28 (2009) 328.

[25] C. Hanisch, W. Haselrieder, A. Kwade, Method for Reclaiming Active Material from a Galvanic Cell, and an Active Material Separation Installation, Particularly an Active Metal Separation Installation, European Patent, EP2742559B1, 2013.

[26] C. Hanisch, W. Haselrieder, A. Kwade, Method for Reclaiming Active Material from a Galvanic Cell, and an Active Material Separation Installation, Particularly an Active Metal Separation Installation, Patent, WO/2013/023640, 2013.

[27] C. Hanisch, T. Loellhoeffel, J. Diekmann, K.J. Markley, W. Haselrieder, A. Kwade, Recycling of lithium-ion batteries: a novel method to separate coating and foil of electrodes, J. Clean. Prod. 108 (2015) 301.

[28] Y. Bai, R. Essehli, C.J. Jafta, K.M. Livingston, I. Belharouak, Recovery of cathode materials and aluminum foil using a green solvent, ACS Sustain. Chem. Eng. 9 (2021) 6048.

[29] M.H.T. Nguyen, E.S. Oh, Application of a new acrylonitrile/butylacrylate water-based binder for negative electrodes of lithium-ion batteries, Electrochem. Commun. 35 (2013) 45.

[30] D. Chen, R. Yi, S. Chen, T. Xu, M.L. Gordin, D.H. Wang, Facile synthesis of graphene-silicon nanocomposites with an advanced binder for high-performance lithium-ion battery anodes, Solid State Ionics 254 (2014) 65.

[31] H.C. Wu, E. Lee, N.L. Wu, T.R. Jow, Effects of current collectors on power performance of $Li_4Ti_5O_{12}$ anode for Li-ion battery, J. Power Sources 197 (2012) 301.

[32] Z. Zheng, Z. Wang, X. Song, S. Xun, V. Battaglia, G. Liu, Biomimetic nanostructuring of copper thin films enhances adhesion to the negative electrode laminate in lithium-ion batteries, ChemSusChem 7 (2014) 2853.

[33] D. Reyter, S. Rousselot, D. Mazouzi, M. Gauthier, P. Moreau, B. Lestriez, D. Guyomard, L. Roue, An electrochemically roughened Cu current collector for Si-based electrode in Li-ion batteries, J. Power Sources 239 (2013) 308.

[34] N. Zhang, Y. Zheng, A. Trifonova, W. Pfleging, Laser structured Cu foil for high-performance lithium-ion battery anodes, J. Appl. Electrochem. 47 (2017) 829.

[35] X. Su, T. Zhang, X. Liang, H. Gao, B.W. Sheldon, Employing nanoscale surface morphologies to improve interfacial adhesion between solid electrolytes and Li ion battery cathodes, Acta Mater. 98 (2015) 175.
[36] C. Yang, D. Zhang, Y. Zhao, Y. Lu, L. Wang, J.B. Goodenough, Nickel foam supported Sn-Co alloy film as anode for lithium ion batteries, J. Power Sources 196 (2011) 10673.
[37] L.B. Hu, F. La Mantia, H. Wu, X. Xie, J. McDonough, M. Pasta, Y. Cui, Lithium-ion textile batteries with large areal mass loading, Adv. Energy Mater. 1 (2011) 1012.
[38] W.R. Hansen, K. Autumn, Evidence for self-cleaning in gecko setae, Proc. Natl. Acad. Sci. U.S.A. 102 (2005) 385.
[39] A. Bietsch, B. Michel, Conformal contact and pattern stability of stamps used for soft lithography, J. Appl. Phys. 88 (2000) 4310.
[40] B. Persson, S. Gorb, The effect of surface roughness on the adhesion of elastic plates with application to biological systems, J. Chem. Phys. 119 (2003) 11437.
[41] S. Vajpayee, K. Khare, S. Yang, C.-Y. Hui, A. Jagota, Adhesion selectivity using rippled surfaces, Adv. Funct. Mater. 21 (2011) 547.
[42] H. Shahsavan, B. Zhao, Conformal adhesion enhancement on biomimetic microstructured surfaces, Langmuir 27 (2011) 7732.
[43] A.K. Singh, Y. Bai, A. Jagota, C.-Y. Hui, Adhesion of microchannel based complementary surfaces, Langmuir 28 (2012) 4213.
[44] F. Cordisco, P.D. Zavattieri, L.G. Hector, A.F. Bower, Toughness of a patterned interface between two elastically dissimilar solids, Eng. Fract. Mech. 96 (2012) 192.
[45] C. Jin, A. Jagota, C.-Y. Hui, Structure and energetics of dislocations at micro-structured complementary interface govern adhesion, Adv. Funct. Mater. 23 (2013) 3452.
[46] C. Jin, K. Khare, S. Vajpayee, S. Yang, A. Jagota, C.-Y. Hui, Adhesive contact between a rippled elastic surface and a rigid spherical indenter: from partial to full contact, Soft Matter 7 (2011) 10728.
[47] B. Bhushan, Biomimetics: lessons from nature—an overview, Phil. Trans. Roy. Soc. A 367 (2009) 1445.
[48] A. Jagota, C.-Y. Hui, N. Glassmaker, T. Tang, Mechanics of bioinspired and biomimetic fibrillar interfaces, MRS Bull. 32 (2007) 492.
[49] C.-Y. Hui, A. Jagota, L. Shen, A. Rajan, N. Glassmaker, T. Tang, Design of bio-inspired fibrillar interfaces for contact and adhesion - theory and experiments, J. Adhes. Sci. Technol. 21 (2007) 1259.
[50] L.F. Boesel, C. Greiner, E. Arzt, A. del Campo, Gecko-inspired surfaces: a path to strong and reversible dry adhesives, Adv. Mater. 22 (2010) 2125.
[51] A.M. Lees, J. Hardie, The organs of adhesion in the aphid Megoura viciae, J. Exp. Biol. 136 (1988) 209.
[52] S.N. Gorb, The design of the fly adhesive pad: distal tenent setae are adapted to the delivery of an adhesive secretion, Proc. Roy. Soc. Lond. B 265 (1998) 747.
[53] T. Eisner, D.J. Aneshansley, Defense by foot adhesion in a beetle (Hemisphaerota cyanea), Proc. Nat. Acad. USA 97 (2000) 6568.
[54] W. Federle, M. Riehle, A.S.G. Curtis, R.J. Full, An integrative study of insect adhesion: mechanics and wet adhesion of pretarsal pads in ants, Integr. Comp. Biol. 42 (2002) 1100.
[55] M.K. Kwak, H.E. Jeong, W.G. Bae, H.S. Jung, K.Y. Suh, Anisotropic adhesion properties of triangular-tip-shaped micropillars, Small 7 (2011) 2296.
[56] K. Autumn, M. Sitti, Y.A. Liang, A.M. Peattie, S. Hansen, S. Sponberg, T.W. Kenny, R. Fearing, R.J. Israelachvili, R.J. Full, Evidence for van der Waals adhesion in gecko setae, Proc. Natl. Acad. Sci. U.S.A. 99 (2002) 12252.
[57] K. Autumn, A. Dittmore, D. Santos, M. Spenko, M. Cutkosky, Frictional adhesion: a new angle on gecko attachment, J. Exp. Biol. 209 (2006) 3569.
[58] J. Lee, R.S. Fearing, Contact self-cleaning of synthetic gecko adhesive from polymer microfibers, Langmuir 24 (2008) 10587.
[59] M.P. Murphy, B. Aksak, M. Sitti, Gecko-Inspired directional and controllable adhesion, Small 5 (2005) 170.

[60] C. Jin, Z. Yang, J. Li, Y. Zheng, W. Pfleging, T. Tang, Bio-inspired interfaces for easy-to-recycle lithium-ion batteries, Extreme Mech. Lett. 34 (2020) 100594.

[61] A. Davoodabadi, J. Li, Y. Liang, R. Wang, H. Zhou, D. Wood, T. Singler, C. Jin, Characterization of surface free energy of composite electrodes for lithium-ion batteries, J. Electrochem. Soc. 165 (2018) A2493.

[62] L.S. de Vasconcelos, R. Xu, J. Li, K. Zhao, Grid indentation analysis of mechanical properties of composite electrodes in Li-ion batteries, Extreme Mech. Lett. 9 (2016) 495.

[63] ASTM Standard D 6862, Standard Test Method for 90 Degree Peel Resistance of Adhesives, ASTM International, West Conshohocken, PA, 2004.

[64] M. Lamblet, E. Verneuil, T. Vilmin, A. Buguin, P. Silberzan, L. Leger, Adhesion enhancement through micropatterning at polydimethylsiloxane − acrylic adhesive interfaces, Langmuir 23 (2007) 6966.

[65] Abaqus FEA (Formerly ABAQUS) Is a Commercial Software Suite for Finite Element Analysis and Computer-Aided Engineering. For More Details See Abaqus 6.14 Analysis User's Guide, Dassault Systèmes Simulia Corp., Providence, RI, USA, 2014.

CHAPTER 8

Materials recycling using pH/thermal-responsive materials

Muhammad Fahad Arain, Arsalan Ahmed and Muhammad Qamar Khan

Department of Textile and Clothing, Faculty of Engineering and Technology, National Textile University Karachi Campus, Karachi, Sindh, Pakistan

1. Introduction

Polymers have been used by humankind for ages in almost every field of life. In the 19th and 20th centuries, research was carried out to study the behavior of various polymeric materials. Advancements led to the investigation of various polymer classes and their applications in daily life as inspired by nature. Devoted research by various scientists and researchers to mimic nature resulted in the development of stimuli-responsive polymeric materials. Materials in this class respond to external environmental factors by changing their physical or chemical properties. The external factors include temperature, pH, mechanical force, and the presence of specific molecules. Stimuli-responsive polymeric materials or smart polymers have been used in several application areas, such as drug delivery [1], ultrafiltration/membranes [2], phase separation [3], food packaging [4], and smart water treatment [5], and are not limited to the research/laboratory level but are also being used in the practical and commercial sectors.

Despite the numerous advantages of nanomaterials and a wide variety of efficient applications developed by researchers, the practical and commercial utilization of nanomaterials is limited. Limitations include the safe use, disposal, and recycling of nanomaterials [6]. Current investigations have shown serious concerns about the toxicity of nanomaterials, highlighting the need for nanomaterial recycling procedures and standards. New technology parameters have been developed to analyze the environmental behavior of novel materials at the nano level. The presence of nanomaterials in wastewater is classified in the following forms: (1) nanotubes, (2) nanomaterials in liquid, (3) solids containing nanomaterials, and (4) integrated solids with nanomaterials [7,8]. Various traditional techniques for separation or recycling require high energy levels for operation, such as centrifugal or solvent separation methods. Therefore, alternative methods have been developed for efficient and cost-effective separation, such as magnetic fields, nanostructured colloidal solvents, and pH/thermal stimuli-responsive nanomaterials. Although several techniques have been developed to recycle nanomaterials, further investigations on the behavior of recycled materials and their intrinsic properties are needed for the efficient design of recycling standards and the use of recycled materials.

This chapter provides a discussion on recent developments in the recycling of pH- and thermal-responsive nanomaterials with specific examples and design methodologies.

2. Recyclable pH-responsive nanomaterials

2.1 Introduction to pH-responsive nanomaterials

The class of pH-based stimulus-responsive materials is either a weak acid (proton accepting) or a weak base (proton-releasing) group within their chemical structure. Such materials are also known as polyelectrolytes, which are responsive to variations in pH. The functional group present in the polymeric material experiences ionization with variations in pH, thus resulting in a change in the physical/chemical structure of the responsive material. The physical properties of such materials can be modified by altering the charges on the polymer backbone [9]. For example, poly (acrylic acid) has a pH of 4; exceeding this value results in the functional group's ionization and expansion of the chemical structure. A widely used pH-responsive material—poly (2-dimethylaminoethyl methacrylate) (PDMAEMA)—responds at a comparatively lower pH value, and on reverting the pH value to the original, the material returns to the original form, thus allowing a reversible process. Due to their pH-responsive behavior, these materials are used in a wide variety of different applications such as water treatment, oil–water separation, drug delivery, etc. There is a high potential of pH-responsive nanomaterials for recycling and functional purposes, as described in the following section of this chapter.

2.2 Recycling and applications

Oil–water separation is a major problem worldwide, and significant research has recently been conducted in this regard. Industrial waste involving oil is frequently released into water without any control, and repeated oil spill incidents have been primary sources of oil–water pollution [10]. Such pollution is harmful to aquatic beings and human health; therefore, it is vital to pay attention and treat oil–water separation as a severe concern. Out of various oil–water pollution, emulsified oil is the most difficult to treat because the presence of different substances results in the formation of stable oil molecule, and smaller oil droplet size restricts the separation of oil and water, especially via traditional techniques [10–12]. Emulsified oils are common in daily life and various sectors such as textiles, foods, mining, and petrochemicals [13]. The conventional technique— for example, the flocculation technique for oil–water separation—has the disadvantages that it is only efficient under certain conditions, is time-consuming, and has low efficiency [14]. Superoleophilic sponges are reportedly effective against floating oil; however, sponges cannot separate the oil–water emulsion [15,16]. Apart from various separation techniques, several oil–water separable materials (synthetic and natural) have been reported in the literature. However, collection of these materials after treatment/separation is still a major concern. To deal with this problem, several researchers have reported the use of membranes for oil–water separation [17,18]. Although membranes have been efficient in separating oil and water, their practical and commercial applications are limited due to a comparatively higher cost, reduced membrane flux efficiency over the service life, and energy consumption.

Recently, magnetic nanoparticles (MNPs) have emerged as an efficient and recyclable material for oil−water separation [19,20]. MNPs control interfacial activity for more efficient interaction, resulting in magnetic oil droplets that can be separated easily with a magnetic field. Hybrid MNPs have also been developed with temperature or pH-sensitive properties and have been efficient against emulsified oil−water separation. Lu et al. [10] synthesized the cost-effective APTES (3-aminopropyl triethoxysilane) coated pH-sensitive synthetic MNPs (Fe_3O_4) using the two-step coating technique. The resultant MNPs were responsive within acidic to neutral pH and recyclable up to nine times without significant loss of oil−water separation efficiency.

Another technique for developing hybrid nanoparticles (NPs) is synthesizing the core-shell structure. The core-shell technique allows the development of the outer layer (shell), which in the present case would consist of a polymeric shell to prevent the (inorganic) core material and provide the possibility of enhancing material properties, such as stimuli-responsiveness. Besides, the controllable interface nature of MNPs allows getting efficient oil−water separation. After completing the oil−water separation, the stimuli-responsive polymer at the outer layer can be recycled completely and prepared for the next cycle of oil−water separation. Furthermore, the addition of certain tethered polymeric materials would also provide the steric repulsion resulting in no aggregation of the particles [21,22]. Several polymers have been reported as the functional material such as poly(N-isopropylacrylamide) [23], poly(2-dimethylaminoethyl methacrylate) [22], chitosan [24,25], polystyrene and poly(butyl acrylate) [26]. Wang et al. [22] synthesized a hybrid core-shell recyclable MNP for emulsified oil−water separation. The experimental work designed the core as a cluster of MNPs and PDMAEMA as a dual stimulus-responsive material for the magnetic field and pH stimuli. The core was synthesized using the solvothermal hydrolysis process, whereas surface-initiated atom transfer radical polymerization (SI-ATRP) of tertiary amine-containing DMAEMA produced the shell material. The as developed core-shell hybrid MNP were reported to be more efficient than the gravity plate separation technique [27] and membranes [28]. The schematic of stimuli-responsive MNP recycling is shown in Fig. 8.1. MNP was added on a certain pH in the diesel-water emulsion, and on the adsorption of MNP at the oil−water interface, a Pickering emulsion was formed. Later, on the magnetic field application, the MNP attached oil particles were attracted to the magnetic source, and the water was separated. On reducing the emulsion's pH to a lower value, the MNP is separated from the oil and then recovered and made ready for reuse for the next separation cycle. The reusability was found to be six times without loss of separation efficiency of the as-prepared core-shell MNP.

The separation performance of APTES (3-aminopropyl triethoxysilane) coated MNP under alkaline conditions is limited, whereas under neutral to acidic conditions, it is based on interfacial activity and electrostatic attraction, respectively. To deal with this limitation, Lu et al. [29] investigated the AEAPTES (containing an additional aminoethyl group

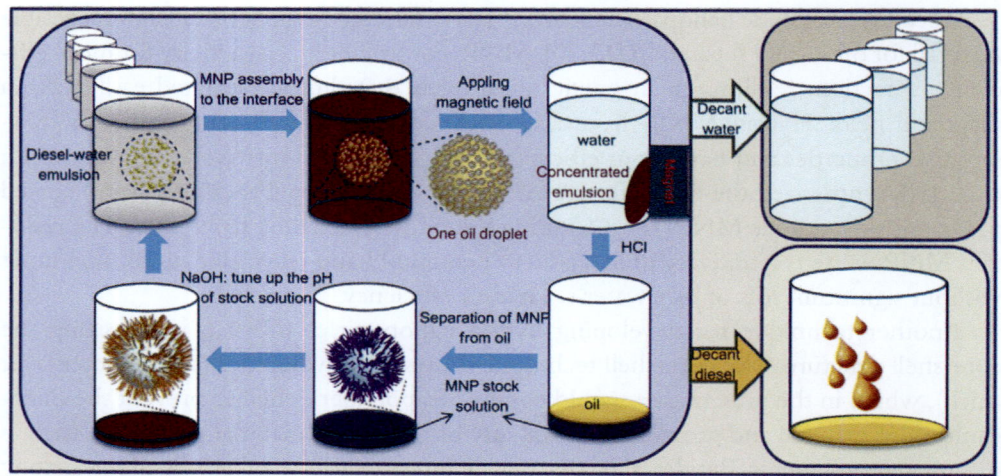

Figure 8.1 Application of pH-responsive magnetic nanoparticles (MNPs) as recyclable stabilizers for oil—water separation [22]. *(Reproduced with permission from Ref. [22]. Copyright 2015 Elsevier.)*

compared with APTES) for surface coating of pH-responsive MNP. AEAPTES MNP was effective under varying pH-responsive values without influencing the efficiency, even in the alkaline condition. In addition, the prepared material was tested to be capable of reuse up to five cycles without affecting the overall oil—water separation property.

NP catalysts having characteristic high surface area and efficiency have been widely researched due to their characteristic properties. However, their practical applications are limited because of additional setup and processing need for separation and recycling [30,31]. Various separation techniques have been developed, such as using the magnetic field, centrifugation, and filtration to solve the issue of separation and recycling; however, such methods require external setup and transfer of the mix from one container to another, loss of material, blockage of filters, and are time-consuming. Therefore, to make the recycling and separation process more convenient, efficient methods are needed to promote a sustainable environment. Yang et al. reported that the pH-responsive Pickering emulsion and biphasic Pickering emulsion, although they have high reaction efficiency but the recovered amount of the organic product, is comparatively low [32]. Another aspect of increasing the recycling efficiency of the organic materials is to break down the Pickering emulsion. Such an approach would require the energy barrier to be broken for detaching the NP from the emulsion, which conversely is related to the oil phase binding energy and is consequently higher than its thermal energy [33]. To overcome the problem, Huang et al. [34] designed a smart Pickering system using pH-responsive NP catalysts. The recycling and separation took place by the addition of acid and via macroscopic phase separation, as shown in Fig. 8.2.

In contrast to the thermal and magnetic-responsive materials, which require high or strong stimulus for the responsive material to react, pH-responsive material requires a comparatively lesser degree of stimulus for the demulsification to occur. Huang et al.

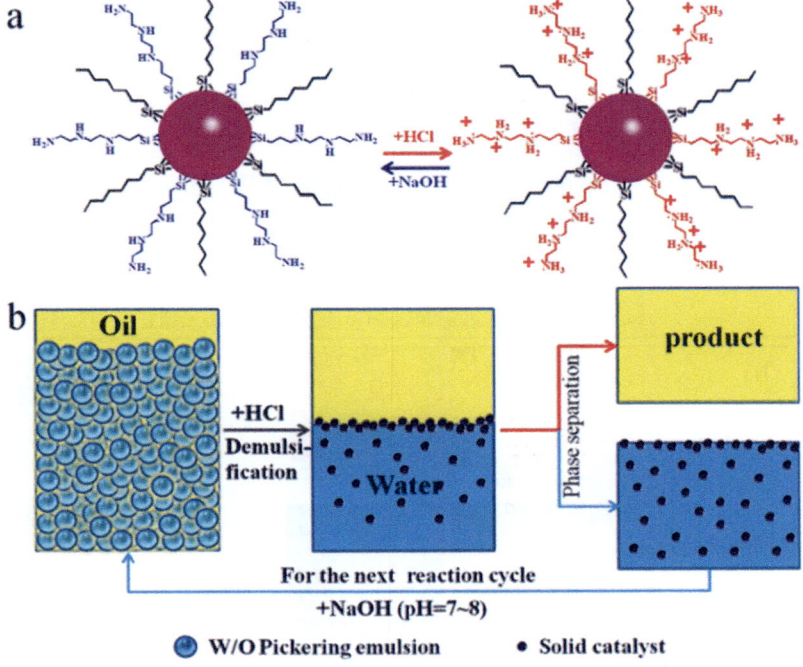

Figure 8.2 Schematic illustration of the pH-switched Pickering emulsion strategy. (A) Protonation–deprotonation of a nanoparticle catalyst. (B) Demulsification, organic product separation and catalyst recycling [34]. *(Reproduced with permission from Ref. [34]. Copyright 2015 RSC.)*

[34] used the Pickering emulsion/organic biphasic system instead of the conventional organic/aqueous biphasic system using silica NP along with the pH-responsive group. The as-prepared material was loaded by Pd NP for organic/biphasic catalysis and demonstrated five times increased reaction rate, and in-situ addition of stimuli-responsive material also showed recyclability 15 times without losing the efficiency. The addition of responsive material allowed the recyclability of the emulsification and demulsification process and because of increased NP surface area, the efficiency was also improved.

Apart from the other silica or organic polymers, titanium dioxide (TiO_2) is a widely used catalyst due to its photocatalytic efficiency and low cost, especially in water-hydrogen splitting [35]. Hao et al. [36] studied the in-situ recycling of a TiO_2 NP catalyst on the pH-responsive oil–water Pickering system. TiO_2 was prepared via the post-grafting method, and its surface wettability was modified using the pH-responsive material. Firstly, the Pickering emulsion was formed with the aid of the TiO_2 acting as a stabilizer. Upon the completion of the reaction, acid is added to break down the emulsion. Consequently, the oil and water phases can be separated using the phase separation technique, whereas the TiO_2 catalyst is available in the recycled form in

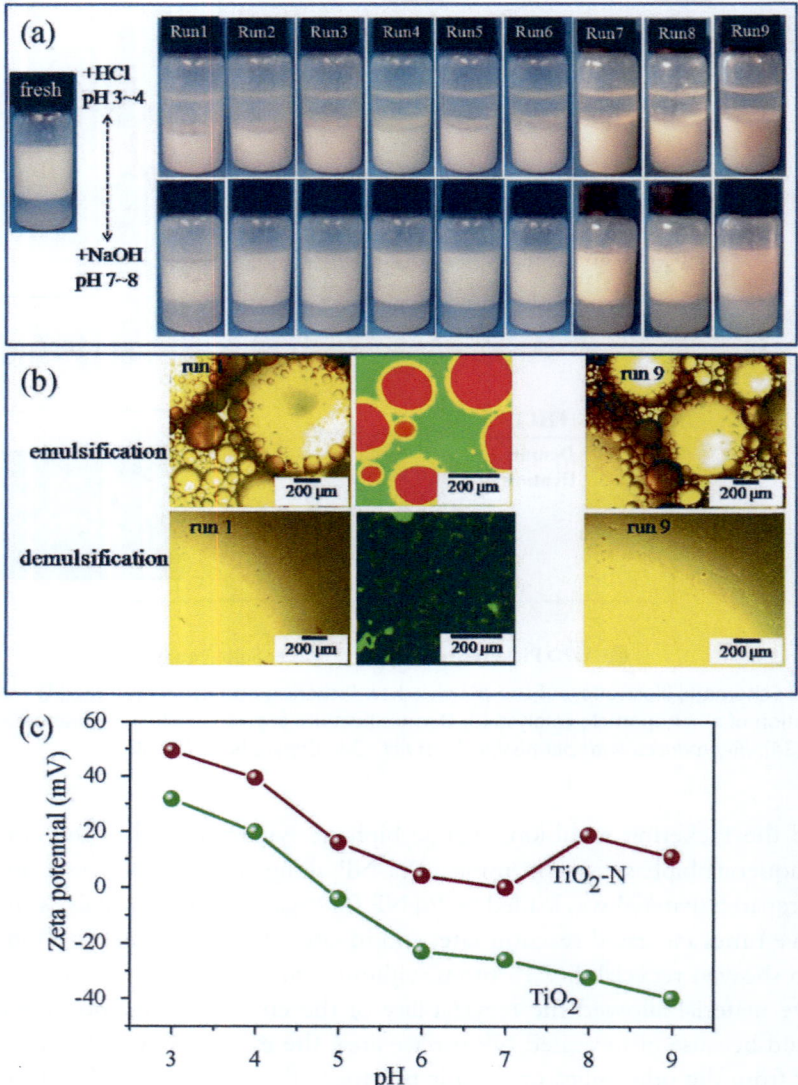

Figure 8.3 pH-triggered switch between emulsification and demulsification systems. (A) Appearance of systems in nine switch runs in response to pH. (B) Optical micrographs for the first and ninth run, and Laser scanning confocal microscopy images for the first run. Red represents the water phase and green is oil droplet. (C) Zeta potentials of TiO_2 and TiO_2–N in water at different pH values [36]. *(Reproduced with permission from Ref. [36]. Copyright 2018 Elsevier.)*

the water phase for reuse. The pH-responsive recycled catalyst, as shown in Fig. 8.3, demonstrated the stability of 10 cycles without losing its efficiency and is dependent on the quantity of aminopropyl group present on the TiO_2 surface.

3. Recyclable thermo-responsive nanomaterials

3.1 Introduction to thermo-responsive nanomaterials

Thermal-responsive nanomaterials respond to the stimulus of temperature variation. Temperature variation is a common phenomenon experienced in daily life, such as the temperature difference in day and night. The widely used class of thermal-responsive nanomaterials are those who experience liquid to liquid phase transition. During the transition, the liquid solution changes its solution, i.e., from clear to a cloudy solution because of nanodroplets' development and the difference in the refractive index of both phases [37]. When a phase transition occurs from the hydrophilic/extended state to the hydrophobic/shrunken state on the temperature above a particular value, this temperature point is known as lower critical solution temperature (LCST). Such a class of thermal-responsive polymers demonstrates a water-soluble state due to hydrogen bonding at the temperature point lower than the LCST value. As the temperature is increased from their LCST point, intramolecular hydrogen bonding occurs along with the change in behavior to hydrophobic [38]. Such a class of thermal-responsive polymers is mostly utilized in biomedical applications (drug delivery, cell culture surfaces, etc. [39,40]) due to their swelling and shrinking phenomena on the stimulus of temperature variation. In this regard, poly(N-isopropyl acrylamide) (PNIPAM) is a widely used thermal-responsive polymer, especially in biomedical applications due to its comparatively lower LCST value (~32°C) and also due to the ability to modify the LCST value to the required application [41]. Besides, the LCST of PNIPAM polymer does not alter with small variation in solution concentration and pH. Another thermal-responsive polymer class is known as the upper critical solution temperature, which demonstrates the reverse phase transition. Apart from PNIPAM, another type of thermal-responsive polymer poly(N-vinyl caprolactam) is also reported to have the LCST value (~31°C) closer to that of PNIPAM, making it another useful polymer for biomedical applications. Although the LCST values for both polymers are closer to each other, PNIPAM is still the widely used thermal-responsive material for biomedical applications. One drawback associated with both of these two polymers is the high glass transition temperature, which leads to hysteresis during the phase transition on thermal stimulus [37]. Apart from the biomedical applications of nano thermal-responsive polymers, they are also being reported efficient for separation application, as discussed in the next section.

3.2 Recycling and applications

One of the methods for oil–water separation using the nano pH-responsive materials is discussed in Section 14.2.2. Another most widely used approach for oil–water separation is the use of recyclable thermo-responsive nanomaterials. Chen et al. [23] synthesized the hybrid MNP containing magnetic core and thermo-responsive outer layer developed through SI-ATRP. The prepared material was tested for its efficiency against toluene–water emulsion, and it was found that the MNP was able to separate the emulsion and recyclable from toluene on the stimulus of temperature, thus promoting the eco-friendly approach. The SI-ATRP technique was effective but costly and complicated. Therefore, to increase the number of practical applications, especially within the industrial sector, Lu et al. [42] investigated the "grafting through" technique as an economical process. The low-density particles produced via the "grafting through" method would also benefit NP accumulation in the emulsion [43]. The MNPs were synthesized by grafting poly (N-isopropylacrylamide) (PNIPAM) and were analyzed for separation in a diesel–water emulsion. The prepared material demonstrated high interfacial activity at room temperature and the strong magnetic response of MNPs for demulsification. The grafted MNPs were able to desorb from the emulsion and were easily separated from the mix at 33°C. Nonetheless, the material was stimuli-responsive to temperature only and had potential recycling of up to seven cycles without any loss of working efficiency.

Various metallic catalysts have been used in advanced materials research, of which the gold (Au) NP is widely used in organic synthesis [44]. Using bare metallic catalysts results in particle agglomeration, thus reducing their efficiency. Therefore, such catalysts are used in other forms such as carbon nanotubes [45] and graphene oxide [46]. Apart from metallic catalysts, the cellulose-based nanocatalyst has also gained significant attention because of its potentially enhanced performance and sustainable characteristics [47]. In addition, hydroxyl groups on the surface provide the added advantage of functionalization. Cellulose has been used as a sponge for its characteristic open-cell structure. However, to enhance service life, improvement in mechanical properties is vitally important. In this regard, several researchers have investigated improving the mechanical performance through additional polymerization and dual crosslinking of cellulose nanofiber with polymeric material [48]. A thermo-responsive catalyst using the core-shell technique for gold NP at the core and thermo-responsive material (PNIPAM) at the shell was proposed by Shuang et al. [49]. However, the prepared material had reduced efficiency after each recycling due to the loss of gold NPs. PNIPAM had been the most widely used thermo-responsive catalyst [50]. Functionalization of PNIPAM with MNP and cellulosic sponge provides dual hydrophilic and hydrophobic functionality to the trigger of temperature response as reported by Ref. [51]. Li et al. [52] investigated the copolymer of PNIPAM and poly(glycidyl methacrylate) on nanocellulose sponge, and the prepared material was tested for catalytic performance and thermal response on the reduction of 4-nitrophenol (4-NP) using $NaBH_4$. The resultant cellulosic sponge had the honeycomb structure with nanopore size and enhanced mechanical properties. In addition, strong catalytic activity for the reduction of 4-NP was shown, and the thermo-responsive material had a recycling life of 22 cycles without a reduction in performance efficiency.

4. Future trends

As discussed in this chapter, there is an excellent potential for pH- and thermal stimuli-responsive nanomaterials with the added advantage of recycling. Besides many other advantages, tuning of the polymeric material to control and engineer responsive behavior provides for its use in various sectors and applications. Therefore, researchers need to further analyze the various possibilities that allow more control of the responsive properties and means of recycling without sacrificing performance efficiency. As more research is conducted in this regard, it will be equally important to develop cost-effective approaches for increasing materials acceptability and practical applications.

5. Conclusion

This chapter discussed recyclable thermal and pH-responsive nanomaterials for various applications. A trigger-in and trigger-out concept is applied to nanomaterials based on external thermal and/or pH stimuli with a main focus on materials recycling. Various methods and application techniques are also covered in this chapter for the efficient functional performance of materials along with their respective recycling efficiency and life. These include magnetic-responsive nanomaterials and core-shell-based design for incorporating stimulus-responsive materials for a required application. Overall, this chapter provides a detailed discussion of various techniques for recycling stimuli-responsive nanomaterials, especially for their application in the separation sector, which would be useful for researchers and industrialists working in similar areas of research.

References

[1] S. Wang, H. Liu, D. Wu, X. Wang, Temperature and pH dual-stimuli-responsive phase-change microcapsules for multipurpose applications in smart drug delivery, J. Colloid Interface Sci. 583 (2020) 470–486.
[2] R. Singh, M.K. Sinha, M.K. Purkait, Stimuli responsive mixed matrix polysulfone ultrafiltration membrane for humic acid and photocatalytic dye removal applications, Separ. Purif. Technol. 250 (2020) 117247.
[3] R. Sepehrifar, R.I. Boysen, B. Danylec, Y. Yang, K. Saito, M.T. Hearn, Design, synthesis and application of a new class of stimuli-responsive separation materials, Anal. Chim. Acta 963 (2017) 153–163.
[4] N. Abu-Thabit, Y. Umar, Z. Sadique, E. Ratemi, A. Ahmad, A.K. Azad, S. Al-Anazi, I. Awan, Optical and pH-responsive nanocomposite film for food packaging application, in: Multidisciplinary Digital Publishing Institute Proceedings, vol. 42, 2019, p. 28.
[5] S. Gao, G. Tang, D. Hua, R. Xiong, J. Han, S. Jiang, Q. Zhang, C. Huang, Stimuli-responsive biobased polymeric systems and their applications, J. Mater. Chem. B 7 (5) (2019) 709–729.
[6] T. Faunce, B. Kolodziejczyk, Nanowaste: Need for Disposal and Recycling Standards, G20 Insights, Policy Era: Agenda 2030, 2017.
[7] J. Bridges, K. Dawson, W. De Jong, T. Jung, A. Proykova, Q. Chaudhry, R. Duncan, E. Gaffet, K. Jensen, W. Kreyling, B. Quinn, Scientific Basis for the Definition of the Term "Nanomaterial", 2012, https://doi.org/10.2772/39703.
[8] D. Živković, L. Balanović, A. Mitovski, N.M. Talijan, N. Štrbac, M. Sokić, D. Manasijević, D. Minić, V. Ćosović, Nanomaterials environmental risks and recycling: actual issues, Recycl. Sustain. Dev. 7 (1) (2014) 1–8.

[9] F.R.-O. Luis García-Fernández, Ana Mora-Boza, pH-responsive polymers: properties, synthesis and applications, in: Smart Polymers and Their Applications, Elsevier, 2014, pp. 45–92.

[10] T. Lü, S. Zhang, D. Qi, D. Zhang, G.F. Vance, H. Zhao, Synthesis of pH-sensitive and recyclable magnetic nanoparticles for efficient separation of emulsified oil from aqueous environments, Appl. Surf. Sci. 396 (2017) 1604–1612.

[11] J. Liu, H. Wang, X. Li, W. Jia, Y. Zhao, S. Ren, Recyclable magnetic graphene oxide for rapid and efficient demulsification of crude oil-in-water emulsion, Fuel 189 (2017) 79–87.

[12] H. Peng, H. Wang, J. Wu, G. Meng, Y. Wang, Y. Shi, Z. Liu, X. Guo, Preparation of superhydrophobic magnetic cellulose sponge for removing oil from water, Ind. Eng. Chem. Res. 55 (3) (2016) 832–838.

[13] R.K. Gupta, G.J. Dunderdale, M.W. England, A. Hozumi, Oil/water separation techniques: a review of recent progresses and future directions, J. Mater. Chem. 5 (31) (2017) 16025–16058.

[14] T. Lü, C. Luo, D. Qi, D. Zhang, H. Zhao, Efficient treatment of emulsified oily wastewater by using amphipathic chitosan-based flocculant, React. Funct. Polym. 139 (2019) 133–141.

[15] F. Beshkar, H. Khojasteh, M. Salavati-Niasari, Recyclable magnetic superhydrophobic straw soot sponge for highly efficient oil/water separation, J. Colloid Interface Sci. 497 (2017) 57–65.

[16] Z.-T. Li, B. Lin, L.-W. Jiang, E.-C. Lin, J. Chen, S.-J. Zhang, Y.-W. Tang, F.-A. He, D.-H. Li, Effective preparation of magnetic superhydrophobic Fe_3O_4/pu sponge for oil-water separation, Appl. Surf. Sci. 427 (2018) 56–64.

[17] Y. Zhao, M. Zhang, Z. Wang, Underwater superoleophobic membrane with enhanced oil–water separation, antimicrobial, and antifouling activities, Adv. Mater. Interfac. 3 (13) (2016) 1500664.

[18] Z. Cheng, C. Li, H. Lai, Y. Du, H. Liu, M. Liu, K. Sun, L. Jin, N. Zhang, L. Jiang, Recycled superwetting nanostructured copper mesh film: toward bidirectional separation of emulsified oil/water mixtures, Adv. Mater. Interfac. 3 (17) (2016) 1600370.

[19] J. Liang, N. Du, S. Song, W. Hou, Magnetic demulsification of diluted crude oil-in-water nanoemulsions using oleic acid-coated magnetite nanoparticles, Colloids Surf. A Physicochem. Eng. Asp. 466 (2015) 197–202.

[20] S. Mirshahghassemi, B. Cai, J.R. Lead, Evaluation of polymer-coated magnetic nanoparticles for oil separation under environmentally relevant conditions: effect of ionic strength and natural organic macromolecules, Environ. Sci. 3 (4) (2016) 780–787.

[21] L.H. Reddy, J.L. Arias, J. Nicolas, P. Couvreur, Magnetic nanoparticles: design and characterization, toxicity and biocompatibility, pharmaceutical and biomedical applications, Chem. Rev. 112 (11) (2012) 5818–5878.

[22] X. Wang, Y. Shi, R.W. Graff, D. Lee, H. Gao, Developing recyclable pH-responsive magnetic nanoparticles for oil–water separation, Polymer 72 (2015) 361–367.

[23] Y. Chen, Y. Bai, S. Chen, J. Ju, Y. Li, T. Wang, Q. Wang, Stimuli-responsive composite particles as solid-stabilizers for effective oil harvesting, ACS Appl. Mater. Interfaces 6 (16) (2014) 13334–13338.

[24] T. Lü, Y. Chen, D. Qi, Z. Cao, D. Zhang, H. Zhao, Treatment of emulsified oil wastewaters by using chitosan grafted magnetic nanoparticles, J. Alloys Compd. 696 (2017) 1205–1212.

[25] T. Lü, S. Zhang, D. Qi, D. Zhang, H. Zhao, Enhanced demulsification from aqueous media by using magnetic chitosan-based flocculant, J. Colloid Interface Sci. 518 (2018) 76–83.

[26] P.M. Reddy, C.-J. Chang, J.-K. Chen, M.-T. Wu, C.-F. Wang, Robust polymer grafted fe3o4 nanospheres for benign removal of oil from water, Appl. Surf. Sci. 368 (2016) 27–35.

[27] Y. Tsujii, K. Ohno, S. Yamamoto, A. Goto, T. Fukuda, Structure and properties of high-density polymer brushes prepared by surface-initiated living radical polymerization, in: Surface-initiated Polymerization I, Springer, 2006, pp. 1–45.

[28] B. Jing, H. Wang, K.-Y. Lin, P.J. McGinn, C. Na, Y. Zhu, A facile method to functionalize engineering solid membrane supports for rapid and efficient oil–water separation, Polymer 54 (21) (2013) 5771–5778.

[29] T. Lü, D. Qi, D. Zhang, S. Lin, Y. Mao, H. Zhao, Facile synthesis of n-(aminoethyl)-aminopropyl functionalized core-shell magnetic nanoparticles for emulsified oil-water separation, J. Alloys Compd. 769 (2018) 858–865.

[30] S. Shylesh, V. Schuenemann, W.R. Thiel, Magnetically separable nanocatalysts: bridges between homogeneous and heterogeneous catalysis, Angew. Chem. Int. Ed. 49 (20) (2010) 3428–3459.

[31] H. Liu, Z. Zhang, H. Yang, F. Cheng, Z. Du, Recycling nanoparticle catalysts without separation based on a pickering emulsion/organic biphasic system, ChemSusChem 7 (7) (2014) 1888–1900.
[32] H. Yang, T. Zhou, W. Zhang, A strategy for separating and recycling solid catalysts based on the pH-triggered pickering-emulsion inversion, Angew. Chem. Int. Ed. 52 (29) (2013) 7455–7459.
[33] K. Du, E. Glogowski, T. Emrick, T.P. Russell, A.D. Dinsmore, Adsorption energy of nano-and microparticles at liquid- liquid interfaces, Langmuir 26 (15) (2010) 12518–12522.
[34] J. Huang, H. Yang, A pH-switched pickering emulsion catalytic system: high reaction efficiency and facile catalyst recycling, Chem. Commun. 51 (34) (2015) 7333–7336.
[35] G. Wang, H. Wang, Y. Ling, Y. Tang, X. Yang, R.C. Fitzmorris, C. Wang, J.Z. Zhang, Y. Li, Hydrogen-treated TiO_2 nanowire arrays for photoelectrochemical water splitting, Nano Lett. 11 (7) (2011) 3026–3033.
[36] Y. Hao, Y. Liu, R. Yang, X. Zhang, J. Liu, H. Yang, A pH-responsive TiO_2-based pickering emulsion system for in situ catalyst recycling, Chin. Chem. Lett. 29 (6) (2018) 778–782.
[37] R. Hoogenboom, Temperature-responsive polymers: properties, synthesis, and applications, in: Smart Polymers and Their Applications, Elsevier, 2019, pp. 13–44.
[38] I. Yildiz, B. Sizirici Yildiz, Applications of thermoresponsive magnetic nanoparticles, J. Nanomater. 2015 (2015).
[39] A.S. Hoffman, Stimuli-responsive polymers: biomedical applications and challenges for clinical translation, Adv. Drug Deliv. Rev. 65 (1) (2013) 10–16.
[40] M.A. Ward, T.K. Georgiou, Thermoresponsive polymers for biomedical applications, Polymers 3 (3) (2011) 1215–1242.
[41] K. Tauer, D. Gau, S. Schulze, A. Völkel, R. Dimova, Thermal property changes of poly (n-isopropylacrylamide) microgel particles and block copolymers, Colloid Polym. Sci. 287 (3) (2009) 299.
[42] T. Lü, S. Zhang, D. Qi, D. Zhang, H. Zhao, Thermosensitive poly (n-isopropylacrylamide)-grafted magnetic nanoparticles for efficient treatment of emulsified oily wastewater, J. Alloys Compd. 688 (2016) 513–520.
[43] L.M. Foster, A.J. Worthen, E.L. Foster, J. Dong, C.M. Roach, A.E. Metaxas, C.D. Hardy, E.S. Larsen, J.A. Bollinger, T.M. Truskett, et al., High interfacial activity of polymers "grafted through" functionalized iron oxide nanoparticle clusters, Langmuir 30 (34) (2014) 10188–10196.
[44] Y. Park, Y. Kim, S. Chang, Transition metal-catalyzed c–h amination: scope, mechanism, and applications, Chem. Rev. 117 (13) (2017) 9247–9301.
[45] Y. Cao, S. Mao, M. Li, Y. Chen, Y. Wang, Metal/porous carbon composites for heterogeneous catalysis: old catalysts with improved performance promoted by n-doping, ACS Catal. 7 (12) (2017) 8090–8112.
[46] Y. Li, X. Fan, J. Qi, J. Ji, S. Wang, G. Zhang, F. Zhang, Palladium nanoparticle-graphene hybrids as active catalysts for the suzuki reaction, Nano Res. 3 (6) (2010) 429–437.
[47] S. Wang, A. Lu, L. Zhang, Recent advances in regenerated cellulose materials, Prog. Polym. Sci. 53 (2016) 169–206.
[48] Y. Li, L. Xu, B. Xu, Z. Mao, H. Xu, Y. Zhong, L. Zhang, B. Wang, X. Sui, Cellulose sponge supported palladium nanoparticles as recyclable cross-coupling catalysts, ACS Appl. Mater. Interfaces 9 (20) (2017) 17155–17162.
[49] S. Wu, J. Dzubiella, J. Kaiser, M. Drechsler, X. Guo, M. Ballauff, Y. Lu, Thermosensitive au-pnipa yolk–shell nanoparticles with tunable selectivity for catalysis, Angew. Chem. Int. Ed. 51 (9) (2012) 2229–2233.
[50] P. Huo, Z. Ye, H. Wang, Q. Guan, Y. Yan, Thermo-responsive pnipam@ AgBr/CSs composite photocatalysts for switchable degradation of tetracycline antibiotics, J. Alloys Compd. 696 (2017) 701–710.
[51] Y. Tu, F. Peng, X. Sui, Y. Men, P.B. White, J.C. van Hest, D.A. Wilson, Self-propelled supramolecular nanomotors with temperature-responsive speed regulation, Nat. Chem. 9 (5) (2017) 480–486.
[52] Y. Li, L. Zhu, B. Wang, Z. Mao, H. Xu, Y. Zhong, L. Zhang, X. Sui, Fabrication of thermoresponsive polymer-functionalized cellulose sponges: flexible porous materials for stimuli-responsive catalytic systems, ACS Appl. Mater. Interfaces 10 (33) (2018) 27831–27839.

CHAPTER 9

Pyrometallurgy-based applications in spent lithium-ion battery recycling

Jiafeng Zhang
National Engineering Laboratory for High Efficiency Recovery of Refractory Nonferrous Metals, School of Metallurgy and Environment, Central South University, Changsha, Hunan, China

1. Introduction

With the increasingly severe energy crisis and environmental pollution in recent years, lithium-ion batteries (LIBs) have been widely applied in new-energy vehicles and energy-storage devices to satisfy growing energy demand. In a lithium-ion power battery, after cyclic charging and discharging, the internal structure of the battery undergoes irreversible changes, resulting in lithium-ion channel blockage and LIB failure. The use cycle of lithium-ion power batteries in new-energy vehicles is 3–5 years, and with the increased demand for and production of lithium-ion power batteries, the volume of scrap LIBs is expected to increase rapidly. The volume of scrap LIBs in China is expected to reach 780,000 t in 2025, with a market size of more than 20 billion yuan. Spent LIBs contain a large number of valuable metals; for example, spent ternary cathode powder contains 2%–5% Li, 5%–12% Ni, 5%–20% Co, and 7%–10% Mn. If these materials are not suitably handled, organic matter, dust, and heavy metal pollution may be generated along with the wasting of resources. In this context, the recycling of spent ternary lithium-ion power battery materials not only saves resources and protects the ecological environment but also promotes the sustainable development of the LIB industry.

Ternary LIBs are generally composed of positive anode materials, cathode materials, electrolytes, collector fluids, diaphragms, and shells. After repeated charging and discharging, the structure of electrode materials is destroyed, resulting in critical capacity decay. With the increasing demand for batteries, the disposal of spent ternary LIBs is expected to witness explosive growth. The large volume of spent ternary LIBs poses a potential threat to the environment, especially from the presence of heavy metals, electrolytes, solvents, and various organic excipients that can be detrimental to the ecosystem and human health if they are discarded without reasonable disposal. Batteries must be recycled for two main reasons: (1) the recycling of valuable materials is of value, especially when the supply of these materials is limited; (2) to ensure sustainability and eliminate the safety hazards associated with disposing used lithium batteries, governments have mandated lithium battery recycling.

From the perspective of economic benefits and environmental protection, the development of recycling processes for spent ternary LIBs has received widespread attention to realize the recycling of valuable metal resources and reduce the impact of solid waste disposal on the environment. Many recycling technologies based on pyrometallurgical processes have been developed for spent LIBs, and an increasing number of companies (Umicore, Sony, OnTo, Accurec, Inmetco, Xstrata, etc.) have commercialized and used pyrometallurgical process. This chapter introduces the main methods and processes for pyroprocessing-based recovery of spent ternary LIBs, including pyropretreatment, high-temperature melting, reduction roasting, and other technologies. Moreover, this chapter summarizes the advantages and disadvantages of various technologies and provides a reference for developing resource-based technology for spent ternary LIBs.

2. Application of pyrometallurgy in pretreatment process

The LIB cathode is composed mainly of cathode active materials ($LiCoO_2$, $LiFePO_4$, $LiCo_xNi_yMn_zO_2$, etc.), a binder, acetylene black, and a collector (aluminum foil). The active material is connected to the aluminum foil through the binder. To obtain pure cathode active material, it is necessary to separate the material from the collector. The commonly used binders include polyvinylidene fluoride (PVDF) and polytetrafluoroethylene. The melting temperature of PVDF is 172°C, and its decomposition reaction occurs at 350–400°C produces HF and its oligomers. The decomposition temperature of acetylene black is 600°C, and the melting point of aluminum foil is 660°C. High-temperature pretreatment is based on the difference in the melting or decomposition temperatures of the three entities to remove the binder and separate the aluminum foil from the active material of the positive electrode. To achieve effective dissociation of electrode materials at lower treatment temperatures, reaction media such as $CaCl_2$, $CaCO_3$ and CaO can be added to the reaction system or a combination of melting agents, such as NaOH–KOH, $NaNO_3$–KNO_3, and $AlCl_3$–NaCl, to enhance the separation effect of positive active materials and aluminum foil, and CaO and other substances can be used to realize heat storage in the reaction process. The addition of the melt agent combination can promote the separation of cathode active material from the aluminum foil through the following mechanism. The added molten salt causes the PVDF to melt in response to the phase-change heat accumulation, thereby separating the cathode active material from the aluminum foil. The mechanism is that the added molten salt causes the PVDF to melt, thereby dissociating the positive active material from the aluminum foil. After vacuum pyrolysis at 450°C, separation of the active material and collector was not significant. In contrast, when the temperature was increased to 600°C, the active material could be easily peeled from the aluminum foil. When the temperature was increased to 700°C, the aluminum foil became brittle, and the active material and aluminum foil were mixed and could not be effectively

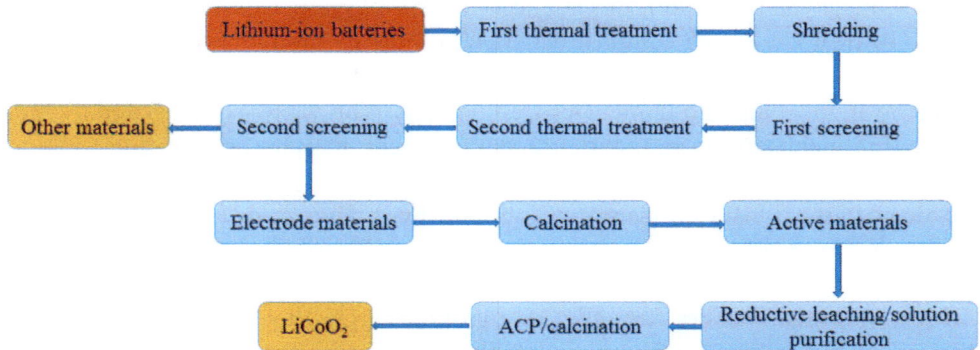

Figure 9.1 Flowchart of proposed thermal treatment for metal recovery.

separated. Lee [1] placed the mechanically treated electrode material in a muffle furnace and first heat-treated it at 100°C–150°C for 1 h. Next, the material was disassembled and sieved in a grinder, and the active material was heat treated at 500°C–900°C for 0.5 to 2 h to remove the carbon and binder. The process flow is shown in Fig. 9.1.

Heat treatment aims to effectively separate the cathode active material from the aluminum foil through the thermal decomposition of the organic binder to reduce the cohesion of the coating active material particles and remove the carbon and PVDF binder from the cathode material. The advantage of heat treatment is that it is simple and easy to implement. Most of the organic binder is removed from the cathode material after heat treatment. However, the heat treatment process is energy intensive and emits toxic gases; therefore, it is necessary to install special equipment to purify the gases and fumes generated by combustion. A high-temperature metallurgical process generally involves the separation and recovery of metals from used LIBs through high-temperature thermal reduction, thermal decomposition, and vacuum metallurgy. This process involves a high chemical conversion rate for valuable elements and a short recovery process and can be easily implemented in industrial applications. The related technologies have been widely studied. However, high-temperature metallurgical methods involve high losses and high energy consumption, generate harmful gases (dioxins, furans, etc.), and have strict processing equipment requirements. Therefore, in the future, it is necessary to adopt methods with a higher recovery rate, lower energy consumption, and fewer environmental hazards to replace the current recycling process for spent LIBs.

3. Pyrometallurgy technology

Pyrometallurgy, also known as dry metallurgy or incineration, removes the organic binder from the electrode material by high-temperature incineration while the metal and its compounds undergo redox reactions to recover the low boiling point metal and its compounds through condensation. The metal in the slag is recovered by sieving, pyrolysis, magnetic separation, or chemical methods. Pyrometallurgical techniques can be used to treat large-scale complex batteries, and the method does not place stringent requirements on the composition of the recovered LIBs. However, high-temperature treatments involve high equipment-processing costs. The waste gas generated by high-temperature incineration can pollute the environment, thereby necessitating additional purification and recovery equipment. The pyrometallurgical recovery processes for valuable metals in spent LIB cathode materials can be divided into high-temperature melting, in situ reduction roasting, chlorination roasting, and sulfidation roasting according to the reaction principle.

Most industrial facilities currently use pyrometallurgical techniques, in which used LIBs are melted at high temperatures (1000°C) by placing them directly in a furnace. In the roasting process, the plastic, electrolyte, and carbonaceous components contained in spent LIBs are decomposed at high temperatures, and valuable metals (Co, Ni, Mn, Cu, etc.) are collected from the melted metals and alloys. This method, as the traditional means of mineral processing, is simple to implement, can be adapted to several raw materials, and involves high production capacity; however, this process results in a high metal loss and low overall recovery efficiency. In addition, the finished roasted alloy requires further purification. Notably, fire-based recovery may lead to a net increase in greenhouse gas emissions and energy consumption.

3.1 High-temperature melting

Pyrometallurgical processes for recovering valuable metals from spent LIBs are classic metal extraction methods based on high-temperature roasting to recover valuable metals from spent LIBs in the form of alloys. For example, Umicore adopts a combination of pyrometallurgical and hydrometallurgical processes to treat spent LIBs. Valuable metals are recovered through a hydrometallurgical process, acid leaching, followed by precipitation in the form of salts. Spent LIBs are placed directly in the smelter without any pretreatment. The plastic, organic solvents, and graphite present in the cell generate significant heat during roasting. The roasted metal components are reduced and converted to alloys. While the alloys of the valued metals (Co, Ni, and Cu) are collected, the oxides of Li, Mn, Fe, and Al are treated as slag. The obtained alloys must be further purified. Cobalt oxide and nickel hydroxide ($Ni(OH)_2$) are obtained via sulfuric acid (H_2SO_4) leaching and solvent extraction. However, this process results in a certain loss of lithium. To reduce lithium loss in the recovery process, researchers have attempted to extract lithium and later recover other valuable metals. Although a large number of lithium batteries can be treated through the high-temperature roasting method, and it is easy to implement, several problems are encountered. For example, the method is energy-intensive, can lead to secondary pollution, and is thus not environmentally friendly. In addition, the loss of several valuable metals during the recycling process is an important issue that must be solved.

In the resource recovery treatment of spent LIBs, high-temperature melting is a typical method based on traditional pyrometallurgy, consisting primarily of three steps: (1) SiO_2–CaO–MgO–Al_2O_3 and reducing agent system slag are added to the spent LIBs and melted at temperatures above 1000°C to convert the valent metals into metal alloys; (2) Co, Ni, Mn, and other target metals are extracted from the polymetallic alloy; and (3) Li-containing slag is recovered. The high-temperature melting method is simple, has a large processing capacity, and does not require pretreatment, which makes it easy to incorporate within the industrial domain. For example, Umicore uses this process by directly placing spent LIBs in a shaft furnace for high-temperature melting and separating them after obtaining a Cu–Co–Ni alloy to realize the recycling of valuable metals. However, the melting slag from this process still contains large amounts of Al_2O_3, Li_2O_3, and MnO, which are usually directly used as construction materials, resulting in wasting of Li and Mn resources.

3.2 Reduction roasting

In the previously mentioned pyrometallurgical process, part of the lithium metal eventually appears in the slag phase by roasting, and a series of subsequent steps is required for further extraction. In recent years, to address these problems, carbothermal reduction has received considerable attention from researchers as a pyrometallurgical method to recover lithium and other metals. In the process of carbothermal reduction, the mixed waste stream is converted to metal oxides, pure metals, or lithium carbonate. In the first step, the lithium carbonate is leached out with water, while the graphite in the leaching residue is burned off, leaving metal oxide as the final residue. In the next step, pure metal, graphite, and lithium carbonate are further separated by wet magnetic separation. However, existing pyrometallurgical technologies face the challenge of reducing energy consumption and meeting the demanding requirements of processing equipment.

In recent years, reduction roasting technology has been widely used to recover valuable metals from spent ternary lithium battery cathode materials. The recovery process involves the following steps: the pretreated spent LIB cathode materials are mixed with a reducing agent (usually C or Al) and placed in a high-temperature environment (>600°C) for the roasting reaction to ensure that the high-valence metal ions contained in these materials, such as Co^{3+}, Mn^{4+}, and Ni^{3+}, are reduced to Co^{2+}. Ni^{3+} is reduced to low-valence compounds, such as Co^{2+}, Ni^{2+}, Mn^{2+}, or metal monomers. The reaction equation of the roasting process is shown as Eq. (9.2).

$$12LiNi_{1/3}Co_{1/3}Mn_{1/3}O_2 + 7C = 6Li_2CO_3 + 4Ni + 4Co + 4MnO + CO_2(g) \quad (9.1)$$

In addition, the method can be applied to treat $LiCoO_2$ scrap. Reduction roasting can achieve efficient recovery of Li and Co from $LiCoO_2$, the cathode material for LIBs, under various reaction atmospheres. The reaction equation of the roasting process is

$$LiCoO_2 + C = Li_2CO_3 + Co + CO_2 \quad (9.2)$$

In a typical pyrometallurgical process, lithium appears in the slag phase and must be further extracted. Carbon thermal reduction has received considerable attention in recent years as a pyrometallurgical method to recover lithium and other metals. Based on conventional in situ reduction roasting, scholars have examined the stepwise reduction roasting process to recover the valuable metals Ni and Co from the spent ternary lithium battery cathode material. This material is mixed well with graphite after pretreatment of the cathode active material of the spent ternary LIB and placed in a tube furnace to realize vacuum-reduction roasting. The metals Ni and Co reduce to metal monomers at 691 and 893°C, respectively, and the magnetic separation method can be used to realize the recovery of Ni and Co. Ni and Co reduce to metal monomers at 691 and 893°C, respectively, and are separated magnetically by exploiting their magnetic properties. In this manner, the selective recovery of Ni and Co can be realized. Other scholars have explored the effect of carbon thermal reduction on battery recycling using an inert gas under vacuum conditions. The researchers proved that graphite powder serves as a reducing agent in the roasting process, and vacuum conditions facilitate a reduced roasting temperature owing to reduced energy consumption. Zhao et al. [2]. exploited the advantages of fast heating by microwave roasting to explore the feasibility of the microwave heating method in the in situ reduction roasting of spent ternary lithium battery cathode materials and noted that the spent ternary battery cathode active material exhibits high wave absorption performance when mixed with carbon at 25–900°C, and the dielectric property of the material increases rapidly from 600°C as the temperature increases, thereby promoting the carbon thermal reduction reaction. When the carbon content reaches 18%, the carbon thermal reduction reaction can be completed, and optimal microwave absorption performance can be obtained when the apparent density of the material is 1.41 g/cm^3. The method is based on microwave heating from the inside of the material to promote reduction roasting. Compared with traditional high-temperature melting technology, the reduction roasting method has a lower heat treatment temperature, and the positive and negative electrode materials of spent LIBs can be effectively utilized, thereby decreasing the treatment cost and achieving selective Li recovery.

3.3 Chlorination roasting

In the method of lithium extraction by chlorination roasting, the lithium in lithium ore is transformed to lithium chloride through chlorination roasting. During roasting, the valuable elements in lithium ore, such as potassium, rubidium, and cesium, are transformed into chlorides to ensure comprehensive extraction. The chlorination roasting method involves a short and closed process, high metal recovery, large equipment capacity, low smelting energy consumption, and no pollution. This method is suitable for processing low-grade lithium ore and extracting lithium with high development prospects. As early as 1817 in Sweden, Smith (J. Smith) used ammonium chloride at 1223 K to roast lithium mica to produce lithium chloride. In India, in 1976, a semiindustrial trial of roasting lithium mica with calcium chloride at 1223 K was conducted to produce lithium chloride, although this method was not aimed at industrial production. In the 1960s, the United States Bureau of Mines performed high-temperature (1143−1813 K) roasting of calcium chloride lithium pyroxene to produce lithium chloride. Notably, due to the high temperature, gaseous alkali metal chlorides are difficult to collect and separate and corrode the equipment. Therefore, this method cannot be used in industrial production. China tested the production of lithium carbonate by roasting lithium mica with calcium chloride at intermediate temperatures (1193−1233 K) in 1976, conducted an industrial test in 1986, and established a workshop for producing lithium carbonate products through this process.

At the roasting temperature, lithium and other alkali metal elements in lithium ore are chlorinated with chlorinating agents. The corresponding chlorides are produced, and various alkali metal compounds are separately extracted from these chlorides. The production of lithium carbonate from lithium mica through medium-temperature chlorination roasting is a typical example of lithium extraction by chlorination roasting. The lithium mica concentrate is mixed with calcium chloride, sodium chloride, and calcium carbonate and roasted at 1193−1233 K to produce solid alkali metal chlorides that are soluble in water. Calcium chloride is used as the main chlorinating agent, sodium chloride is used as the mineralizing and partial chlorinating agent, and calcium carbonate is used as the slagging agent to produce water-insoluble $CaO-SiO_2$, $CaO-Al_2O_3-2SiO_2$, and CaF_2. The generated carbon dioxide, water vapor, and trace chlorides are volatilized into furnace gas.

Chlorination roasting involves adding a chlorinating agent during the roasting process. The chlorinating agent decomposes during the roasting process at high temperatures to produce gases, such as HCl or Cl_2, that convert the stable state metal into active metal chlorides, which are subsequently separated and recovered using the difference in the water solubility of the metal chlorides. Fan et al. [3] used NH_4Cl as the chlorinating agent, mixed the spent $LiCoO_2$ cathode material with NH_4Cl in a mortar, and heated and calcined the material to convert the cathode active material into soluble ammonium chloride salt and chloride salt. After the reaction, the roasted products were leached out in water. The experimental results showed that the leaching rates of Co and Li were more than 99% at a calcination temperature of 350°C with a mass ratio of NH_4Cl to cathode active material of 2:1 and a reaction time of 20 min. The leaching rates of Li and Mn in $LiMn_2O_4$ were 97.99% and 95.27%, respectively, and the leaching rates of Li, Mn, Co, and Ni in ternary lithium batteries were more than 90%. Compared with sulfurization roasting, the reaction temperature of chlorination roasting is lower, and the efficiency is higher. Moreover, no ionic impurities are introduced, and the leaching rate is higher. However, the reaction process produces HCl gas, which is corrosive to equipment and poses a serious environmental pollution risk.

3.4 Sulfate roasting

Sulfate roasting, a type of salt roasting, has been widely used in ore fossil pyrometallurgy. The main purpose is to recover valuable metals by converting them into soluble salts by mixing the added cosolvent with the raw material after roasting. Scholars have performed mixed roasting of used LIBs by adding various types of sulfates to ensure that lithium ions are present in the form of sulfates, and the calcination products can be separated and recovered through water leaching and chemical precipitation. The structure of the cathode material is destroyed at a lower temperature under the action of simple salt roasting. This process exhibits notable potential for industrial applications. Other valuable metals are recovered in the form of oxides, sulfides, etc. This method can be applied to NCM cathode material as well as lithium cobaltate.

Since the reduction roasting temperature is usually more than 600°C, to reduce the treatment temperature and energy consumption while enhancing the recovery efficiency, several scholars have proposed the method of sulfidation roasting to recover valuable metals in spent LIB cathode materials. The sulfurization roasting method involves mixing the spent LIB cathode material with sulfides and roasting it at a high temperature (200–600°C) to convert the metal oxides in the cathode active material into water-soluble sulfate, thereby facilitating subsequent separation and recovery. This method uses sulfide to destroy the structure of the cathode material of LIBs to ensure that valuable metal ions, such as lithium, nickel, cobalt, and manganese, can be recovered. The additives include sulfuric acid, sulfate, sulfide, and singlet sulfur. Lin et al. [4] effectively recovered valuable metals by mixing sulfuric acid and pretreated used LIBs, followed by roasting and water leaching to separate lithium ions from nickel-cobalt-manganese. The roasting mechanism is presented in the following equation, and the thermodynamics of the roasting process are illustrated in Fig. 9.2:

$$4LiMO_2 + 6H_2SO_4 = 2Li_2SO_4 + 4MSO_4 + 6H_2O + O_2 \qquad (9.3)$$

The transformation of $LiCoO_2$, a spent LIB cathode material, in sulfurization roasting has been studied. As the reaction temperature increased, $K_2S_2O_7$ gradually decomposed, metal sulfates, such as $KLiSO_4$, $K_2Co_2(SO_4)_3$ and K_2SO_4, appeared in the roasted products, and the reaction finished when the temperature increased to 500°C. Yu et al. [5] roasted pretreated spent ternary battery cathode active materials with $NaHSO_4-H_2O$, $KHSO_4$, $Na_2S_2O_7$, and other sulfur-containing salts and noted that sodium salt exhibited the best sulfidation effect. Moreover, the roasted slag could yield Co, Ni, Mn, and Li after leaching when roasted at 600°C for 0.5 h, and the molar ratio of the cathode active material to $NaHSO_4-H_2O$ was 1:3. The recoveries of Co, Ni, Mn, and Li were close to 100%. The reaction products were leached out with 60 mL of distilled water in a water bath at 60°C for 2 h. The leaching recovery efficiency of Ni, Co, and Mn was more than 98%.

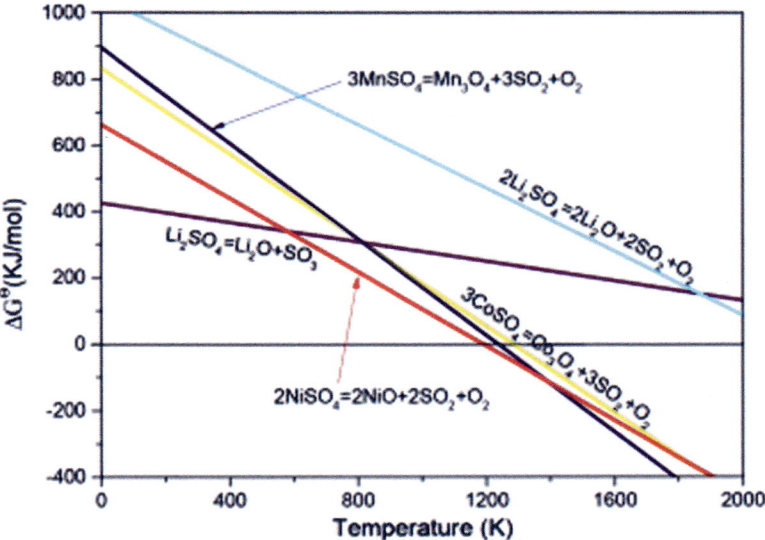

Figure 9.2 The relationship between ΔG and temperature for different reactions, calculated by HSC 6.0.

In contrast to the conventional sulfide roasting method, Lin et al. [6] incorporated the principles of green chemistry into pyrometallurgy and proposed the use of H_2SO_4 as the sulfiding agent instead of conventional inorganic sulfate for sulfide roasting. In this study, the material was first mixed with H_2SO_4 solution and aged in a muffle furnace at 120°C for 720 min. Part of the $LiCoO_2$ was converted to Li_2SO_4 and $CoSO_4$. The remaining $LiCoO_2$ and $CoSO_4$ reacted to yield Co^{2+}, Co^{3+}, and Li_2SO_4, and the reaction equation is shown in Eq. (9.4):

$$2CoSO_4 + 4LiCoO_2 = 2Li_2SO_4 + 6CoO + O_2 \tag{9.4}$$

The total reaction between H_2SO_4 and $LiCoO_2$ is expressed as Eq. (9.5):

$$12LiCoO_2 + 6H_2SO_4 = 6Li_2SO_4 + 4Co_3O_4 + 6H_2O + O_2 \tag{9.5}$$

The results show that 99.3% of Li is converted to Li_2SO_4, and 98.7% of Co is converted to Co_3O_4, and Li and Co can subsequently be recovered by water leaching. Compared with traditional sulfide roasting, this method can effectively reduce the introduction of ionic impurities, the roasted products can be selectively recovered through water leaching, and SO_2 gas can be eliminated in the tail gas by controlling the molar ratio of $LiCoO_2$ to H_2SO_4 at 2:1. However, the mixed roasting temperature is as high as 800°C, which results in high energy consumption, and the powder is prone to sintering during the high-temperature reaction. Therefore, certain pretreatment of the roasted solid-phase products must be performed before the metal leaching reaction.

Pyrometallurgy can be applied to the treatment of large-scale complex batteries, and the composition of recovered lithium batteries does not involve stringent requirements; however, high-temperature treatment involves higher equipment processing costs. The waste gas generated by high-temperature incineration can pollute the environment, necessitating additional purification and recovery equipment.

4. Summary and prospects

Pyrometallurgical processes are designed to recover or extract valuable metals from lithium scrap through physical or chemical transformation at high temperatures. Conventional pyrometallurgical processes typically involve roasting at approximately 1000°C, and the roasting products obtained by this method are Co, Fe, and Ni-based alloys; however, this method results in lithium metal entering the roasted slag, which then must be leached and extracted. The basic principle of reduction roasting is to separate and recover metals by converting high-valence metal compounds into low-valence substances by doping with reducing substances, such as carbon under vacuum or an inert atmosphere. This method has received widespread attention in recent years. In contrast, sulfation roasting is performed by doping sulfate, sulfide, or sulfur monomers to destroy the structure of lithium battery cathode materials to obtain soluble lithium salts to realize lithium extraction. Subsequently, valuable metals, such as nickel, cobalt, and manganese, can be recovered from the roasted slag. LIBs used in first-generation electric vehicles are reaching the end of their useful lives; therefore, it is necessary to actively develop waste management systems.

Since the commercialization of LIBs, the recycling treatment of spent LIBs has received extensive research attention worldwide, and considerable research has been performed. Traditional high-temperature chemical conversion methods to realize the resource recovery of used LIBs have focused on the recovery of the most valuable metals in batteries, such as cobalt and nickel. Although existing recovery technology can realize the selective recovery of lithium, the recovery rate is low, and lithium-containing material is often wasted as slag during the melting or roasting process. Improving the selective recovery of lithium resources in the resource recovery process of used LIBs is the focus of current and future research. To recover valuable elements from cathode active materials, batteries must be separated and crushed, and the main challenge is ensuring safe and effective automated dismantling, especially for the large number of power batteries expected to appear in the future. Risks of fires, explosions, and other safety accidents in the dismantling process must be reduced while ensuring automation efficiency. Existing high-temperature chemical conversion recycling technology is only one step in the recycling of used LIBs. After high-temperature treatment, water or chemical reagents must be introduced to gently leach valuable metals. Therefore, future research should focus on reducing the discharge of waste liquid in the recycling process.

When the cathode active material is repaired and regenerated through the sintering process, toxic and harmful gases are generated by decomposition of the battery shell and organic materials, which are harmful to the environment and human body. Avoidance of this potential secondary pollution is a key consideration in recycling. The high energy consumption associated with high temperatures is a key issue that limits the recovery process of high-temperature chemical conversion methods. Vacuum metallurgy technology can reduce the roasting temperature to a certain extent; however, because this method is implemented under vacuum conditions, it imposes high requirements for equipment, scale, investment, and operational control. Reducing the roasting temperature and energy consumption through effective intensification is an important research direction for ensuring effective LIB recovery by high-temperature chemical conversion methods.

As an important technical means of regenerating spent LIB resources, pyrotechnology has been widely used to dissociate spent LIB cathode materials, recover valuable metals, and regenerate cathode active materials. Compared with wet recycling technology, thermal technology involves higher energy consumption; however, the operating process is short, and the equipment covers a small area. High-temperature dissociation of battery cathode material is based on using high temperatures to decompose and remove the binder to obtain a purer cathode active material. The high-temperature smelting and recovery technology of valuable metals mainly includes high-temperature smelting, in situ reduction roasting, chlorination roasting, and sulfidation roasting. Of these, high-temperature smelting technology involves the highest operating temperature and cannot achieve effective Li resource recovery. However, the associated process flow is short, and the method is easy to implement. In situ reduction roasting can fully exploit the positive/anode materials of spent LIBs and achieve selective Li recovery. The reaction temperature of chlorination and sulfidation roasting techniques is relatively low (usually 200–600 °C), and valuable metals, such as Ni, Co, Mn, and Li, can be recovered efficiently. Direct high-temperature regeneration of the cathode active material can effectively shorten the resource regeneration process of spent LIB cathode material, considerably reduce the production cost, and increase the efficiency of resource reuse, especially for lithium iron phosphate batteries, which have a relatively low recovery value. However, existing directly regenerated LIBs do not satisfy the performance requirements of automotive power batteries. Due to the presence of electrolytes, binders, and other substances in spent LIBs, the high-temperature heat treatment process can produce HF, PF_5, and other toxic and harmful gases, and in the recycling process, chlorination and sulfidation roasting produce SO_x, HCl, and other acidic gases. These gases can lead to equipment corrosion and environmental pollution. Therefore, effectively reducing the reaction roasting temperature while avoiding the introduction of ionic impurities, designing an environmentally friendly recovery process combined with green chemistry principles, and avoiding the generation of secondary pollution are future research directions for the pyrogenic resource recovery technology of spent LIBs.

References

[1] C.K. Lee, K.I. Rhee, Reductive leaching of cathodic active materials from lithium ion battery wastes, Hydrometallurgy 68 (1) (2003) 5–10.
[2] Y. Zhao, B. Liu, L. Zhang, et al., Microwave-absorbing properties of cathode material during reduction roasting for spent lithium-ion battery recycling, J. Hazard Mater. 384 (121487) (2020).
[3] E. Fan, L. Li, J. Lin, et al., Low-temperature molten-salt-assisted recovery of valuable metals from spent lithium-ion batteries, ACS Sustain. Chem. Eng. 7 (19) (2019) 16144–16150.
[4] J. Lin, L. Li, E. Fan, et al., Conversion mechanisms of selective extraction of lithium from spent lithium-ion batteries by sulfation roasting, ACS Appl. Mater. Interfaces (2020).
[5] Y. Yu, D. Wang, H. Chen, et al., Mechanism of lithium and cobalt recovery from spent lithium-ion batteries by sulfation roasting process, Chem. Res. Chin. Univ. 36 (5) (2019) 908–914.
[6] J. Lin, C. Liu, H. Cao, et al., Environmentally benign process for selective recovery of valuable metals from spent lithium-ion batteries by using conventional sulfation roasting, Green Chem. 21 (21) (2019) 5904–5913.

CHAPTER 10

Application of hydrometallurgy in spent lithium-ion battery recycling

Jiafeng Zhang
National Engineering Laboratory for High Efficiency Recovery of Refractory Nonferrous Metals, School of Metallurgy and Environment, Central South University, Changsha, Hunan, China

1. Introduction
1.1 Concept of hydrometallurgical recovery

The process of extracting metals from aqueous solutions is defined as hydrometallurgical recovery. The process includes valuable metal extraction from waste raw materials via aqueous solution leaching, purification of aqueous solutions containing valuable metals, separation of similar elements, and precipitation of metals or metal compounds from aqueous solutions. In particular, the hydrometallurgical recovery of spent lithium-ion batteries refers to extracting and recycling rare metals, e.g., Li, Ni, Co, and Mn, from cathode materials, depending on the various properties of relevant metals in aqueous solutions. Fig. 10.1 displays the process flow of the hydrometallurgical recovery of waste lithium-ion batteries.

Hydrometallurgical recovery has gradually become the mainstream approach in the waste lithium-ion battery disposal and resource recycling industry due to its universal applicability and economical nature. At present, more than 80% of waste batteries are hydrometallurgically recovered.

The process of hydrometallurgical recovery can be divided into two processes according to the concept of traditional hydrometallurgical extraction:
- Leaching: transforming the solid (alloy state) metals in prerecovered waste lithium-ion batteries to the solution ionic state.
- Precipitation: separating the tailored metal from solution by modifying the solubility product.

1.1.1 Leaching

Leaching refers to a key operation in hydrometallurgical recovery. An appropriate leaching agent is used to selectively react with a member of the solid waste battery to dissolve it and form a solution containing the metal to be recovered, which is preliminarily separated from other insoluble components. Effective leaching requires the selection of an appropriate leaching agent according to the chemical characteristics of the raw material and determination of the leaching system from the study of the leaching chemistry.

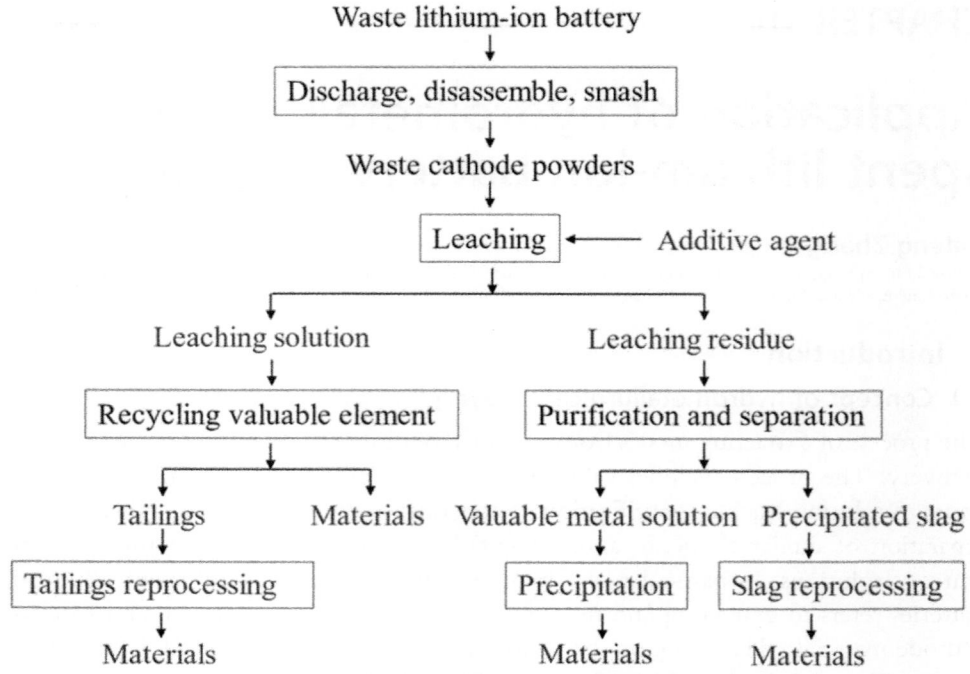

Figure 10.1 Process flow of hydrometallurgical recovery.

According to the different leaching agents and systems, the leaching process of waste lithium-ion battery recycling can be divided into the following three types:

(i) Simple dissolution reaction: This process primarily includes the dissolution of spent cathode oxide materials in an n acid or alkali. For example, $LiFePO_4$ can be spontaneously dissolved in acidic solutions with H^+ concentrations greater than 0.5 mol/L. In this case, all valuable elements, such as lithium, iron, and phosphorus, efficiently leach into the solution. $LiNi_xCo_yMn_{1-x-y}O_2$ can be dissolved with a leaching efficiency of 70% by acid leaching. This method is convenient and exhibits low complexity, although the potential environmental hazards associated with the high-concentration acid (alkali) wastewater and low leaching efficiency limit its practical application.

(ii) Redox reaction: The valence state of a valence metallic element or an element closely related to a valence metallic element is changed, accompanied by electron transport. This process can be divided into the oxidation of olivine structure compounds (e.g., $LiFePO_4$ and $LiMnPO_4$) and the reduction in layered structure compounds (e.g., $LiNi_xCo_yMn_{1-x-y}O_2$ and $LiNi_xCo_yAl_{1-x-y}O_2$). The redox implementation process contains chemical additive redox, electrochemical redox, and microbial redox. Redox leaching systems are the primary leaching technology in industrial recovery owing to their high efficiency and economical nature.

(iii) Complexing reaction: Tailored metal ions exist as ions in a solution in the environment in which they should be precipitated due to complexation with certain substances. The most important complexing reaction occurs between the transition metal ions Ni^{2+} (Ni^{3+}), Co^{2+} (Ni^{3+}), and NH^{4+}.

The leaching reaction and system are often complicated in the actual leaching process, including two single reactions or multiple reactions. For example, in the ammonia leaching of Li $Ni_xCo_yMn_{1-x-y}O_2$ materials, efficient leaching results from the reduction reaction between the metal ion and reducing agent and complexation with ammonium ions. Furthermore, in the study of leaching chemistry, it is necessary to focus on the influence of the reaction of nontarget metal elements with the leaching agent on the main leaching reaction and propose a rational design to inhibit these adverse side reactions.

1.1.2 Precipitation

Precipitation is an essential separation and purification method in hydrometallurgy. Valuable metals in a solid state from the solution are precipitated as a single or mixed salt to achieve purification and separation. In practice, the precipitation process is included in all industrial production processes involving hydrometallurgical processes owing to its simple operation, low cost, and small investment. Precipitation aims to introduce appropriate measures to supersaturate a substance in the solution to achieve the precipitation of tailored elements. Table 10.1 lists the solubility values of the sediments involved in the hydrometallurgical recovery process of waste lithium-ion batteries.

According to the different sequences of valuable metal extraction, the precipitation methods can be divided into two types:

(i) Impurities are removed from the solution, and valuable metals remain. In the recovery of waste lithium-ion batteries, impurities in the solution are usually precipitated first to ensure that the solution contains only one or more target elements for extraction. Next, a product with high purity is obtained at a high extraction rate by precipitating the target elements. For example, in a system involving high-concentration sulfuric acid leaching to recover $LiFePO_4$, the leaching solution contains Li^+, Fe^{2+}, P^{5+}, Al^{3+}, Cu^{2+}, and other elements. To increase the recovery rate and product purity of lithium, impurities, such as Fe^{2+}, P^{5+}, Al^{3+}, and Cu^{2+}, are precipitated through chemical precipitation to obtain a solution containing only lithium. This mode can maximize the extraction effect of target elements and obtain greater economic benefits; however, the solutions obtained from front-end leaching systems must exhibit a high concentration of target elements in the solution and the presence of elemental impurities that are few and can be easily separated.

(ii) The target elements are precipitated and directly separated from the leaching solution while ensuring that the elemental impurities remain in the solution. Usually, this method requires a low content of elemental impurities in the solution or a conspicuous concentration difference between the elemental impurities and target

Table 10.1 Solubility of common metallic compounds in water.

Materials	0°C	10°C	20°C	30°C	40°C	50°C	60°C	70°C	80°C	90°C	100°C
$Co(BrO_3)_2$	91.9		45.5								
$CoBr_2$	135	162	112	128	163		227		241		257
$Co(ClO_3)_2$			180	195	214		316				
$CoCl_2$	43.5	47.7	52.9	59.7	69.5		93.8		97.6	101	106
CoF_2			1.36								
$Co(IO_3)_2$					0.88		0.82		0.73		0.7
CoI_2			203								
$Co(OH)_2$			6.3×10^{-6}								
$Co(NO_3)_2$	84	89.6	97.4	111	125		174		204	300	
$Co(NO_2)_2$	7.6×10^{-2}	0.24	0.4	0.61	0.85						
$Co(ClO_4)_2$			104								
$CoSO_4$	25.5	30.5	36.1	42	48.8		55		53.8	45.3	38.9
$LiC_2H_3O_2$	31.2	35.1	40.8	50.6	68.6						
LiN_3	61.3	64.2	67.2	71.2	75.4		86.6				100
$LiC_7H_5O_2$	38.9	41.6	44.7	53.8							
$LiBrO_3$	154	166	179	198	221		269		308	329	355
$LiBr$	143	147	160	183	211		223		245		266
Li_2CO_3	1.54	1.43	1.33	1.26	1.17	1.08	1.01		0.85		0.72
$LiClO_3$	241	283	372	488	604		777				
$LiCl$	69.2	74.5	83.5	86.2	89.8		98.4		112	121	128
$Li_2CrO_4 \cdot 2H_2O$				151							
$Li_2Cr_2O_7 \cdot 2H_2O$			142								
LiH_2PO_4	126										
LiF			0.16								
$Li_2SiF_6 \cdot 2H_2O$			73								
$LiHCO_2$	32.3	35.7	39.3	44.1	49.5		64.7		92.7	116	138
Li_2HPO_3	4.43			9.97	7.61		7.11				6.03
$LiOH$	12.7	12.7	12.8	12.9	13	13.3	13.8		15.3		17.5
LiI	151	157	165	171	179		202		435	440	481
Li_2MoO_4	82.6		79.5	79.5	78						73.9

Compound									
LiNO$_3$	53.4	60.8	70.1	138	152	175	233	272	324
LiNO$_2$	70.9	82.5	96.8	114	133	177			
Li$_2$C$_2$O$_4$			8						
LiClO$_4$	42.7	49	56.1	63.6	72.3	92.3	128	151	
LiMnO$_4$			71.4						
Li$_3$PO$_4$			0.039						
Li$_2$Se			57.7						
Li$_2$SeO$_3$	25	23.3	21.5	19.6	17.9	14.7	11.9	11.1	9.9
Li$_2$SO$_4$	36.1	35.5	34.8	34.2	33.7	32.6	31.4	30.9	
Li$_2$C$_4$H$_4$O$_6$	42	31.8	27.1	26.6	27.2	29.5			
LiSCN			114	131	153				
LiVO$_3$	2.5		4.82	6.28	4.38	2.67			
AlCl$_3$	43.9	44.9	45.8	46.6	47.3	48.1	48.6		49
AlF$_3$	0.56	0.56	0.67	0.78	0.91	1.1	1.32		1.72
Al(NO$_3$)$_3$	60	66.7	73.9	81.8	88.7	106	132	153	160
Al(ClO$_4$)$_3$	122	128	133						
Al$_2$(SO$_4$)$_3$	31.2	33.5	36.4	40.4	45.8	59.2	73	80.8	89
Al(OH)$_3$			1×10^{-7}						
MnBr$_2$	127	136	147	157	169	197	225	226	228
MnCO$_3$			4.877×10^{-5}						
MnCl$_2$	63.4	68.1	73.9	80.8	88.5	109	113	114	115
Mn$_2$Fe(CN)$_6$			1.882×10^{-3}						
MnF$_2$			10.6		0.67	0.44			0.48
MnSiF$_6$•6H$_2$O			140						
Mn(OH)$_2$			3.221×10^{-4}						
Mn(NO$_3$)$_2$	102	118	139	206					
MnC$_2$O$_4$•2H$_2$O	2×10^{-2}	2.4×10^{-2}	2.8×10^{-2}	3.3×10^{-2}					
MnSO$_4$	52.9	59.7	62.9	62.9	60	53.6	45.6	40.9	35.3
Ni(BrO$_3$)$_2$•6H$_2$O			28						
NiBr$_2$	113	122	131	138	144	153		154	155
NiCO$_3$			9.643×10^{-4}						

Continued

Table 10.1 Solubility of common metallic compounds in water.—cont'd

Materials	0°C	10°C	20°C	30°C	40°C	50°C	60°C	70°C	80°C	90°C	100°C
$Ni(ClO_3)_2$	111	120	133	155	181		221		308		
$NiCl_2$	53.4	56.3	66.8	70.6	73.2		81.2		86.6		87.6
NiF_2			2.56								
$Ni(IO_3)_2$	0.74		6.2×10^{-2}	1.43			2.56			2.59	
NiI_2	124	135	148	161	174		184		187	188	
$Ni(OH)_2$			5.7×10^{-6}								
$Ni(NO_3)_2$	79.2		94.2	105	119		158		187	188	
$Ni(ClO_4)_2$	105	107	110	113	117						
$Ni_2P_2O_7$			1.017×10^{-3}								
$NiSO_4 \cdot 6H_2O$			44.4	46.6	49.2		55.6		64.5	70.1	76.7
$FeCO_3$			6.554×10^{-5}								
$FeCl_2$	49.7	59	62.5	66.7	70		78.3		88.7	92.3	94.9
$FeSiF_6$	72.1	74.4	77				84		88		100
$Fe(OH)_2$		8×10^{-3}	5.255×10^{-5}								
$Fe(NO_3)_2$	113		28.8	48							
$FeSO_4$					73.3		101	79.9	68.3	57.8	
$FeAsO_4$			1.47×10^{-9}								
$FeCl_3$	74.4	91.8	107								
FeF_3			9.1×10^{-2}								
$Fe(OH)_3$	2.097×10^{-9}		4.785×10^{-9}								
$Fe(IO_3)_3$			0.36								
$Fe(NO_3)_3$	112	138			175						
$Fe(ClO_4)_3$	289		368	422	478		772				
$Fe_2(SO_4)_3$	440										
$CuCl$			9.9×10^{-3}								
$CuCN$			1.602×10^{-9}								
$CuOH$			8.055×10^{-7}								
CuI			1.997×10^{-5}								
Cu_2S			1.361×10^{-15}								
$CuSCN$			8.427×10^{-7}								

Compound									
CuBr$_2$	107	116	126	128	131		104	108	120
CuCO$_3$			1.462×10^{-4}						
Cu(ClO$_3$)$_2$	68.6	70.9	242						
CuCl$_2$			73	77.3	87.6	96.5			
CuCrO$_4$			3.407×10^{-2}						
CuF$_2$	73.5	76.5	7.5×10^{-2}				93.2		
CuSiF$_6$			81.6	84.1	91.2				
Cu(HCO$_2$)$_2$			12.5						
Cu(OH)$_2$			1.722×10^{-6}						
Cu(IO3)$_2$	0.109								
Cu(NO3)$_2$	83.5	100	125	156	163	182	208	222	247
Cu(ClO4)$_2$				146					
CuSeO$_4$	12	14.5	17.5	21	25.2	36.5	53.7		
CuSeO$_3$			2.761×10^{-3}						
CuSO$_4$•5H$_2$O	23.1	27.5	32	37.8	44.6	61.8	83.8		114
CuS			2.4×10^{-17}						

element (the concentration of the target element must be more than 100 times that of the impurities). A representative example is a leaching solution containing only Li^+, Ni^{2+}, and Co^{2+} recovered by the ammonia leaching of waste $LiNi_xCo_yMn_{1-x-y}O_2$ cathode materials.

1.2 Advantages of hydrometallurgical recovery

Hydrometallurgical and pyrometallurgical recovery techniques include nearly all waste lithium-ion battery recycling technologies. Compared with pyrometallurgical recovery, hydrometallurgical recovery is the primary market for waste battery recovery because of its high applicability, recovery effect, and economic benefits. In practice, the hydrometallurgical recovery of waste lithium-ion batteries can be regarded as an important branch of hydrometallurgy. Therefore, hydrometallurgical recovery has many advantages similar to those of hydrometallurgy, which can be summarized as follows:

- Pure metals can be directly obtained from the leaching solution—for instance, through the displacement precipitation of copper and electrodeposition of elemental nickel—and the corresponding process can be adjusted to satisfy the frequently changing requirements for the final product.
- Hydrometallurgical recovery is mostly performed at room temperature, and thus, a large amount of fuel is not consumed, in contrast to pyrometallurgical recovery.
- In hydrometallurgical recovery, equipment corrosion is negligible, and the equipment service life is long.
- The composition of leached slag is simpler than that of roasted slag, resulting in an easier and more economical disposal process.
- A large amount of electrolyte is attached to the waste cathode material, and the aqueous fluoride solution produced by hydrometallurgical recovery is easier to manage than fluoride gas.
- Hydrometallurgical recovery can be performed to simultaneously process different types of lithium-ion batteries, whereas pyrometallurgical recovery can recycle only a single waste battery cathode material with the same composition.
- From an economic viewpoint, the unit cost of hydrometallurgical recovery is the same for equipment of different sizes. Nevertheless, building large roasters is considerably more economical than building several small roasters to handle the same volume in the pyrometallurgical process. Therefore, establishing a hydrometallurgical recovery industry line is more flexible in practical applications.
- Hydrometallurgical recovery corresponds to a high degree of automation and low labor costs.

1.3 The status of technological research on hydrometallurgical recovery

Because of its simple process, high product purity, and significant economic benefits, the hydrometallurgical recovery of waste lithium-ion batteries has emerged as a focus of

Table 10.2 Waste lithium-ion battery hydrometallurgical recycling industry at home and abroad.

Company	Headquarters location	Lithium recovery product	Other products
Gem	China	Li_2CO_3	$NiSO_4$, $CoSO_4$, $MnSO_4$
GreenM	China	$LiNi_xCo_yMn_{1-x-y}O_2$	CoO, Cu, Al
Brunp	China	$LiNi_xCo_yMn_{1-x-y}O_2$	$NiSO_4$, $CoSO_4$, $MnSO_4$
AEA	UK	LiOH	Co_2O_3, Cu, Al
IME	Germany	Li_2CO_3	Al, Fe, Ni, cobalt alloy
Recupyl	France	Li_2CO_3, Li_3PO_4	Cu, alloy
Mitsubishi	Japan	$LiCoO_2$	Cu

research. At present, many domestic and foreign companies have established and are operating industrial lines for hydrometallurgical processes to recycle waste lithium-ion batteries. Table 10.2 lists the main domestic and foreign companies involved in the hydrometallurgical recycling of lithium-ion batteries.

AEA (UK): The waste lithium-ion battery is crushed at low temperatures, and the steel is separated. Acetonitrile is added as an organic solvent to extract the electrolyte, n-methylpyrrolidone (NMP) is used as a solvent to extract adhesive (PVDF), and the solid is sorted to obtain Cu, Al, and plastic. The solid residue is dissolved using H_2SO_4 and recovered through electrodeposition to obtain Co_2O_3.

Recupyl (France): Under the protection of an inert mixed gas, the waste lithium-ion battery is broken down, and the paper, plastic, steel, and copper are separated by magnetic separation. Certain metal ions are leached out with LiOH solution, and the insoluble matter is leached out with sulfuric acid. The precipitation of Cu and other metals is realized by adding Na_2CO_3. After filtration, the $Co(OH)_2$ precipitate and Li_2SO_4 solution are obtained by adding NaClO to the filtrate solution for oxidation treatment, and the Li_2CO_3 precipitate is obtained by blowing CO_2 into the solution containing Li.

Mitsubishi (Japan): The waste lithium-ion battery is disassembled after being frozen with liquid nitrogen, and the plastic is separated. Subsequently, the steel is extracted by crushing, magnetic separation, and water washing, and finally, the copper foil is obtained by vibration separation. The remaining materials are successively dissolved with an acid and alkali to recover Li_2CO_3 and metal oxide.

IME (Germany): After sorting the battery shell and electrode material, the electrode material is placed in the reaction tank and heated to 250°C to recover the electrolyte after volatilization and reflux cooling. Subsequently, the powder is crushed, screened, and separated by magnetic separation. The serrated classifier separates large particles (mainly containing Fe and Ni) and small particles (mainly containing Al and electrode material). Co alloy is prepared by melting small particles in an electric arc furnace. Li_2CO_3 is prepared by the hydrometallurgical dissolution of flue ash and slag.

2. Leaching technology

Waste lithium-ion batteries are mainly composed of metal shells, cathode materials, anode materials, other high-value recovery components and electrolytes, diaphragms, and other harmful components. Usually, waste lithium-ion batteries are subjected to discharg`ing, crushing, sorting, and crushing to obtain a mixed powder of cathode and anode materials. Tables 10.3 and 10.4 list the main elemental contents of the pretreatment powder for waste $LiFePO_4$ and $LiNi_xCo_yMn_{1-x-y}O_2$ lithium-ion batteries.

Li, Ni, Co, Mn, and Cu are the target elements requiring recovery, P must be reasonably aggregated to address potential environmental hazards, and Al and Fe can be selectively recovered according to the economic benefits of recovery. According to the different leaching agents, the leaching of waste lithium-ion batteries can be divided into five kinds of leaching systems: acid, alkali, redox, electrochemical, and microbial.

Due to the various types and compositions of lithium-ion batteries prepared by different manufacturers, mixed recycling of $LiFePO_4$ and $LiNi_xCo_yMn_{1-x-y}O_2$ batteries often occurs in the hydrometallurgical recovery process, which further increases the types of elemental impurities and difficulty of recycling. At present, ensuring the consistency of the elemental composition of the same batch of waste batteries is a crucial problem in hydrometallurgical recovery. Promoting company-based integration of lithium-ion battery manufacturing and recycling and printing an "ID card" on every lithium-ion battery are viable solutions. However, these measures are implemented by only a few companies and have not been popularized in the lithium-ion battery industry.

2.1 Acid leaching process

Lithium, nickel, cobalt, manganese, copper, aluminum, and iron are the main metal elements involved in the hydrometallurgical recycling of waste lithium-ion batteries. When not present as an insoluble salt with other anions (precipitation in only the hydroxide form), metal ions can remain in an aqueous solution within a certain acidic pH

Table 10.3 Main elemental contents (wt%) of pretreatment powder of waste $LiFePO_4$ lithium-ion batteries.

Element	Li	Fe	P	Al	Cu	C
Content (wt%)	2.76	18.74	10.41	1.85	4.01	29.58

Table 10.4 Main elemental content (wt%) of pretreatment powder of waste $LiNi_{0.6}Co_{0.2}Mn_{0.2}O_2$ lithium-ion batteries.

Element	Li	Ni	Co	Mn	Al	Cu	C
Content (wt%)	4.17	19.84	6.52	6.59	2.16	5.06	25.19

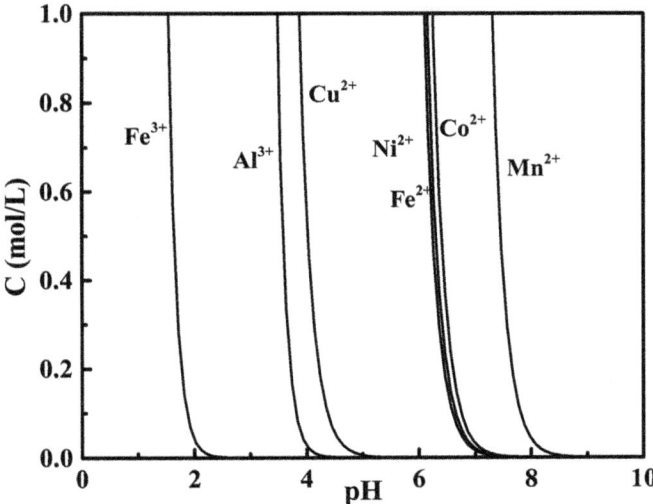

Figure 10.2 Concentration of metal ions in aqueous solution under various pH conditions.

range, as shown in Fig. 10.2. Therefore, lithium metal exists as Li^+ in aqueous solutions, the solubility of LiOH is large, and Li^+ is not precipitated in acidic solutions. Nickel, cobalt, and manganese account for a large proportion of active substances in cathode materials owing to their relatively large atomic mass, thus accounting for the largest recovery value from the hydrometallurgical recycling of $LiNi_xCo_yMn_{1-x-y}O_2$ lithium-ion batteries. Ni(III), Co(III), and Mn(III) are the main valence states in solid cathode material and are reduced to Ni^{2+}, Co^{2+}, and Mn^{2+}, respectively, in the following reduction leaching process. Fig. 10.2 shows that Ni^{2+}, Co^{2+}, and Mn^{2+} are precipitated when the pH of the aqueous solution is more than 7, 7, and 8, respectively. The hydrometallurgical recovery of waste lithium-ion batteries involves two main sources of iron elements: the residual ferroalloy shell and the iron in the cathode material of $LiFePO_4$. Iron has two stable valence states in water solution, divalent and trivalent. Compared with Fe^{2+}, Fe^{3+} can be precipitated at a lower pH value (pH less than 2). Therefore, in the subsequent impurity removal process, all Fe^{2+} in the solution is oxidized to Fe^{3+}, and the iron element in the solution is removed by adjusting the pH. Small amounts of the cathode and anode current collectors are pulverized into powder during precrushing and passed to the leaching system. The aluminum foil is spontaneously transformed into Al^{3+} in an acidic solution environment. The copper foil does not react with acid because of its low metal activity. However, when an oxidizer is added to the leaching system, the copper foil oxidizes to Cu^{2+}. The Al^{3+} and Cu^{2+} in the solution begin to precipitate in the form of hydroxides after the solution pH has increased to 4.

2.1.1 Inorganic acid

Inorganic acid leaching refers to leaching valuable metals by integrating all the main metal elements in waste lithium-ion batteries into the solution through hydrochloric, sulfuric, or nitric acid. Figs. 10.3 and 10.4 present the respective $LiFePO_4$ and $LiNi_xCo_yMn_{1-x-y}O_2$ leaching efficiencies of various leaching systems. In a sulfuric acid $LiFePO_4$ leaching system, the lithium, iron, and phosphorus can be fully extracted from the powder into the solution, and similar laws can be observed for hydrochloric and nitric acid leaching systems. The reaction equation is shown as Eq. (10.1). This inorganic acid leaching system has been widely applied to recover waste $LiFePO_4$ because of its low price. However, a concentration of more than 1 mol/L of hydrogen ions is required in the inorganic acid leaching system; this can corrode the equipment and produce a large amount of acid wastewater. Notably, inorganic acid leaching systems without additives cannot effectively extract the target metal in the recovery of waste $LiNi_xCo_yMn_{1-x-y}O_2$ lithium-ion batteries because of the high bond energy between the trivalent nickel and cobalt and oxygen atoms. The reaction equations are shown as Eqs. (10.2)–(10.4).

$$2LiFePO_4 + 2H^+ + \frac{1}{2}O_2 = 2Li^+ + 2Fe^{2+} + 2PO_4^{3-} + H_2O \quad (10.1)$$

$$2LiNi_xCo_yMn_{1-x-y}O_2 + 6H^+ = 2Li^+ + 2xNi^{2+} + 2yCo^{2+} + 2(1-x-y)Mn^{2+} + 3H_2O + \frac{1}{2}O_2 \quad (10.2)$$

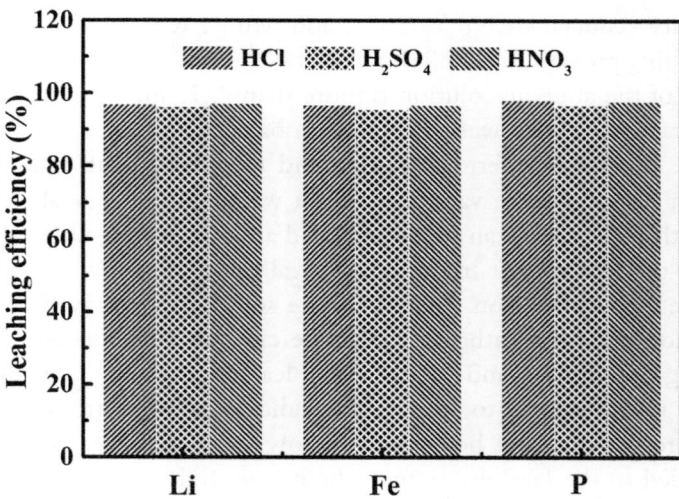

Figure 10.3 Leaching efficiency of $LiFePO_4$ lithium-ion batteries for various inorganic acid leaching wastes.

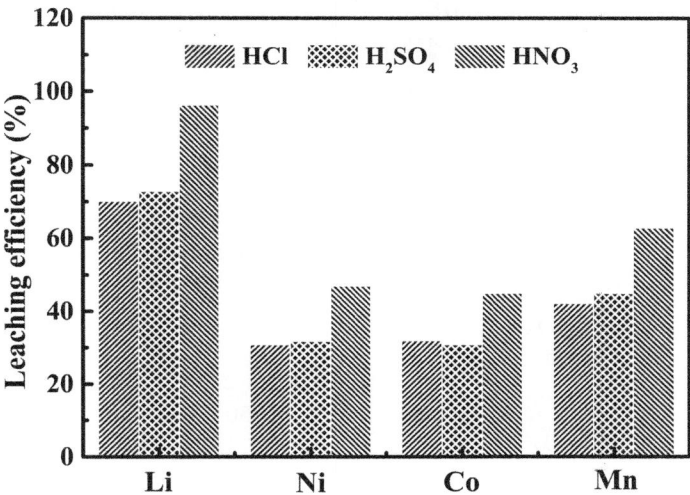

Figure 10.4 Leaching efficiency of $LiNi_xCo_yMn_{1-x-y}O_2$ lithium-ion batteries for various inorganic acid leaching wastes.

$$2LiNi_xCo_yMn_{1-x-y}O_2 + (6-4y)H^+ =$$
$$2Li^+ + 2xNi^{2+} + yCo_2O_3 + 2(1-x-y)Mn^{2+} + (3-2y)H_2O + \frac{1-y}{2}O_2 \tag{10.3}$$

$$2LiNi_xCo_yMn_{1-x-y}O_2 + (6-6x)H^+ =$$
$$(2-2x)Li^+ + xLi_2Ni_2O_4 + 2yCo^{2+} + 2(1-x-y)Mn^{2+}$$
$$+ (3-3x)H_2O + \frac{1-x}{2}O_2 \tag{10.4}$$

To solve the problem of a single inorganic acid leaching system being unable to effectively recover waste $LiNi_xCo_yMn_{1-x-y}O_2$ materials, a small amount of reducing agent is often added to the leaching system to increase its leaching efficiency. Common leaching reducing agents include H_2O_2, Na_2SO_3, glucose, $NaBH_4$, and Fe^{2+}, and the reaction equation is shown as Eq. (10.5):

$$LiNi_xCo_yMn_{1-x-y}O_2 + 3H^+ + \frac{1}{2}H_2O_2 = Li^+ + xNi^{2+} + yCo^{2+}$$
$$+ (1-x-y)Mn^{2+} + 2H_2O + \frac{1}{2}O_2 \tag{10.5}$$

According to current research results, the leaching concentration, amount of reducing agent, leaching temperature, leaching time, solid−liquid ratio and other factors considerably influence leaching efficiency. Table 10.5 lists the leaching rates of various metals in different inorganic acid leaching systems.

In practice, hydrochloric acid is highly corrosive and volatile and can easily pollute the air in the workshop. Consequently, leaching equipment and workshops must be anticorrosive. Nitric acid, due to its high price, is associated with low economic benefits.

Table 10.5 Leaching efficiencies of spent lithium-ion materials in inorganic acid leaching system.

Materials	Leaching agent	Temperature (°C)	S/L ratio (g/L)	Time (h)	Leaching rate (%)
$LiCoO_2$	2 M H_2SO_4 + 2.0 vol% H_2O_2	60	33	2	Co: 96.3, Li: 87.5
$LiCoO_2$	2 M H_2SO_4 + 5.0 vol% H_2O_2	80	50	1	Co: >99, Li: >99
$LiCoO_2$	2 M H_2SO_4 + 5.0 vol% H_2O_2	75	100	1	Li: 99.1, Co: 70.0
$LiCoO_2$	2 M H_2SO_4 + 8.0 vol% H_2O_2	75	50	1	Co: 98
$LiCoO_2$	2 M H_2SO_4 + 6.0 vol% H_2O_2	60	100	1	Co: 98, Li: 97
$LiCoO_2$	3 M H_2SO_4 + 0.4 g/g cellulose	95	25	2	Co: 54; Li: 100
$LiCoO_2$	3 M H_2SO_4 + 0.4 g/g glucose	95	25	2	Co: 96, Li: 100
$LiCoO_2$	3 M H_2SO_4 + 0.4 g/g glucose	95	25	2	Co: 98, Li: 96
$LiCoO_2$	1.0 M HNO_3 + 1.7 vol% H_2O_2	80	10	0.5	Co: ~99, Li: ~99
$LiCoO_2$	0.7 M H_3PO_4 + 4 vol% H_2O_2	75	50	1	Co: 99.7, Li: 99.9
$LiNi_xCo_yMn_{1-x-y}O_2$	1 M H_2SO_4 + 5 wt% H_2O_2	50	50	4	Li: 94.5, Co: 79.2, Ni: 96.4, Mn: 84.6
$LiNi_xCo_yMn_{1-x-y}O_2$	2 M H_2SO_4 + 6 wt% H_2O_2	100	100	1	Co: 99
$LiNi_xCo_yMn_{1-x-y}O_2$	4 M H_2SO_4 + 10 wt% H_2O_2	100	100	2	Co: 95, Li: 96
$LiNi_xCo_yMn_{1-x-y}O_2$	1 M H_2SO_4 + 0.78 wt% $NaHSO_3$	20	20	4	Li: 96.7, Co: 91.6, Ni: 96.4, Mn: 87.9

Therefore, sulfuric acid is often chosen as the leaching acid in industrial hydrometallurgical waste lithium-ion battery recycling.

2.1.2 Organic acid

Organic acid leaching refers to the process in which the chemical bonds between the atoms of the waste active material of lithium-ion batteries are broken, and the powders gradually decompose into various stable metal ions in the reducing organic acid aqueous solution. For organic acid molecules, the chemical bond between the carbon atom connected with $-OH$ or $-COOH$ and another connected carbon atom is easy to break, resulting in strong reducibility. Reductive organic acids can overcome the disadvantage of insufficient reductive ability when the inorganic acid leaches the active substance of waste $LiNi_xCo_yMn_{1-x-y}O_2$ lithium-ion batteries and are thus often applied in the study of hydrometallurgical recovery. At present, citric, tartaric, oxalic, malic, aspartic, methanesulfonic, formic, acetic, and ascorbic organic acids have a nearly 100% leaching effect in the hydrometallurgical recovery of waste $LiNi_xCo_yMn_{1-x-y}O_2$ materials. The relevant leaching conditions and leaching efficiency are listed in Table 10.6.

Although organic leaching technology has only recently been developed, it has emerged as a popular research topic in the domain of hydrometallurgical recovery due to its high leaching efficiency. The advantages and disadvantages of organic acid leaching are as follows:

Advantages:
(i) Most organic acids are mild acids, and thus, the pH of the obtained leaching solution is not extremely low. In the subsequent removal of iron, copper, and other impurities by adjusting the pH, alkaline additive use can be considerably reduced, which can decrease the reclamation cost.
(ii) Organic acids are composed of carbon, hydrogen, and oxygen and do not generate Cl_2, SO_2, or other harmful substances.
(iii) Organic acids are usually strongly reductive and can enable the recycling of waste $LiNi_xCo_yMn_{1-x-y}O_2$ materials without a reductant.

Disadvantages:
(i) The price of organic acids is generally extremely high, generally 5–20 times that of inorganic acids.
(ii) The leaching rates of organic acids are generally lower than those of inorganic acids.
(iii) The solid–liquid ratio (usually less than 50 g/L) is a key feature of the inorganic acid leaching system of hydrometallurgical recovery of waste lithium-ion batteries. However, the organic acid leaching process usually requires a lower solid-liquid ratio, resulting in the need for a large amount of aqueous solution.

2.2 Alkaline leaching process

Ammonia leaching is often applied in metallurgical processes to extract nickel and cobalt from ores. The specific principle involves alkaline (pH of 8 to 11) solutions, with most metal ions precipitated in the form of hydroxides. However, when the solution contains NH_4^+,

Table 10.6 Leaching efficiencies of spent lithium-ion materials in organic acid leaching system.

Materials	Leaching agent	Temperature (°C)	S/L ratio (g/L)	Time (min)	Leaching rate (%)
$LiCoO_2$	1.5 M succinic acid + 4 vol% H_2O_2	70	15	40	Co: 100, Li: 96
$LiCoO_2$	1.25 M ascorbic acid	70	25	20	Co: 94.8, Li: 98.5
$LiCoO_2$	1.25 M citric acid + 1.0 vol% H_2O_2	90	20	30	Co: 91, Li: 99
$LiCoO_2$	n(citric acid)/n($LiCoO_2$) = 4:1 + 1.0 vol% H_2O_2	90	15	300	Co: 99.07
$LiCoO_2$	1 M $H_2C_2O_4$	80	50	120	Li: 98
$LiCoO_2$	2 M H_3Cit + 0.6 g H_2O_2/g cathode materials	70	50	80	Co: 98, Li: 99
$LiCoO_2$	1.25 M citric acid + 1 vol% H_2O_2	90	20	30	Co: >90, Li: ~100
$LiCoO_2$	1.5 M malic acid + 2 vol% H_2O_2	90	20	40	Co: >90, Li: ~100
$LiCoO_2$	1.5 M aspartic acid + 4 vol% H_2O_2	90	10	120	Co: ~60, Li: ~60
$LiCoO_2$	n(formic acid):n($LiCoO_2$) = 10:1	60	20	20	Co: 99.96, Li: 99.9
$LiCoO_2$	1.5 M DL-malic acid + 2.0 vol% H_2O_2	90	20	40	Co: 93, Li: 94
$LiNi_xCo_yMn_{1-x-y}O_2$	1.2 M DL-malic acid + 1.5 vol% H_2O_2	90	40	30	Li: 98.9, Co: 94.3, Ni: 95.1, Mn: 96.4
$LiNi_xCo_yMn_{1-x-y}O_2$	1 M acetic acid and 6 vol % H_2O_2	70	20	60	Li: 98.4, Co: 97.7, Ni: 97.3, Mn: 97.1
$LiNi_xCo_yMn_{1-x-y}O_2$	2 M maleic acid and 4 vol % H_2O_2	70	20	60	Li: 98.2, Co: 98.4, Ni: 98.1, Mn: 98.1

Table 10.6 Leaching efficiencies of spent lithium-ion materials in organic acid leaching system.—cont'd

Materials	Leaching agent	Temperature (°C)	S/L ratio (g/L)	Time (min)	Leaching rate (%)
$LiNi_xCo_yMn_{1-x-y}O_2$	1 M citric acid + 12 vol % H_2O_2	60	80	40	Total metals: >98
$LiNi_xCo_yMn_{1-x-y}O_2$	0.5 M citric acid + 1.5 vol % H_2O_2	90	40	60	Li: 99.1, Co: 99.8, Ni: 98.7, Mn: 95.2
$LiNi_xCo_yMn_{1-x-y}O_2$	3.5 M acetic acid + 4 vol % H_2O_2	60	20	60	Li: 99.97, Co: 93.6, Ni: 92.7, Mn: 96.3

cobalt, nickel, and NH_4^+ can be combined to form stable complex compounds. The complex generally exhibits a steady state with ions in the solution owing to the large solubility product. When ammonia ions are introduced as complexes in the hydrometallurgical alkaline leaching recovery of waste lithium-ion batteries, the leaching solution contains only lithium, nickel, cobalt and copper, which have the largest recovery value, and the other metal ions precipitate as hydroxides into the leaching residue. The leaching system avoids the introduction of elemental impurities in the leaching solution, thereby considerably reducing the follow-up processes of impurity removal and metal extraction. Under this scenario, lithium cannot be precipitated from the aqueous solution because of the large solubility product of LiOH, and nickel, cobalt, and copper are in equilibrium in the aqueous solution due to the formation of the corresponding stable complexes with NH_4^+. Table 10.7 presents the stability constants of various nickel, cobalt, and copper complexes.

The stability constant β for $Ni(NH_3)_6^{2+}$ can be defined as follows:

$$\lg\beta = \frac{Ni(NH_3)_6^{2+}}{Ni^{2+} \cdot (NH_3)^6} \tag{10.6}$$

Table 10.7 Stability constants of various nickel, cobalt, and copper complexes at 298 K.

Complexes	$\lg\beta_i$	Complexes	$\lg\beta_i$	Complexes	$\lg\beta_i$
$Cu(NH_3)_2^+$	10.58	$Ni(NH_3)^{2+}$	2.80	$Co(NH_3)^{2+}$	2.05
$Cu(NH_3)^{2+}$	4.12	$Ni(NH_3)_2^{2+}$	5.04	$Co(NH_3)_2^{2+}$	3.62
$Cu(NH_3)_2^{2+}$	7.63	$Ni(NH_3)_3^{2+}$	6.77	$Co(NH_3)_3^{2+}$	4.61
$Cu(NH_3)_3^{2+}$	10.51	$Ni(NH_3)_4^{2+}$	7.96	$Co(NH_3)_4^{2+}$	5.31
$Cu(NH_3)_4^{2+}$	12.60	$Ni(NH_3)_5^{2+}$	8.71	$Co(NH_3)_5^{2+}$	5.43
$Cu(NH_3)_5^{2+}$	12.43	$Ni(NH_3)_6^{2+}$	8.74	$Co(NH_3)_6^{2+}$	4.75
$Cu(NH_3)(OH)^+$	14.9	$Ni(OH)_2$	8.55	$Co(OH)_2$	5.20
$Cu(NH_3)(OH)_3^-$	16.3	—	—	—	—
$Cu(NH_3)_2(OH)_2$	15.7	—	—	—	—
$Cu(OH)_2$	12.8	—	—	—	—

2.2.1 Nickel and cobalt mechanism

In the ammonia leaching system for recycling waste $LiNi_xCo_yMn_{1-x-y}O_2$ lithium-ion batteries, ammonia is used as a complex compound to realize the separation of nickel and cobalt from other metal ions. To increase the leaching efficiency of nickel and cobalt, reducing agents, such as H_2O_2, Na_2SO_3, and $NaHSO_3$, are often added to the solution. Fig. 10.5 illustrates the influence of various reducing agents on the leaching efficiency of the target extraction metal. The leaching conditions are 30 wt% ammonia, 0.1 mol/L reducing agent concentration, 20 g/L solid—liquid ratio, leaching time of 6 h, and leaching temperature of 40°C. The specific reaction is shown in Eqs. (10.7)—(10.9). Nickel and cobalt form complexes with ammonia molecules, thereby maintaining equilibrium in an aqueous solution. The specific reaction is shown in Eqs. (10.10)—(10.21).

$$LiNi_xCo_yMn_{1-x-y}O_2 + H_2O + \frac{1}{2}H_2O_2 = Li^+ + xNi^{2+} + yCo^{2+} \\ + (1-x-y)Mn^{2+} + 3OH^- + \frac{1}{2}O_2 \quad (10.7)$$

$$LiNi_xCo_yMn_{1-x-y}O_2 + \frac{3}{2}H_2O + \frac{1}{2}SO_3^{2-} = Li^+ + xNi^{2+} + yCo^{2+} \\ + (1-x-y)Mn^{2+} + 3OH^- + \frac{1}{2}SO_4^{2-} \quad (10.8)$$

$$LiNi_xCo_yMn_{1-x-y}O_2 + \frac{3}{2}H_2O + \frac{1}{2}HSO_3^- = Li^+ + xNi^{2+} + yCo^{2+} \\ + (1-x-y)Mn^{2+} + \frac{7}{2}OH^- + \frac{1}{2}SO_4^{2-} \quad (10.9)$$

$$Ni^{2+} + NH_3 = Ni(NH_3)^{2+} \quad (10.10)$$

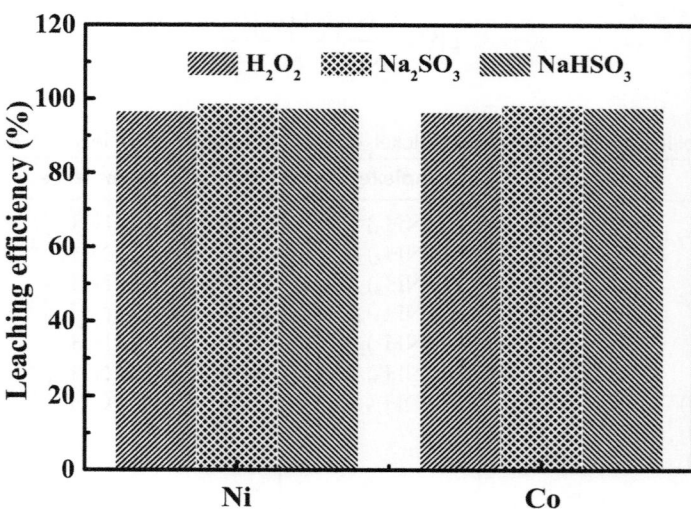

Figure 10.5 Influence of various reducing agents on leaching efficiency of Ni and Co.

$$Ni^{2+} + 2NH_3 = Ni(NH_3)_2^{2+} \tag{10.11}$$

$$Ni^{2+} + 3NH_3 = Ni(NH_3)_3^{2+} \tag{10.12}$$

$$Ni^{2+} + 4NH_3 = Ni(NH_3)_4^{2+} \tag{10.13}$$

$$Ni^{2+} + 5NH_3 = Ni(NH_3)_5^{2+} \tag{10.14}$$

$$Ni^{2+} + 6NH_3 = Ni(NH_3)_6^{2+} \tag{10.15}$$

$$Co^{2+} + NH_3 = Co(NH_3)^{2+} \tag{10.16}$$

$$Co^{2+} + 2NH_3 = Co(NH_3)_2^{2+} \tag{10.17}$$

$$Co^{2+} + 3NH_3 = Co(NH_3)_3^{2+} \tag{10.18}$$

$$Co^{2+} + 4NH_3 = Co(NH_3)_4^{2+} \tag{10.19}$$

$$Co^{2+} + 5NH_3 = Co(NH_3)_5^{2+} \tag{10.20}$$

$$Co^{2+} + 6NH_3 = Co(NH_3)_6^{2+} \tag{10.21}$$

2.2.2 Copper mechanism

The copper leaching mechanism is different from that of nickel and cobalt. Copper is oxidized from Cu(0+) to Cu^+ or Cu^{2+} under the action of oxygen because leaching is performed in air. The reaction is shown in Eqs. (10.22) and (10.23). The reaction speed is relatively low because of the slow kinetic process. Reducing the stirring speed, reaction temperature, and reaction time can effectively reduce the leaching efficiency of copper, thereby decreasing the content of copper impurities in the leaching solution.

$$Cu + \frac{1}{4}O_2 + \frac{1}{2}H_2O = Cu^+ + OH^- \tag{10.22}$$

$$Cu + \frac{1}{2}O_2 + H_2O = Cu^{2+} + 2OH^- \tag{10.23}$$

Cu^+ and Cu^{2+} can form complexes with ammonia molecules, thereby maintaining equilibrium in an aqueous solution. The specific reaction is shown in Eqs. (10.24)–(10.32).

$$Cu^+ + 2NH_3 = Cu(NH_3)_2^+ \tag{10.24}$$

$$Cu^{2+} + NH_3 = Cu(NH_3)^{2+} \tag{10.25}$$

$$Cu^{2+} + 2NH_3 = Cu(NH_3)_2^{2+} \tag{10.26}$$

$$Cu^{2+} + 3NH_3 = Cu(NH_3)_3^{2+} \tag{10.27}$$

$$Cu^{2+} + 4NH_3 = Cu(NH_3)_4^{2+} \tag{10.28}$$

$$Cu^{2+} + 5NH_3 = Cu(NH_3)_5^{2+} \qquad (10.29)$$

$$Cu^{2+} + NH_3 + OH^- = Ni(NH_3)OH^+ \qquad (10.30)$$

$$Cu^{2+} + 5NH_3 + 2OH^- = Ni(NH_3)_5(OH)_2 \qquad (10.31)$$

$$Cu^{2+} + 5NH_3 + 3OH^- = Ni(NH_3)_5(OH)^- \qquad (10.32)$$

2.2.3 Lithium mechanism

With the breaking of the metal–oxygen bond and collapse of the layered structure of the $LiNi_xCo_yMn_{1-x-y}O_2$ material, lithium ions enter the solution without participating or undergoing any reaction or change in valence. In the ammonia leaching recovery system of waste lithium-ion batteries, the maximum pH of ammonia water is 11.58, and the pH of the leaching solution must be maintained at less than 12. Table 10.8 lists the hydroxide concentrations required to initiate and complete the precipitation of LiOH. Lithium ions cannot be precipitated from the ammonia leaching solution due to the high solubility of LiOH, although the lithium ions extracted from the layered structure do not form a stable complex compound with ammonia molecules.

2.2.4 Manganese mechanism

The leaching mechanism of manganese is similar to that of nickel and cobalt, as shown in Eqs. (10.7)–(10.9). In an ammonia solution, the complex formed by manganese and ammonia exhibits limited stability and is prone to precipitation. Table 10.9 lists the stability constants of the manganese–ammonia complex. The stability constants of $Mn(NH_3)^{2+}$ and $Mn(NH_3)_2^{2+}$ are only 0.8 and 1.3, respectively, indicating that the forward reaction constants of Eqs. (10.33) and (10.34) are low, with a single manganese ion constituting a dominant proportion of the solution. In the solution of the alkali leaching system, due to the high pH of the solution, manganese combines with the hydroxide to

Table 10.8 Hydroxide concentration required to initiate and complete the precipitation of LiOH.

Material	Solubility	Initiate OH⁻	Complete OH⁻
LiOH	12.84	2.86×10^2 mol/L	2.86×10^6 mol/L

Table 10.9 Stability constants of various manganese complexes at 298 K.

Complexes	lgβᵢ
$Mn(NH_3)^{2+}$	0.8
$Mn(NH_3)_2^{2+}$	1.3

form a manganese hydroxide precipitate, as shown in Fig. 10.2, resulting in the separation of manganese ions from the solution.

$$Mn^{2+} + NH_3 = Mn(NH_3)^{2+} \quad (10.33)$$

$$Mn^{2+} + 2NH_3 = Mn(NH_3)_2^{2+} \quad (10.34)$$

2.2.5 Iron and aluminum mechanism

Fe and Al are elemental impurities in the recycling process of waste $LiNi_xCo_yMn_{1-x-y}O_2$ lithium-ion batteries. In the leaching process, the iron shell and aluminum current collector powders are oxidized to Fe^{2+}, Fe^{3+}, and Al^{3+} by O_2; the specific reaction is shown in Eqs. (10.35)–(10.37). In the ammonia leaching recovery system, complexation between Fe^{2+} (Fe^{3+}), Al^{3+}, and molecules does not occur, thus promoting the precipitation reaction for the hydroxide form.

$$Fe + \frac{1}{2}O_2 + H_2O = Fe^{2+} + 2OH^- = Fe(OH)_2 \quad (10.35)$$

$$Fe + \frac{3}{4}O_2 + \frac{3}{2}H_2O = Fe^{3+} + 3OH^- = Fe(OH)_3 \quad (10.36)$$

$$Al + \frac{3}{4}O_2 + \frac{3}{2}H_2O = Al^{3+} + 3OH^- = Al(OH)_3 \quad (10.37)$$

2.3 Redox leaching process

In recent years, redox leaching has gradually developed as a lithium-ion battery recycling leaching method. In this method, a single oxidant (or reducing agent) is used to realize the oxidation (reduction) leaching of $LiFePO_4$ ($LiNi_xCo_yMn_{1-x-y}O_2$) to obtain a leaching solution containing only lithium and leaching residue, including other elemental impurities. This approach efficiently shortens the process of lithium recovery. The main principle of the oxidation leaching of $LiFePO_4$ involves an oxidant in the solution to oxidize $LiFePO_4$ to $FePO_4$ and release lithium ions that combine with the oxidant or its derived anion to form soluble lithium salts. Because no hydrogen ion is introduced into the leaching system, the solution pH is not low—generally greater than 5. In this case, $FePO_4$, Cu^{2+}, Al^{3+}, and other impurities cannot be maintained in the ionic state in the solution balance, and only soluble lithium salt is dissolved in the solution, thereby achieving the separation of lithium and other elements. Similarly, owing to the high pH of the $LiNi_xCo_yMn_{1-x-y}O_2$ leaching system, the leaching solution contains only lithium ions, which enables the efficient recovery of lithium. This recycling process considerably shortens the process of lithium recovery, and the whole process is environmentally friendly.

2.3.1 Oxidization leaching

The oxidation leaching process refers to the leaching recovery system of LiFePO$_4$ based on the oxidation of LiFePO$_4$ to FePO$_4$. The commonly studied oxidation leaching systems include mild S$_2$O$_8^{2-}$, Fe^{3+}, and ClO$^-$ leaching systems, and the specific reaction processes are shown in Eqs. (10.38)–(10.41). Furthermore, several strong oxidants, including MnO$_4^-$, ClO$_3^-$, and ClO$_4^-$, exhibit excellent lithium extraction efficiency for the oxidative leaching recovery of LiFePO$_4$; however, the strong oxidizing properties and high costs limit the application of such oxidants in research and industrial domains.

$$LiFePO_4 + \frac{1}{2}S_2O_8^{2-} = FePO_4 + Li^+ + SO_4^{2-} \qquad (10.38)$$

$$LiFePO_4 + Fe^{3+} + 2OH^- = FePO_4 + Li^+ + Fe(OH)_2 \qquad (10.39)$$

$$LiFePO_4 + Fe^{3+} + 2OH^- + \frac{1}{4}O_2 + \frac{1}{2}H_2O = FePO_4 + Li^+ + Fe(OH)_3 \qquad (10.40)$$

$$LiFePO_4 + \frac{1}{2}H_2O + ClO^- = FePO_4 + Li^+ + 2OH^- \qquad (10.41)$$

In recent years, the oxidation leaching recovery of LiFePO$_4$ has attracted considerable research attention. The advantages and disadvantages of this method are as follows:

Advantages:
(i) Oxidation leaching can enable one-step separation of lithium from iron, aluminum, copper, and other impurities.
(ii) No acid wastewater is produced in the process of oxidation leaching, and thus, the process does not pose a threat to the environment.
(iii) Oxidation leaching systems are not highly corrosive, leading to less stringent requirements for equipment.

Disadvantages:
(i) The oxidant in the oxidation leaching system is more expensive than inorganic acid, resulting in a low economic benefit.
(ii) Generally, the kinetic process of oxidation leaching is slow, and the reaction time is significant.

2.3.2 Reduction leaching

Similar to oxidation leaching, reduction leaching generally refers to the hydrometallurgical recovery of LiNi$_x$Co$_y$Mn$_{1-x-y}$O$_2$ materials by transitioning metals from high to low states. Commonly used reduction leaching systems include SO$_3^{2-}$ and Fe^{2+} leaching systems. The specific reaction is shown in Eqs. (10.42) and (10.43):

$$LiNi_xCo_yMn_{1-x-y}O_2 + \frac{1}{2}SO_3^{2-} + 3H^+ = Li^+ + xNi^{2+} + yCo^{2+}$$
$$+ (1-x-y)Mn^{2+} + \frac{1}{2}SO_4^{2-} + \frac{3}{2}H_2O \qquad (10.42)$$

$$\begin{aligned}\text{LiNi}_x\text{Co}_y\text{Mn}_{1-x-y}\text{O}_2 + \text{Fe}^{2+} + 4\text{H}^+ &= \text{Li}^+ + x\text{Ni}^{2+} + y\text{Co}^{2+} \\ &+ (1-x-y)\text{Mn}^{2+} + \text{Fe}^{3+} + 2\text{H}_2\text{O}\end{aligned}$$
(10.43)

In practice, the leaching solution obtained by recycling $\text{LiNi}_x\text{Co}_y\text{Mn}_{1-x-y}\text{O}_2$ materials via reduction leaching often contains a large volume of Ni^{2+}, Co^{2+}, and Mn^{2+}, limiting the direct separation of lithium and other elements. In the reduction leaching process to recover $\text{LiNi}_x\text{Co}_y\text{Mn}_{1-x-y}\text{O}_2$ materials, the commonly applied sulfites and divalent iron salts are usually acidic salts, resulting in a leaching solution with a low pH. According to Fig. 10.2, the pH of the deposited transition metals (Ni, Co, and Mn) in the solution is greater than 7, and as a result, large amounts of Ni^{2+}, Co^{2+}, and Mn^{2+} can be found in the leaching solution. In terms of existing separation technologies, effective methods for separating lithium, nickel, cobalt, and manganese ions in solution are lacking. Therefore, the leaching of Ni^{2+}, Co^{2+}, and Mn^{2+} during lithium extraction increases the difficulty of subsequent element separation and extraction steps.

2.4 Electrochemical leaching process

According to the Eh−pH diagram of the hydrometallurgy process, the extraction and separation of elements depend not only on the change in the solution pH, which promotes precipitation, but also on the increase in the external potential of the solution, which changes the state of the ion. At present, water solution systems are the most commonly used electrochemical recovery systems because of their convenient operation mechanisms, nonstringent equipment requirements, easy automation, high productivity, safety, and high economic value. In the design of electrochemical systems, the cathode material of waste lithium-ion batteries is usually used as the anode, and inert materials, such as graphite and platinum sheets, are used as the cathode. The anode (waste cathode material) undergoes oxidation under the action of an external potential, while the cathode (inert material) undergoes reduction and releases H_2. The commonly used electrolytes can be divided into neutral electrolytes (sodium chloride solution) and acidic electrolytes (ammonium sulfate solution). Notably, only a few researchers have focused on alkaline electrolyte electrochemical recovery systems. In addition, only limited research has been conducted on molten salt electrochemical systems, ionic liquid electrochemical systems, and other targeted recovery techniques of waste lithium-ion batteries. The electrochemical recovery of waste lithium-ion battery cathode materials involves a high current efficiency and simple operating conditions. The separation efficiency of different metal ions can be controlled by changing the magnitude of the applied potential. Domestic and foreign researchers have successfully applied this method to recycle current mainstream lithium-ion batteries, such as waste LiFePO_4 batteries, waste LiCoO_2 batteries, and waste $\text{LiNi}_x\text{Co}_y\text{Mn}_{1-x-y}\text{O}_2$ batteries, and summarized the optimal experimental conditions. In this context, the electrochemical leaching process of water solutions is a key research topic for the hydrometallurgical recovery of waste lithium-ion battery cathode materials.

In the process of electrochemical recovery, due to the limitation of the water decomposition voltage (when the applied voltage exceeds 1.23 V, the electrolytic anode begins to release O_2), only the oxidation of Fe^{2+} to Fe^{3+} and reduction in Ni^{3+} (Co^{3+}, Mn^{3+}) to Ni^{4+} (Co^{3+}, Mn^{3+}) can be realized. The specific reaction is presented in Eqs. (10.44)–(10.46). Therefore, lithium ions emerge from the anode and enter the solution as ions under the action of an external potential, forming a lithium–ion–external electrolyte solution. Other elements in the lithium-ion battery, such as iron, nickel, cobalt, manganese, and aluminum, continue to adhere to the anode, enabling the separation of lithium and other metallic impurities.

$$LiFePO_4 - \delta e^- = \delta Li^+ + Li_{1-\delta}FePO_4 \quad (10.44)$$

$$LiCoO_2 - \delta e^- = \delta Li^+ + Li_{1-\delta}CoO_2 \quad (10.45)$$

$$LiNi_xCo_yMn_{1-x-y}O_2 - \delta e^- = \delta Li^+ + Li_{1-\delta}Ni_xCo_yMn_{1-x-y}O_2 \quad (10.46)$$

2.5 Microbial leaching process

In the microbial leaching process, microorganisms separate valuable metal elements from waste lithium-ion battery powder after pretreatment. Recycling waste lithium-ion batteries by biological methods has the advantages of low acid consumption, low cost, and simple operation. The process also has disadvantages, such as a considerable leaching time, use of bacteria that are difficult to culture, pollution, and difficult separation of the leached solution ions. Thus, it is important to study the microorganism-based leaching of waste lithium-ion batteries under the present scenario involving resource shortages and increasingly severe environmental pollution.

In 1955, Zimmerle applied for the first bioleaching patent, corresponding to the first application of bioleaching technologies in industrial production. At present, microbial leaching technology can be divided into three types according to the type of reaction:

(i) Direct acid leaching: minerals are leached by microorganisms through the acids produced by bacterial metabolism.
(ii) Direct bacterial leaching: bacteria directly use minerals as nutrients and leach minerals through their own metabolism.
(iii) Indirect bacterial leaching: bacteria leach minerals by using other elements as nutrients and produce other substances with redox properties through their own metabolism.

Notably, the development of microbial leaching for waste lithium-ion batteries has been slow, and only a few studies have demonstrated the feasibility of this approach.

Table 10.10 Leaching bacteria and their main properties.

Bacteria	Physiological reaction	Optimum pH	Optimum temperature
Thiobacillus ferrooxidans	$Fe^{2+} \rightarrow Fe^{3+}$	1.8~2.5	30~35°C
Acidithiobacillus thiooxidans	$Fe^{2+} \rightarrow Fe^{3+}$	2~2.5	28~30°C
Leptospirillum ferrooxidans	$Fe^{2+} \rightarrow Fe^{3+}$	2.5~3.0	45~50°C
Sulfolobus	$Fe^{2+} \rightarrow Fe^{3+}$	2~3	55~80°C
Acidianus	$Fe^{2+} \rightarrow Fe^{3+}$	1.5~2	45~75°C

Table 10.10 lists certain microorganisms that can be used to leach waste lithium-ion batteries and the optimum conditions for leaching.

In addition, many bacteria can leach oxide ore during microbial metallurgy, and the mechanism has been explained. For example, the fungi *Penicillium notatum* and *Aspergillus niger* can be used to extract lithium from $LiAlSi_2O_6$, and heterotrophic microorganisms, such as *Bacillus megaterium, B. circulans, Pseudomonas,* and *A. niger*, can be used to extract nickel from silicon magnesium nickel slag (garnierite-SiO_2). The reasonable application of these microbial leaching methods to waste lithium-ion battery leaching systems may be a popular research topic for lithium-ion battery recycling in the future.

3. Separation and extraction technologies

After pretreatment and the leaching process of waste lithium-ion batteries, a leaching solution containing various ions (Li^+, Ni^{2+}, Co^{2+}, Mn^{2+}, Fe^{2+}, Fe^{3+}, Al^{3+}, Cu^{2+}) can be obtained. Considering the most complex leaching solution: acid leaching + redox additive leaching system as an example, this section describes the separation and extraction of various elements from waste $LiFePO_4$ and $LiNi_xCo_yMn_{1-x-y}O_2$ lithium-ion battery leaching solutions. Tables 10.11 and 10.12 present the contents of major elements in the leaching solution (the experimental conditions are 1 mol/L H_2SO_4 +5 wt% H_2O_2, 200 g/L, 60°C, and 60 min).

The simplest separation method is to separate each element and prepare a high-purity compound salt; however, this method requires a specific extraction process for each ion, which is difficult to implement and has a low recovery efficiency. According to the final function of different elements, researchers have proposed the simultaneous extraction of

Table 10.11 Main elemental contents (wt%) of pretreatment powder of waste $LiFePO_4$ lithium-ion batteries.

Element	Li	Fe	P	Al	Cu	F
Content (g/L)	5.49	37.41	20.71	3.51	7.86	0.014

Table 10.12 Main elemental contents (wt%) of pretreatment powder of waste $LiNi_{0.6}Co_{0.2}Mn_{0.2}O_2$ lithium-ion batteries.

Element	Li	Ni	Co	Mn	Al	Cu	F
Content (g/L)	8.26	39.53	12.91	13.01	4.28	10.01	0.021

multiple elements to form compound salts containing multiple target elements. The extraction process is as follows:

(i) Through ion exchange, precipitation, extraction, membrane separation, and other methods, each ion is separated and resynthesized into the corresponding industrial raw materials.

(ii) For the leaching solution of $LiFePO_4$, the ratio of Fe and P in the solution and pH of the solution are adjusted to ensure the prioritized removal of Al, Cu, and other elements such that Fe and P are completely precipitated in the form of $FePO_4$, and the leaching solution containing only lithium is obtained. The obtained $FePO_4$ and lithium salt are resynthesized using a high-temperature solid-state method to prepare $LiFePO_4$ as the cathode material of lithium-ion batteries. This method involves a high recovery with reasonable material properties and high economic benefits and has emerged as the main method for the hydrometallurgical recovery of waste $LiFePO_4$ batteries.

(iii) For the leaching solution of $LiFePO_4$, after removing Al, Cu, and F, the ratio of Li, Fe, and P in the solution and pH are directly adjusted, and the cathode material $LiFePO_4$ for lithium-ion batteries is directly prepared using a hydrothermal method or spray pyrolysis method. This method corresponds to the highest recovery rate of lithium and the shortest technological process; however, this method is not suitable for industrial production due to the extremely harsh synthetic conditions of $LiFePO_4$.

(iv) In the leaching solution of $LiNi_xCo_yMn_{1-x-y}O_2$ materials, lithium, nickel, cobalt, and manganese exhibit high recovery values and can be directly prepared as new cathode materials for lithium-ion batteries. Therefore, Al^{3+}, Fe^{2+}, Fe^{3+}, and Cu^{2+} are first removed from the leaching solution, and the $Ni_xCo_yMn_{1-x-y}O_2$ precursor and lithium-containing solution are prepared. The obtained $Ni_xCo_yMn_{1-x-y}O_2$ and lithium salts can be resynthesized into $LiNi_xCo_yMn_{1-x-y}O_2$ cathode materials through high-temperature sintering. This method is similar to that specified in category 2 and is presently the mainstream method for recycling waste $LiNi_xCo_yMn_{1-x-y}O_2$ batteries.

(v) Similarly, for the leaching solution of $LiNi_xCo_yMn_{1-x-y}O_2$ materials, after removing Al^{3+}, Fe^{2+}, Fe^{3+} and Cu^{2+}, the ratio of Li^+, Ni^{2+}, Co^{2+}, Mn^{2+} in the solution and the pH are directly adjusted, and the cathode material $LiNi_xCo_yMn_{1-x-y}O_2$ for lithium-ion batteries is directly prepared using a hydrothermal method or spray pyrolysis method. Compared with the direct synthesis method of LiFePO$_4$ pertaining to category 3, the synthesis conditions of LiNi$_x$Co$_y$Mn$_{1-x-y}$O$_2$ materials in this category are more demanding.

The five categories only correspond to leaching solutions containing a variety of ions obtained from the leaching system of acid leaching + redox additives. Other superior solutions may exist for different leaching solutions. For example, the leaching solution obtained by recycling waste LiNi$_x$Co$_y$Mn$_{1-x-y}$O$_2$ materials in an alkaline leaching system only contains Li$^+$, Ni^{2+}, and Co^{2+}. Only a small part of the category 4 process, that is, the preparation of nickel-cobalt oxide precursor and precipitation of lithium salt, can be implemented to realize the separation and extraction of ions in the leaching solution. Therefore, this section does not describe the specific recycling process of the leaching solution and only elaborates upon a part of the process, such as iron removal and precursor preparation. Readers can refer to the abovementioned five categories and different characteristics of leaching liquids to perform the corresponding modifications and select the appropriate extraction process.

3.1 Impurity removal

In the leaching solution of waste lithium-ion batteries, except for the elements that can be resynthesized as new cathode materials, other elements can be considered elemental impurities, mainly Fe, Al, Cu, and F.

3.1.1 Copper removal
3.1.1.1 Copper removal by replacement precipitation

The principle of replacement precipitation is that a more active metal (more negative potential) can reduce an inert metal (more positive potential) to metal, leading to precipitation in the metal ion solution. In removing copper from solution, the commonly used active metals are Ni, Fe, Zn, and Mg. The reaction equations are as follows:

$$Ni + Cu^{2+} = Cu + Ni^{2+} \tag{10.47}$$

$$Fe + Cu^{2+} = Cu + Fe^{2+} \tag{10.48}$$

$$Zn + Cu^{2+} = Cu + Zn^{2+} \tag{10.49}$$

$$Mg + Cu^{2+} = Cu + Mg^{2+} \tag{10.50}$$

To avoid introducing other impurities, Fe and Ni are active metals commonly used for removing impurities from the leaching solution of waste lithium-ion batteries. Iron is the self-contained ionic impurity in the solution, and nickel is the target metal.

Copper removal via replacement precipitation is implemented by many companies. To ensure high copper removal efficiency, the following points must be considered:

(i) The active metals must have high activity and purity.
(ii) The solution must be maintained between 60°C and 80°C, and the pH should be less than 3.5 during copper replacement removal.
(iii) The precipitation equipment should be sealed to prevent reoxidation of displaced copper.

3.1.1.2 Copper removal by sulfide precipitation

The sulfide precipitation method is based on the principle that the sulfides of many elements do not easily dissolve in water. Therefore, when the corresponding metal ions exist in the solution, S^{2-} is added, and the following precipitation reactions occur:

$$2M^{n+} + nS^{2-} = M_2S_n \tag{10.51}$$

The solubility product $Ksp = [Mn^+]^2[S^{2-}]^n$. The solubility product of several sulfides is shown in Table 10.13.

In the process of impurity removal of the leaching solution of waste lithium-ion batteries, H_2S is generally applied to a precipitator. The solution temperature in this process is generally greater than 60°C, and the pH is controlled to be between 1.8 and 2.5. The precipitating equation is as follows:

$$Cu^{2+} + H_2S = CuS + 2H^+ \tag{10.52}$$

3.1.1.3 Copper removal by extraction reaction

The extraction reaction refers to the process in which an extraction agent dissolved in an organic solvent reacts with metal ions in an aqueous solution to form compounds dissolved in an organic solvent. Because the organic solvent is not miscible with the aqueous solution, which is separate and known as the organic phase, the relative aqueous solution is named the aqueous phase. The extraction process can be divided into simple complex extraction and chelate extraction according to the types of complexes produced. The complexes and chelate extraction agents commonly used in the copper extraction process are shown in Table 10.14.

Copper is an inexpensive metal, but the extraction process to remove impurities from the leaching solution of waste lithium-ion batteries must be performed at a low cost to realize industrial application. Chelating extractants are the most commonly used

Table 10.13 Solubility product of several sulfides at 25°C.

Sulfide	K_{sp}	lgK_{sp}	Sulfide	K_{sp}	lgK_{sp}
MnS	2.8×10^{-13}	-12.25	FeS	4.9×10^{-18}	-17.31
NiS	2.8×10^{-21}	-20.55	CuS	8.9×10^{-36}	-35.05
CoS	1.8×10^{-22}	-21.74	Cu_2S	2.0×10^{-47}	-46.70

Table 10.14 Commonly used complexes and chelate extraction agents.

Type	Code	Type	Code
Complex extraction agent	P 205 Cyanex 272 Cyanex 302 Cyanex 301 LIX 1104	Chelate extraction agent	Kelex 100 LIX 54 XI 51 CLX 50 LIX 64 LIX 65N LIX 70

extractants because they can extract copper in acidic solutions without additional pH adjustment and are less expensive than other agents.

In general, due to the low copper content in the leaching solution (approximately 1–3 g/L), the copper concentration of the organic phase after extraction can reach 20–40 g/L. To increase the extraction efficiency of copper, multistage extraction is usually performed to reverse-extract the organic phase.

3.1.2 Iron removal
3.1.2.1 Iron removal using iron hydroxide

Iron hydroxide precipitation is a precipitation removal method widely used in hydrometallurgical processes. It is a simple operation that does not introduce additional ionic impurities and has a high economic benefit. The iron element in the solution usually has two bivalent and trivalent forms, and the trivalent iron ion easily forms insoluble iron hydroxide with the hydroxide in the solution due to its high potential and strong polarization ability.

In the leaching solution of waste lithium-ion batteries, iron ions exist as bivalent and trivalent ions. A certain amount of oxidant, such as hydrogen peroxide, is first added to the solution. All the iron in the solution oxidizes to the trivalent form and precipitates as iron hydroxide.

$Fe(OH)_3$ precipitated using the iron hydroxide precipitation method usually behaves as a colloid, which considerably increases the operational difficulty of the subsequent solid–liquid separation. When the concentration of iron is excessively high, the separation process may be interrupted.

3.1.2.2 Iron removal using ihleite

The chemical formula of ihleite is $A_2O \cdot 3Fe_2O_3 \cdot 4SO_4 \cdot 6H_2O$, where A represents monovalent cations, such as K^+, Na^+, and NH^{4+}. For example, the chemical formula of jarosite is $KFe_3(SO_4)_2(OH)_6$.

The ihleite method for iron removal involves precipitating Fe^{3+} in a weak acid sulfate solution (or chloride solution containing sulfate). For example, the reaction equation of jarosite precipitation is shown in Eqs. (10.53)–(10.55):

$$Fe^{3+} + SO_4^{2-} + H_2O = Fe(OH)SO_4 + H^+ \quad (10.53)$$

$$4Fe(OH)SO_4 + 4H_2O = 2Fe_2(OH)_4SO_4 + 2H_2SO_4 \quad (10.54)$$

$$Fe(OH)SO_4 + Fe_2(OH)_4SO_4 + Na^+ + H_2O = NaFe_3(SO_4)_2(OH)_6 + H^+ \quad (10.55)$$

In retreating the waste lithium-ion battery leaching solution, it is usually necessary to add an appropriate amount of an oxidizing agent, such as MnO_2, H_2O_2, or O_3, to oxidize all the Fe^{2+} in the solution to Fe^{3+} and precipitate it through ihleite precipitation. The results show that jarosite formed with K^+ is the most stable among many monovalent cations.

The formation of ihleite is usually accompanied by the generation of sulfuric acid, which must be neutralized to continue the precipitation reaction. Usually, an appropriate amount of ammonium sulfate is added as a pH buffer.

3.1.2.3 Iron removal by goethite method

Goethite is composed of α-$Fe_2O_3 \cdot H_2O$ or α-FeOOH. Goethite is an inorganic polymer, and the units represented by the molecular formula FeOOH cannot exist independently. The molecular formula for goethite must be expressed as α-$[FeOOH]_n$, where n is a large number.

The removal of iron by goethite precipitation is aimed at Fe^{2+} in the solution. In the process of removing iron from the leaching solution of waste lithium-ion batteries, it is necessary to add an appropriate amount of reducing agent, usually sulfite, to reduce all the Fe^{3+} in the solution to Fe^{2+}. Subsequently, Fe^{2+} is reoxidized to Fe^{3+} in air, and the oxidation reaction equation is described as follows:

$$4H^+ + 4Fe^{2+} + O_2 = 4Fe^{3+} + 2H_2O \quad (10.56)$$

The oxidized removal of iron by the goethite process includes two closely related reactions, specifically, the oxidation of Fe^{2+} and hydrolysis of Fe^{3+}. The hydrolysis of Fe^{3+} reduces the concentration of OH in the solution, thereby increasing the concentration of acid in the system. Therefore, the total reaction of the oxidation precipitation can be expressed as

$$\frac{1}{2}O_2 + 2Fe^{2+} + 3H_2O = 2FeOOH + 4H^+ \quad (10.57)$$

The main advantage of goethite precipitation for iron removal is that precipitation slag with a high filtration performance can be obtained without adding other alkali metal cations, the slag amount is less than that of the ihleite method, and the iron content is relatively high. However, in the process of precipitation, the anions and cations in the solution critically pollute the precipitate, which reduces the value of iron slag as a byproduct.

3.1.3 Aluminum removal

Aluminum removal based on aluminum hydroxide:

The aluminum hydroxide precipitation method is currently the main method for aluminum removal because of its simple operation, high economy, and easy-to-achieve subsequent solid—liquid separation. When the pH is 3.47, Al(OH)$_3$ begins to precipitate. When the pH is 5.81, Al(OH)$_3$ completely precipitates. When the pH is 12, Al(OH)$_3$ begins to dissolve and forms AlO$_2^-$.

In the process of impurity removal from the leaching solution of waste lithium-ion batteries, the leaching solution of LiFePO$_4$ only contains Li$^+$ and Al^{3+} (Fe^{2+}, Fe^{3+}, and Cu^{2+} have already been removed). Therefore, the pH of the aluminum removal solution can be controlled at 6—10. In the case of the LiNi$_x$Co$_y$Mn$_{1-x-y}$O$_2$ leaching solution, due to a large volume of nickel, cobalt, and manganese ions in the solution, it is necessary to ensure that Ni^{2+}, Co^{2+}, and Mn^{2+} do not precipitate when the aluminum is completely removed. Therefore, the pH of the aluminum removal solution should be controlled at six to seven.

3.1.4 Fluorine removal

To avoid the introduction of other impurities, the F removal associated with lime milk, which is commonly used in hydrometallurgy, cannot be applied to remove F from the lithium-ion battery leaching solution. This paper introduces a method of silica gel fluoride removal, the basic principle of which is as follows: in an acidic solution, fluorine in hydrofluoric acid molecular state and silica gel are polymerized and adsorbed on silica gel. In neutral or alkaline solutions, hydrofluoric acid and silica gel do not react, and only physical adsorption occurs. Therefore, after washing, silica gel can be defluorinated to realize regeneration.

3.2 Precursor preparation

The leaching solution after impurity removal contains only lithium, nickel, manganese, and cobalt (lithium, iron, phosphate) metal ions, which are necessary elements for the repreparation of cathode materials from lithium-ion batteries. Therefore, it is important to study methods to efficiently prepare lithium-ion battery cathode materials to ensure a high economy and resource circulation.

3.2.1 Lithium iron phosphate preparation

The efficient recovery of FePO$_4$ from the leaching solution containing Li$^+$, Fe^{2+}, Fe3+, and P^{5+} ensures the removal of impurities and realizes the recovery and value-added reuse of iron and phosphorus. Iron in FePO$_4$ is trivalent iron, and dihydrate is the dominant state at room temperature and pressure. The K$_{SP}$ of FePO$_4$•2H$_2$O is 9.91 × 10^{-16}.

In the leaching solution of waste LiFePO$_4$, iron usually exists in the form of Fe^{2+}. Therefore, in the process of precipitation of FePO$_4$, a certain amount of iron salt or phosphorus salt is added to equalize the concentrations of iron and phosphorus in the solution, a certain amount of NaOH is added to adjust the pH of the solution to 4.5—5.5, and H_2O_2 is added to oxidize the Fe^{2+} in the solution to precipitate $FePO_4$.

3.2.2 Lithium nickel-cobalt-manganate preparation

As described in Section 3, two methods can be used to treat the $LiNi_xCo_yMn_{1-x-y}O_2$ battery leaching solution. One method is to obtain $NiSO_4$, $CoSO_4$, $MnSO_4$, and lithium-containing solutions through extraction, membrane separation, and precipitation. The other technique is to directly separate the lithium in the solution and prepare the remaining nickel-cobalt-manganese ion solution as the precursor of $Ni_xCo_yMn_{1-x-y}O_2$. The first method involves a lengthy process and low economic benefit, although the operational difficulty is low and industrial production can be easily achieved. The second method involves a short process; however, the lithium separation effect is inferior, and the technical barrier is large, which renders it challenging to achieve industrial production.

3.2.2.1 Separation and extraction of nickel, cobalt, and manganese

(i) Mn precipitation via ammonification:

As described in Section 2.2.4, nickel and cobalt in ammonia-containing solutions form stable complexes with ammonia molecules, and thus, neither cobalt nor nickel ions can be precipitated even in alkaline solutions with pH values of 8—11. Conversely, manganese is difficult to balance effectively in the solution due to the low stability of the complex formed with ammonia molecules, resulting in the combination of manganese and hydroxide to form $Mn(OH)_2$ precipitates.

(ii) Ni and Co separation via extraction

The commonly used extractants to separate Ni^{2+} and Co^{2+} are P507, N235, N263, N509, and N510.

P507 (Ethyl hexyl phosphonic acid monoethyl hexyl ester) and the synthesized fatty acids are acidic extractants, and the reaction process pertains to cation exchange. P507 can effectively extract Co^{2+} in an acidic solution. During extraction, Co^{2+} enters the extraction solution after reverse extraction, while Li^+ and Ni^{2+} remain in the raffinate. In addition, N235 (trialkyl amine, tertiary amine) and N263 (methyl trialkyl amine, quaternary amine chloride) exhibit similar regularities in alkaline solutions.

N509 (5,8 diethyl-7 carbonyl-6-dodecyl ketone) and N510 (2 carbonyl-5-sec-octyl-dibenzophenone) are chelating extractants. Specifically, compounds with chelating rings are formed during extraction. Both N509 and N510 can effectively extract Ni^{2+} in acidic solutions and separate Ni^{2+} from Li^+ and Co^{2+}.

3.2.2.2 Prioritized separation of lithium and preparation of $Ni_xCo_yMn_{1-x-y}O_2$

(i) Li separation via extraction

The solvent extraction method of lithium ions has received considerable attention in the processing of lithium brine. Laboratory extractants include phosphorus extractants, amine extractants, diketones, ketones, alcohols, and ethers. Diisobutyl ketone has been used in the United States to extract lithium from brine with an extraction rate of 80%. In China, a 20% 503%-20% TBP (tributyl phosphate)-60% 200 kerosene system has been used to extract lithium ions with an extraction efficiency of 90%. With improvements, the extraction rate of lithium reached 99.1%, the purity of anhydrous lithium chloride reached 98.8%, and the total recovery rate of lithium was 98%.

(ii) Li separation via ion-exchange membranes

Ion exchange membranes can be divided into anion, cation, and dual ion-exchange membranes according to the type of ion exchange.

In the membrane separation process of lithium ions, the adopted selective permeation membrane is usually a monovalent cation-selective permeation membrane. This membrane allows the passage of only monovalent cations and not divalent and high-valence ions. The basic principle is that the surface of the positive film is coated with a thin negative film layer, which has a fixed charge group with a positive charge, and the repulsive force of the divalent cation is greater than that of the monovalent cation. Under the action of an electric field, monovalent cations can pass through this barrier layer and enter the positive membrane, but divalent cations cannot pass through this barrier layer.

3.3 Lithium salt extraction

The precipitation method makes it difficult to extract lithium salt from the solution of the crystallization precipitation method. Among common lithium salts, Li_2CO_3 is the least soluble (Table 10.1), with a solubility of 1.25 at 25°C, compared with 25.6 for Li_2SO_4 and 45.8 for LiCl. In addition, among the alkali metal carbonates, Li_2CO_3 is the least soluble, and other alkali metals generally do not precipitate with Li_2CO_3.

The solubility of lithium carbonate exhibits a negative temperature coefficient; that is, solubility decreases with increasing temperature. For example, the solubility of Li_2CO_3 is only 0.72 at 100°C. Therefore, the precipitation of lithium carbonate usually involves adding saturated Na_2CO_3 solution to the solution, and the process is implemented at a high temperature to enhance lithium recovery.

Further reading

[1] H. Liu, et al., Hydrometallurgy—Leaching Technology [M], Metallurgical Industry Press, Beijing, China, 2016.
[2] J. Chen, et al., Handbook of Hydrometallurgy [M], Metallurgical Industry Press, Beijing, China, 2005.
[3] R. Peng, et al., Nickel Metallurgy [M], Central South University Press, Changsha, China, 2005.

[4] T. Zhu, et al., Extraction and Ion Exchange [M], Metallurgical Industry Press, Beijing, China, 2005.
[5] X. Zhai, et al., Reduction and Precipitation [M], Metallurgical Industry Press, Beijing, China, 2008.
[6] T. Zhu, et al., Modern Copper Hydrometallurgy [M], Metallurgical Industry Press, Beijing, China, 2008.
[7] M. Tang, et al., Fundamental and Technology of Complex Metallurgy [M], Central South University Press, Changsha, China, 2011.
[8] R. Ma, et al., Principle on Hydrometallurgy [M], Metallurgical Industry Press, Beijing, China, 2007.
[9] Q. Tian, et al., Study on Preparation of Cobalt Oxide by Precipitation-Thermolysis Process of Oxalic Acid without Ammonia [M], Metallurgical Industry Press, Beijing, China, 2015.
[10] L. Chai, et al., Metallurgical Environmental Engineering [M], Sciences Press, Beijing, China, 2010.
[11] J. Li, et al., Waste Battery Management and Recycling [M], Chemical Industry Press, Beijing, China, 2005.
[12] X. Yang, et al., Microbial Hydrometallurgy [M], Metallurgical Industry Press, Beijing, China, 2003.
[13] S. Luo, et al., Leaching Technology of Lithium Cobalt Oxide from Waste Lithium-Ion Batteries [M], Metallurgical Industry Press, Beijing, China, 2014.
[14] H. Yang et al., Bacteriometallurgy [M], Chemical Industry Press, Beijing, China, 2006.
[15] C. Tang, et al., Metallurgical Electrochemistry Principle [M], Metallurgical Industry Press, Beijing, China, 2013.
[16] Z. Fang, et al., Leaching [M], Metallurgical Industry Press, Beijing, China, 2007.
[17] P. Guo, et al., Efficient Utilization of Metallurgical Resources [M], Metallurgical Industry Press, Beijing, China, 2012.
[18] M. Han, et al., Metallurgical Principle [M], Metallurgical Industry Press, Beijing, China, 2008.
[19] H. Li, et al., Metallurgical Principle [M], Sciences Press, Beijing, China, 2015.
[20] H. Guo, et al., Metallurgical Physical Chemistry Course [M], Metallurgical Industry Press, Beijing, China, 2008.
[21] F. Sheng, et al., Metallurgical Physical Chemistry [M], Higher Education Press, Beijing, China, 2017.
[22] G. Xie, et al., Thermodynamic Principle and Application of Metallurgical Solution [M], Sciences Press, Beijing, China, 2013a.

CHAPTER 11

Biohydrometallurgical recycling approaches for returning valuable metals to the battery production cycle

Tannaz Naseri[1], Vahid Beigi[1], Ashkan Namdar[2], Arnavaz Keikavousi Behbahan[3] and Seyyed Mohammad Mousavi[1,4]

[1]Biotechnology Group, Chemical Engineering Department, Tarbiat Modares University, Tehran, Iran; [2]Faculty of Materials Science and Engineering, Khajeh Nasir Toosi University of Technology, Tehran, Iran; [3]Department of Chemistry, Sharif University of Technology, Tehran, Iran; [4]Modares Environmental Research Institute, Tarbiat Modares University, Tehran, Iran

1. Introduction

Batteries play a vital role in today's energy-dependent world. Batteries are electrochemical devices with a wide range of sizes, power, energy, weights, safety, and shapes that convert chemical energy into electrical energy. Nowadays, there is enormous demand for batteries due to developments in technology, industry, and communication [1]. Generally speaking, batteries are divided into rechargeables, such as nickel—metal hydride (Ni—MH), nickel—cadmium (Ni—Cd), and lithium-ion batteries (LIBs), and non-rechargeables, such as alkaline and zinc batteries [2]. LIBs have many advantages over Ni—Cd and Ni—MH batteries, such as high voltage per cell, low self-discharge, good cycle life, weak memory impact, long storage life, and high energy density [3]. Thanks to these merits, LIBs have been broadly used in numerous devices such as laptops, cameras, remotes, mobile phones, consumer electronics, and electric vehicles (EVs) [3].

Over the past decade, EVs have developed rapidly— more than one million EVs worldwide were sold in 2017, and the global stock of EVs reached 5 million units in 2018 [4,5]. Global EV sales were anticipated to reach 6.9 million units in 2020. Multiple studies report different predicted lifetimes of the LIBs in EVs. Depending on their type, EV LIBs typically last 8—10 years. According to these statistics, 0.33 to 4 million metric tons (MT) of LIBs will have been used in EVs from 2015 to 2040 [6]. It is predicted that when EVs reach their end of life, the generated, spent LIBs will weigh 1.25 MT from five million EVs [4].

The growth in spent LIBs comes with various hazardous characteristics, such as their ability to spontaneously ignite or release toxic chemicals under landfill conditions and thus cause serious tolls on humans and the environment [7]. According to the Environmental Protection Agency, spent batteries are hazardous materials [8]. The European Union (EU) imposes Directive 2006/66/EC, requiring at least 50% of spent battery material to be recycled [9].

The directive also required that all states intensify their spent battery recycling rates from 25% in 2012 to 45% in 2016. In addition, waste electrical and electronic equipment (WEEE) Directive 2012/19/EU was announced, requiring that 75% of spent mobile phones containing LIBs had to be reused and recycled by 2018 [10]. With these and future changes, proper recycling methods are needed for spent LIBs. This chapter discusses the recovery of spent LIBs and the reuse of valuable materials from these wastes.

1.1 Spent lithium-ion batteries as a precious secondary resource

Lithium (Li) and cobalt (Co) are important metals with limited sources of about 64 million and 7.1 million tons (as metal), respectively [11]. Li and Co laterites' geological occurrence is quite uneven, and they are listed as vital raw materials by both the United Nations Environment Programme and the EU [12,13]. Despite their limited relative crustal abundance (26 ppm for Co and 17 ppm for Li), they have widespread applications in industries, including catalysts, magnets, electronics, ceramics, drugs, and manufacturing lubricating greases [14]. In addition, about 25–35% of the global production of Li and Co is applied in LIB manufacturing [15,16]. LIBs typically contain precious materials including Co = 5%–20%, nickel (Ni) = 5%–10%, Li = 5%–7%, and other metals—manganese (Mn), copper (Cu), aluminum (Al), iron (Fe), etc. = 5%–10% [17].

Based on the increasing annual production of LIBs (at least 20 million LIBs each year) and confined ore resources, it is estimated that in less than 60 years, all the Co that exists on the earth will have been consumed [16]. This situation sounds an alarm for us to look for a new way to secure resources while respecting the environment. Since the percentage of Li in LIBs is approximately 10 times higher than that in mineral ores [18], one of the best ways to secure more metals is to stop their endless extraction from mines and increase the recycling of end-of-life electronic wastes (e-waste).

1.2 The environmental aspects of inappropriate spent lithium-ion battery management

LIBs, including a cathode, an anode, an electrolyte, a separator, and some other elements, generally contain an overwhelming mass of heavy metals and toxic flammable electrolytes that threaten human life and the ecosystem [19,20]. Heavy metals usually have long half-lives, and alone, they can be harmful to human beings. For instance, Co and Co-containing powder in contact with skin can cause dermatitis. Mn can cause chronic and acute poisoning [20]. Spent LIBs are classified as non-environmentally friendly wastes because of their high concentrations of heavy metals, such as Li, Co, Cu, Ni, and lead (Pb) [2,5].

Depositing spent LIBs in landfills results in environmental pollution, such as toxicity to underground water bodies, unless huge landfill areas are considered, which seems to be an expensive resolution [5]. The mechanism by which water bodies receive toxic runs is the leaching potential of the chemicals used in making LIBs. Water that enters landfills acts as a solvent for these chemicals (usually heavy metals and inorganic elements in the case of LIBs) [6]. Burning LIBs would also be problematic because of the atmospheric pollution that would result, as burning LIBs would generate an appreciable number of toxic gases, particularly hydrofluoric acid. Thus, a cost-effective and environmentally friendly way to recover LIBs is needed [21].

2. Approaches for recycling spent lithium-ion batteries

Several methods can be applied to extract metals from spent batteries: pyrometallurgy, hydrometallurgy, and biohydrometallurgy [22]. The pyrometallurgical process has been a traditional method of using heat to recover metals from e-waste or minerals for the past 2 decades [23]. However, pyrometallurgical processes require high temperatures that lead to increased energy consumption [20]. In addition, these processes emit poisonous gases such as furans and dioxins [3].

Hydrometallurgical methods are much more accurate and predictable than the pyrometallurgical method. The main step in hydrometallurgical methods is washing solids with a set of acids [3]. Therefore, hydrometallurgical methods are correlated with environmental pollutants because of the consumption of chemicals and dangerous reagents and the considerable volumes of by-products that are produced [23].

Biohydrometallurgy is a new, effective, economical, and environmentally friendly method for recycling spent LIBs [20]. Biohydrometallurgical methods are classified into three categories according to the type of reaction between metals and microorganisms: biosorption, bioaccumulation, and bioleaching. Metals extraction from aqueous solutions using biosorption and bioaccumulation has received considerable attention [24]. Bioleaching is one of the most important fields of biohydrometallurgy, in which microbes dissolve metal ions from the solid waste matrix [25]. The advantages of this method are lower energy consumption, lower costs, and a reduction in harmful by-products [20].

3. Types of biohydrometallurgical methods

3.1 Biosorption

Biosorption is a passive physicochemical and metabolically independent method between heavy metal components (contaminants) and biomass in which many different mechanisms can contribute to heavy metals removal. Some possible mechanisms involved in biosorption are absorption, adsorption, ion exchange, surface complexation, and precipitation [26]. Biosorption has several advantages over conventional methods, including economical operating costs, reduction in the volume of chemicals or biological sludge, reduced adsorbents, possible metal recovery, and high-performance detoxification of streams [27].

To clarify the mechanisms, we shall divide them into two sequential steps. The first step is cell interactions, followed by the second step, cell accumulation and surface microprecipitation [28]. As far as heavy metals toxicity is concerned, dead biomass is a better choice because living biomass can be affected by toxic metals. Numerous parameters can affect biosorption efficiency. Structure is one crucial aspect to be considered, especially the structure of the cell surface and cell wall. Functional groups like carboxylate, hydroxyl, sulfate, phosphate, and amino are present in the biosorption process [29,30].

In addition, functionally, the temperature, pH, initial metal concentration, and oxygen content of the environment play a huge role in the biosorption of heavy metals [28]. Chemical and genetic enhancement of dead or living cell surfaces and walls would also improve the sorption capacity of cells [28]. To sum up, biosorption has these advantages: (1) selective sorption of precious metals, (2) efficiency, (3) temperature- and pH-controllable conditions, (4) the ability to harvest metals, and (5) regenerability of living cells [28].

3.2 Bioaccumulation

Bioaccumulation is considered a specific kind of biosorption that has been narrowed down to living biomass. Bioaccumulation is usually undertaken to remove metals from a solution by intracellular and extracellular uptake. Passive uptake can also be involved (biosorption). The process of bioaccumulation is usually known as a means of protection against metal toxicity, where physiochemical conditions (namely, Eh and pH) can affect the intensity of toxicity [26,31]. This process often involves two steps. The first step, which is metabolism- and energy-independent, is relatively quick and identical to biosorption. The second step, which is metabolism- and energy-dependent, is the transport of ion metals into the cytoplasm [32].

It is worth mentioning that some filamentous fungi such as *Aspergillus* and *Penicillium* species have a great capacity for accumulating metals from their media [33]. To sum up, the mechanism of bioaccumulation involves three events: (1) extracellular binding of metabolic and nonmetabolic metal ions (2) intracellular binding of metabolic metals (e.g., K, Fe, Mg, Mo, traces of Cu, and Ni) and their transfer inside the cytoplasm, and (3) intracellular binding of large amounts of nonmetabolic metals (e.g., Co, Ni, Cu, Cd, and Ag) and their transfer inside the cytoplasm [31].

3.3 Bioleaching

Bioleaching was used industrially for decades without knowing the specific functions of the microorganisms in metal dissolution, until 1961, when *Acidithiobacillus ferrooxidans,* an iron-oxidizing bacteria (ferrous ions (Fe^{2+}) to ferric ions (Fe^{3+})), was found in leachates. Since 1980, the bioleaching technique has been implemented in the mining industry better understand the performance of microorganisms [22,34]. As a result, bioleaching has proven to be one of the most appropriate technologies for processing metals from low-grade ores and WEEE, which contain many metals in their zero valence metallic, combined with other metals in waste matrices [30,35].

Bioleaching refers to the process of solubilizing metal cations from insoluble materials. Bioleaching is defined as the dissolution of metal oxides with microbial metabolites or enzymatic processes on solid waste into the leaching medium as metal cations so that the metal ions are removed from the solid waste after filtering water through it [31,36]. The role of microbial metabolites in the bioleaching process is through ligand- and proton-promoted mechanisms [37].

The bioleaching method can be conducted three ways: one step, two-step, and spent medium [36,38]. The first two approaches depend on biomass exposure to solid materials. The one-step method is the classical method, wherein waste material (LIB powder) and microbial inclusion (spore suspension) are added simultaneously to the leaching medium. In the two-step method, LIB powder is added after the microorganism enters the logarithmic phase (maximum growth) in the culture medium. It has been reported that this approach is more practical for metal dissolution [2,36].

In the spent medium process, the microorganism is inoculated in the absence of LIB powder until it reaches the stationary phase (the end of its logarithmic phase). Then, the biomass is separated by centrifugation and filtration. Finally, the leaching process is initiated by adding LIB powder to the biomass-free medium with various metabolites [36,39]. The bioleaching method chosen affects metal recovery. Other factors also strongly affect the bioleaching process, such as microorganism type, pH, particle size, pulp density, redox potential, and CO_2 and O_2 concentrations [36]. Therefore, in developing bioleaching performance, these operating conditions need to be optimized.

- **pH**

pH value is an essential factor in the bioleaching of solid wastes. Proper adjustment of pH value is essential for bacterial growth, activity, and metal dissolution [40]. For iron- and sulfur-oxidizing bacteria, the optimum pH is in the range of 2—2.5; when the pH is higher than 2.5, the Fe^{3+} concentration decreases, and the formation of jarosite precipitation occurs, in which the bioleaching rate decreases [22]. In contrast, for the most commonly used genus, *Penicillium*, the optimum pH is between 2 and 8, since organic acid excitation by fungi is influenced by the composition of the culture medium [41].

- **Temperature**

The impact of temperature on the bioleaching process has been broadly investigated. In general, the dissolution of metals at higher temperatures occurs more rapidly because of increased kinetic rates. Temperature affects bacterial activity, and high temperatures can cause the cessation of bacterial activity. In the case of sulfur-oxidizing bacteria, the optimum temperature is in the range of 30°C—35°C [40]. At temperatures above 50°C, thermophilic bacteria can be applied in the bioleaching process [42]. Since the metals dissolution process is mainly exothermic, it is crucial to control the temperature when the bioleaching method is used [22].

- **Particle size**

   The particle size of solid waste is another vital factor in metals bioleaching. The smaller the particle size is, the greater the surface area available for mass transfer and the higher the efficiency of the bioleaching process that can occur. It is worth mentioning that sometimes a very fine particle (less than 45 μm) can damage microbial cells and result in a lower bioleaching rate [22]. For example, with a small particle size, oxalic acid secretion was reduced because of toxic particles on the fungal structure that led to reduced biomass production [41].

- **Nutrient**

   The microbial species in the bioleaching process require different substrates [22]. There is a correlation between the nutrients and metabolism of microbial strains [40]. Chemolithoautotrophic bacteria use sulfur and Fe^{2+} (inorganic compounds) as their energy source. Generally speaking, they use CO_2 as their carbon source [22]. In addition, organic compounds (e.g., yeast extract, sucrose, glucose) are used by heterotrophs or mixotroph microorganisms in the culture medium [40].

- **O_2 and CO_2**

   The amount of dissolved O_2 and CO_2 plays an essential role in the bioleaching process. Dissolved O_2 is supplied by aeration, stirring, and shaking [22]. Oxidation of sulfur and iron requires high levels of dissolved O_2. Decreased levels of available O_2 lead to reduced sulfuric acid production, which in turn reduces metal recovery. In addition, heterotrophic microorganisms need O_2 to produce metabolites from organic compounds [40]. Although the effect of CO_2 on bacterial growth has been less studied due to the difficulty of measuring dissolved CO_2, the optimal CO_2 concentration for oxidation of Fe^{2+} has been reported to be between 17 and 7% (v/v) [22].

4. Microbes in bioleaching

The bioleaching process is connected with the activities of microorganisms. Three main groups of microorganisms are commonly applied in bioleaching: autotrophic bacteria (i.e., sulfur and iron oxidizers), heterotrophic bacteria, and fungi [35]. State-of-the-art research has been conducted on the removal of metals from spent LIBs via these microorganisms [2,20,43]. Fig. 11.1 summarizes the bioleaching process for secondary wastes using autotrophic and heterotrophic microorganisms of secondary resources.

4.1 Chemolithotrophic autotrophic bacteria

Autotrophic bacteria do not need an organic carbon source for growth. They can tolerate high concentrations of metal ions, grow at an acidic pH of 1 to 3, and can grow under aerobic conditions, and as a result, they are the most widely used bacteria in bioleaching processes [44]. They are gram-negative, non-spore-forming rods [22]. These bacteria grow at 30°C–35°C and use inorganic components such as ferrous iron (Fe^{2+}) and

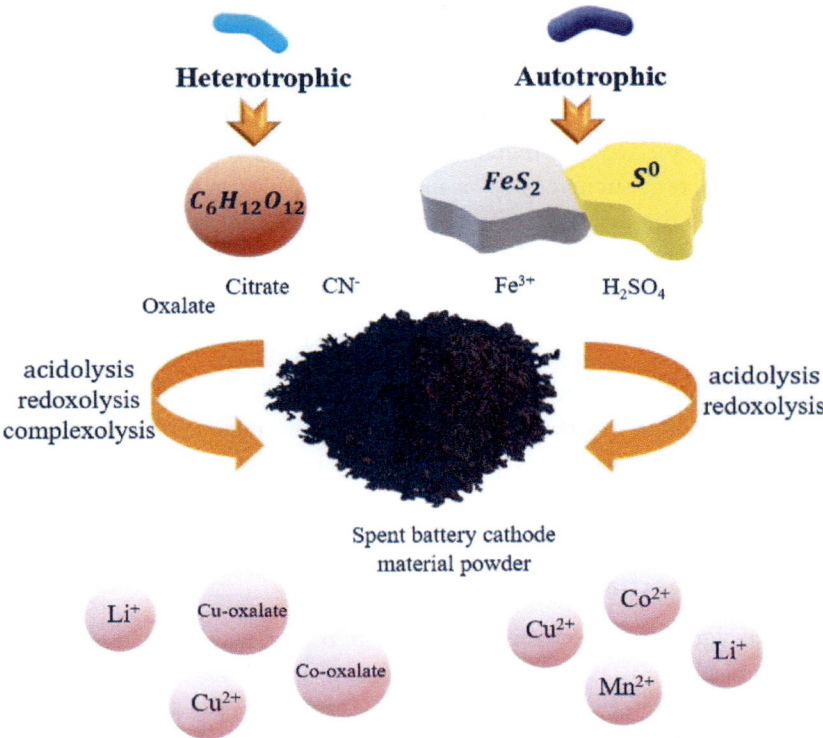

Figure 11.1 Heterotrophic and autotrophic bioleaching process for secondary materials. *(Modified and redrawn after [30].)*

elemental sulfur (S^0) as carbon sources [44]. The two most important uses of these microorganisms are the oxidation of Fe^{2+} to Fe^{3+} and the conversion of S^0 to H_2SO_4 [45].

Acidithiobacillus species such as *A. prosperus, A. caldus, A. thiooxidans, A. ferrooxidans, A. concretivorus*, and *A. albertis* are the most extensively studied autotrophic acidophilic species in spent battery bioleaching [36]. Among these, *A. ferrooxidans* and *Leptospirillum ferrooxidans* as iron-oxidizing bacteria, and *A. thiooxidans* as sulfur-oxidizing bacteria, are the most important bacteria in metals dissolution [46].

4.2 Heterotrophic bacteria and fungi

Heterotrophic bacteria use hydrogen and organic carbon as energy sources and produce metabolic by-products such as organic acids (e.g., lactic, citric, oxalic, and gluconic acids), enzymes, etc. [22]. These by-products can solubilize metals through acidolysis (protonic reaction through acid production), complexolysis (forming soluble complexes of metal), or redoxolysis (oxidation-reduction reactions thanks to the enzymatic reduction of oxidized complexes) [23].

Researchers has been conducted on the bioleaching of valuable metals from solid waste using cyanide and organic acid-generating microorganisms [47]. Various fungi such as *Aspergillus niger, Penicillium chrysogenum, Penicillium simplicissimum* and bacteria such as *Gluconobacter oxydans, Pseudomonas* strains, and *Chromobacterium violaceum* have been applied in the recovery of metals from various solid wastes [30].

Cyanogenic bacteria such as *Pseudomonas* strains, *C. violaceum*, and *Bacillus megaterium* produce cyanide by the carboxylation of glycine in the stationary phase [48]. According to Eq. (11.1), two forms of cyanide production are possible in an equilibrium reversible reaction: (1) cyanide anion (CN^-) and (2) HCN [49,50]:

$$HCN \leftrightarrow H^+ + CN^- \tag{11.1}$$

Various valuable metals and platinum group metals, including Au, Ag, Pt, Pd, Rh, and Ru, which are often not reachable by mineral acids, make a complex form with cyanide. Therefore, complexolysis is the only mechanism in the cyanogenic bioleaching process [44].

5. Bioleaching mechanism

In general, two significant mechanisms are used in microbial bioleaching, direct (contact bioleaching) and indirect (a noncontact mechanism) [51]. The direct mechanism occurs in one-step and two-step bioleaching processes; conversely, the indirect mechanism occurs during spent medium bioleaching. The microorganisms working within these two mechanisms can dissolve heavy metals from solid materials [33]. Fig. 11.2 demonstrates the direct and indirect mechanisms of microbial bioleaching.

5.1 Direct or contact mechanism

In the direct mechanism, microorganisms attack and destroy solid waste directly by physical contact with its surfaces [22]. Metals solubilization occurs through electrochemical reactions and can cause electron transfer from solid minerals [35]. It is noteworthy that the extracellular polymeric substances (EPSs) of iron- and sulfur-oxidizing bacteria cause cells to attach to the surfaces of solid materials and fill the space between the solid material and bacterial outer membrane [53]. Enzymatic attack and electrochemical interaction result from direct microbial attack through EPS, leading to metals dissolution [23]. In addition, the rate of biooxidation reaction inside the EPS layer reaction space is faster and more effective than the bulk [54].

In bioleaching by iron-oxidizing bacteria, Fe^{2+} and Fe^{3+} are absorbed into the EPS, leading to an increased concentration of these ions and improving the Fe^{2+}/Fe^{3+} cycle inside the aggregates (cell—EPS—solid material), causing a more effectual reductive attack

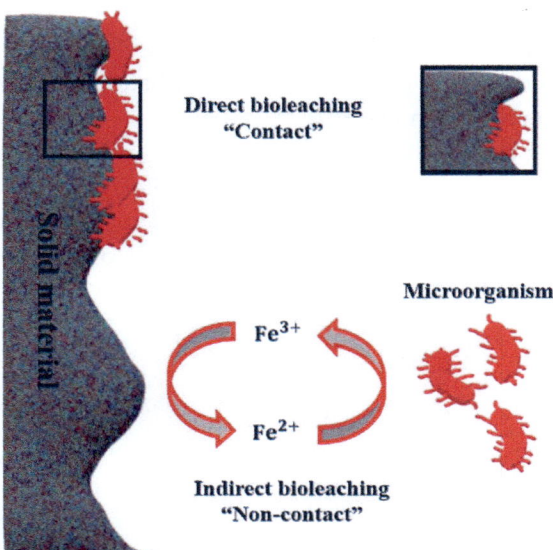

Figure 11.2 Direct and indirect mechanisms in the bioleaching processes for heavy metals. *(Modified and redrawn after [52].)*

on metals through the direct mechanism. In other words, Fe^{3+} in the EPS layer (Fe^{3+}-glucuronic acids complex) reduces to Fe^{2+}, and this electron transfer between the solid material and EPS layer causes the metals dissolution. This has been confirmed by Wang et al. [55]; the author investigated the effect of EPS in microorganism attachment on the LIB cathode material. The formation of aggregates and electron transfer inside them leads to a better reduction attack, resulting in increased metals recovery. In the direct mechanism, the electrostatic forces among bacteria and cathode particles occur in the EPS layer through a minor electrostatic force and a major hydrophobic force. The effective dissolution of Co, Ni, and Mn occurs via direct contact between bacterial cells and the LIB cathode [55].

5.2 Indirect or noncontact mechanism

In the indirect mechanism, microorganisms have no physical contact with solid surfaces. Instead, this mechanism involves the generation of various organic acids and metabolites by fungi and bacteria. The metabolites produced by the microorganisms then react with waste particles to dissolve metals [35]. In many cases, because of the physical contact of microorganisms with solid surfaces, the direct mechanism receives more attention than the indirect mechanism [23]. In general, the indirect mechanism is performed in three ways: (1) secretion of organic or inorganic acids (proton formation), (2) redox reactions, and (3) secretion of ligands that form complexes with metals [22]. Generally speaking, the indirect mechanism occurs in spent-medium approaches [33].

5.2.1 Biogenic ferric sulfate and biogenic sulfuric acid mechanism

In the indirect mechanism, iron- and sulfur-oxidizing bacteria convert Fe^{2+} and S^0 to Fe^{3+} and H_2SO_4, respectively. In the bioleaching of metals from spent batteries, these two intermediates are used for metals solubilization [44].

In metals extraction by iron-oxidizing bacteria, Fe^{2+} is oxidized to Fe^{3+} as shown in Eq. (11.2). Fe^{3+} is an oxidizing agent that attacks the solid material (LIB powder) and reacts with metal compounds. Based on Eqs. (11.3)–(11.5), the reduction process of Fe^{3+} leads to proton (H^+) production by which oxygen atoms in metal oxides are attacked and metal components are dissolved [20,25]:

$$4Fe^{2+} + O_2 + 4H^+ \xrightarrow{Microorganism} 4Fe^{3+} + 2H_2O \tag{11.2}$$

$$Fe^{3+} + H_2O \to Fe(OH)^{2+} + H^+ \tag{11.3}$$

$$Fe^{3+} + 2H_2O \to \quad + H^+ \tag{11.4}$$

$$Fe^{3+} + 3H_2O \to Fe(OH)_3 + 3H^+ \tag{11.5}$$

Naseri et al. [20] applied the bioleaching method of spent coin cells with $LiMn_2O_4$ cathode material by *A. ferrooxidans*. The experiment was done at various pulp densities (10–40 g/L). According to Eq. (11.6), the strain-produced proton releases Li and Mn from the cathode matrix. In addition, Fe^{2+} in the presence of the proton reacts with MnO_2 through a chemical reaction—see Eq. (11.7)—and reduces Mn oxide (insoluble form) into Mn^{2+} (soluble form) [20]:

$$4LiMn_2O_4 + 4H^+ \to 2Li^+ + 3MnO_2 + Mn^{2+} + 2H_2O \tag{11.6}$$

$$MnO_2 + 2Fe^{2+} + 4H^+ \to 2Fe^{3+} Mn^{2+} + 2H_2O \tag{11.7}$$

Regarding $LiNi_xCo_yMn_{1-x-y}O_2$ cathode material, metals such as Ni, Co, and Mn exist in their insoluble form with a high valence state. Through biochemical reactions due to the Fe^{2+}/Fe^{3+} cycle—see Eqs. (11.3)–(11.5)—insoluble high valence Co^{3+}, Mn^{4+}, and Ni^{3+} can be converted into soluble-form Ni^{2+}, Co^{2+}, and Mn^{2+}, as shown in Eqs. (11.8)–(11.10) [56,57]:

$$M^{4+}/M^{3+}(M = Mn, Co, Ni) + Fe^{2+} \to M^{2+}(MO) + Fe^{3+} \tag{11.8}$$

$$MO + 2H^+ \to M^{2+} + H_2O \tag{11.9}$$

$$6LiNi_{1/3}Co_{1/3}Mn_{1/3}O_2 + 6FeSO_4 + 12H^+ \to \\ 3Fe_2(SO_4)_3 + 2MnSO_4 + 2NiSO_4 + 2CoSO_4 + 3Li_2SO_4 + 12H_2O \tag{11.10}$$

Zeng et al. [58] reported the bioleaching process of the Co and Li in $LiCoO_2$ by *A. ferrooxidans*. Based on Eq. (11.11), proton generation from the reduction of Fe^{3+} caused leaching of Co and Li [59]:

$$2LiCoO_2 + 6H^+ \to 2Co^{2+} + 2Li^+ + 3H_2O + 0.5O_2 \tag{11.11}$$

On the other hand, sulfur-oxidizing bacteria metabolize S^0 to H_2SO_4—Eq. (11.12) [2]—and then the biogenic sulfuric acid produced from this metabolite leaches metals from LIBs through proton attack, as shown in Eq. (11.13):

$$S^0 + H_2O + 1.5O_2 \rightarrow H_2SO_4 \qquad (11.12)$$

$$H_2SO_4 + MO \rightarrow MSO_4 + H_2O \quad (M = \text{Metal}) \qquad (11.13)$$

As mentioned, $LiMn_2O_4$ can react with protons and produce MnO_2—see Eq. (11.6). In addition, S^0 in the sulfur-oxidizing bacteria system can respond with MnO_2 and produce Mn^{2+}—Eq. (11.14) [2]. This was proved by Naseri et al. [2]; the authors found that with biogenic sulfuric acid, the removal of Co, Mn, and Li reached 60%, 20%, and 99%, respectively [2]:

$$MnO_2 + S^0 + 2H_2O \rightarrow SO_4^{2-} + 3Mn^{2+} + 4OH^- \qquad (11.14)$$

A perfect example of the indirect bioleaching process by *A. thiooxidans* is the recovery of Co and Li from $LiCoO_2$. According to Eq. (11.15), non-water-soluble Co^{3+} converts to water-soluble Co^{2+} through indirect bioleaching [60]:

$$4LiCoO_2 + 3H_2SO_4 \rightarrow Co_3O_4(s) + 2Li_2SO_4(aq) + CoSO_4(aq) + 3H_2O + O_2 \qquad (11.15)$$

5.2.2 Biogenic organic acids

Heterotrophic microorganisms can generate citric, gluconic fumaric, oxalic, lactic, pyruvic, malic, succinic, and other organic acids. Their acidity is highly related to their functional groups (e.g., carboxylic or hydroxyl). The number of functional groups is not equal in organic acid molecules. These acids are generally classified based on carbon chains, saturation, and their functional groups [61].

Table 11.1 shows the acid dissociation constants (pKa), formulae, dissociation reactions, and metallic complexes of organic acids. pKa values are correlated with the ionic strength of organic acids. The smaller the pKa value, the stronger the acid is [62]. In addition, organic acids with more hydroxyl groups in their structure react faster, and those with carboxyl groups are beneficial for metal mobilization owing to the formation of stable ligands [63]. Table 11.1 also shows the complexation reactions of organic acids with metals (Li, Co, Mn, and Ni) in spent LIBs. Many studies have shown that organic acids can be applied as green and eco-friendly recycling agents [62].

Eqs. (11.16)–(11.19) list the leaching reactions between metal ions of $LiCoO_2$ cathodic material and organic acid.

Malic acid [64]:

$$\begin{aligned} 4LiCoO_2(S) + 12C_4H_6O_5(aq) &\rightarrow 4LiC_4H_5O_5(aq) \\ &+ 4Co(C_4H_6O_5)_2(aq) + 6H_2O + O_2 \end{aligned} \qquad (11.16)$$

Table 11.1 Complexation reactions of organic acids with Ka values [62].

Organic acid	Reaction	pKa	No.	Metallic complexes	No.
Citric acid	$C_6H_8O_7 \rightarrow C_6H_7O_7^- + H^+$	3.09	(1)	$n[C_6H_7O_7^-]+M^{n+} \rightarrow M[C_6H_7O_7]_n$	(13)
	$C_6H_7O_7^- \rightarrow C_6H_6O_7^{2-} + H^+$	4.75	(2)	$n[C_6H_6O_7^{2-}]+2M^{n+} \rightarrow M_2[C_6H_6O_7]_n$	(14)
	$C_6H_7O_7^{2-} \rightarrow C_6H_5O_7^{3-} + H^+$	6.4	(3)	$n[C_6H_5O_7^{3-}]+3M^{n+} \rightarrow M_3[C_6H_5O_7]_n$	(15)
Oxalic acid	$C_2H_2O_4 \rightarrow C_2HO_4^- + H^+$	1.25	(4)	$n[C_2HO_4^-]+M^{n+} \rightarrow M[C_2HO_4]_n$	(16)
	$C_2HO_4^- \rightarrow C_2O_4^{2-} + H^+$	4.14	(5)	$n[C_2O_4^{2-}]+2M^{n+} \rightarrow M_2[C_2O_4]_n$	(17)
Malic acid	$C_4H_6O_5 \rightarrow C_4H_5O_5^- + H^+$	3.4	(6)	$n[C_4H_5O_5^-]+M^{n+} \rightarrow M[C_4H_5O_5]_n$	(18)
	$C_4H_5O_5^- \rightarrow C_4H_4O_5^{2-} + H^+$	5.11	(7)	$n[C_4H_4O_5^{2-}]+2M^{n+} \rightarrow M_2[C_4H_4O_5]_n$	(19)
Gluconic acid	$C_6H_{12}O_7 \rightarrow C_6H_{11}O_7^- + H^+$	3.86	(8)	$n[C_6H_{11}O_7^-]+M^{n+} \rightarrow M[C_6H_{11}O_7]_n$	(20)
Succinic acid	$C_4H_6O_4 \rightarrow C_4H_5O_4^- + H^+$	4.22	(9)	$n[C_4H_5O_4^-]+M^{n+} \rightarrow M[C_4H_5O_4]_n$	(21)
	$C_4H_5O_4^- \rightarrow C_4H_4O_4^{2-} + H^+$	5.7	(10)	$n[C_4H_4O_4^{2-}]+2M^{n+} \rightarrow M_2[C_4H_4O_4]_n$	(22)
Tartaric acid	$C_4H_6O_6 \rightarrow C_4H_5O_6^- + H^+$	3.03	(11)	$n[C_4H_5O_6^-]+M^{n+} \rightarrow M[C_4H_5O_6]_n$	(23)
	$C_4H_5O_6^- \rightarrow C_4H_4O_6^{2-} + H^+$	4.37	(12)	$n[C_4H_4O_6^{2-}]+2M^{n+} \rightarrow M_2[C_4H_4O_6]_n$	(24)

$M = Co^{2+}, Li^+, Mn^{2+}, Ni^{2+}$.

$$4LiCoO_2(S) + 12C_4H_5O_5^-(aq) + 4Li^+(aq) + 4Co^{2+}(aq) \rightarrow$$
$$4Li_2C_4H_4O_5(aq) + 8CoC_4H_4O_5(aq) + 6H_2O + O_2 \quad (11.17)$$

Oxalic acid [65]:

$$2LiCoO_2(S) + 7C_2H_2O_4(aq) \rightarrow 4LiC_2HO_4(aq)$$
$$+ 2Co(C_2HO_4)_2(aq) + 4H_2O + 2CO_2 \quad (11.18)$$

$$2LiCoO_2(S) + 4C_2H_2O_4(aq) \rightarrow 4Li_2C_2O_4(aq) + 2CoC_2O_4(S) + 4H_2O + 2CO_2 \quad (11.19)$$

According to Eq. (11.19), oxalic acid negatively affects Co recovery from spent LIBs, and insoluble Co oxalate is formed [65].

Microorganisms like *A. niger* can produce various organic acids and extract over 90% of metals [66]. The factors affecting organic acid production are the carbon source, nitrogen and phosphate limitations, the pH, aeration, trace elements, and microorganism morphology [67]. Rasoulnia et al. [39] investigated how the phosphorus source affects organic acid production by *G. oxydans* and *Streptomyces pilosus* and the bioleaching of base metals and rare earth elements from Ni−MH batteries. They used three different phosphorus sources [$Ca_3(PO_4)_2$, KH_2PO_4, and K_2HPO_4]. The authors claimed that the phosphorus source affects organic acid production by bacteria and the bioleaching of metals. Among the various phosphorus sources, K_2HPO_4 at a concentration of 27 mM for *S. pilosus* led to the production of 10.7 mM pyruvic acid and at a concentration of 2.7 mM for *G. oxydans* led to the production of 45 mM gluconic acid. In two-step bioleaching, 34% Mn, 88% Fe, 41.5% Co, 18.5% Ni, 21% Cu, 12% Zn, and 6.5% total rare earth elements were recovered, and in spent media bioleaching, 35% Mn, 68% Fe, 35.5% Co, 16.5% Ni, 20% Cu, 9% Zn, and 9% of rare earth elements were recovered [39].

5.3 Mixed culture

A mixed culture of microorganisms (autotrophs and heterotrophs) can improve LIB bioleaching efficiency compared with the pure culture of a single microorganism. In addition, synergistic effects occur between microorganisms during bioleaching [68]. For example, according to Eq. (11.20), the formation of hydroxyl sulfate precipitation, known as jarosite ($MFe_3(SO_4)_2(OH)_6$, where M = Na^+, Ag^+, K^+, NH_4^+, or H_3O^+), is deposited on the solid surface and reduces metals extraction during the bioleaching process [20]. The mixed microorganism of sulfur-oxidizing bacteria decreases the pH by producing H_2SO_4. The jarosite formation is reduced by low pH and improves oxidant diffusion to the solid surface to increase metals extraction [68]:

$$3Fe^{3+} + K^+ + 2SO_4^{2-} + 6H_2O \rightarrow KFe_3(SO_4)_2(OH)_6 + 6H^+ \quad (11.20)$$

The mixed culture of various microorganisms, from iron- and sulfur-oxidizing bacteria, can be advantageous for bacterial activity and growth. For example, *A. thiooxidans*, by producing H_2SO_4, can provide a desirable condition for the growth of *A. ferrooxidans* [69].

In addition, mixed culture can affect two critical factors in the bioleaching process: the culture medium's initial pH and redox potential. It was believed that the initial pH values decreased and the redox potential obtained ultimate potential (550–600 mV) faster with mixed rather than pure culture [70]. It was reported that high fluctuations in ORP in mixed culture systems caused a reduction in some metals such as insoluble Mn^{4+} to soluble Mn^{2+} during the bioleaching process. Furthermore, the lower pH values of the mixed culture system lead to effective production of the Fe^{2+}/Fe^{3+} cycle, which strongly increases the Mn^{4+} reduction rate [59,71].

A mixture of thermophilic bacteria (*A. caldus*, *Leptospirillum ferriphilum*, *Sulfobacillus* spp. and *Ferroplasma* spp.) for LIB bioleaching have been studied by Ghassa et al. [72]. The authors investigated the potential of a mixed consortia of moderate thermophilic bacteria and the effect of their energy sources (S^0 and $FeSO_4.7H_2O$) on Co, Li, and Ni recovery at 45°C. The results showed that with increasing S^0 concentration, Li extraction increased due to a more acidic environment; conversely, by adding $FeSO_4.7H_2O$, Co and Ni extraction rates increased significantly [72].

Heydarian et al. [73] implemented two-step bioleaching of laptop LIBs using a mixture of *A. ferrooxidans* and *A. thiooxidans*. The initial pH, initial S^0 and $FeSO_4.7H_2O$, and different inoculum ratios of the two consortia were investigated and optimized. Results showed that *A. thiooxidans* maintained high acidity in the culture medium. This condition was favorable for Li recovery; conversely, the synergistic effect of acidic and reduction dissolution for Co and Ni recovery was preferred. Under optimal conditions (pH = 1.5, $FeSO_4.7H_2O$ = 36.7 g/L, S^0 = 5 g/L) and an inoculum ratio of 3/2 for *A. ferrooxidans* and *A. thiooxidans*, the maximum recovery of Li, Ni, and Co was 99.2%, 89.4%, and 50%, respectively [73].

The concentration of mixed energy sources also affected LIB bioleaching performance. In this study, a mixed consortium of *Alicyclobacillus* sp. as sulfur-oxidizing bacteria and *Sulfobacillus* sp. as iron-oxidizing bacteria was studied by Niu et al. [74] for Co and Li recovery from LIBs. they reported that under a pulp density of 2%, 4.0 g/L S^0, and 4.0 g/L FeS_2 as energy sources, maximum extraction increases from 57% to 62% for Li and from 21% to 32% for Co. The authors reported that a greater concentration of FeS_2 did not lead to higher metals extraction [74].

6. A brief summary of spent battery bioleaching

Table 11.2 summarizes spent battery bioleaching including the microorganism type, spent battery type, method applied, recovery process condition, and metal recovery rate.

Table 11.2 Summary of bioleaching processes of spent batteries using various microorganisms.

Batteries	Bleaching method	Microorganism	Agent	Condition Temperature	Pulp density	Stirring speed	Recovery (%)	Reference
LIBs	Two-step	*Acidithiobacillus ferrooxidans*	H_2SO_4 and Fe^{3+}	30°C	100 g/L	160	Ni: 90 Mn: 92 Co: 82 Li: 89	[57]
	Two-step	*A. ferrooxidans*	H_2SO_4 and Fe^{3+}	30°C	100 g/L	160	Co: 94 Li: 60	[75]
	Two-step	*Leptospirillum ferrooxidans Sulfobacillus thermosulfidooxidans*	H_2SO_4, Fe^{3+}, and exogenous glutathione	NM	5%	NM	Li: 98.1 Co: 96.3	[76]
	Two-step	Sulfur-oxidizing bacteria	H_2SO_4 and $S_2O_3^{2-}$	30°C	400 g/L	120	Co: 91.45 Li: 93.64 Mn: 87.9	[77]
	One-step	*Lysinibacillus* sp.	Citric acid hybrid combination	37°C	5 g/L	130	Li: 25	[78]
	One-step	*Leptospirillum feriphilum S. thermosulfidooxidans*	H_2SO_4 and Fe^{3+}	42°C	15 g/L	180	Li: 100 Co: 99.3	[79]
	Spent-medium	*A. ferrooxidans Acidithiobacillus thiooxidans*	Fe^{3+} with 100 (mM) H_2SO_4	30°C	10%	250	Li: 60.0 Ni: 48.7 Mn: 81.8 Cu: 74.4 Co: 53.2	[80]
	Spent-medium	*Aspergillus niger*	Citric acid gluconic acid Oxalic acid Malic acid	30°C	1%	130	Cu: 100 Li: 95 Mn: 70 Al: 65 Co: 45 Ni: 38	[81]
	Spent-medium	*A. ferrooxidans A. thiooxidans*	H_2SO_4 and Fe^{3+}	30°C	1%	120	Li: 95 Mn: 96	[56]

Continued

Table 11.2 Summary of bioleaching processes of spent batteries using various microorganisms.—cont'd

Batteries	Bleaching method	Microorganism	Agent	Condition			Recovery (%)	Reference
				Temperature	Pulp density	Stirring speed		
Ni–Cd	Spent-medium	A. niger KUC5254 Aspergillus tubingensis KUC5037	Oxalic acid citric acid	27°C	0.1%	150	Ni, Cd, and Zn: >95	[82]
	One-step	A. ferrooxidans	H_2SO_4 and Fe^{3+}	30°C	1%	–	Ni: 51	[83]
Ni–MH and Ni–Cd		A. ferrooxidans	H_2SO_4 and Fe^{3+}	30°C	10 g/L	160	Ni: 87 Co: 93.7 Cd: 67	[84]
Zn–Mn	One-step	Sulfur-oxidizing bacteria Iron-oxidizing bacteria Sulfur- and iron-oxidizing bacteria	H_2SO_4 and Fe^3	35°C	8%	–	Zn: 50	[85]
			H_2SO_4 and Fe^3	30°C	10%	–	Mn: 50.5	
	Spent-medium	A. ferrooxidans	H_2SO_4 and Fe^{3+}	30°C	10 g/L	140	Mn: 99 Zn: 53	[86]
	Spent-medium	Aspergillus fumigatus KUC1520 Aspergillus flavipes KUC5033 Aspergillus japonicus KUC5035 A. tubingensis KUC5037 Aspergillus versicolor KUC5201	Oxalic acid citric acid	27°C	0.1%	150	Zn and Mn: >90	[82]
Ag–Zn	Two-step	A. thiooxidans L. ferriphilum	H_2SO_4 and Fe^{3+}	35°C	10%	120	Mn: 62.4 Zn: 62.5	[87]
	Spent-medium	A. ferrooxidans	H_2SO_4 and Fe^{3+}	30°C	0.2%	150	Ag: 98	[88]

7. Reuse of valuable metals in battery production

After recovering heavy metals from spent LIBs, the main aim is to achieve pure metal or chemical material from the leachate to reuse in the battery production cycle. Solid residues from the recovery process are removed by filtration, and the filtrate is the leaching liquor containing the various metals. How to separate these metals is a crucial issue.

Solvent extraction and chemical precipitation methodologies are commonly applied to separate valuable metals from leachate. For efficient production, the metal ions in leachate are usually abundant and not certainly enriched. Recently, the innovative regeneration technique for recycling cathode materials has gained notice as a relevant means of resynthesizing them directly from leachate. It is worth mentioning that cathode material synthesis and raw chemical component processes are integrated, which leads to some increased complexity and energy effectiveness [89].

7.1 Separation process
7.1.1 Chemical precipitation

Chemical precipitation is a way to remove metals after the leaching process. Chemical precipitation involves adding a chemical called a precipitant to a solution, resulting in a chemical reaction that converts undesired dissolved components into solid particles. Afterward, the particles can be aggregated by chemical coagulation and separated by filtration [90].

Table 11.3 shows various precipitants employed for the chemical precipitation of various metals depending on their chemical features. Carbonates, hydroxides, sulfides, phosphates, oxalates, strong oxidants, and chelating precipitants are employed the most. Several chemical reagents are introduced in the following section to analyze current chemical precipitation processes [105].

- **Hydroxide precipitation**

Hydroxid precipitation is a general chemical precipitation method used to remove dissolved heavy metals. In this process, an alkaline agent is used to raise the pH of leachate and decrease the solubility of metal ions, thus precipitating them out of the solvent [105]. The optimum pH at which hydroxide precipitation occurs depends on the type of metal. Some valuable metals in LIBs, including Ni, Co, and Mn, can be precipitated simply by adjusting the pH to 8, 10, or 12, respectively, demonstrating that pH control is the most critical parameter in hydroxide precipitation. Low pH values are typically used to remove impurity ions, including Fe, Al, and Cu in leachate. In contrast, low-value metals like Mn/Ca/Mg are extracted by increasing the pH with NaOH solution to values of about 12 [106]. In addition, essential materials like new cathodes can be generated from leaching solutions by utilizing NaOH for the coprecipitation of valuable metals with the formula $Ni_xCo_yMn_z(OH)_2$ [107].

Table 11.3 Summary of chemical precipitating processes.

Chemical regents	Metal	Precipitate	pH	References
NaOH	Co	$Co(OH)_2$	10	[91]
	Mn	$Mn(OH)_2$	12	[92]
	Ni	$Ni(OH)_2$	8	[93]
	Zn/Ni/Co	$Zn(OH)_2/Ni(OH)_2/Co(OH)_2$	7	[94]
	Ni/Co/Mn	$Ni_xCo_yMn_z(OH)_2$	11	[1,95]
	Fe/Al/Cu	$Fe(OH)_3/Al(OH)_3/Cu(OH)_2$	5.3	[94,96]
	Mn/Ca/Mg	$Mn(OH)_2/Ca(OH)_2/Mg(OH)_2$	12	[94]
Na_2CO_3	Co	$CoCO_3$	9–10	[97]
	Mn	$MnCO_3$	—	[98]
	Ni	$NiCO_3$	9	[99]
	Li	Li_2CO_3	—	[97,98]
	Ni/Co/Mn	$(Ni_xCo_yMn_z)CO_3$	7.5–8	[100]
$H_2C_2O_4/(NH_4)_2C_2O_4$	Co	$CoC_2O_4 \cdot 2H_2O$	1.5–2	[101]
$Na_2C_2O_4$	Mg/Ca	CaC_2O_4/MgC_2O_4	—	[102]
H_2S	Fe	FeS_2	—	[103]
$(NH_4)_2S$	Co	CoS	6	[98]
$C_4H_8N_2O_2$	Ni	$NiC_8H_{15}N_4O_4$	8	[104]
Na_3PO_4	Li	Li_3PO_4	—	[101]

- **Carbonate precipitation**

    Another widely used precipitant is sodium carbonate (Na_2CO_3), which can be used to remove metals either by direct precipitation or by hydroxide to carbonate conversion using carbon dioxide. Similar to NaOH, valuable metals like Ni, Co, and Mn can be precipitated using sodium carbonate at an optimized pH of around 9—10. It is worth noting that metal carbonate precipitates present good chemical stability and easier filtration with milder pH situations [108]. Sodium carbonate can also form coprecipitates of important metals with the formula $(Ni_xCo_yMn_z)CO_3$, which can be applied as the precursor for cathode materials preparation. But because various metals can be precipitated with sodium carbonate, the contamination of produced precursors can affect its electrochemical performance [105].

- **Oxalate precipitation**

    As a green precipitation agent, oxalate is found in various chemical formulas, including sodium oxalate ($Na_2C_2O_4$), ammonium oxalate (($NH_4)_2C_2O_4$), and oxalic acid ($H_2C_2O_4$). This natural low-toxicity precipitant is a metabolite of organisms and is mainly used to extract Co [109]. Unlike Ni, Li, and Mn, Co can be efficiently precipitated using oxalate with the CoC_2O_4 formula and high purity [110].

    Biswal et al. [36] investigated chemical precipitation methods for Li and Co extraction from an LIB bioleaching solution. First, to remove suspended particles from leachate liquor, the leachate was purified by centrifuging at 10,000 g and filtrating with a 0.45 μM PES filter. For the separation of Co, three chemical reagents were applied—sodium sulfide (Na_2S), sodium hydroxide (NaOH), and sodium oxalate ($Na_2C_2O_4$)—to precipitate Co as Co sulfide (CoS), Co hydroxide ($Co(OH)_2$), and Co oxalate ($CoC_2O_4.2H_2O$), respectively. It is worth noting that for ($Co(OH)_2$) precipitation, the leachate was adjusted to pH 12 using 1 M NaOH solution. For Li precipitation as Li carbonate (Li_2CO_3), sodium carbonate (Na_2CO_3) was applied at pH 12. The precipitation reactions of Co and Li extraction using various chemical reagents are shown in Eqs. (11.21)—(11.24). The results show that coprecipitation with Na_2S and NaOH was 100%, while 87.7% recovery was achieved using $Na_2C_2O_4$. For Li employing Na_2CO_3, 73.6% Li extraction observed [36]:

$$Co^{2+}(aq) + Na_2S \rightarrow CoS\ (S) + 2Na^+(aq) \tag{11.21}$$

$$Co^{2+}(aq) + 2NaOH \rightarrow Co(OH)_2(S) + 2Na^+(aq) \tag{11.22}$$

$$Co^{2+}(aq) + Na_2C_2O_4 + 2H_2O \rightarrow CoC_2O_4.2H_2O\ (S) + 2Na^+(aq) \tag{11.23}$$

$$2Li^+(aq) + Na_2CO_3 \rightarrow Li_2CO_3(S) + 2Na^+(aq) \tag{11.24}$$

7.1.2 Solvent extraction

Solvent extraction (SX), or liquid—liquid extraction, is an efficient separation method for extracting various metals from leachate. SX occurs as a solute is distributed between two immiscible liquid phases. The ratio of distribution indicates the successfulness of separation [111]. SX techniques have a broad range of applications, from large-scale industrial separation to pharmaceutical and biochemical industries and waste treatment. Metals recovery by SX techniques is simple, speedy, and broad, and it does not require much complex equipment [112]. Most metal ions can be recovered by SX. The most popular media is sulfate media, despite the success of chloride-based media [113]. SX usually operates as its molecules replace metal complexes. As a result, metallic ions become more lipophilic [111]. As shown in Table 11.4, various SX agents are used to separate valuable metals (Co^{2+}, Mn^{2+}, Ni^{2+}) and impurity ions (Al^{3+}, Fe^{3+}) from leachate solution [113].

Separating Co, Ni, and Mn from LIBs could become a rugged cliff to climb. Di-(2-ethylhexyl) phosphoric acid (D2EHPA), bis-(2,4,4-trimethyl-pentyl) phosphonic acid (Cyanex 272/Cyanex 301), and 2-ethylhexyl phosphonic acid mono-2-ethylhexyl ester (PC88A) are of greatest interest in this case. Cyanex 272 is the best synergic option for selectively separating Co from complicated leachate solution with other metal ions such as Ni and Mn [114]. For instance, Cyanex 272 can efficiently separate more than 90% of Co^{2+} ions from leachate containing Ni, Li, and Mn at room temperature with an optimum pH range of 4.5—5.5 and a short removal time of less than 20 min [115]. Cyanex 301 selectively separates Co and Ni from leachate containing Mn ions. These solvents are best used in acidic media [114]. In acidic media, Co extraction with D2EHPA starts at a pH over 5.5, and it increases slowly with increases in pH [116].

Table 11.4 Summary of various solvent extraction reagents for separation of specific metals [113].

Element	Solvating agent
Al/Fe	DNNSA + EHPA from nitrate
Cd	DEHPA + MEHPA with TBP for separation of Cd—Zn
Co/Ni	LIX 63 + Cyanex272
	Separation of Ni/Co from nitrate with a carboxylic acid (4-tert-butylbenzoic acid) + pyridine (4-(5-nonyl) pyridine in xylene
	Versatic acid + 4-nonyl pyridine for bioleaching solution containing high Ca
Cu	Cu from chloride using mixed extractants, e.g., Alamine 336 + LIX 54; Acorga CLX50 + LIX 54
	Cu from pickling bath with Cyanex 302 + LIX 860
Fe	Primene JM-T + EHPA
	MEHPA + Primene better but not for stripping
	DEHPA + Cyanex 923 for In and Fe from sulfate to provide easy stripping
	Selective Fe extraction from Zn liquors with derivatives of aminomethylene phosphonic acid

The extractants of D2EHPA and PC88A are frequently applied to separate Mn ions from leachate. Chen et al. [117] studied the separation of Mn^{2+} ions from LIB leachate solution using 20 vol.% D2EHPA with a 70% saponification rate. About 95% of Mn can be separated under conditions of removal with an equilibrium pH of 4, a volume ratio of the aqueous and organic phases (A/O) = 1, and a contact time of 5 min at room temperature [117]. Alamine and hexaacetatocalix(6)arene are some other Co separating agents [118]. pH control is an aid to separation, as its contribution to replacing water molecules is an essential factor.

- **Separation of impurity ions**

It is well known that the leachate of spent batteries may be polluted by impurity ions, such as Fe, Al, and Cu ions. These impurity ions are produced during the recovery process [119]. For example, Fe ions usually generate metallic shells of spent batteries during mechanical operations (e.g., crushing and grinding). Al impurity ions are produced when cathode materials are peeled off from their wastes coated with Al foils. Cu ions can be separated either as valuable metals or as impurity ions for the purification process [120]. In this case, 10% PC88A was adopted as an organic extractant to remove Al ions [121]. In addition, mixed organic extractants of 7% Ionquest 801 and 2% Acorga M5640 were adopted for the selective recovery of Al, Fe, and Cu ions from leachate. These mixtures can be separated into approximately 100% Al, Fe, and Cu under the optimal conditions of pH = 4, T = 40°C, and O/A = 1:2 [122]. It is worth mentioning that the stripping of impurity ions from loaded organics is very challenging, for a relatively high concentration of sulfuric acid (e.g., 3 M) will be needed for adequate stripping of impurity ions from the impurity ions loaded organics [123].

7.2 Regeneration process

Regeneration is a novel strategy for synthesizing cathode material from spent LIB leaching liquor and has many advantages, such as reduced greenhouse gas emissions and the economic benefits from extracting high-value-added products. Sol-gel and coprecipitation are two main ways of regenerating cathode material. Analyzing the concentration of every single metal in leachate before regeneration is necessary, and then by adding certain chemical compounds, each one is adjusted to the desired value.

In coprecipitation cases, different coprecipitators (hydroxide or carbonate compounds) are added to the leachate solution according to the molar ratio of metals in cathode material coprecipitate all essential metal ions and obtain precursor precipitants [124]. This precipitant is subjected to various sequential methods such as purification, drying, and calcining to achieve the essential precursor ($Co_xNi_yMn_zO$). Finally, the precursor is used to form cathode materials. Different factors during the coprecipitation process affect the quality of the final product, such as the pH value of the leachate, temperature, metal ion concentration, and stirring speed. This method has been widely applied to regenerate cathode materials that display regular morphology, fine particles, and uniform distribution of elements, thereby presenting excellent electrochemical performance. However, this method still has some deficiencies, such as high-cost complex procedures [125].

Recently, a coprecipitation method using hydroxide oxalate, and carbonate was investigated to regenerate cathode materials with good crystal structure and electrochemical performance [126,127]. Song et al. [128] studied a novel strategy for bioleaching Zn—Mn batteries and coprecipitation at 30°C along with boiling reflux at 100°C. The authors claimed that this novel multistep process was low-cost, energy-saving, and eco-friendly. A maximum recovery of 100% for Zn and 89% for Mn was obtained using a mixed culture of *Sulfobacillus* sp. and *Alicyclobacillus* sp. 50 mL of bioleaching liquor injected into 500 mL three-neck flask connecting a motor-driven tetrafluoroethylene agitator through the middle neck, located in a constant temperature oil-bath. Then, NaOH, $NaHCO_3$, and NH_4HCO_3 solution (6.0 mol/L) as a coprecipitator was added at 30°C and 350 rpm until the pH value of the liquor increased from 2.0 to 7.0 to initiate the coprecipitation process. The results showed that NaOH was the best coprecipitator in comparison with $NaHCO_3$ and NH_4HCO_3. Due to the effectiveness of increasing the pH value of the second coprecipitation on crystal and magnetization, different amounts of NaOH under the same condition (e.g., 6.0 mol/L, 30°C, and 350 rpm) were continuously added to start the second coprecipitation until the pH value increased from 7.0 to 13. For improved magnetization, the boiling reflux (at 100°C and time set at 0, 3, and 5 h) was achieved after 2-hour incubation of the second coprecipitation at pH 13.0. The authors claimed this process produced Zn—Mn ferrite soft magnetic materials with the highest saturation magnetization measured at 102 emu/g [128].

The sol-gel process, also known as chemical solution precipitation, is widely used in engineering to synthesize a variety of nanostructures. This process involves a series of irreversible chemical reactions that ultimately lead to the production of the final product. These reactions convert the first homogeneous solution molecules into an infinite, heavy, three-dimensional polymer molecule called gel [129].

The sol-gel process has been broadly used to generate cathode materials such as $LiCoO_2$ [130,131] and $LiNi_xCo_yMn_zO_2$ [132,133]. It is worth noting that the sol-gel method is used for acid leaching [125], especially organic acid leaching of cathode material, because the organic components can serve as chelates [134]. This method typically has three primary steps: (1) complex agents are added to the leachate and generate a hydrolytic reaction and polymerization to form sol, (2) a gel with a spatial structure is produced, and (3) the cathode material is regenerated after the sintering process [125].

In this case, Wu et al. [135] investigated the regeneration of $LiCo_{1/3}Ni_{1/3}Mn_{1/3}O_2$ cathode material from citric acid and H_2O_2 leaching solution. The authors claimed that this method exhibited good electrochemical performance of the regenerated cathode material. First, spent LIBs were pretreated to form the active cathode material. After the leaching of valuable metals, corresponding metal salts were added to the liquor to adjust the molar ratio. Finally, the liquor was dried to form a transparent gel, and the product was generated during heat treatment [136].

The sol-gel method can completely employ metal ions in the leachate and mix them homogeneously at the molecular level [137]. In addition, during the process, the electrode presents the same electrochemical and morphology features as the freshly fabricated one. Nevertheless, the operational method of sol-gel is more complicated, and the preparation time is extended by significant energy consumption and generation costs [138]. Hence, this method only employs the basic investigation of cathode materials because the lengthy preparation time restricts its large-scale application in industry. It is worth noting the Li^+ in the leachate can be reused as a Li source without a separation process, and the regenerated products have smaller particle sizes (100–300 nm) than those in coprecipitation methods [137].

8. Conclusion and perspectives

This chapter provides a comprehensive overview of biohydrometallurgy recovery of various metals from spent batteries, provides a general introduction, and investigates the bioleaching process. Depending on the factors above, several conclusions can be obtained regarding the biohydrometallurgical processes involved with bioleaching, metal ion removal and separation from leachate solution, and cathode materials regeneration. The bioleaching approach is a revolutionary solution and promising future technology to address the historical problems related to traditional metal recovery techniques. Although this method was introduced as a lab-scale method in the past, the increasing acceptance of the bioleaching approach as a commercial technology has made it a practical and realistic economic and environmental option for metal recovery. However, despite the widespread development of methods for this new technology, some significant challenges remain for the economical use of bioleaching processes, including slow process kinetics, the ratio of solid waste to liquid, new procedures for microorganism development and metabolite generation increment, the preparation of simple bioreactors for microbial culture and growth, and economical microorganism nutrient sourcing. First, after pretreating spent batteries, the bioleaching process results in the efficient removal of cathode materials. For separation and recovery after the bioleaching process and the development of effective approaches such as chemical precipitation, SX should be adopted. It is highly recommended that a combination of several separation methods is investigated for improving the quality of recovered metals. Finally, the regeneration of cathode materials is an essential part of the recovery process. Coprecipitation, sol-gel, and direct regeneration are the most efficient methods in this field. As a result, the current situation of biohydrometallurgical approaches within a global scenario is a promising and commercializing approach that should be more seriously considered.

Acknowledgments

This study was financially supported by Tarbiat Modares University under grant number IG-39701. The authors thank Iran's National Elites Foundation for supporting part of this research (Project Number 521402).

References

[1] E. Gratz, Q. Sa, D. Apelian, Y. Wang, A closed loop process for recycling spent lithium ion batteries, J. Power Sources 262 (2014) 255–262.
[2] T. Naseri, N. Bahaloo-Horeh, S.M. Mousavi, Environmentally friendly recovery of valuable metals from spent coin cells through two-step bioleaching using *Acidithiobacillus thiooxidans*, J. Environ. Manag. 235 (2019) 357–367.
[3] B. Huang, Z. Pan, X. Su, L. An, Recycling of lithium-ion batteries: recent advances and perspectives, J. Power Sources 399 (2018) 274–286.
[4] Y. Hua, X. Liu, S. Zhou, Y. Huang, H. Ling, S. Yang, Toward sustainable reuse of retired lithium-ion batteries from electric vehicles, Resour. Conserv. Recycl. (2020) 105249.
[5] M. Sethurajan, S. Gaydardzhiev, Bioprocessing of spent lithium ion batteries for critical metals recovery—a review, Resour. Conserv. Recycl. 165 (2021) 105225.
[6] K.M. Winslow, S.J. Laux, T.G. Townsend, A review on the growing concern and potential management strategies of waste lithium-ion batteries, Resour. Conserv. Recycl. 129 (2018) 263–277.
[7] P. Meshram, A. Mishra, R. Sahu, Environmental impact of spent lithium ion batteries and green recycling perspectives by organic acids—a review, Chemosphere 242 (2020) 125291.
[8] M. Rinne, H. Elomaa, A. Porvali, M. Lundström, Simulation-based life cycle assessment for hydrometallurgical recycling of mixed LIB and NiMH waste, Resour. Conserv. Recycl. 170 (2021) 105586.
[9] E. Asadi Dalini, G. Karimi, S. Zandevakili, M. Goodarzi, A review on environmental, economic and hydrometallurgical processes of recycling spent lithium-ion batteries, Miner. Process. Extr. Metall. Rev. (2020) 1–22.
[10] W. Zhang, C. Xu, W. He, G. Li, J. Huang, A review on management of spent lithium ion batteries and strategy for resource recycling of all components from them, Waste Manag. Res. 36 (2) (2018) 99–112.
[11] P. Meshram, B. Pandey, Perspective of availability and sustainable recycling prospects of metals in rechargeable batteries—a resource overview, Resour. Pol. 60 (2019) 9–22.
[12] M. Buchert, D. Schüler, D. Bleher, Critical Metals for Sustainable Technologies and Their Recycling Potential, United Nations Environment Programme (UNEP) and Oko-Institute, Nairobi, Kenya, 2009.
[13] D. ENTR, Report on Critical Raw Materials for the EU, Ares (2015), 2014, p. 1819503.
[14] D.E. Garrett, Handbook of Lithium and Natural Calcium Chloride, first ed., 2004.
[15] I. Santana, T. Moreira, M. Lelis, M. Freitas, Photocatalytic properties of Co_3O_4/$LiCoO_2$ recycled from spent lithium-ion batteries using citric acid as leaching agent, Mater. Chem. Phys. 190 (2017) 38–44.
[16] B. Swain, Recovery and recycling of lithium: a review, Separ. Purif. Technol. 172 (2017) 388–403.
[17] J. Ordoñez, E.J. Gago, A. Girard, Processes and technologies for the recycling and recovery of spent lithium-ion batteries, Renew. Sustain. Energy Rev. 60 (2016) 195–205.
[18] J. Zhao, X. Qu, J. Qu, B. Zhang, Z. Ning, H. Xie, X. Zhou, Q. Song, P. Xing, H. Yin, Extraction of Co and Li_2CO_3 from cathode materials of spent lithium-ion batteries through a combined acid-leaching and electro-deoxidation approach, J. Hazard Mater. 379 (2019) 120817.
[19] X. Zheng, Z. Zhu, X. Lin, Y. Zhang, Y. He, H. Cao, Z. Sun, A mini-review on metal recycling from spent lithium ion batteries, Engineering 4 (3) (2018) 361–370.
[20] T. Naseri, N. Bahaloo-Horeh, S.M. Mousavi, Bacterial leaching as a green approach for typical metals recovery from end-of-life coin cells batteries, J. Clean. Prod. 220 (2019) 483–492.

[21] T. Naseri, N. Bahaloo-Horeh, S.M. Mousavi, Two-step bioleaching of Li, Co and Mn from spent lithium-ion coin cells batteries using Acidithiobacillus ferrooxidans, Iran. J. Health Environ. (2018) 11.

[22] T. Gu, S.O. Rastegar, S.M. Mousavi, M. Li, M. Zhou, Advances in bioleaching for recovery of metals and bioremediation of fuel ash and sewage sludge, Bioresour. Technol. 261 (2018) 428–440.

[23] F. Anjum, M. Shahid, A. Akcil, Biohydrometallurgy techniques of low grade ores: a review on black shale, Hydrometallurgy 117 (2012) 1–12.

[24] S. Kathi, A. Padmavathy, E-waste: global scenario, constituents, and biological strategies for remediation, in: Electronic Waste Pollution, 2019, pp. 75–96.

[25] F. Pourhossein, S.M. Mousavi, Enhancement of copper, nickel, and gallium recovery from LED waste by adaptation of *Acidithiobacillus ferrooxidans*, Waste Manag. 79 (2018) 98–108.

[26] S. Bindschedler, T.Q.T.V. Bouquet, D. Job, E. Joseph, P. Junier, Fungal biorecovery of gold from e-waste, Adv. Appl. Microbiol. 99 (2017) 53–81.

[27] A.A. Juwarkar, S.K. Singh, A. Mudhoo, A comprehensive overview of elements in bioremediation, Rev. Environ. Sci. Biotechnol. 9 (3) (2010) 215–288.

[28] H. Qin, T. Hu, Y. Zhai, N. Lu, J. Aliyeva, The improved methods of heavy metals removal by biosorbents: a review, Environ. Pollut. 258 (2020) 113777.

[29] T. Viraraghavan, A. Srinivasan, Fungal Biosorption and Biosorbents. Microbial Biosorption of Metals, 2011, pp. 143–158.

[30] A. Isıldar, E.D. van Hullebusch, M. Lenz, G. Du Laing, A. Marra, A. Cesaro, S. Panda, A. Akcil, M.A. Kucuker, K. Kuchta, Biotechnological strategies for the recovery of valuable and critical raw materials from waste electrical and electronic equipment (WEEE)—a review, J. Hazard Mater. 362 (2019) 467–481.

[31] M.E. Hoque, O.J. Philip, Biotechnological recovery of heavy metals from secondary sources—an overview, Mater. Sci. Eng. C 31 (2) (2011) 57–66.

[32] A. Hansda, V. Kumar, A comparative review towards potential of microbial cells for heavy metal removal with emphasis on biosorption and bioaccumulation, World J. Microbiol. Biotechnol. 32 (10) (2016) 1–14.

[33] I. Asghari, S. Mousavi, F. Amiri, S. Tavassoli, Bioleaching of spent refinery catalysts: a review, J. Ind. Eng. Chem. 19 (4) (2013) 1069–1081.

[34] D. Mishra, D.-J. Kim, J.-G. Ahn, Y.-H. Rhee, Bioleaching: a microbial process of metal recovery; a review, Met. Mater. Int. 11 (3) (2005) 249–256.

[35] M. Jafari, H. Abdollahi, S.Z. Shafaei, M. Gharabaghi, H. Jafari, A. Akcil, S. Panda, Acidophilic bioleaching: a review on the process and effect of organic–inorganic reagents and materials on its efficiency, Miner. Process. Extr. Metall. Rev. 40 (2) (2019) 87–107.

[36] B.K. Biswal, U.U. Jadhav, M. Madhaiyan, L. Ji, E.H. Yang, B. Cao, Biological leaching and chemical precipitation methods for recovery of Co and Li from spent lithium-ion batteries, ACS Sustain. Chem. Eng. 6 (9) (2018) 12343–12352.

[37] A. Potysz, E.D. van Hullebusch, J. Kierczak, Perspectives regarding the use of metallurgical slags as secondary metal resources—a review of bioleaching approaches, J. Environ. Manag. 219 (2018) 138–152.

[38] A.K. Awasthi, M. Hasan, Y.K. Mishra, A.K. Pandey, B.N. Tiwary, R.C. Kuhad, V.K. Gupta, V.K. Thakur, Environmentally sound system for E-waste: biotechnological perspectives, Curr. Res. Biotechnol. 1 (2019) 58–64.

[39] P. Rasoulnia, R. Barthen, K. Valtonen, A.M. Lakaniemi, Impacts of phosphorous source on organic acid production and heterotrophic bioleaching of rare earth elements and base metals from spent nickel-metal-hydride batteries, Waste Biomass Valorization (2021) 1–15.

[40] V. Fonti, A. Dell'Anno, F. Beolchini, Does bioleaching represent a biotechnological strategy for remediation of contaminated sediments? Sci. Total Environ. 563 (2016) 302–319.

[41] I. Asghari, S. Mousavi, Effects of key parameters in recycling of metals from petroleum refinery waste catalysts in bioleaching process, Rev. Environ. Sci. Biotechnol. 13 (2) (2014) 139–161.

[42] K. Bosecker, Bioleaching: metal solubilization by microorganisms, FEMS Microbiol. Rev. 20 (3–4) (1997) 591–604.

[43] Z. Kazemian, M. Larypoor, R. Marandi, Evaluation of myco-leaching potential of valuable metals from spent lithium battery by *Penicillium chrysogenum* and *Aspergillus niger*, Int. J. Environ. Anal. Chem. (2020) 1–14.

[44] S. Ilyas, J c Lee, Biometallurgical recovery of metals from waste electrical and electronic equipment: a review, ChemBioEng Rev. 1 (4) (2014) 148–169.

[45] F. Zhao, S. Wang, Bioleaching of electronic waste using extreme acidophiles, in: Electronic Waste Management and Treatment Technology, 2019, pp. 153–174.

[46] G.R. Marlenne, M.V. Fernanda, G.R.A. Norma, Acidithiobacillus thiooxidans DSM 26636: an alternative for the bioleaching of metallic burrs, Catalysts 10 (11) (2020) 1230.

[47] A. Isıldar, J. van de Vossenberg, E.R. Rene, E.D. van Hullebusch, P.N. Lens, Biorecovery of metals from electronic waste, in: Sustainable Heavy Metal Remediation, 2017, pp. 241–278.

[48] J.D. García-García, R. Sánchez-Thomas, R. Moreno-Sánchez, Bio-recovery of non-essential heavy metals by intra-and extracellular mechanisms in free-living microorganisms, Biotechnol. Adv. 34 (5) (2016) 859–873.

[49] F. Pourhossein, S.M. Mousavi, F. Beolchini, M.L. Martire, Novel green hybrid acidic-cyanide bio-leaching applied for high recovery of precious and critical metals from spent light emitting diode lamps, J. Clean. Prod. 298 (2021) 126714.

[50] Y. Lu, Z. Xu, Precious metals recovery from waste printed circuit boards: a review for current status and perspective, Resour. Conserv. Recycl. 113 (2016) 28–39.

[51] R. Marcincakova, J. Kadukova, A. Mrazikova, O. Velgosova, A. Luptakova, S. Ubaldini, Metal bio-leaching from spent lithium-ion batteries using acidophilic bacterial strains, Inz. Miner. 17 (1) (2016) 117–120.

[52] A. Habibi, S. Shamshiri Kourdestani, M. Hadadi, Biohydrometallurgy as an environmentally friendly approach in metals recovery from electrical waste: a review, Waste Manag. Res. 38 (3) (2020) 232–244.

[53] W. Sand, T. Gehrke, Extracellular polymeric substances mediate bioleaching/biocorrosion via inter-facial processes involving iron (III) ions and acidophilic bacteria, Res. Microbiol. 157 (1) (2006) 49–56.

[54] D. Mishra, Y.H. Rhee, Microbial leaching of metals from solid industrial wastes, J. Microbiol. 52 (1) (2014) 1–7.

[55] J. Wang, B. Tian, Y. Bao, C. Qian, Y. Yang, T. Niu, B. Xin, Functional exploration of extracellular polymeric substances (EPS) in the bioleaching of obsolete electric vehicle $LiNi_xCo_yMn1_{-x-y}O_2$ Li-ion batteries, J. Hazard Mater. 354 (2018) 250–257.

[56] Y. Xin, X. Guo, S. Chen, J. Wang, F. Wu, B. Xin, Bioleaching of valuable metals Li, Co, Ni and Mn from spent electric vehicle Li-ion batteries for the purpose of recovery, J. Clean. Prod. 116 (2016) 249–258.

[57] J. Jegan Roy, M. Srinivasan, B. Cao, Bioleaching as an eco-friendly approach for metal recovery from spent NMC-based lithium-ion batteries at a high pulp density, ACS Sustain. Chem. Eng. 9 (8) (2021) 3060–3069.

[58] G. Zeng, S. Luo, X. Deng, L. Li, C. Au, Influence of silver ions on bioleaching of cobalt from spent lithium batteries, Miner. Eng. 49 (2013) 40–44.

[59] B. Xin, W. Jiang, H. Aslam, K. Zhang, C. Liu, R. Wang, Y. Wang, Bioleaching of zinc and manganese from spent Zn–Mn batteries and mechanism exploration, Bioresour. Technol. 106 (2012) 147–153.

[60] A. Pathak, L. Morrison, M.G. Healy, Catalytic potential of selected metal ions for bioleaching, and potential techno-economic and environmental issues: a critical review, Bioresour. Technol. 229 (2017) 211–221.

[61] M.M. Theron, J.R. Lues, Organic Acids and Food Preservation, CRC Press, 2010.

[62] R. Golmohammadzadeh, F. Faraji, F. Rashchi, Recovery of lithium and cobalt from spent lithium ion batteries (LIBs) using organic acids as leaching reagents: a review, Resour. Conserv. Recycl. 136 (2018) 418–435.

[63] Y. Yan, J. Gao, J. Wu, B. Li, Effects of Inorganic and Organic Acids on Heavy Metals Leaching in Contaminated Sediment. An Interdisciplinary Response to Mine Water Challenges, China University of Mining and Technology Press, Xuzhou, 2014.
[64] L. Li, J. Ge, R. Chen, F. Wu, S. Chen, X. Zhang, Environmental friendly leaching reagent for cobalt and lithium recovery from spent lithium-ion batteries, Waste Manag. 30 (12) (2010) 2615−2621.
[65] L. Sun, K. Qiu, Organic oxalate as leachant and precipitant for the recovery of valuable metals from spent lithium-ion batteries, Waste Manag. 32 (8) (2012) 1575−1582.
[66] K. Pollmann, S. Kutschke, S. Matys, J. Raff, G. Hlawacek, F.L. Lederer, Bio-recycling of metals: recycling of technical products using biological applications, Biotechnol. Adv. 36 (4) (2018) 1048−1062.
[67] B. Max, J.M. Salgado, N. Rodríguez, S. Cortés, A. Converti, J.M. Domínguez, Biotechnological production of citric acid, Braz. J. Microbiol. 41 (4) (2010) 862−875.
[68] H. Srichandan, R.K. Mohapatra, P.K. Parhi, S. Mishra, Bioleaching approach for extraction of metal values from secondary solid wastes: a critical review, Hydrometallurgy 189 (2019) 105122.
[69] M. Baniasadi, F. Vakilchap, N. Bahaloo-Horeh, S.M. Mousavi, S. Farnaud, Advances in bioleaching as a sustainable method for metal recovery from e-waste: a review, J. Ind. Eng. Chem. 76 (2019) 75−90.
[70] H. Liu, G. Gu, Y. Xu, Surface properties of pyrite in the course of bioleaching by pure culture of *Acidithiobacillus ferrooxidans* and a mixed culture of *Acidithiobacillus ferrooxidans* and *Acidithiobacillus thiooxidans*, Hydrometallurgy 108 (1−2) (2011) 143−148.
[71] B. Xin, D. Zhang, X. Zhang, Y. Xia, F. Wu, S. Chen, L. Li, Bioleaching mechanism of Co and Li from spent lithium-ion battery by the mixed culture of acidophilic sulfur-oxidizing and iron-oxidizing bacteria, Bioresour. Technol. 100 (24) (2009) 6163−6169.
[72] S. Ghassa, A. Farzanegan, M. Gharabaghi, H. Abdollahi, Novel bioleaching of waste lithium ion batteries by mixed moderate thermophilic microorganisms, using iron scrap as energy source and reducing agent, Hydrometallurgy 197 (2020) 105465.
[73] A. Heydarian, S.M. Mousavi, F. Vakilchap, M. Baniasadi, Application of a mixed culture of adapted acidophilic bacteria in two-step bioleaching of spent lithium-ion laptop batteries, J. Power Sources 378 (2018) 19−30.
[74] Z. Niu, Y. Zou, B. Xin, S. Chen, C. Liu, Y. Li, Process controls for improving bioleaching performance of both Li and Co from spent lithium ion batteries at high pulp density and its thermodynamics and kinetics exploration, Chemosphere 109 (2014) 92−98.
[75] J.J. Roy, S. Madhavi, Cao B Metal extraction from spent lithium-ion batteries (LIBs) at high pulp density by environmentally friendly bioleaching process, J. Clean. Prod. 280 (2021) 124242.
[76] X. Liu, H. Liu, W. Wu, X. Zhang, T. Gu, M. Zhu, W. Tan, Oxidative stress induced by metal ions in bioleaching of $LiCoO_2$ by an acidophilic microbial consortium, Front. Microbiol. 10 (2020) 3058.
[77] T. Huang, L. Liu, S. Zhang, Recovery of cobalt, lithium, and manganese from the cathode active materials of spent lithium-ion batteries in a bio-electro-hydrometallurgical process, Hydrometallurgy 188 (2019) 101−111.
[78] T. Dolker, D. Pant, Chemical-biological hybrid systems for the metal recovery from waste lithium ion battery, J. Environ. Manag. 248 (2019) 109270.
[79] W. Wu, X. Liu, X. Zhang, X. Li, Y. Qiu, M. Zhu, W. Tan, Mechanism underlying the bioleaching process of $LiCoO_2$ by sulfur-oxidizing and iron-oxidizing bacteria, J. Biosci. Bioeng. 128 (3) (2019) 344−354.
[80] N.J. Boxall, K.Y. Cheng, W. Bruckard, A.H. Kaksonen, Application of indirect non-contact bioleaching for extracting metals from waste lithium-ion batteries, J. Hazard Mater. 360 (2018) 504−511.
[81] N.B. Horeh, S. Mousavi, S. Shojaosadati, Bioleaching of valuable metals from spent lithium-ion mobile phone batteries using *Aspergillus niger*, J. Power Sources 320 (2016) 257−266.
[82] M.J. Kim, J.Y. Seo, Y.S. Choi, G.H. Kim, Bioleaching of spent Zn−Mn or Ni−Cd batteries by *Aspergillus* species, Waste Manag. 51 (2016) 168−173.

[83] O. Velgosová, J. Kaduková, R. Marinčáková, A. Mražíková, L. Fröhlich, The role of main leaching agents responsible for Ni bioleaching from spent Ni-Cd batteries, Separ. Sci. Technol. 49 (3) (2014) 438–444.

[84] M.I. Bajestani, S. Mousavi, S. Shojaosadati, Bioleaching of heavy metals from spent household batteries using *Acidithiobacillus ferrooxidans*: statistical evaluation and optimization, Separ. Purif. Technol. 132 (2014) 309–316.

[85] C. Ruhatiya, S. Shaosen, C.T. Wang, A. Jishnu, Y. Bhalerao, Optimization of process conditions for maximum metal recovery from spent zinc-manganese batteries: illustration of statistical based automated neural network approach, Energy Storage 2 (3) (2020) e111.

[86] M.S. Sadeghabad, N. Bahaloo-Horeh, S.M. Mousavi, Using bacterial culture supernatant for extraction of manganese and zinc from waste alkaline button-cell batteries, Hydrometallurgy 188 (2019) 81–91.

[87] Z. Niu, Q. Huang, J. Wang, Y. Yang, B. Xin, S. Chen, Metallic ions catalysis for improving bioleaching yield of Zn and Mn from spent Zn-Mn batteries at high pulp density of 10%, J. Hazard Mater. 298 (2015) 170–177.

[88] U. Jadhav, H. Hocheng, Extraction of silver from spent silver oxide–zinc button cells by using *Acidithiobacillus ferrooxidans* culture supernatant, J. Clean. Prod. 44 (2013) 39–44.

[89] B. Swain, Cost effective recovery of lithium from lithium ion battery by reverse osmosis and precipitation: a perspective, J. Chem. Technol. Biotechnol. 93 (2) (2018) 311–319.

[90] L.K. Wang, Y.T. Hung, N.K. Shammas, Physicochemical Treatment Processes 2005th, 744, Humana, 2005.

[91] Y. Zheng, W. Song, W.T. Mo, L. Zhou, J.W. Liu, Lithium fluoride recovery from cathode material of spent lithium-ion battery, RSC Adv. 8 (16) (2018) 8990–8998.

[92] R.C. Wang, Y.C. Lin, S.H. Wu, A novel recovery process of metal values from the cathode active materials of the lithium-ion secondary batteries, Hydrometallurgy 99 (3–4) (2009) 194–201.

[93] M. Joulié, R. Laucournet, E. Billy, Hydrometallurgical process for the recovery of high value metals from spent lithium nickel cobalt aluminum oxide based lithium-ion batteries, J. Power Sources 247 (2014) 551–555.

[94] Y. Song, Z. Zhao, Recovery of lithium from spent lithium-ion batteries using precipitation and electrodialysis techniques, Separ. Purif. Technol. 206 (2018) 335–342.

[95] H. Zou, E. Gratz, D. Apelian, Y. Wang, A novel method to recycle mixed cathode materials for lithium ion batteries, Green Chem. 15 (5) (2013) 1183–1191.

[96] L. Chen, X. Tang, Y. Zhang, L. Li, Z. Zeng, Y. Zhang, Process for the recovery of cobalt oxalate from spent lithium-ion batteries, Hydrometallurgy 108 (1–2) (2011) 80–86.

[97] G. Granata, E. Moscardini, F. Pagnanelli, F. Trabucco, L. Toro, Product recovery from Li-ion battery wastes coming from an industrial pre-treatment plant: lab scale tests and process simulations, J. Power Sources 206 (2012) 393–401.

[98] S. Natarajan, A.B. Boricha, H.C. Bajaj, Recovery of value-added products from cathode and anode material of spent lithium-ion batteries, Waste Manag. 77 (2018) 455–465.

[99] P. Meshram, B. Pandey, T. Mankhand, Hydrometallurgical processing of spent lithium ion batteries (LIBs) in the presence of a reducing agent with emphasis on kinetics of leaching, Chem. Eng. J. 281 (2015) 418–427.

[100] Y. Weng, S. Xu, G. Huang, C. Jiang, Synthesis and performance of Li $[(Ni_{1/3}Co_{1/3}Mn_{1/3})_{1-x}Mg_x]O_2$ prepared from spent lithium ion batteries, J. Hazard Mater. 246 (2013) 163–172.

[101] X. Chen, T. Zhou, J. Kong, H. Fang, Y. Chen, Separation and recovery of metal values from leach liquor of waste lithium nickel cobalt manganese oxide based cathodes, Separ. Purif. Technol. 141 (2015) 76–83.

[102] J.W. An, D.J. Kang, K.T. Tran, M.J. Kim, T. Lim, T. Tran, Recovery of lithium from Uyuni salar brine, Hydrometallurgy 117 (2012) 64–70.

[103] J. Pakarinen, E. Paatero, Recovery of manganese from iron containing sulfate solutions by precipitation, Miner. Eng. 24 (13) (2011) 1421–1429.

[104] X. Chen, Y. Chen, T. Zhou, D. Liu, H. Hu, S. Fan, Hydrometallurgical recovery of metal values from sulfuric acid leaching liquor of spent lithium-ion batteries, Waste Manag. 38 (2015) 349–356.

[105] M. Sethurajan, Metallurgical Sludges, Bio/Leaching and Heavy Metals Recovery (Zn, Cu). Paris Est, 2015.
[106] J. Li, P. Shi, Z. Wang, Y. Chen, C.-C. Chang, A combined recovery process of metals in spent lithium-ion batteries, Chemosphere 77 (8) (2009) 1132–1136.
[107] S. Castillo, F. Ansart, C. Laberty-Robert, J. Portal, Advances in the recovering of spent lithium battery compounds, J. Power Sources 112 (1) (2002) 247–254.
[108] A. Kowalski, O.N. Center, Metals removal to low levels using chemical precipitants, in: Proc., Proceedings of AESF/EPA Conference for Environmental Excellence, Orlando, Florida, 2002, pp. 143–150.
[109] X. Zhang, Y. Bian, S. Xu, E. Fan, Q. Xue, Y. Guan, F. Wu, L. Li, R. Chen, Innovative application of acid leaching to regenerate Li $(Ni_{1/3}Co_{1/3}Mn_{1/3})$ O_2 cathodes from spent lithium-ion batteries, ACS Sustain. Chem. Eng. 6 (5) (2018) 5959–5968.
[110] F. Ibis, P. Dhand, S. Suleymanli, A.E. van der Heijden, H.J. Kramer, H.B. Eral, A combined experimental and modelling study on solubility of calcium oxalate monohydrate at physiologically relevant pH and temperatures, Crystals 10 (10) (2020) 924.
[111] Y. El-Nadi, Solvent extraction and its applications on ore processing and recovery of metals: classical approach, Separ. Purif. Rev. 46 (3) (2017) 195–215.
[112] V.S. Kislik, Solvent Extraction: Classical and Novel Approaches 1st, 576, Elsevier, 2012.
[113] G.M. Ritcey, Solvent extraction in hydrometallurgy: present and future, Tsinghua Sci. Technol. 11 (2) (2006) 137–152.
[114] S.H. Joo, S.M. Shin, D. Shin, C. Oh, J.P. Wang, Extractive separation studies of manganese from spent lithium battery leachate using mixture of PC88A and Versatic 10 acid in kerosene, Hydrometallurgy 156 (2015) 136–141.
[115] X. Chen, L. Cao, D. Kang, J. Li, S. Li, X. Wu, Hydrometallurgical Processes for Valuable Metals Recycling from Spent Lithium-Ion Batteries. Recycling of Spent Lithium-Ion Batteries, 2019, pp. 93–139.
[116] D.P. Mantuano, G. Dorella, R.C.A. Elias, M.B. Mansur, Analysis of a hydrometallurgical route to recover base metals from spent rechargeable batteries by liquid–liquid extraction with Cyanex 272, J. Power Sources 159 (2) (2006) 1510–1518.
[117] X. Chen, T. Zhou, Hydrometallurgical process for the recovery of metal values from spent lithium-ion batteries in citric acid media, Waste Manag. Res. 32 (11) (2014) 1083–1093.
[118] B. Swain, S.S. Cho, G.H. Lee, C.G. Lee, S. Uhm, Extraction/separation of cobalt by solvent extraction: a review, Appl. Chem. Eng. 26 (6) (2015) 631–639.
[119] F. Peng, D. Mu, R. Li, Y. Liu, Y. Ji, C. Dai, F. Ding, Impurity removal with highly selective and efficient methods and the recycling of transition metals from spent lithium-ion batteries, RSC Adv. 9 (38) (2019) 21922–21930.
[120] J. Ren, R. Li, Y. Liu, Y. Cheng, D. Mu, R. Zheng, J. Liu, C. Dai, The impact of aluminum impurity on the regenerated lithium nickel cobalt manganese oxide cathode materials from spent LIBs, New J. Chem. 41 (19) (2017) 10959–10965.
[121] T. Suzuki, T. Nakamura, Y. Inoue, M. Niinae, J. Shibata, A hydrometallurgical process for the separation of aluminum, cobalt, copper and lithium in acidic sulfate media, Separ. Purif. Technol. 98 (2012) 396–401.
[122] Y. Pranolo, W. Zhang, C. Cheng, Recovery of metals from spent lithium-ion battery leach solutions with a mixed solvent extractant system, Hydrometallurgy 102 (1–4) (2010) 37–42.
[123] O.A.E.N. Desouky, Liquid-liquid Extraction of Rare Earth Elements from Sulfuric Acid, University of Leeds, 2006.
[124] S. Refly, O. Floweri, T.R. Mayangsari, A.H. Aimon, F. Iskandar, Green recycle processing of cathode active material from $LiNi_{1/3}Co_{1/3}Mn_{1/3}O_2$ (NCM 111) battery waste through citric acid leaching and oxalate co-precipitation process, Mater. Today Proc. 44 (2021) 3378–3380.
[125] Y. Zhao, X. Yuan, L. Jiang, J. Wen, H. Wang, R. Guan, J. Zhang, G. Zeng, Regeneration and reutilization of cathode materials from spent lithium-ion batteries, Chem. Eng. J. 383 (2020) 123089.
[126] Q. Sa, E. Gratz, M. He, W. Lu, D. Apelian, Y. Wang, Synthesis of high performance $LiNi_{1/3}Mn_{1/3}Co_{1/3}O_2$ from lithium ion battery recovery stream, J. Power Sources 282 (2015) 140–145.

[127] Q. Sa, E. Gratz, J.A. Heelan, S. Ma, D. Apelian, Y. Wang, Synthesis of diverse LiNi x Mn y Co z O 2 cathode materials from lithium ion battery recovery stream, J. Sustain. Metall. 2 (3) (2016) 248–256.

[128] Y. Song, Q. Huang, Z. Niu, J. Ma, B. Xin, S. Chen, J. Dai, R. Wang, Preparation of Zn–Mn ferrite from spent Zn–Mn batteries using a novel multi-step process of bioleaching and co-precipitation and boiling reflux, Hydrometallurgy 153 (2015) 66–73.

[129] R.X. Che, H. Gao, H.B. Zhao, J.X. Fang, Developing history and present situation of sol-gel science, J. Yunnan Univ. S (2005).

[130] C.K. Lee, K.I. Rhee, Preparation of $LiCoO_2$ from spent lithium-ion batteries, J. Power Sources 109 (1) (2002) 17–21.

[131] L. Li, R. Chen, X. Zhang, F. Wu, J. Ge, M. Xie, Preparation and electrochemical properties of re-synthesized LiCoO 2 from spent lithium-ion batteries, Chin. Sci. Bull. 57 (32) (2012) 4188–4194.

[132] L. Li, Y. Bian, X. Zhang, Q. Xue, E. Fan, F. Wu, R. Chen, Economical recycling process for spent lithium-ion batteries and macro-and micro-scale mechanistic study, J. Power Sources 377 (2018) 70–79.

[133] L. Yao, H. Yao, G. Xi, Y. Feng, Recycling and synthesis of LiNi 1/3 Co 1/3 Mn 1/3 O 2 from waste lithium ion batteries using d, l-malic acid, RSC Adv. 6 (22) (2016) 17947–17954.

[134] J. Wang, Z. Guo, Hydrometallurgically Recycling Spent Lithium-Ion Batteries. Recycling of Spent Lithium-Ion Batteries, 2019, pp. 27–55.

[135] J. Wu, J. Lin, E. Fan, R. Chen, F. Wu, L. Li, Sustainable regeneration of high-performance $Li_{1-x}Na_xCoO_2$ from cathode materials in spent lithium-ion batteries, ACS Appl. Energy Mater. 4 (2021) 2607–2615.

[136] L. Yao, Y. Feng, G. Xi, A new method for the synthesis of LiNi 1/3 Co 1/3 Mn 1/3 O 2 from waste lithium ion batteries, RSC Adv. 5 (55) (2015) 44107–44114.

[137] X. Zhang, L. Li, E. Fan, Q. Xue, Y. Bian, F. Wu, R. Chen, Toward sustainable and systematic recycling of spent rechargeable batteries, Chem. Soc. Rev. 47 (19) (2018) 7239–7302.

[138] L. Li, X. Zhang, M. Li, R. Chen, F. Wu, K. Amine, J. Lu, The recycling of spent lithium-ion batteries: a review of current processes and technologies, Electrochem. Energy Rev. 1 (4) (2018) 461–482.

CHAPTER 12

Technologies for separating nanomaterials from spent lithium-ion batteries

Jiafeng Zhang

National Engineering Laboratory for High Efficiency Recovery of Refractory Nonferrous Metals, School of Metallurgy and Environment, Central South University, Changsha, Hunan, China

1. Introduction

Separating a reasonable level of electrode nanomaterials from spent lithium-ion batteries (LIBs) is a key prerequisite for increasing the efficiency of nanoelectrode material recovery. Domestic and international research on the separation of electrode nanomaterials has focused on three main steps: (1) Pretreatment of spent LIBs: pretreatment generally includes discharge treatment, mechanical disassembly, and crushing. (2) Separation of electrode materials from collector fluid: commonly used methods include pyrolysis, alkali dissolution, and mechanical separation. (3) Separation among nanomaterials: this step pertains to separating cathode nanomaterials from anode nanomaterials [1].

For the effective liberation of nanomaterials from LIBs, we describe the structure and main components of LIBs. An LIB consists of anode material, cathode material, a shell, dispersion mass, aluminum foil, copper foil, and a diaphragm. The specific mass percentages are listed in Table 12.1. Table 12.2 lists the main LIB components along with their chemical properties and potential hazards [2,3].

1.1 Cathode

The LIB cathode is usually prepared by mixing the cathode active material, conductive agent, and binder, coating this mixture onto an aluminum foil collector, and compacting, rolling, and cutting the resulting entity. Commonly used conductive agents are conductive carbon black and acetylene black, with a binder of polyvinylidene fluoride (PVDF). According to the different active substances of the cathode, LIB cathode materials include lithium cobaltate ($LiCoO_2$), lithium manganate ($LiMnO_2$), lithium iron phosphate

Table 12.1 Typical lithium-ion battery composition.

Components	Casing	Cathode materials	Anode materials	Electrolyte	Copper foil	Aluminum foil	Separator
wt%	30	25	15	11	8	7	4

Table 12.2 Chemical properties and potential hazards of lithium-ion battery components.

Composition	Component name	Chemical property	Potential hazards
Cathode materials	$LiNi_xCo_yMn_{1-x-y}O_2$	Reacts with water, acid, or strong oxidizing agents; decomposes by heat to produce toxic lithium, nickel, and other oxides	Heavy metal nickel, cobalt, and manganese pollution. Raises environmental pH
	$LiCoO_2$	Reacts with water, acid, strong oxidizer; decomposes by heat to produce toxic lithium and cobalt oxides	Heavy metal cobalt contamination. Raises environmental pH
	$LiMnO_2$	Reacts easily with organic, oxidizing, and reducing agents to produce toxic substances	Heavy metal manganese pollution. Raises environmental pH
Anode materials	Graphite	Combustion produces CO, CO_2, dust	Dust pollution
Electrolyte	$LiPF_6$	Strong corrosiveness, decomposition to HF from contact with water, combustion to produce P_2O_5, etc.	Fluorine contamination
Organic solvents	Vinyl carbonate	Reacts with acids, bases, enhanced oxidizers, and reducing agents to produce aldehydes and acids	Aldehyde, organic pollution
	Propylene carbonate	Reacts with strong oxidizers, water, and air; decomposes by heat to produce aldehyde, ketone, etc. Harmful gas combustion can cause explosions	Ketone, aldehyde, organic pollution
	Dimethyl carbonate	Reacts violently with water, strong acids, strong bases, and strong reducing agents and hydrolyzes to produce methanol	Methanol, organic pollution
	Diethyl carbonate	Reacts violently with strong oxidizers, water, strong acids, strong bases, and burns to produce CO, CO_2	Alcohol and other organic pollution

cathode (LiFePO$_4$), nickel—cobalt—manganese ternary cathode (LiNi$_x$Co$_y$Mn$_{1-x-y}$O$_2$), nickel—cobalt—aluminum ternary material (LiNi$_x$Co$_y$Al$_{1-x-y}$O$_2$), and lithium-rich manganese-based material. As the parts that store energy in LIBs, the performance and cost of cathode materials directly determine the performance and cost of LIBs. Lithium-cobalt acid cathode materials are usually used in 3C applications, such as mobile phones and laptops. Lithium iron phosphate materials and nickel—cobalt—manganese ternary materials are usually used in the power sector—for instance, in new energy vehicles [3].

1.2 Anode

The LIB anode usually consists of an even mixture of anode material, a conductive agent, and a binder coated onto a copper foil collector and vacuum dried, compacted, rolled, and cut. LIB anodes are usually made of carbon-based materials. The layered carbon material can absorb lithium ions when charging, thus forming LiC$_6$ and enabling lithium ions to pass to the cathode more easily when discharging. The duration of the cyclic process is reduced, and the volume of the battery material is not significantly changed, thereby facilitating the realization of efficient LIB cycle performance. Commonly used carbon-based anode materials include natural graphite, multiplier graphite, energy graphite, silicon-carbon anode materials, and hard carbon materials. In addition, lithium titanate (Li$_4$Ti$_5$O$_{12}$), Sn-based intermetallic compounds, metal nitride lithium manganese nitride (Li$_7$MnN$_4$), silicon dioxide, and other materials can be used [4].

1.3 Diaphragm

The main function of the diaphragm is to separate the anode and cathode of the battery to prevent short-circuiting due to cathode and anode contact. Moreover, the diaphragm facilitates the transport of electrolyte ions. To this end, the diaphragm must exhibit high electrical insulation, corrosion resistance, moisture absorption capacity, mechanical properties, thermal stability, spatial stability, and flatness. In addition, the diaphragm must exhibit a certain pore size and porosity to ensure low resistance and high ionic conductivity with high permeability to lithium ions. Marketed diaphragms consist of a base material, usually polyolefin based on polyethylene, polypropylene, and a ceramic coating such as alumina. In recent years, new types of diaphragms, such as those made from PVDF as the base polymer, and diaphragm materials, such as cellulose composite membranes, have been developed [5].

1.4 Electrolyte

As the "flowing blood" the electrolyte plays a vital role in LIBs, providing a venue for electrochemical reactions in such batteries and serving as a carrier for the diffusion of lithium ions between electrical stages. The salts commonly involved in LIB electrolytes are $LiPF_6$, $LiClO_4$, $LiBF_4$, $Li(SO_2CF_3)_3$, and $LiCF_3SO_3$. Commonly used solvents include ethyl carbonate, dimethyl carbonate (DMC), propylene carbonate, and dimethyl sulfoxide (DMSO). Typically, additives, such as solid electrolyte interface improvers, cathode protectors, $LiPF_6$ stabilizers, flame retardants, and lithium-ion deposition improvers, are added to the electrolyte to enhance the physicochemical and electrochemical properties of LIBs [5].

1.5 Other structural components

LIB structural components include connecting and packaging parts. The structural parts of soft-pack batteries are aluminum plastic films and pole lugs. The structural parts of aluminum-shell batteries are internal connecting tapes, internal protective films, external insulating films, shells, top covers, and pole pillars. In addition, safety protection devices are typically set on the top cover of aluminum shells, such as overcurrent fuses, overvoltage protection, and safety valves.

1.6 Failure mechanisms

An extensive understanding of LIB failure mechanisms can facilitate safer and more efficient recycling of used LIBs. LIBs are rechargeable, have a specific service life, and are subject to abnormal factors that may lead to failures such as battery expansion, short-circuiting, performance degradation, and electrolyte leakage.

Therefore, from a structural perspective, the main causes of rechargeable battery failure can be analyzed as failure in one of the following four areas—positive electrode, negative electrode, electrolyte, or diaphragm—as shown in Fig. 12.1. End-of-life for rechargeable batteries may be associated with problems such as electrolyte decomposition, deterioration, diaphragm aging, punctures, and blockages. In addition, the potential causes of positive and negative electrode failure considerably influence how these batteries are recycled as well as the process design. Specific failure mechanisms are classified as follows:

➢ *Structural failure of the electrode material and particle cracking*
➢ *Corrosion of the collector fluid and decomposition of the binder*
➢ *Dissolution of transition metals in the cathode material*
➢ *Overgrowth of the solid electrolyte interface phase on the cathode surface, resulting in the depletion of Li from the internal circulation system of the cell*
➢ *Growth of lithium dendrites on the negative electrode, leading to short circuits during overcharging and overdischarging and even to battery explosions*

Understanding the causes of capacity decay in electrode materials can promote the design of targeted recovery methods. Capacity degradation results from two basic principles, structural changes during cycling and chemical decomposition/dissolution reactions. Therefore, an appropriate recycling method must be selected to ensure the

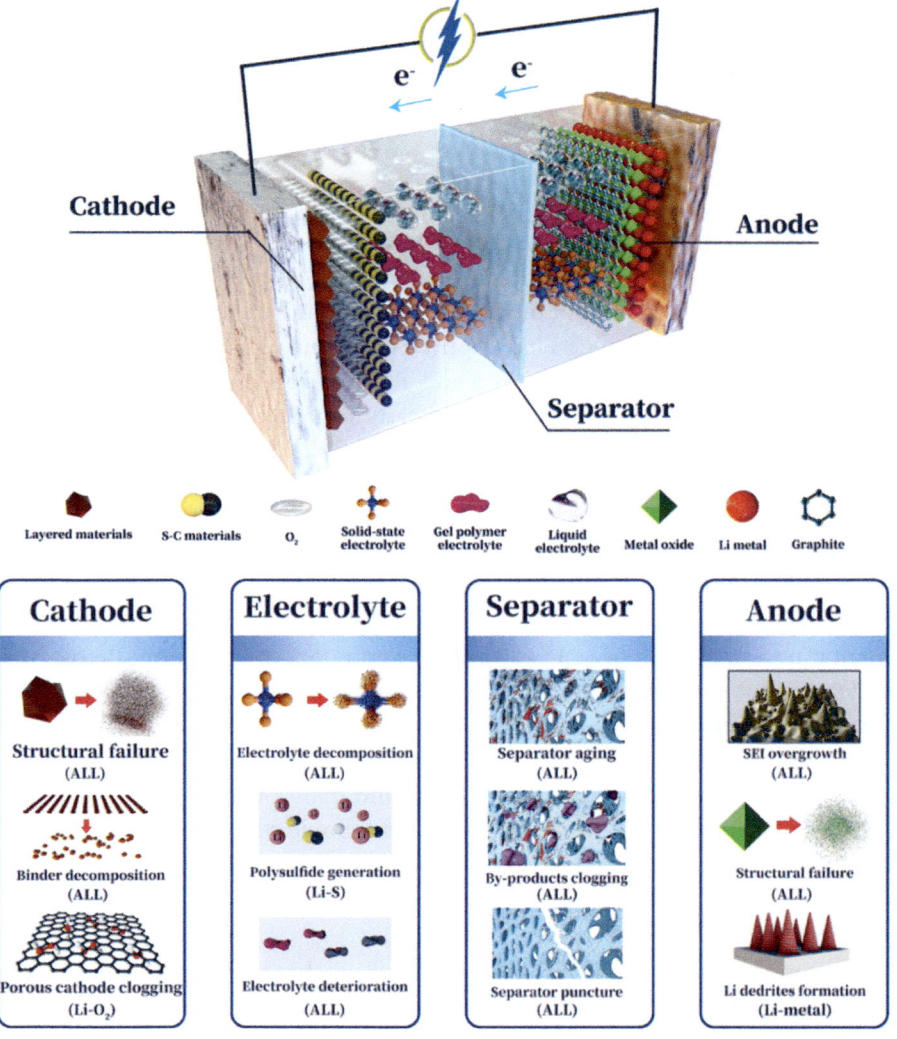

Figure 12.1 Schematic diagram of four aspects of existing degradation mechanisms for rechargeable batteries [6].

maximum recovery of used electrode materials, especially positive electrode materials. Undamaged electrode materials and other components can be separated or directly recovered using simple physical methods. In the case of battery failure due to a lack of metal ions in the cathode material, the cathode material can be repaired and directly regenerated by adding metal ions and subjecting the material to heat treatment to restore its electrochemical properties. Batteries that fail due to severe damage to the cathode material can be recovered by material extraction techniques, such as wetting and pyrometallurgy. Moreover, failed negative electrode materials can be recovered and regenerated with targeted treatment processes [6,7].

2. Pretreatment process

Although LIBs are not treated and disposed of as hazardous waste, they may be hazardous to the environment without proper treatment and disposal. Used LIBs contain copper, cobalt, nickel, manganese, lithium, aluminum, and other metal elements, as well as electrolytes, organic solvents, and byproducts of the battery charging and discharging cycle, all of which are hazardous to the environment. First, copper, cobalt, and nickel are heavy metals that can directly contaminate soil and water sources and ultimately endanger human health through bioconcentration. Second, $LiPF_6$ in the electrolyte is highly corrosive and produces HF when it meets water and P_2O_5 when it burns—i.e., toxic, harmful gases. Finally, organic solvents, such as ethyl carbonate, DMC, propylene carbonate, and DMSO do not easily degrade in nature and eventually contaminate water, air, and soil [5].

Based on the complex composition of LIBs, it is necessary to fine-tune the process of stripping valuable items to facilitate subsequent nanomaterial recovery. Considering the potential dangers of recycling LIBs, including fire, explosion, and chemical hazards, pretreatment is necessary to ensure safe and effective component separation. Pretreatment includes deactivation treatment, disassembly, and shredding. To minimize the risks associated with high voltages and the high reactivity of components, deactivation is conducted. Existing deactivation methods include discharging, freezing in liquid nitrogen, or treating cells in an inert atmosphere. In addition, the storage and transport of used LIBs are associated with potential hazards. Used LIBs requiring disposal should be deactivated before transport and storage to release all energy and render the battery inactive [8,9].

2.1 Deactivation treatment

The common deactivation methods are electrical discharge treatment, liquid nitrogen freezing, and dismantling under an inert atmosphere. As the latter two methods are expensive, the discharge treatment method has attracted more research attention. The purpose of discharge treatment is to reduce the residual voltage of LIBs to a safe range (<0.6 V). Used LIBs exhibit a large amount of residual voltage; thus, to ensure the safety of personnel and equipment, such batteries must be discharged to ensure that the residual voltage is within the safe range. None of the existing studies have focused on the discharge of used LIBs. In the following, we summarize several LIB discharge methods.

The most widely used method for discharging LIBs is salt solution immersion. Commonly used salt solutions include sodium chloride, sodium sulfate, and sodium hydroxide solutions. Of these, sodium chloride solution immersion is the most commonly applied method. Specifically, used LIBs are soaked in 5%—10% NaCl salt solution for approximately 48 h until the residual voltage of the battery can satisfy the requirements for safe dismantling. This method involves an extremely low discharge rate, and more than 24 h are generally required to achieve a residual voltage of less than 1 V. Moreover, difficult-to-remove chloride ions may be introduced in the leach solution, which can negatively affect subsequent decontamination and product recovery stages.

Sodium hydroxide solution may be used instead of sodium chloride solution in discharge treatment. Specifically, soaking is performed in 0.1–1 mol/L NaOH solution at room temperature for 1–3 h to discharge the battery. Notably, OH^- in aqueous solution is more difficult to discharge than Cl^-. Consequently, using NaOH solution is expected to reduce the discharge rate of the battery and weaken the discharge effect. Moreover, while treating soft-pack batteries, NaOH may induce corrosion of the aluminum case. To avoid introducing ionic impurities and increase the discharge efficiency, a reducing salt immersion discharge method has been developed, in which used LIBs are immersed in reducing solutions, such as 5%–20% sodium sulfide, at room temperature. The battery discharge time to 0 V can be reduced by approximately 6 h, and no secondary pollution is generated.

In addition, researchers have proposed that waste batteries to be treated should be discharged in a solution with 1 mol/L $CuCl_2$ solution as the electrolyte and 0.1 g/L conductive carbon black Super-P as the conductive agent after the positive and negative electrodes are led out with wires using a steel needle. The discharge time of this method is within 10 h, corresponding to an increased discharge rate; however, the positive and negative electrodes need to be led out with wires during the discharge process, which is highly tedious. Some people have proposed that by placing the battery directly in an electrolyte containing copper sulfate and ascorbic acid, the voltage of the used battery can be reduced from 3.5 V to less than 1 V after 5 h of immersion.

The discharge process is significant in ensuring the safety of subsequent recycling processes. Regardless of how the discharge process is implemented, leaking cells are present in spent LIBs. These leaks can contaminate the electrolyte used in the discharge process, and the battery is likely to react with the electrolyte solution to produce toxic and harmful substances during discharge. Thus, this process generates highly polluting wastewater and exhaust gas. None of the existing studies have focused on addressing these polluting substances. Management of the harmful wastewater and exhaust gas generated during discharge is a difficult aspect of recycling spent LIBs [10–12].

2.2 Mechanical dismantling and shredding

Discharged spent LIBs must be dismantled or mechanically crushed before they can be subjected to subsequent processes. Some people have crushed square LIBs using a VM-20-type vertical high-speed rotary crusher. Electrode active substances, copper foil, and aluminum foil were obtained as granular products that could be sieved to initially separate the various spent LIB substances. Some people have used an HYP-250 angle-cut crusher to perform the coarse crushing of spent LIBs, which were later finely crushed using an FL-150 universal crusher. A combination of coarse and fine crushing processes was used to treat the spent LIBs. Finally, $LiCoO_2$ and Co were enriched, and Co recovery was calculated to be up to 87.50% for the whole process. Some people have used a toothed shear roll crusher to coarsely crush spent LIBs and a vertical impact crusher to finely crush the obtained batteries. The authors studied the sorting and enrichment behavior of valuable components during crushing, established an evaluation method for the selective crushing of LIBs, and used a self-designed crusher for efficient crushing and dissociation of spent LIBs. Spent LIBs contain toxic and hazardous substances, such as binders and electrolytes, and the continuous release of these substances during crushing likely poses a threat to the environment and operation workers. Therefore, in recent years, most scholars have recommended the removal of organic substances before crushing [13].

3. Separation of nanomaterials from the collector fluid

Nanomaterials generally adhere to collector fluid and are difficult to separate from it after mechanical disassembly and fragmentation due to the durable adhesion provided by organic binders. Common methods used to separate nanomaterials from collector fluid include pyrolysis, alkali dissolution, and mechanical treatment.

3.1 Thermal treatment

In the thermal treatment approach, an organic binder is thermally decomposed to reduce the cohesion of the active material particles in the coating and remove the binder from the cathode material, thereby effectively separating the nanomaterials from the collector fluid. The PVDF binder in the positive electrode and conductive agent start to decompose at approximately 350°C and 600°C, respectively. Researchers have used pyrolysis to separate aluminum foil from the rest of the cathode material; however, when the temperature reached 700°C, the aluminum foil melted, and the active material could not be separated from the foil. The binder and carbonaceous conductor were removed by controlling the pyrolysis temperature, and the aluminum foil was completely separated from the active material after heating at 600°C for 15 min. In addition, the heat treatment process changed the molecular structure of the cathode material, reducing the charge of transition metal ions in the cathode material and facilitating the recovery of valuable metals in the subsequent leaching process. The thermal behavior of cathode active powder was characterized by Thermo Gravimetry-Differential Scanning Calorimeter (TG-DSC). In the TG-DSC curve shown in Fig. 12.2, a significant weight loss was observed between 519.44°C and 548.77°C, indicating the rapid decomposition of PVDF and carbon. Researchers have used vacuum pyrolysis to pretreat LIB cathode materials. When the system pressure was less than 1 kPa with pyrolysis at 600°C for 30 min, the organic

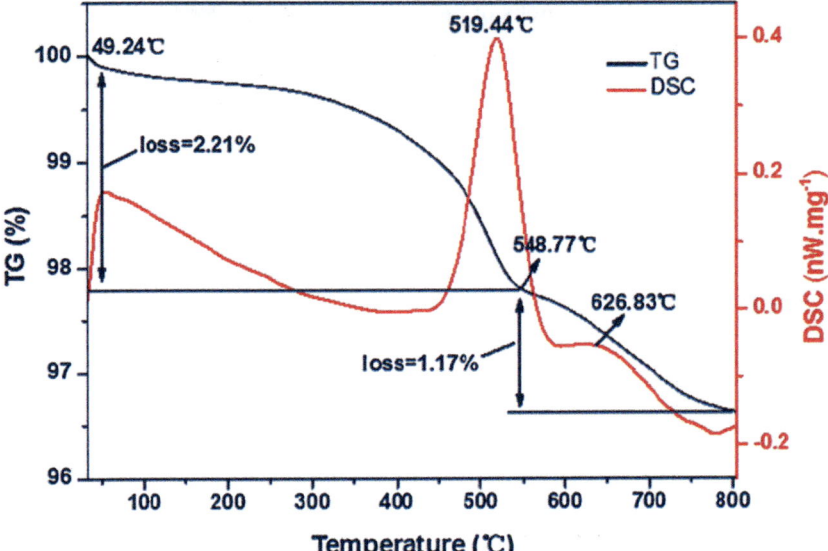

Figure 12.2 Thermo Gravimetry-Differential Scanning Calorimeter (TG-DSC) curves of active cathode powder calcined under various temperatures [14].

binder could be removed, and most of the cathode active material was separated from the aluminum foil, which remained intact.

The core problem in separating aluminum foil from the active material by pyrolysis is controlling the temperature to avoid melting the aluminum foil while the conductive agent and binder are being completely decomposed. The pyrolysis process is simple and suitable for large-scale industrial production; however, energy consumption is high, and the production process involves the emission of toxic and harmful gases that generate secondary pollution, thereby necessitating the installation of special equipment to purify the gases and fumes generated by combustion [15–17].

The results of various pretreatment processes for waste LIBs are summarized in Table 12.3.

3.2 Solvent soaking
3.2.1 Alkali soaking method
The alkali leaching method uses a sodium hydroxide solution to leach the aluminum foil from the positive electrode material, thereby separating the active material from the aluminum foil. Some people have used NaOH to separate the cathode active material from the aluminum foil and investigated the effect of NaOH concentration (1%–15% (w/w)) on the aluminum foil. Some people have used NaOH to dissolve aluminum foil to avoid introducing Al^{3+} into the subsequent separation step. When the liquid–solid ratio was 10:1, the amount of NaOH was 5% (mass fraction), and the required

Table 12.3 Summary of research results of pretreatment processes for spent lithium-ion batteries [18].

Methods	Solution	Temperature/ °C	Time/ min	Separation efficiencies
Alkali dissolving	2 mol/L NaOH	25	120	—
Alkali dissolving	5 wt% NaOH	25	240	99.99%
Alkali dissolving	5 mol/L NaOH	25	30	100%
Dissolution process	NMP	25	60	68%
Dissolution process	NMP	70	90	99%
Thermal treatment	—	600	15	—
Thermal treatment	—	300	60	—
Thermal treatment	—	600	30	—

NMP, N-methyl-pyrrolidone.

time was 4 h. At room temperature, 99.9% of the aluminum was dissolved, corresponding to a satisfactory separation effect. The reaction of the dissolution of aluminum foil in NaOH is presented as Eq. (12.1):

$$2Al\ (s) + 2NaOH\ (aq) + 6H_2O \rightarrow 2Na[Al(OH)_4]\ (aq) + 3H_2\ (g) \quad (12.1)$$

In this manner, the alkali leaching method can effectively separate aluminum foil from other cathode substances. Alkali leaching is a simple, low-cost operation; however, the strong alkali can corrode equipment, resulting in stringent equipment requirements. In addition, other positive substances, such as binders and conductive agents, cannot be separated from active substances, which limits subsequent separation and recovery [19,20].

3.2.2 Organic solvent soaking method

Bonding agents are organic solvents and can be dissolved in specific organic solvents to separate the collecting fluid and active substance. PVDF is a polar reagent and the most widely used bonding agent. Organic solvents can dissolve the organic binder in the electrode material via the principle of "similarity solubility" and other nanomaterials are separated from the collector. In this manner, nanoelectrode material can be separated from the collector. Commonly used organic solvents include N-methyl-pyrrolidone (NMP), N-dimethylacetamide (DMAC), N-dimethylformamide (DMF), and DMSO.

Some researchers have studied the solubility of PVDF in NMP, DMF, and DMAC at various temperatures and observed that DMF almost completely dissolves the binder PVDF in the lithium-ion cathode after 2 h of immersion at 65°C, thereby separating the aluminum foil from the other cathode materials. Some researchers have proposed an ultrasonic-assisted dissolution process by which the cathode material is immersed in NMP and sonicated for 3 min at room temperature to separate the cathodic active material from the aluminum collector, with a separation efficiency of 99%. The effective separation

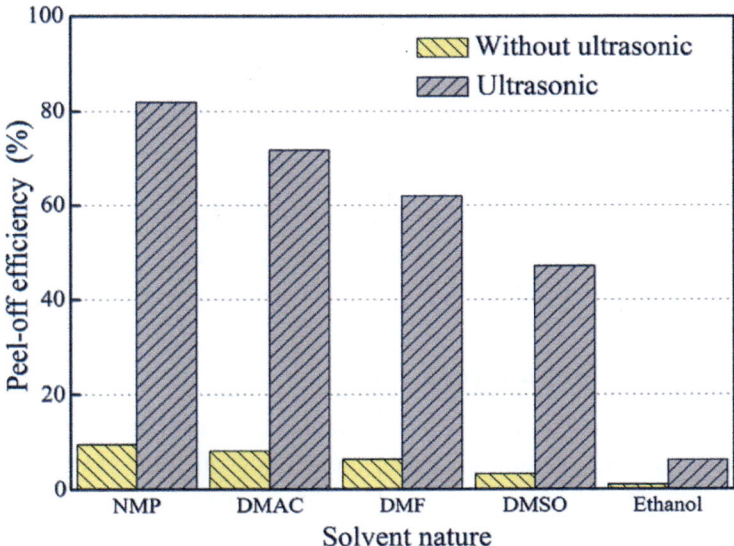

Figure 12.3 Separation efficiency of cathode materials using various solvents [21].

of the aluminum foil and cathode material could be attributed to the combined effect of PVDF dissolution and ultrasonic cavitation, as shown in Fig. 12.3. The separation efficiency of various solvents for the cathode material exhibited the following order: NMP > DMAC > DMF > DMSO > ethanol. The experimental results showed that the separation efficiency of aluminum foil from the cathode material reached 99% under the effect of NMP at a temperature of 70°C, ultrasonic power of 240 W, and 90 min of ultrasonic treatment, and the aluminum was recovered in metallic form with high purity.

Thus, NMP can effectively separate aluminum foil from the positive active material under ultrasonication, and the solvent can be distilled, enabling the recycling of the NMP. However, organic solvents cannot remove all impurities, recovered cathode active material requires further calcination to burn off residues, such as PVDF, and organic solvents are expensive and not suitable for large-scale recycling processes [21,22].

3.3 Mechanical treatment

In the mechanical separation method, the various materials in spent LIBs are sorted according to their physical properties, such as particle size, density, magnetism, electrical conductivity, elasticity, surface wettability, and color. Given their varying physical properties, aluminum foil, copper foil, steel, separators, cathodes, and anodes can be easily separated by crushing and screening. Gravity separation is based on the principle that mixtures with different sizes and densities exhibit different movements in certain separation media. Materials with the same particle size but a different composition, separated by screening and sieving, can be effectively sorted using gravity separation methods based on density differences. The low-density fraction consists mainly of separators, plastics, and aluminum foil. Mechanical separation can be generally divided into three steps: a basic grinding breakdown, a sieving process to further enrich the material, and a fine crushing process to separate the nanomaterial from the collector fluid and increase the nanomaterial recovery rate.

The mechanical separation method is usually combined with thermal treatment to efficiently separate the nanomaterials from the collector fluid. The method has a high processing efficiency and is suitable for large-scale industrial applications [13].

4. Separation of nanomaterials

The primary battery nanomaterials can be divided into two categories, carbon-based anode materials and metal oxide materials. Depending on their physical and chemical properties, these materials can be separated. The commonly used physical separation methods are flotation and electrostatic separation. In general, these methods do not damage nanomaterial structures; however, the purity of the products obtained after sorting is not high, and the added value of the products is small. Traditional wet metallurgy is the commonly used chemical method. Wet metallurgy usually destroys the structure of metal compound nanomaterials, and the obtained products are carbon-based nanomaterials and high-purity, metal-added products, such as cobalt sulfate, nickel sulfate, and lithium carbonate. This aspect is extensively described in the section on wet metallurgy.

4.1 Flotation separation methods

Cathodes and anodes can be separated by a flotation process based on differences in wettability. Specifically, graphite is hydrophobic, while cathode materials are hydrophilic.

Some researchers have first crushed spent LIBs using a vertical shear crusher, and after crushing, the waste LIBs have been graded using a wind shaker and vibrating screen. The grading yielded light products (positive and negative electrode isolation materials), metal products (copper and aluminum), and electrode materials. The resulting electrode material was heat-treated to remove organic matter and separated from the lithium-cobalt oxide and graphite material by flotation. Under optimal flotation conditions, the recovered cobalt oxide product contained more than 93% lithium and cobalt. Some researchers have crushed pretreated waste LIBs with an impact crusher and sieved out materials smaller than 0.25 mm, which were modified using Fenton's reagent and separated by flotation to recover C. Moreover, the authors experimentally analyzed the effect of Fenton's reagent on the modification of the active material.

Some researchers have separated $LiCoO_2$ powder from C powder using centrifugal separation equipment, as shown in Fig. 12.4. The slurry is poured through an inlet

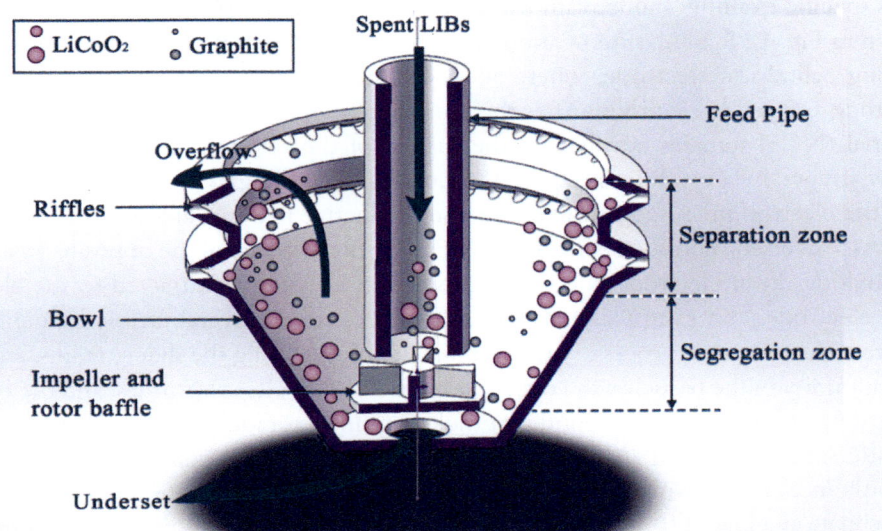

Figure 12.4 Schematic diagram of separation process of Falcon series centrifugal concentrator [23].

into the bottom of a high-speed rotating drum. Under the action of a centrifugal force 50—200 times greater than gravity, the slurry is thrown onto the inner surface of the drum, forming a thin hydrated film. Under the action of the powerful centrifugal force field, the heavy mineral particles can overcome water back pressure and settle in the drum tank. During the separation of LIBs, it is difficult to overcome water back pressure due to the small size of the light graphite particles. The denser lithium cobaltate particles settle in the cell. Moreover, the graphite particles are discharged from the drum as tailings. At the end of the sorting process, the concentrate is flushed out using high-pressure water to collect lithium cobaltate particles.

4.2 Electrostatic separation

The various electrical properties of carbon-based anode and metal oxide cathode materials can be exploited to separate them via electrostatic separation. The electrostatic separation method shows promise because the size of each particle charged or polarized under the action of an external electric field is different, resulting in different trajectories of their movement. The electric field around the particles can be generated in various ways, such as electrical induction, contact galvanization, and ion bombardment. In the ion bombardment method, the electrical charge is generated by the ionization of air caused by the discharge from a corona electrode connected to a high-voltage DC power supply.

A specific example can explain this aspect. In a roller electrostatic separator plant, as shown in Fig. 12.5, a mixture of anode and cathode materials is fed using a feeder into a rotating cylindrical electrode, where an electric field is generated between the rolling electrode and active electrode. After the ions have been bombarded, the nonconductive material (NCF) remains attached to the surface of the rotating drum electrode and is finally scraped by a brush into the NCF collector. Since the gravitational effect is greater than the electron microscopic force, the nonconductive material falls into the NCF collector. Conversely, conductive materials (CFs) acquire a charge of the opposite polarity to the rotating drum electrode due to electrostatic induction, are attracted to the electrostatic electrode, and eventually fall into the CF collector. Several factors influence the electrostatic separation process, with the most significant being the shape, radius, and surface humidity of the particles; the radius and speed of the roller electrodes; and the type of corona electrode. This study found that the optimum separation conditions for this electrostatic separator, when used with battery materials, were a roller rotation speed of 20 rpm, an electrode voltage of 25 kV, an electrostatic electrode distance of 6 cm, and an inclination angle of 0° for the deflector plate. These conditions produced conductive and nonconductive fractions containing 98.98 wt% metallic material and 99.6 wt% polymer, respectively [24].

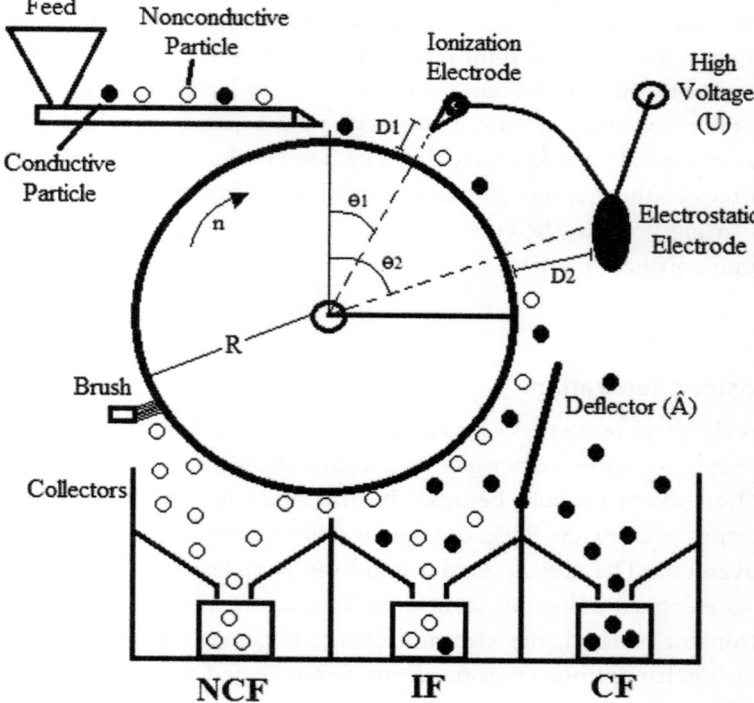

Figure 12.5 Schematic illustration of operation of roll-type electrostatic separator [24].

References

[1] A. Chagnes, B. Pospiech, A brief review on hydrometallurgical technologies for recycling spent lithium-ion batteries, J. Chem. Technol. Biotechnol. 88 (2013) 1191–1199.
[2] S. Kim, J. Kim, S. Kim, et al., Electrochemical lithium recovery and organic pollutant removal from industrial wastewater of a battery recycling plant, Environ. Sci. Water Res. Technol. 4 (2018) 175–182.
[3] A. Yamada, M. Hosoya, S.C. Chung, et al., Olivine-type cathodes achievement and problems, J. Power Sources 119 (2003) 232–238.
[4] C. Liu, F. Li, L.P. Ma, et al., Advanced materials for energy storage, Adv. Mater. 22 (2010) E28–E62.
[5] W. Lv, Z. Wang, H. Cao, et al., A critical review and analysis on the recycling of spent lithium-ion batteries, ACS Sustain. Chem. Eng. 6 (2) (2018) 1504–1521.
[6] E. Fan, L. Li, et al., Sustainable recycling technology for Li-ion batteries and beyond: challenges and future prospects, Chem. Rev. 120 (2020) 7020–7063.
[7] B. Scrosati, J. Garche, Lithium batteries: status, prospects and future, J. Power Sources 195 (2010) 2419–2430.
[8] S. Ojanen, M. Lundstrom, A. Santasalo-Aarnio, R. SernaGuerrero, Challenging the concept of electrochemical discharge using salt solutions for lithium-ion batteries recycling, Waste Manag. 76 (2018) 242–249.
[9] J. Yu, Y. He, Z. Ge, H. Li, W. Xie, S. Wang, A promising physical method for recovery of $LiCoO_2$ and graphite from spent lithium-ion batteries: grinding flotation, Separ. Purif. Technol. 190 (2018) 45–52.
[10] F. Wang, T. Zhang, Y. He, Y. Zhao, S. Wang, G. Zhang, Y. Zhang, Y. Feng, Recovery of valuable materials from spent lithium ion batteries by mechanical separation and thermal treatment, J. Clean. Prod. 185 (2018) 646–652.
[11] J. Li, G. Wang, Z. Xu, Generation and detection of metal ions and volatile organic compounds (VOCs) emissions from the pretreatment processes for recycling spent lithium-ion batteries, Waste Manag. 52 (2016) 221–227.
[12] X. Wang, G. Gaustad, C.W. Babbitt, Targeting high value metals in lithium-ion battery recycling via shredding and size-based separation, Waste Manag. 51 (2016) 204–213.
[13] L. Li, J. Ge, F. Wu, et al., Recovery of cobalt and lithium from spent lithium ion batteries using organic citric acid as leachant, J. Hazard Mater. 176 (2010) 288–293.
[14] Y. Chen, N. Liu, F. Hu, et al., Thermal treatment and ammoniacal leaching for the recovery of valuable metals from spent lithium-ion batteries, Waste Manag. (2018) 469.
[15] D.D.S. Leite, P.L.G. Carvalho, L.R.D. Lemos, et al., Hydrometallurgical separation of copper and cobalt from lithium-ion batteries using aqueous two-phase systems, Hydrometallurgy 169 (2017) 245–252.
[16] L. Sun, K. Qiu, Vacuum pyrolysis and hydrometallurgical process for the recovery of valuable metals from spent lithium-ion batteries, J. Hazard Mater. 194 (2011) 378–384.
[17] Y. Yang, G. Huang, S. Xu, Y. He, X. Liu, Thermal treatment process for the recovery of valuable metals from spent lithium-ion batteries, Hydrometallurgy 165 (2016) 390–396.
[18] Q.H. Tian, A.L. Zou, et al., Research progress on recycling technology of cathode materials for spent ternary lithium-ion batteries, Mater. Rep. 35 (2021) 1.
[19] L. Li, Y. Bian, X. Zhang, Y. Guan, E. Fan, F. Wu, R. Chen, Process for recycling mixed-cathode materials from spent lithium-ion batteries and kinetics of leaching, Waste Manag. 71 (2018) 362–371.
[20] L. Li, E. Fan, Y. Guan, X. Zhang, Q. Xue, L. Wei, F. Wu, R. Chen, Sustainable recovery of cathode materials from spent lithium-ion batteries using lactic acid leaching system, ACS Sustain. Chem. Eng. 5 (2017) 5224–5233.
[21] L.-P. He, S.-Y. Sun, et al., Recovery of cathode materials and Al from spent lithium-ion batteries by ultrasonic cleaning, Waste Manag. 46 (2015) 523–528.

[22] L. Li, R. Chen, F. Sun, et al., Preparation of LiCoO$_2$ films from spent lithium-ion batteries by a combined recycling process, Hydrometallurgy 108 (2011) 220–225.
[23] Y. Zhang, Y. He, T. Zhang, et al., Application of falcon centrifuge in the recycling of electrode materials from spent lithium ion batteries, J. Clean. Prod. 202 (2018) 736–747.
[24] A.V.M. Silveira, M.P. Santana, E.H. Tanabe, et al., Recovery of valuable materials from spent lithium ion batteries using electrostatic separation, Int. J. Miner. Process. 169 (2017) 91–98.

CHAPTER 13

Separating battery nano/microelectrode active materials with the physical method

Hammad Al-Shammari[1] and Siamak Farhad[2]

[1]Department of Mechanical Engineering, Jouf University, Sakaka, Saudi Arabia; [2]Advanced Energy & Manufacturing Laboratory, Mechanical Engineering Department, University of Akron, Akron, OH, United States

1. Introduction

In 1991, SONY Company launched lithium-ion batteries (LIBs) because of their advantages of a low self-discharge rate, a high capacity to net weight, and excellent cycle life. Since then, LIB technology has advanced significantly, and a range of chemistries have been introduced to the market. LIBs are used in many broad applications, such as portable electronic devices, electric vehicles [1,2], electric aircraft ([3–6] and [7,8] two articles), and renewable storage [9,10]. The rapid growth in LIB production will lead to a rapid increase in the inventory of retired LIBs [11]. The spent materials will effectuate an influx of hazardous waste in landfills if no appropriate action is taken [12,13]. The increase in the disposal rate of nonrecycled LIBs will harm the environment because of their toxic and harmful ingredients [14]. Additionally, the valuable materials waste in spent LIBs necessitates the exploration and extraction of new materials, causing further pollution. A study shows Li, Ni, Co, and Mn in the cathode of a typical electric vehicle has cathode chemistry of $LiNi_{1/3}Co_{1/3}Mn_{1/3}O_2$ of 3.5, 10.9, 10.9, and 9.8 kg, respectively [15]. These high quantities of potential environmental impacts emphasize the necessity to recycle retired LIBs. If spent materials collected from LIBs are separated and appropriately regenerated, they can be reused to build new LIBs. Recycling and keeping LIB materials within the closed-loop of build–recycle–rebuild will improve the safety of nations that depend on importing these materials to make LIBs.

The three current methods used to recycle spent LIBs are pyrometallurgy, hydrometallurgy, and direct physical [16]. Despite the tremendous benefits of these recycling methods, each has some limitations and disadvantages [17]. The pyrometallurgy method is most adopted by recycling companies because pretreatment is unnecessary. The pyrometallurgy method is typically performed by heating the materials of retired LIBs above their melting points. The pyrometallurgy target is to recover valuable metals such as cobalt and nickel. Regardless of the noticeable economic benefit, the increasing cost of required energy and the environmental risk of residual gases that may exhaust into the atmosphere have limited its use in the wide range of recycling industries [18,19].

Hydrometallurgy has a higher recovery efficiency than pyrometallurgy because it generates a larger number of recycled elements. Yet not all components are recovered [20]. In this method, the first step is a complete discharge of spent LIBs to avoid explosion during the process. Second, all spent LIBs are crushed in a hydraulic chamber until converted to a powder called filter cake. Then, the filter cake materials are dissolved in an aqueous acid solution containing reagents. After that, the resulting metal-rich solution is further treated by solvent extraction to recover the valuable metals. Although this method has higher efficiency than pyrometallurgy, it may generate toxic, environmentally harmful substances. This procedure may not be economically efficient because of its massive consumption of chemical reagents [21]. Both methods mentioned above focus only on recovering valuable metals from spent materials, such as Mn, Ni, and Co. In some cases, the remaining elements are disposed of as waste.

The direct physical recycling method for spent LIB materials has attracted more attention recently because it does not change the morphology of the electrode active materials and provides the opportunity for their direct regeneration and use in making new LIBs. Thus, this is a more economical LIB recycling process. In an experiment [22], researchers regenerated $LiNi_{1/3}Co_{1/3}Mn_{1/3}O_2$ (NMC-111) from spent cathode scrap. The regeneration process to fix the material defects was conducted by hydrothermal treatment followed by short thermal annealing to restore the missing lithium into the spent materials. The results indicated that the discharge capacity of the regenerated NMC is 158.4 mAh/g, which is very close to the capacity of fresh NMC. Another experiment was conducted to regenerate $LiNi_{0.8}Co_{0.15}Al_{0.05}O_2$ (NCA) cathode material from retired lithium-ion batteries. First, the cathode materials were separated from the current collector using the acid-leaching technique. Then, the cathode material was synthesized by adding lithium carbonate in a ratio of 1:1.1 at a calcination temperature of 800°C for 15 h. The results showed that the specific capacity was 248.7 mAh/g for the charge phase, and the discharge-specific capacity was 162 mAh/g, which is very close to the performance of new NCA [19,23].

Several studies have been conducted by researchers based on the advantages of direct physical recycling, such as its low emissions, low energy consumption, and simplicity, compared with pyrometallurgical recycling to regenerate and make new LIBs from spent electrode materials [24—26]. However, most of these studies have focused on one type of LIB chemistry with known active material. In practice, all LIB chemistries are recycled together in recycling facilities, and the output of these facilities is a filter cake that contains a mixture of different anode and cathode active materials. It is not practical and economical for recycling facilities to sort spent LIBs based on their chemistries and manufacturing to achieve a filter cake that contains only one anode and one cathode active material. We can count at least two reasons for this practical limitation. One reason is that battery manufacturers do not share the details of battery materials data with recycling facilities to enable effective LIB sorting. The other reason is that the recycling facilities cannot store a LIB with specific chemistry until enough similar LIBs are collected and then recycled together. For these reasons, all spent LIBs are recycled together once they arrive at the recycling facility without considering LIB chemistry and manufacturer. To address this practical limitation, a process to separate different filter cake materials should be added to the direct physical method. Therefore, this chapter focuses on introducing a separation process for the filter cake obtained from spent LIBs with varying chemistries. For more information, readers can refer to other publications of the authors ([27] and [28]). In this chapter shows that the separation of electrode active materials from spent LIBs can be accomplished rapidly and effectively (high purity) by a physical method with minimum damage to materials morphology and composition. The separation process targets the recovery of each type of spent LIB material (e.g., LFP, LMO, NCA, NMC-111, LCO, and graphite) and prepares it for the regeneration process. Regardless of the enormous differences in materials' size distributions, as each density value is fixed, it is possible to separate materials by adopting Stokes' law. The proposed process adopts only a physical separation model based on density differences with no need for thermal or chemical treatments, which avoids the emission of byproduct gases that pose significant risks to the environment. This proposed process separates the filter cake materials from spent LIBs, focusing on cathode active materials that contain the most valuable compounds and elements.

The separation process is validated by running an experiment in which the mixture types of cathode materials (LMO, NMC, NCA, LCO, and LFP) and the anode materials (Graphite) were mixed homogeneously before separation. Finally, scanning electron microscope (SEM)/EDX and X-ray diffraction (XRD) analyses were conducted to validate the success of the separation procedure. The results prove the complete separation of materials in a short time.

2. Materials

Fresh LIB cathode and anode active materials were purchased from MTI Corporation with the specifications shown in Table 13.1. Sodium polytungstate (SPT), used as a heavy liquid in the first part of the experiment, was purchased from GEOLIQUIDS, INC. for $0.617/g (the price is for a small amount for the lab) with the specifications shown in Table S.1 in the Supplementary Materials. The Clerici solution used as a heavy liquid in the second part of the experiment was purchased from AMERICAN ELEMENTS for $14/g (the price is for a small amount for the lab) with the specification shown in Table S.1 in the Supplementary Materials. Both heavy liquids were diluted by distilled water to adjust their densities. A mini centrifuge device was used for the separation process. A set of centrifugal tubes with a 1.5 mL capacity and an approximate height of 2 cm were used in the experiments. A viscometer device was used to measure the viscosity of heavy liquids. A Hitachi TM3030 SEM and XRD instruments recorded the materials' images and identified their components before and after the separation process.

Table 13.1 Density and typical particle size of lithium-ion battery active materials (these values are the input data for the modeling).

Material	Theoretical density (kg/m^3)	Particle size		
		D_{10} (μm)	D_{50} (μm)	D_{90} (μm)
LiFePO$_4$ (LFP)	3600	1.0	3.5	15.0
LiMn$_2$O$_4$ (LMO)	4280	5.0 (D_3)	24.5	27.0 (D_{97})
LiNi$_{0.8}$Co$_{0.15}$Al$_{0.05}$O$_2$ (NCA)	4450	7.3	13.6	26.8
LiNi$_{0.33}$Co$_{0.33}$Mn$_{0.33}$O$_2$ (NMC-111)	4770	5.0	10.5	20
LiCoO$_2$ (LCO)	5050	6.8	12.0	20.8
Graphite/mesophase carbon microbeads	2200	8.1	17.6	33.1

3. Modeling

3.1 Separation process

A typical practical treatment in the recycling industry is to crush all types of spent LIBs together without sorting the spent materials based on their chemistry. Before crushing starts, it is essential to discharge the spent batteries of the remaining energy that may be existed in the spent materials to avoid any safety issues during the crushing process [29]. However, the discharging step is usually not performed in LIB recycling facilities because of the complexity of the process. The materials of spent LIBs in recycling facilities typically contain several components [30]. For example, the casing (e.g., steel); the separator (e.g., plastic); cathode active materials (e.g., LCO, NMC-111, LMO, LFP, and NCA); anode active materials (e.g., graphite); current collectors (e.g., Al and Cu); electrolyte (e.g., LiPF$_6$) [31]; and impurities (e.g., dirt) [32,33].

According to the flowchart in Fig. 13.1, to regenerate a new LIB from spent materials, the spent components must be separated, recovered, and then regenerated as a new LIB. Therefore, the second step after crushing is to dissolve the spent materials in an aqueous solution such as N-methyl pyrrolidone (NMP) to break down the joint between the active materials and the current collectors. The anode and cathode active materials are attached to the current collectors by a joint with binder materials (PVDF) [34]. Therefore, dissolving is vital to breaking down the joint of the electrode materials. After the batteries are crushed, ground, and sieved, the plastic, steel, aluminum, and copper are preliminarily separated. Separation of the casing and separator is not a concern because the casing is usually made of metals, and the separator is made of plastic. In this case, a magnetic attraction method can be applied to recover the casing material (e.g., steel). For the plastic separator, separation by density is recommended since plastic density (920−1200 kg/m^3) is much lower than other materials. Therefore, by making a heavy liquid with a density of 1300 kg/m^3, the plastic can be easily separated from the filter

Separating battery nano/microelectrode active materials with the physical method 267

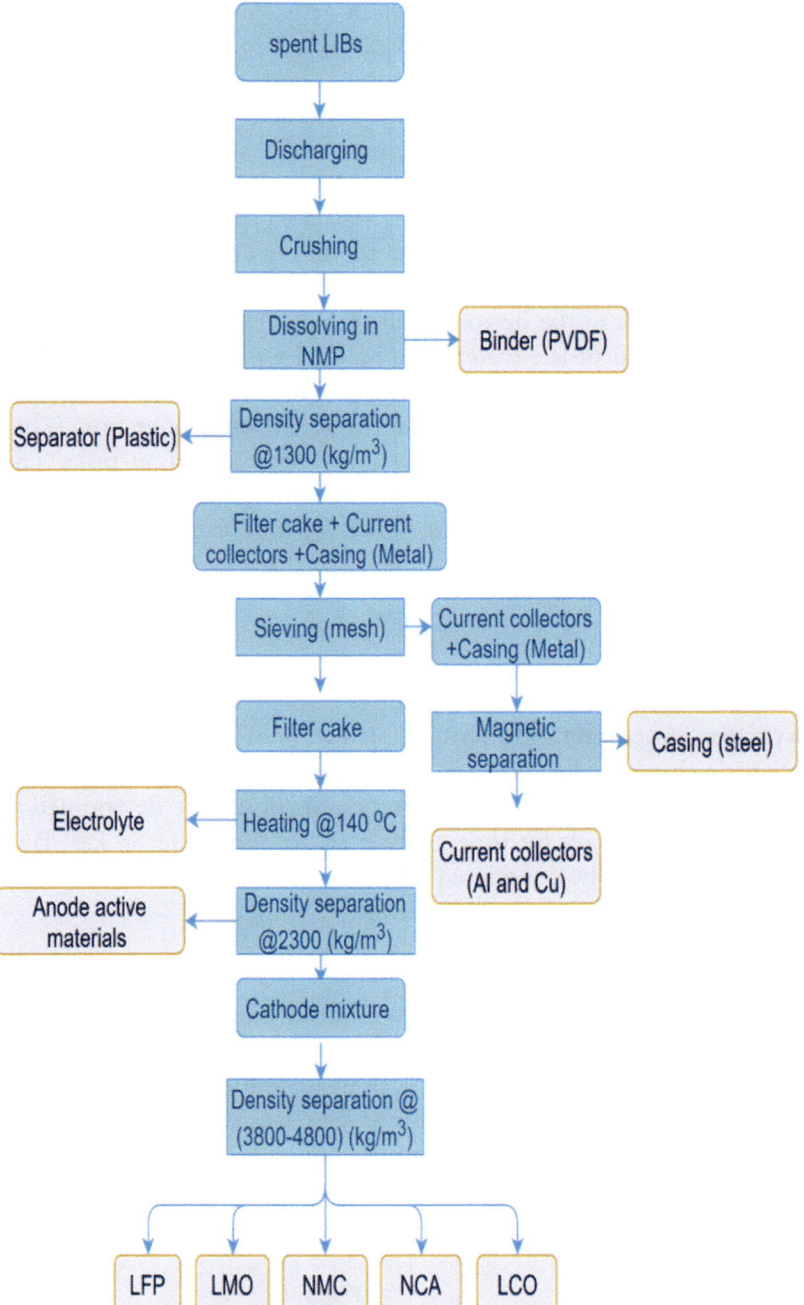

Figure 13.1 Flowchart of the separation process for spent materials obtained from retired lithium-ion batteries.

cake. In this step, the recovered current collectors (e.g., Al and Cu), steel, and plastic have the advantage that they can be used in LIBs a second time.

The most expensive materials in LIBs are the cathode active materials, as they contain valuable elements such as Co, Ni, Li, and Mn. These active materials come with different densities and size distributions, making separation a challenging process. This chapter applies Stokes' law for mineral separation to address this problem. Through this model, the spent LIB components can be effectively separated based on density differences. To illustrate, particles with a density lower than the heavy liquid's density will float on the heavy liquid's surface. On the other hand, particles with a density higher than the heavy liquid's density will sink to the bottom of the sample.

The size distribution of the materials is only an essential factor when gravitational force is applied. In a separation with only gravitational force, the separation process may take a long time because of gravitational acceleration limitation. This particular work is interested in completing the separation process in a short time. Therefore, an external centrifugal force is applied. By doing so, the required time for the separation process is reduced significantly compared with gravitational force alone. In addition, because the centrifugal force is much greater than the gravitational force, the effect of gravitational force can be neglected.

3.2 Stokes' law

Stokes' law describes the movement of particles throughout a liquid medium. Applying this theory in the separation model enables calculation of the terminal velocity of a small spherical particle that has fallen in a dense medium, as illustrated in Fig. 13.2. By calculating the speed, each particle's direction and destination were determined after multiplying the terminal velocity by time. However, to adopt the Stokes' law equation, some assumptions should be considered [35]:
- The particles should have a small diameter (<1 mm).
- The flow of the viscous layers should be laminar.
- The particles should be spherical.

This study validated all assumptions by measuring particle sizes and shapes via SEM images, as shown in Fig S.1 in the Supplementary Materials.

According to the schematic in Fig. 13.2, three main forces, gravitational (F_G), buoyancy (F_B), and drag (F_D) apply to a falling or floating particle. If the particle is assumed to be spherical, these forces are obtained from the following equations:

$$F_G = \frac{1}{6}\pi d^3 \rho_s g \tag{13.1}$$

$$F_B = \frac{1}{6}\pi d^3 \rho_f g \tag{13.2}$$

$$F_D = 3\pi\mu d v_t \tag{13.3}$$

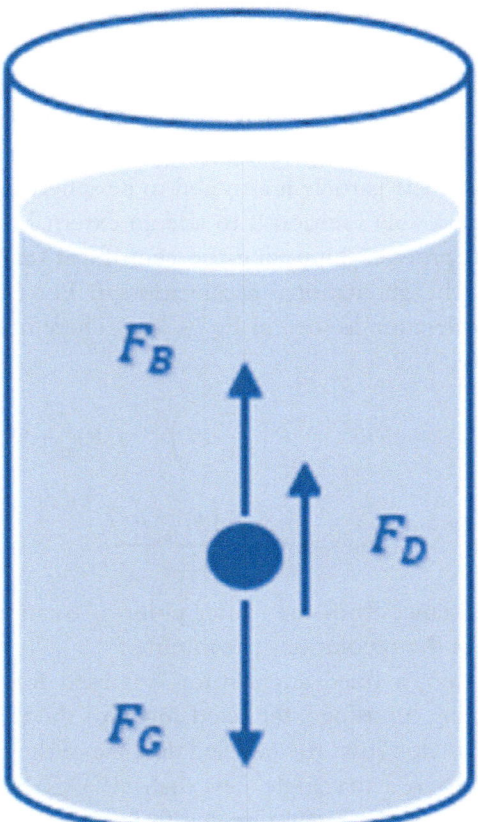

Figure 13.2 The schematic of a fallen particle in a liquid.

where g is the gravitational acceleration (9.8 m/s^2), d is the particle diameter (m), μ is the viscosity of the liquid (kg/m.s), ρ_s is the density of the particle (kg/m^3), ρ_f is the density of the liquid (kg/m^3), and v_t is the terminal velocity of the particle.

Eq. (13.3) demonstrates the drag force applied to small spherical particles (low Reynolds number) in the liquid; if the particles are assumed to move with constant velocity, $F_G - F_B - F_D = 0$. Solving this equation yields the terminal velocity of each particle, as in Eq. (13.4):

$$v_t = \frac{gd^2\left(\rho_s - \rho_f\right)}{18\mu} \tag{13.4}$$

In Eq. (13.4), the velocity goes downwards if $\rho_s > \rho_f$ and upwards if $\rho_s < \rho_f$. As seen in this equation, the terminal velocity of solid particles is a function of the square of the particle diameter. Thus, the terminal velocity of particles with sizes in micrometer is very small. To increase the terminal speed of the microparticles, the difference between the density of particles and liquid should increase or the viscosity of fluid should decrease. However, the range of variation of these two parameters may not be enough to meaningfully increase the terminal velocity. The shape factor is also essential to the separation process because the particle is assumed to be spherical. However, in practice, it is not spherical. One possible solution is to add an external force, such as centrifugal force, to the separation process. Supposing the centrifugal force is much greater than the gravitational force, the gravitational acceleration in Eq. (13.4) can be substituted with the centrifugal acceleration as seen in Eq. (13.5). Therefore, Eq. (13.6) is obtained for the terminal velocity:

$$a_C = R\omega^2 = R\left(\frac{2\pi}{60}N\right)^2 = R\frac{\pi^2}{900}N^2 \tag{13.5}$$

$$v_t = \frac{RN^2 d^2 \left(\rho_s - \rho_f\right)}{1641\mu} \tag{13.6}$$

where R is the radial distance from the center point of rotation (m), ω is the angular velocity (rad/s), and N is the revolutions per minute.

As shown in Fig. 13.3, a fixed angle rotor was used for the separation process. Compared with a swinging centrifuge, the fixed angle has the advantage of reduced separation time because its angle allows the traveled distance of the particles to be shortened [36]. Since the rotor has an acute angle (less than 90°), R had to be replaced by $\ln(R_{max}/R_{min})$, where R_{min} is the radial distance from the center point of the rotation axis to the center point of the top of the sample, and R_{max} is the radial distance from the center point of the rotation axis to the center point of the bottom of the sample. Therefore, the final form of the separation equation will be as in Eq. (13.7):

$$v_t = \frac{\ln\left(\frac{R_{max}}{R_{min}}\right) N^2 d^2 \left(\rho_s - \rho_f\right)}{1641\mu} \tag{13.7}$$

3.3 Heavy liquid selection

Stokes' law can interpret the mechanical separation of minerals. This theory explains the phenomenon of the movement of a small spherical particle through a dense liquid. The particle's motion is caused by the density change between the particle and the liquid medium, as explained in the previous section. By adopting the Stokes' law equation, the calculation of the terminal velocity of each particle was executed. Regarding the separation medium (heavy liquid), the experiment's first challenge was to find a medium with

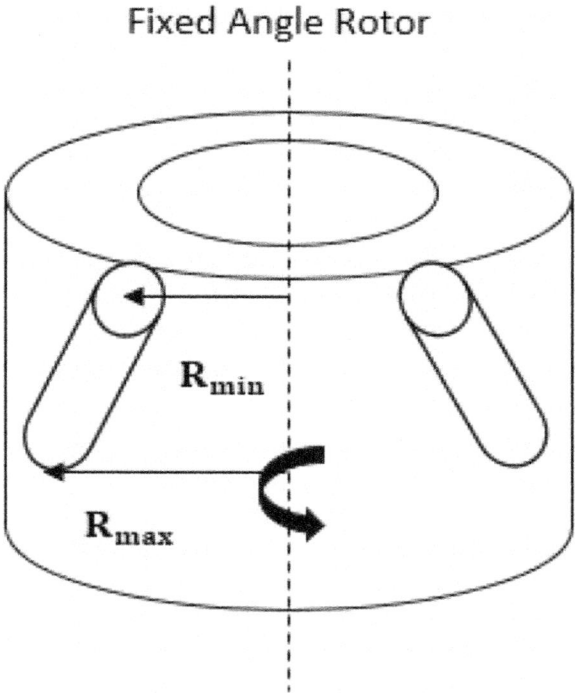

Figure 13.3 The schematic for a centrifugal device with a fixed angle rotor.

specifications such as low toxicity, controllable density-viscosity ratio, and reusability. These factors should be considered to avoid any safety issues and maintain continuity in the separation process. For this purpose, an investigation was performed of the commercially available heavy liquids in the minerals separation industry. The investigation results are listed in Table S.1 in the Supplementary Materials.

The properties of the heavy liquid should be considered when controlling the separation process. The viscosity of the heavy liquid is an essential property second only to density. When only gravitational force is applied, the viscosity should be as low as possible to shorten the procedure's time. However, viscosity is less critical when using centrifugal force as the external force.

The heavy liquids used in this experiment were SPT and Clerici solution. SPT is a heavy liquid that is widely implemented in mineral industries because of its thermal and chemical stability and low toxicity. The density of this liquid can be adjusted by dilution with distilled water or by evaporation. The maximum density for SPT at room temperature is 2890 kg/m^3. However, by increasing the temperature to 90°C, the saturated solution can reach a maximum density of 3000 kg/m^3. The properties of the heavy liquid, such as viscosity and density, are the major contributing factors in the separation process. Regarding this matter, a lab study was conducted to measure the viscosity change from increasing the density of the SPT. The result of the experiment is shown in Fig S.1 in the Supplementary Materials. The results indicate that when the density increases, the viscosity increases as well. This increase in the viscosity slowed the movements of the particles, which caused an increase in the processing time. Therefore, an optimized procedure that considers the density and viscosity is recommended to obtain a smooth separation over a short period.

Clerici solution heavy liquid was also used in the experiment. This solution is considered the heaviest water-based liquid with a density of 4360 kg/m^3 at room temperature. The density of the solution can be controlled by dilution with water (1000–4360 kg/m^3). However, by increasing the temperature to 90°C, the solution can reach a density of 5000 kg/m^3. Studying the relation between density and viscosity indicates that the viscosity of this heavy liquid decreases slightly with increasing temperature. However, the viscosity was set at room temperature for the maximum possible value of 0.005 kg/ms. Despite the advantages of this solution, it is considered toxic. Therefore, in this experiment, safety tools such as masks, gloves, and facial protectors were used and are strongly recommended for lab-scale research. In addition, heavy liquids are costly, and the ability to recover them from the water is of interest in this research to increase the practicality of the separation process. Therefore, we have proposed a simple model for the continuous separation and recovery of heavy liquids in the following section.

3.4 Industry-scale separation process

In this study, the separation was performed in a lab using a mini-centrifugal device with a maximum speed of 10,000 rpm. The separated materials were removed manually from the bottom and the top of the falcon tubes. For industrial applications, this lab-scale process should be scaled up. Therefore, in this section, a simple process for large-scale separation is proposed for potential use in LIB recycling facilities. This process design should be considered a preliminary design for the large-scale process, and more studies are needed to achieve the final design. The objective is to design a process to minimize the consumption of heavy liquids. This is because of the high cost of heavy liquids and to increase the economic viability of the process.

The schematic of the suggested process design, which is a batch process, is shown in Fig. 13.4. The process steps involved are the following: (1) the mixed filter cake and heavy liquid are transferred by gravity force to stainless steel centrifugal tubes, (2) the centrifugal tubes with a butterfly valve at the middle of each tube are rotated until solid particles that are lighter and heavier than the heavy liquid are moved to the top and bottom of the tubes, (3) the tube butterfly valves are closed, and the tube contents at the top

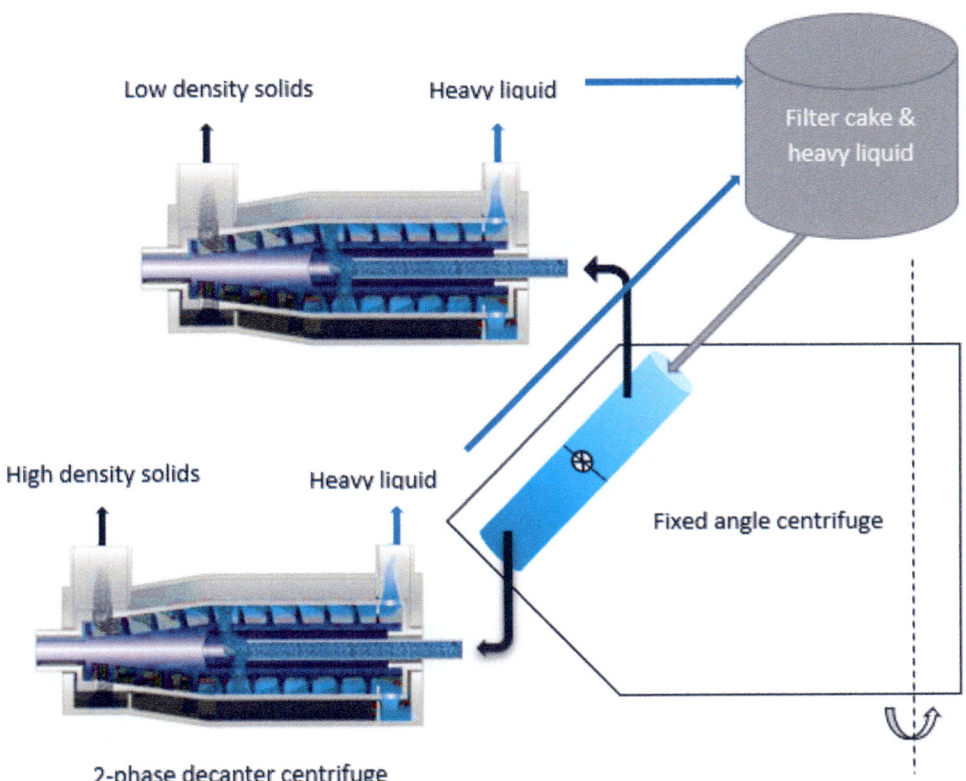

Figure 13.4 Schematic of proposed industry-scale separation process.

and bottom sections are transferred to two-phase decanter centrifuges, (4) the solid particles are separated from the heavy liquid in the decanter centrifuges, (5) solid particles are moved to a tank for washing with water (not shown in Fig. 13.4) to ready them for the regeneration process, and (6) the recovered heavy liquid is returned to the filter cake tank and mixed with new filter cake, and the process continues from step (1). A two-phase decanter centrifuge was used in this process because it has shown effective results for separating solid particles from liquid [37–39].

The heavy liquids we adopted in this study are soluble in water, as mentioned in Table S.1 in the Supplementary Materials. Therefore, any small amount of heavy liquid in the separated solid particles can be collected after washing the solid particles in water. The obtained heavy liquid can be recovered and transferred back to the filter cake tank by water evaporation. Thus, it is expected that the heavy liquid will remain in a closed loop in the proposed separation process.

4. Results and discussion

In this study, separation of the filter cake based on density differences was investigated theoretically and experimentally. The separation model based on Stokes' law was generated to separate the different cathode materials from each other and the anode material with high purity and minimal time. The separation by density differences is complex because of each material's wide variety of particle sizes. For example, the SEM test has shown that the fine powder of LCO cathode materials is distributed in a range of particle sizes of 6.8–20.8 μm with an average of 12 μm, as seen in Fig S.2 in the Supplementary Materials. When two materials have different size distributions, the particles of lower density and larger size are accompanied by the particles of higher density and smaller size after separation. Therefore, centrifugal force was applied to avoid the effect of size distribution and the geometry of the particles, focusing only on the separation based on density differences. By doing so, a tremendous amount of force is applied to the particle in two directions. One is parallel to the direction of the gravitational force, and the second is the opposite. Thus, due to the vast differences between the centrifugal and gravitational forces (e.g., centrifugal is 200 times larger), the gravitational force is neglected. Therefore, by centrifugation using a heavy liquid as a medium between the two materials, spent materials with different densities were successfully separated from each other regardless of their variable size distributions.

4.1 Separation model

As modeled by Stokes' law, cathode and anode materials with different types and size distributions can be separated by applying a centrifugation force and adding a heavy liquid as a separation medium. The model was used to calculate the optimum properties of heavy liquid and the time required for the operating process. In this research, the focus is to choose the lowest possible density for the heavy liquid to increase the separation efficiency and reduce the cost of the heavy liquid. Therefore, the most common anode (graphite) and cathode mixture (LFP, LMO, NCA, NMC-111, and LCO) can be separated in five steps, as shown in Fig. 13.5. It is noted that nanostructured $Li_4Ti_5O_{12}$ (LTO) ([40] and [41]) as the anode active material may be found in some retired LIBs. However, the number of these LIBs is presently not significant compared with retired LIBs that use graphite anode.

4.1.1 Step 1: separation of anode and cathode materials

The separation between the anode and a mixture of five cathode materials can be obtained by adding a heavy liquid with a density of 2300 kg/m^3; a centrifugation speed of 10,000 rpm resulted in the anode material floating to the top of the tube (graphite) and the five cathode materials (LFP, LMO, NCA, NMC-111, and LCO) sinking to the bottom. In this step, the graphite anode material was separated from the five cathodes

Separating battery nano/microelectrode active materials with the physical method 275

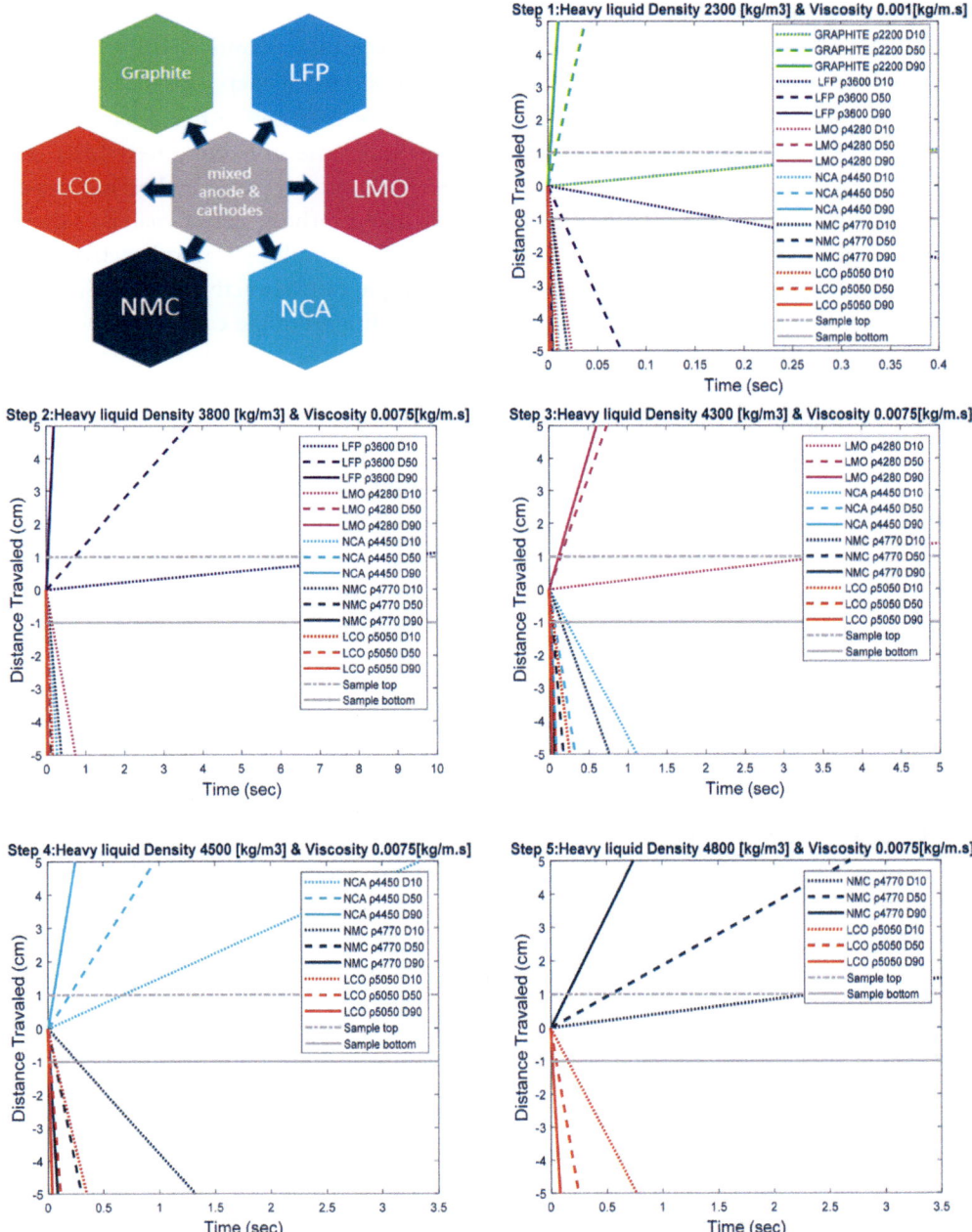

Figure 13.5 Separation process for the mixture of anode and cathode active materials.

in a short time as seen in Fig. 13.5 (step 1). SPT is used as the heavy liquid for this separation due to its density, which can be adjusted between the densities of the anode and cathode materials ($\rho_{SPT} = 1000-3100$ kg/m^3). Therefore, this step allows complete recovery of the anode active material from the cathode materials. After separation, the anode active material can be sent for regeneration and reuse in making new LIBs. The regeneration of the anode active material can potentially be economical when more spent LIBs are available for mass recycling [42]. The separated cathode active materials can be either sent for regeneration and reuse for LIBs made from multiple cathode active materials or sent for separation. For LIBs made from multiple cathode active materials, readers are referred to the study done by Ref. [43]. Regenerating multiple cathode active materials together without separation can reduce the recycling cost; however, controlling the battery performance is challenging. That is why we recommend separating different cathode active materials using steps 2—5.

4.1.2 Step 2: separation of LFP from cathode mixture

In this step, LFP can be recovered from the cathode mixture using a heavy liquid with a density of 3800 kg/m^3 and a centrifugation speed of 10,000 rpm. As a result, the LFP cathode material will rise to the surface, and the other cathode materials (LMO, NCA, NMC-111, and LCO) will sink to the bottom of the tube. In this step, Clerici solution ($\rho_{Clerici} = 1000-4360$ kg/m^3 at room temperature) is used as a medium for separation due to its density, which can be adjusted between the LFP density and those of the remaining cathode mixture materials. Therefore, LFP can be separated in a short period using this step (less than 10 s).

4.1.3 Step 3: separation of LMO from cathode mixture

In this step, LMO can be separated from the cathode mixture using a heavy liquid with a density of 4300 kg/m^3 and a centrifugation speed of 10,000 rpm. As a result, the LMO cathode will rise to the surface, and the other cathode mixture materials (NCA, NMC-111, and LCO) will sink to the bottom of the tube. The heavy liquid suggested in this step is the Clerici solution due to its density, which can be controlled up to 4360 kg/m^3 at room temperature. As shown in Fig. 13.5 (step 3), LMO is separated from the cathode mixture in a short period (less than 5 s).

4.1.4 Step 4: separation of NCA from cathode mixture

In this step, the NCA can be recovered from the cathode mixture using a heavy liquid with a density of 4500 kg/m^3 and a centrifugation speed of 10,000 rpm. The heavy liquid in this step is the Clerici solution. It is noted that the density of this solution increases by temperature and can reach from 4360 kg/m^3 at the room temperature to about 5000 kg/m^3 at elevated temperatures. By adjusting the density of Clerici solution to

4500 kg/m³ (controlled by temperature), NCA will rise to the surface of the tube, and other cathode materials (NMC-111 and LCO) will sink to the bottom. In this step, the time required to separate the NCA is less than 3.5 s.

4.1.5 Step 5: separation of lithium cobalt oxide and lithium nickel cobalt manganese oxide

Finally, in step 5, NMC and LCO cathode materials are separated using a heavy liquid with a density of 4800 kg/m³ and a centrifugation speed of 10,000 rpm. By adjusting the density of the Clerici solution to 4800 kg/m³ at a temperature of about 90°C, the NMC-111 cathode material rises to the surface and the LCO sinks to the bottom of the tube. In this step, the time required for separation is less than 3.5 s.

The separation time is controlled by two variables: the heavy liquid's density and the centrifugal speed. The choice of an appropriate heavy liquid with the desired density to separate different cathode active materials was the major challenge in this work due to the lack of information in the literature. The only available heavy liquid that we could find to separate cathode active materials is the Clerici solution. This solution is the heaviest water-based available heavy liquid on the market. However, it is toxic. Therefore, this heavy liquid should be used carefully. In the lab, we used appropriate safety tools suggested by the supplier, such as gloves, masks, and face protectors. We also did the majority of the work inside the fume hood. For the industry-scale separation proposed in Section 3.4, special attention should be given to the seals of rotary equipment and connectors to prevent any leakage of this toxic heavy liquid.

Centrifuge speed is another important factor in the separation process. By increasing the rotational speed, the materials are separated faster. In the lab, the rotational speed was adjusted to 10,000 rpm, which is the maximum speed of the small centrifugal device used in this study. However, in large-scale industry separation, the centrifugation size and speed should be adjusted to achieve the desired results.

4.2 Validation of separation of mixture of anode and cathode active materials

The separation process using heavy liquids was modeled to (1) separate the graphite anode from a mixture of two cathode active materials of LFP and LCO, and (2) separate LFP and LCO from each other. The theoretical density of the graphite anode active material is 2200 kg/m³, and the theoretical densities of the LCO and LFP cathode materials are 5050 kg/m³ and 3600 kg/m³, respectively. The modeling results are presented in Fig. 13.6. As seen in Fig. 13.6A, if STP solution with a density of 2300 kg/m³ is used as the heavy liquid to separate the graphite anode from the mixture of LFP and LCO cathode materials, the separation of the graphite occurs in less than 0.2 s if the rotational speed of the centrifuge device is adjusted to 10,000 rpm. The separation of LFP and LCO cathode active materials can be also done in about 10 s (see Fig. 13.6B) when Clerici

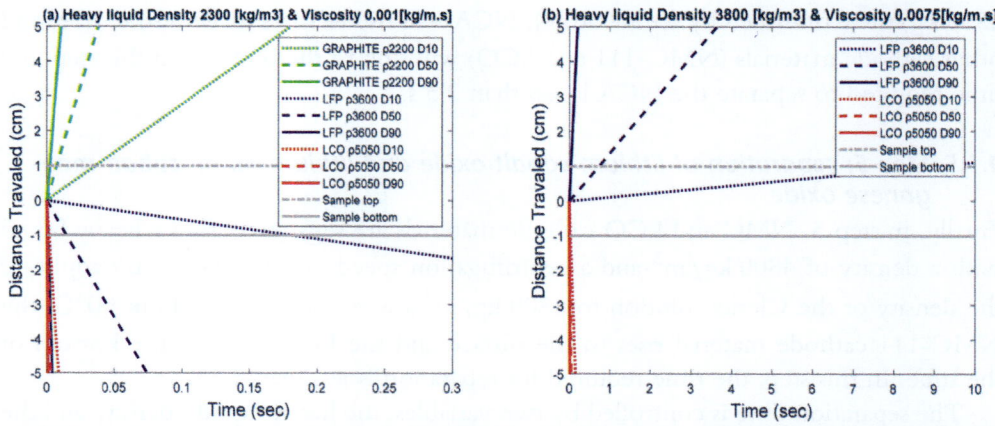

Figure 13.6 The modeling results for (A) separation of the graphite anode and the mixture of LFP and LCO cathode using SPT heavy liquid, and (B) separation of LFP and LCO using Clerici solution heavy liquid. The centrifugation speed is 10,000 rpm.

solution with an adjusted density of 3800 kg/m^3 is used as the heavy liquid in the centrifuge device with a rotational speed of 10,000 rpm.

4.2.1 Experiment for separating anode and cathode mixture

The first part of the experiment separated the graphite anode particles from the mixture of LFP and LCO cathode particles. For this purpose, first, 250 mg mixture of LFP and LCO cathode active materials and 250 mg graphite anode active material were mixed to make a homogeneous mixture in a centrifugal falcon tube with the volume and height of 1.5 mL and 2 cm, respectively. Then, SPT heavy liquid with a density of 2300 kg/m^3 was added to the falcon tube. A centrifugal force with a rotating speed of 10,000 rpm was applied for a time calculated via the mathematical separation model. The model results in Fig. 13.6A showed that about 0.2 s is enough for the separation of the graphite. However, since the particle size of the LFP material is small (see Table 13.1), the simulation shows that it takes about 10 s for these particles to settle down. Since it takes about 3 s for the centrifuge device to reach the steady-state rotational speed of 10,000 rpm, we run the device for about 13 s. As shown in Fig. 13.7A, this time is enough to separate the graphite anode from the mixture of LCO and LFP cathode materials. The results of the SEM/EDX test confirm that the material collected from the top of the tube in Fig. 13.7A is graphite. Thus, there is a good agreement between the model and experiment. We have also examined the flotation time of the graphite anode material in the SPT heavy liquid at the centrifugation speed of 10,000. As shown in Fig. 13.7B, the graphite moved to the top of the falcon tube as soon as the centrifuge device reached the steady-state speed of 10,000, about 3 s. This also indicates a good agreement between the model and experiment.

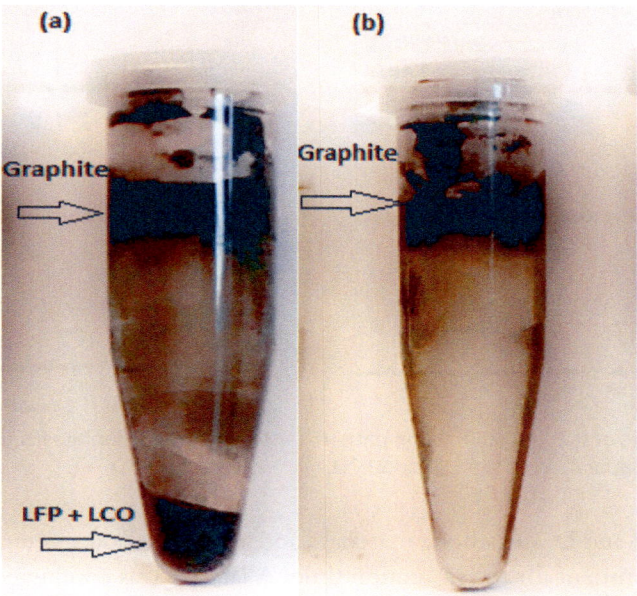

Figure 13.7 (A) Image of the separation between graphite anode and the mixture of LFP and LCO cathode materials, and (B) Image of the flotation of the graphite anode material by centrifugation using the SPT heavy liquid.

It is noted that the tubes used in the experiments were made of plastic. That is why, as seen in Fig. 13.7, a small amount of graphite adheres to the outer surface of the tube and makes it a little dirty. It seems that if stainless steel were used to make the tube, the amount of adhered graphite would be reduced due to the reduction of static electricity on the tube surface.

4.2.2 Experiment for separating of lithium iron phosphate and lithium cobalt oxide

The second part of the experiment was conducted to separate the two cathode active materials of LFP and LCO from each other after the graphite anode was separated. For this purpose, the separated mixture of LFP and LCO was collected from the bottom of the tube in Fig. 13.7A, washed with water, and transferred to a new centrifugal falcon tube. Then, Clerici solution heavy liquid with a density of 3800 kg/m^3 was added to the falcon tube. A centrifugal force with a rotating speed of 10,000 rpm was applied for a time calculated via the mathematical separation model. The model results in Fig. 13.6B showed that about 10 s is enough to bring the LFP particles to the top of the tube. Since it takes about 3 s for the centrifuge device to reach the steady-state rotational speed of 10,000 rpm, we run the device for about 13 s. We found that this is enough time to

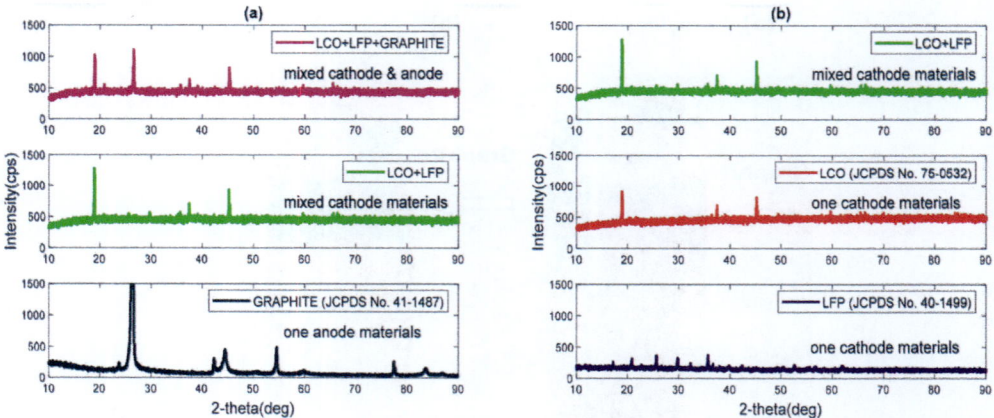

Figure 13.8 Images of the X-ray diffraction pattern. (A) separation of graphite from the mixture of LFP and LCO, and (B) separation of the LFP and LCO.

separate the LFP and LCO cathode materials, and there is a good agreement between the model and experiment for the separation time. A lab test was conducted to measure the Clerici solution's viscosity at room and elevated temperatures. The results showed that the Clerici solution's viscosity at room temperature is 0.0075 kg/ms (water 0.0009 kg/ms), and as the temperature increases, the viscosity decreases slightly. However, in the model, we assumed that viscosity is constant and equals 0.0075 kg/ms. This may cause discrepancies between the model and experiment when the temperature of the Clerici solution increases.

Fig. 13.8 shows the XRD images of the three materials of graphite, LFP, and LCO before and after the separation. By comparing the spikes in the graphs in part (a) of Fig. 13.8, it is evident that the graphite anode material has been separated from the mixture of the LFP and LCO cathode material. Similarly, comparing the spikes in the graphs in part (b) proves that the mixture of LFP and LCO has been separated. Therefore, the separation of the active materials with heavy liquid is effective in terms of the purity of the obtained material.

4.3 Validation of separation of mixture of cathode active materials

In the previous experiment, it was proved that a mixture of anode and cathode materials can be rapidly and effectively separated. It was also proved that cathode active materials with a relatively large difference in their theoretical density, such as LFP ($\rho = 3600$ kg/m^3) and LCO ($\rho = 5050$ kg/m^3), could be separated rapidly and effectively. Here, we focus on cathode active materials with various theoretical densities to prove the validity of the separation for materials having smaller differences in their theoretical densities.

4.3.1 Separation of lithium manganese oxide and lithium nickel cobalt aluminum oxide

In this experiment, the LMO ($\rho = 4280 \text{ kg/m}^3$) and NCA ($\rho = 4450 \text{ kg/m}^3$) cathode active materials were mixed homogeneously in a falcon tube. Then, the Clerici solution heavy liquid was added to the falcon tube and the rotational speed of the centrifuge was adjusted to 10,000 rpm. The density of the Clerici solution was adjusted to 4400 kg/m^3. As shown in Fig. 13.9A, the model predicts the separation is completed in less than 3.5 s. Considering the 3 s time that the centrifuge requires to reach the steady-state speed of 10,000 rpm, as seen in Fig. 13.9B, the separation of LMO and NCA occurred in about 6 s in the experiment.

4.3.2 Separation of lithium nickel cobalt manganese oxide-111 and lithium cobalt oxide

In this experiment, the targeted materials for the separation were the NMC-111 ($\rho = 4770 \text{ kg/m}^3$) and LCO ($\rho = 5050 \text{ kg/m}^3$) cathode active materials. These two materials were mixed in a falcon tube. Then, the Clerici solution heavy liquid was added, and the rotational speed of the centrifuge was adjusted to 10,000 rpm. The density of Clerici solution was adjusted to 4800 kg/m^3. For this purpose, the temperature of the heavy liquid was increased to 90°C. As shown in Fig. 13.10A, the model predicts the separation is completed in less than 2.5 s. Considering the 3 s time that the centrifuge requires to reach the steady-state speed of 10,000 rpm, as seen in Fig. 13.10B, the separation of LMO and NCA occurred in about 5 s in the experiment.

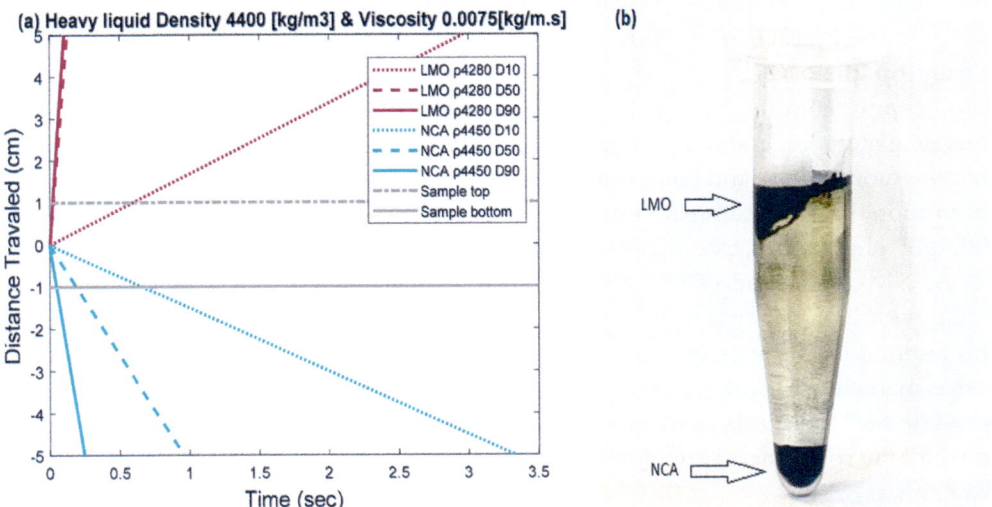

Figure 13.9 (A) The model prediction for separation of LMO and NCA using the Clerici solution and centrifugation speed of 10,000 rpm, and (B) image of the separated LMO and NCA.

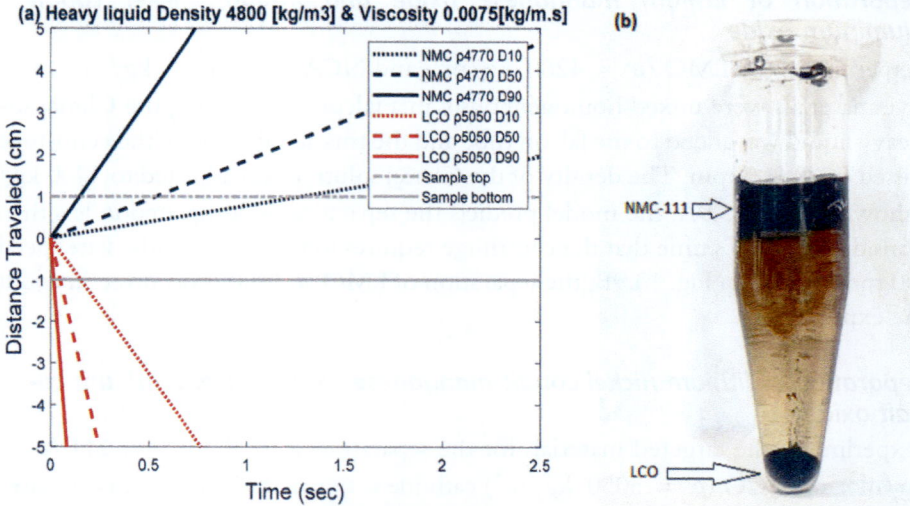

Figure 13.10 (A) The model prediction for separation of NMC-111 and LCO using the Clerici solution and centrifugation speed of 10,000 rpm, and (B) image of the separated NMC-111 and LCO in the centrifuge.

5. Conclusions

Recycling LIBs by the direct physical recycling method is potentially more economical and promising than pyrometallurgical and hydrometallurgical recycling methods. However, since all types of LIBs are usually recycled together in LIB recycling facilities, one of the technical barriers of this method is separating the various electrode active materials from the output filter cake product. Since the main goal of this recycling method is to keep the morphology of the electrode active materials unchanged to enable their direct regeneration and reuse in new LIBs, the separation process should be designed so that the morphology and composition of the active material do not change. This chapter proposed a physical separation method based on using heavy liquids for separating different electrode active materials, including the graphite anode and LFP, LMO, NCA, NMC-111, and LCO cathode active materials. It was shown that a rapid separation with high purity is technically possible for these materials. It was also shown that this method does not change the morphology and composition of the spent electrode active materials. A simple mathematical model was developed for minerals separation based on Stokes' law that can be used to evaluate the effects of the density of heavy liquids and the rotational speed of the centrifuge on the separation time of electrode active materials.

Nomenclature

N Centrifugal rotation (rpm)
PP Polyethylene
R Radius from the center of the rotation (m)
SPT Sodium Polytungstate
F_B Buoyancy force (N)
F_C Rotational force (N)
F_D Drag force (N)
F_G Gravitational force (N)
R_{max} Radius from center to the sample's bottom (m)
R_{min} Radius from center to the sample's top (m)
v_t Terminal velocity of the sphere (m/s)
r Radius of the sphere (m)
ρ_s Density of the sphere (kg/m^3)
ρ_f Density of the medium (kg/m^3)
ω Angular velocity (rad/s)
μ Viscosity of the fluid (kg/m s)

Abbreviations

LCO lithium cobalt oxide
LFP lithium iron phosphate
LIBs lithium ion batteries
LMO lithium manganese oxide
NCA lithium nickel cobalt aluminum oxide
NCM lithium nickel cobalt manganese oxide

References

[1] A.H. Mohammed, M. Alhadri, W. Zakri, H. Aliniagerdroudbari, R. Esmaeeli, S.R. Hashemi, G. Nadkarni, S. Farhad, Design and comparison of cooling plates for a prismatic lithium-ion battery for electrified vehicles, in: SAE Technical Paper, 2018-01-1188, 2018, https://doi.org/10.4271/2018-01-1188.

[2] E. Samadani, S. Farhad, S. Panchal, R. Fraser, M. Fowler, Modeling and evaluation of Li-ion battery performance based on the electric vehicle field tests, in: SAE Technical Paper, 2014, https://doi.org/10.4271/2014-01-1848.

[3] M. Alhadri, R. Esmaeeli, A.H. Mohammed, W. Zakri, S.R. Hashemi, H. Aliniagerdroudbari, H. Barua, S. Farhad, Studying the degradation of lithium-ion batteries using an empirical model for aircraft applications, in: ASME Power & Energy Conference, Lake Buena, FL, United States, June 25–27, 2018, https://doi.org/10.1115/POWER2018-7428.

[4] S.R. Hashemi, R. Esmaeeli, H. Aliniagerdroudbari, M. Alhadri, H. Al-Shammari, A. Mahajan, S. Farhad, New intelligent battery management system for drones, in: Proceeding of the ASME 2019, Internal Mechanical Engineering Congress and Exposition (IMECE), Salt Lake City, UT, USA, November 2019, https://doi.org/10.1115/IMECE2019-10479.

[5] S.R. Hashemi, R. Esmaeeli, A. Nazari, H. Aliniagerdroudbari, M. Alhadri, W. Zakri, A.H. Mohammed, A. Mahajan, S. Farhad, A fast diagnosis methodology for typical faults of a

lithium-ion battery in electric and hybrid electric aircraft, ASME J. Electrochem. Energy Conv. Storage 17 (1) (2020), https://doi.org/10.1115/1.4044956.
[6] S.R. Hashemi, A. Nazari, R. Esmaeeli, H. Aliniagerdroudbari, M. Alhadri, W. Zakri, A.H. Mohammed, A. Mahajan, S. Farhad, Fast fault diagnosis of a lithium-ion battery for hybrid electric aircraft, in: ASME Power & Energy Conference, Lake Buena, FL, United States, June 25–27, 2018, https://doi.org/10.1115/POWER2018-7476.
[7] S.R. Hashemi, A. Bahadoran Baghbadorani, R. Esmaeeli, A. Mahajan, S. Farhad, Machine learning-based model for lithium-ion batteries in BMS of electric/hybrid electric aircraft, Int. J. Energy Res. 45 (2021), https://doi.org/10.1002/er.6197. Accepted Manuscript.
[8] S.R. Hashemi, A.M. Mahajan, S. Farhad, Online estimation of battery model parameters and state of health in electric and hybrid aircraft application, Energy 229 (15) (2021) 120699, https://doi.org/10.1016/j.energy.2021.120699.
[9] M. Alhadri, W. Zakri, R. Esmaeeli, A.H. Mohammed, S.R. Hashemi, H. Barua, H. Aliniagerdroudbari, S. Farhad, "Analysis of second-life of a lithium-ion battery in an energy storage system connected to a wind turbine, in: 2019 IEEE Power and Energy Conference at Illinois (PECI), 1–8, 2019, https://doi.org/10.1109/PECI.2019.8698782.
[10] J. Xie, K. Huang, Z. Nie, W. Yuan, X. Wang, Q. Song, X. Zhang, C. Zhang, J. Wang, J.C. Crittenden, An effective process for the recovery of valuable metals from cathode material of lithium-ion batteries by mechanochemical reduction, Resour. Conserv. Recycl. 168 (2020), https://doi.org/10.1016/j.resconrec.2020.105261.
[11] X. Wang, G. Gaustad, C.W. Babbitt, K. Richa, Economies of scale for future lithium-ion battery recycling infrastructure, Resour. Conserv. Recycl. 83 (2014), https://doi.org/10.1016/j.resconrec.2013.11.009.
[12] I. Kay, R. Esmaeeli, S.R. Hashemi, A. Mahajan, S. Farhad, Recycling Li-ion batteries: robot disassembly of electric vehicle battery systems, in: Proceeding of the ASME 2019, Internal Mechanical Engineering Congress and Exposition (IMECE), Salt Lake City, UT, USA, November 2019, https://doi.org/10.1115/IMECE2019-11949.
[13] S.M. Shin, N.H. Kim, J.S. Sohn, D.H. Yang, Y.H. Kim, Development of a metal recovery process from Li-ion battery wastes, Hydrometallurgy 79 (2005) 172–181, https://doi.org/10.1016/j.hydromet.2005.06.004.
[14] Y. Zhang, Y. Wang, H. Zhang, Y. Li, Z. Zhang, W. Zhang, Recycling spent lithium-ion battery as adsorbents to remove aqueous heavy metals: adsorption kinetics, isotherms, and regeneration assessment, Resour. Conserv. Recycl. 156 (2020), https://doi.org/10.1016/j.resconrec.2020.104688.
[15] J. Diekmann, C. Hanisch, L. Froböse, G. Schälicke, T. Loellhoeffel, A.-S. Fölster, A. Kwade, Ecological recycling of lithium-ion batteries from electric vehicles with focus on mechanical processes, J. Electrochem. Soc. 164 (2017) A6184–A6191, https://doi.org/10.1149/2.0271701jes.
[16] L. Li, Y. Bian, X. Zhang, Y. Guan, E. Fan, F. Wu, R. Chen, Process for recycling mixed-cathode materials from spent lithium-ion batteries and kinetics of leaching, Waste Manag. 71 (2018) 362–371, https://doi.org/10.1016/j.wasman.2017.10.028.
[17] X. Zhang, Y. Xie, X. Lin, H. Li, H. Cao, An overview on the processes and technologies for recycling cathodic active materials from spent lithium-ion batteries, J. Mater. Cycles Waste Manag. 15 (2013), https://doi.org/10.1007/s10163-013-0140-y.
[18] X. Chen, Y. Chen, T. Zhou, D. Liu, H. Hu, S. Fan, Hydrometallurgical recovery of metal values from sulfuric acid leaching liquor of spent lithium-ion batteries, Waste Manag. 38 (2015) 349–356, https://doi.org/10.1016/j.wasman.2014.12.023.
[19] B. Wang, X.Y. Lin, Y. Tang, Q. Wang, M.K.H. Leung, X.Y. Lu, Recycling $LiCoO_2$ with methanesulfonic acid for regeneration of lithium-ion battery electrode materials, J. Power Sources 436 (2019), https://doi.org/10.1016/j.jpowsour.2019.226828.
[20] Y. Yao, M. Zhu, Z. Zhao, B. Tong, Y. Fan, Z. Hua, Hydrometallurgical processes for recycling spent lithium-ion batteries: a critical review, ACS Sustain. Chem. Eng. 6 (2018), https://doi.org/10.1021/acssuschemeng.8b03545.
[21] A. Chagnes, B. Pospiech, A brief review on hydrometallurgical technologies for recycling spent lithium-ion batteries, J. Chem. Technol. Biotechnol. 88 (2013), https://doi.org/10.1002/jctb.4053.

[22] Y. Shi, G. Chen, F. Liu, X. Yue, Z. Chen, Resolving the compositional and structural defects of degraded LiNi$_x$Co$_y$Mn$_z$O$_2$ particles to directly regenerate high-performance lithium-ion battery cathodes, ACS Energy Lett. 3 (2018), https://doi.org/10.1021/acsenergylett.8b00833.

[23] Y. Wang, L. Ma, X. Xi, Z. Nie, Y. Zhang, X. Wen, Z. Lyu, Regeneration and characterization of LiNi0.8Co0.15Al0.05O$_2$ cathode material from spent power lithium-ion batteries, Waste Manag. 95 (2019), https://doi.org/10.1016/j.wasman.2019.06.013.

[24] S. Chen, T. He, Y. Lu, Y. Su, J. Tian, N. Li, G. Chen, L. Bao, F. Wu, Renovation of LiCoO$_2$ with outstanding cycling stability by thermal treatment with Li$_2$CO$_3$ from spent Li-ion batteries, J. Energy Storage 8 (2016), https://doi.org/10.1016/j.est.2016.10.008.

[25] M.J. Ganter, B.J. Landi, C.W. Babbitt, A. Anctil, G. Gaustad, Cathode refunctionalization as a lithium ion battery recycling alternative, J. Power Sources 256 (2014), https://doi.org/10.1016/j.jpowsour.2014.01.078.

[26] X. Song, T. Hu, C. Liang, H.L. Long, L. Zhou, W. Song, L. You, Z.S. Wu, J.W. Liu, Direct regeneration of cathode materials from spent lithium iron phosphate batteries using a solid phase sintering method, RSC Adv. 7 (2017), https://doi.org/10.1039/C6RA27210J.

[27] H. Al-Shammari, S. Farhad, Regeneration of cathode mixture active materials obtained from recycled lithium ion batteries, in: SAE Technical Paper 2020-01-0864, 2020, https://doi.org/10.4271/2020-01-0864.

[28] H. Al-Shammari, R. Esmaeeli, H. Aliniagerdroudbari, M. Alhadri, S.R. Hashemi, H. Zarrin, S. Farhad, Recycling lithium-ion battery: mechanical separation of mixed cathode active materials, in: ASME International Mechanical Engineering Congress and Exposition, Proceedings (IMECE), American Society of Mechanical Engineers (ASME), 2019, https://doi.org/10.1115/IMECE2019-10755.

[29] W. Lv, Z. Wang, H. Cao, Y. Sun, Y. Zhang, Z. Sun, A critical review and analysis on the recycling of spent lithium-ion batteries, ACS Sustain. Chem. Eng. 6 (2018), https://doi.org/10.1021/acssuschemeng.7b03811.

[30] E. Foremen, W. Zakri, M.H. Sanatimoghaddam, A. Modjtahedi, S. Pathak, N.G. Garafolo, S. Farhad, A review of inactive materials and components of flexible lithium-ion batteries, Adv. Sustain. Syst. 1 (11) (2017), https://doi.org/10.1002/adsu.201700061.

[31] H. Zarrin, S. Farhad, F. Hamdullahpur, V. Chabot, A. Yu, M. Fowler, Z. Chen, Effects of diffusive charge transfer and salt concentration gradient in electrolyte on Li-ion battery energy and power densities, Electrochim. Acta 125 (2014) 117–123, https://doi.org/10.1016/j.electacta.2014.01.022.

[32] R. Golmohammadzadeh, F. Faraji, F. Rashchi, Recovery of lithium and cobalt from spent lithium ion batteries (LIBs) using organic acids as leaching reagents: a review, Resour. Conserv. Recycl. 136 (2018), https://doi.org/10.1016/j.resconrec.2018.04.024.

[33] N.B. Horeh, S.M. Mousavi, S.A. Shojaosadati, Bioleaching of valuable metals from spent lithium-ion mobile phone batteries using Aspergillus Niger, J. Power Sources 320 (2016), https://doi.org/10.1016/j.jpowsour.2016.04.104.

[34] J. Ordoñez, E.J. Gago, A. Girard, Processes and technologies for the recycling and recovery of spent lithium-ion batteries, Renew. Sustain. Energy Rev. 60 (2016), https://doi.org/10.1016/j.rser.2015.12.363.

[35] A. Terfous, A. Hazzab, A. Ghenaim, Predicting the drag coefficient and settling velocity of spherical particles, Powder Technol. 239 (2013), https://doi.org/10.1016/j.powtec.2013.01.052.

[36] K. Ohlendieck, S.E. Harding, Centrifugation and ultracentrifugation, in: Wilson and Walker's Principles and Techniques of Biochemistry and Molecular Biology. Cambridge University Press. https://doi.org/10.1017/9781316677056.014.

[37] C. Bai, H. Park, L. Wang, Modelling solid-liquid separation and particle size classification in decanter centrifuges, Separ. Purif. Technol. 263 (2021), https://doi.org/10.1016/j.seppur.2021.118408.

[38] G.A. Lyons, A. Cathcart, J.P. Frost, M. Wills, C. Johnston, R. Ramsey, B. Smyth, Review of two mechanical separation technologies for the sustainable management of agricultural phosphorus in nutrient-vulnerable zones, Agronomy 11 (2021), https://doi.org/10.3390/agronomy11050836.

[39] T. Schubert, A. Meric, R. Boom, J. Hinrichs, Z. Atamer, Application of a decanter centrifuge for casein fractionation on pilot scale: effect of operational parameters on total solid, purity and yield in solid discharge, Int. Dairy J. 84 (2018), https://doi.org/10.1016/j.idairyj.2018.04.002.

[40] A.G. Kashkooli, G. Lui, S. Farhad, D.U. Lee, K. Feng, A. Yu, Z. Chen, Nano-particle size effect on the performance of $Li_4Ti_5O_{12}$ spinel, Electrochim. Acta 196 (2016) 33–40, https://doi.org/10.1016/j.electacta.2016.02.153.

[41] A.G. Kashkooli, E. Foreman, S. Farhad, D. Un Lee, W. Ahn, K. Feng, V. De Andrade, Z. Chen, Morphological and electrochemical characterization of a nanostructure $Li_4Ti_5O_{12}$ electrode using multiple imaging mode synchrotron X-ray computed tomography, J. Electrochem. Soc. 164 (12) (2017) A2861–A2871, https://doi.org/10.1149/2.0101713jes.

[42] X. Ma, M. Chen, B. Chen, Z. Meng, Y. Wang, High-performance graphite recovered from spent lithium-ion batteries, ACS Sustain. Chem. Eng. 7 (2019), https://doi.org/10.1021/acssuschemeng.9b05003.

[43] S.B. Chikkannanavar, D.M. Bernardi, L. Liu, A review of blended cathode materials for use in Li-ion batteries, J. Power Sources 248 (2014), https://doi.org/10.1016/j.jpowsour.2013.09.052.

SECTION 3

Recycling battery materials

CHAPTER 14

Recycling battery anode materials

Xing Ou and Long Ye
National Engineering Laboratory for High Efficiency Recovery of Refractory Nonferrous Metals, School of Metallurgy and Environment, Central South University, Changsha, Hunan, China

1. Anode materials in lithium-ion batteries

An ideal anode material should meet the following seven conditions at a minimum:
  (i) low chemical potential (compared with the counterpart electrode material);
  (ii) high specific capacity;
  (iii) channels for de/intercalation of Li^+, and high coulomb efficiency;
  (iv) good electronic and ionic conductivity;
  (v) good stability and compatibility with electrolytes;
  (vi) abundant resources, low cost, and simple manufacturing process;
  (vii) safe, green, and pollution-free.

Finding a perfect material that satisfies the above conditions is presently quite difficult. High energy density, good safety performance, low cost, and easy availability are the top priorities for selecting anode materials. According to the compositions, anode materials for lithium-ion batteries (LIBs) are divided into two categories, graphite and nongraphite. Graphite materials can be classified as natural graphite, artificial graphite, mesocarbon microbeads (MCMBs), soft carbon (such as coke), hard carbon, carbon nanotubes, graphene, carbon fiber, and so on. Nongraphite materials include silicon-based materials, nitride materials, tin-based materials, lithium titanate, and alloy materials. Specific data for the major characteristics and differences among these materials are provided in Table 14.1. Practically, graphite materials are commercially dominant as anode materials for LIBs because of their low cost, abundance, and favorable electrochemical performance. Thus, the following discussion of recycling objects primarily revolves around graphite materials.

2. The imperative of anode materials recycling

Given the rapid development of new energy vehicles and renewable energy equipment in recent years, explosive growth has occurred in the demand for LIBs, the core component providing energy storage for these products. Moreover, as a novel and efficient energy conversion and storage technology, LIB applications are expected to drive the energy supply revolution, no matter whether in military, telecommunications, electric power, or other fields. Statistically, global shipments of power batteries totaled

Table 14.1 Specific data for major characteristics of various anode materials.

Material	Li	C	$Li_4Ti_5O_{12}$	Si	Sn	Sb	Al	Mg
Density (g/cm^3)	0.53	2.25	3.5	2.33	7.29	6.7	2.7	1.3
Compound with Li	Li	LiC_6	$Li_7Ti_5O_{12}$	$Li_{4.4}Si$	$Li_{4.4}Sn$	Li_3Sb	LiAl	Li_3Mg
Mass specific capacity (mAh/g)	362	372	175	4200	994	660	993	3350
Volume specific capacity (mAh/cm^3)	2047	837	613	9786	7246	4422	2681	4355
Volumetric change	100	12	1	320	260	200	96	100
Potential versus Li^+/Li	0	0.05	1.6	0.4	0.6	0.9	0.3	0.1

128 GWh in 2019, up 19.6% year-on-year, making them the fastest-growing segment in the power, consumer, and energy-storage sectors. Global new energy vehicle sales totaled about 2.21 million units in 2019, up 15% year-on-year. The new energy vehicle industry is predicted to continuously drive sustained development of the LIB market. In light of this growth, recycling spent LIBs at the back end of the industry chain will be indispensable for future development.

Historically, the recycling process has focused on recycling the most valuable objects, giving higher priority to cathode materials owing to their scarcer ingredients. However, with a growing accumulation and volume of spent LIBs, every component becomes valuable in some way, especially overlooked, low-value anode materials. From an economics standpoint, although anode materials do not have a value comparable to that of cathode materials, they account for roughly 25%—28% of total battery production costs. Therefore, a reasonable and efficient recycling process is needed to realize the maximum use of anode materials as secondary resources. Furthermore, long-term cycling causes damage to electrode structural integrity, resulting in the mingling of electrolytes and the production of byproducts that are quite harmful to the environment if not properly handled. Hence, it is not difficult to conclude that anode materials should be given more attention in the recycling process.

Instead of being stacked or discarded, waste anode materials have potential value in reuse. For example, MCMBs, spherical lamellar particles with an isotropic microstructure, are the most common anode materials in LIBs. The excellent structural characteristic of MCMBs originates from high-temperature treatment (2800°C) in the manufacturing process, which helps reduce the negative structural influence of long-term cycling [1,2]. Thus, MCMBs are practical for possible reuse in LIBs as long as their impurities are removed. Moreover, the negative characteristic of waste MCMBs can sometimes be a positive factor from a different perspective. The structural damage resulting from long-term cycling is reflected in the reduced interlayered bonding force, but such damage is beneficial when stripping graphite layers to prepare value-added graphene material. In conclusion, whether it is recycled back into battery materials to form a closed-loop industrial chain or transformed into materials with other functions, reusing waste anode materials is both valuable and meaningful.

3. Graphite materials in lithium-ion batteries

3.1 Why do we choose graphite?

Graphite remains unbeaten as an anode material in most commercial LIBs because of its three distinguishing characteristics. First, its good electrochemical performance is highly attached to its unique structure. The graphite is a stratified structure with a layer spacing of 0.335 nm. The carbon atoms in the same layer are bonded by sp^2 hybridization, and the graphite layers are bonded by van der Waals forces. The carbon atoms on each layer are arranged in a six-membered ring that extends indefinitely in two dimensions. The layered structure of graphite provides convenient channels for the de/intercalation of Li^+. Second, graphite has an obvious low potential charge-and-discharge platform (0.01–0.2 V). Most chemical reactions for the de/intercalation of Li^+ happen within that voltage region. Specifically, the charge-and-discharge platform corresponds to the formation and decomposition of the graphite interlayer compound LiC_6, which is conducive to providing high and stable working voltage for LIBs. The theoretical capacity of graphite is 372 mAh/g, but in practice, the specific capacity is 330–370 mAh/g. The third reason for the attractiveness of graphite, of course, is its incomparable advantages in cost and resource abundance.

Without a doubt, however, graphite material has some defects. During the charge–discharge process, it easily reacts with electrolytes to generate a solid electrolyte interface (SEI) that reduces coulomb efficiency in the first cycling. In addition, co-embedding with the organic electrolyte solvent leads to expansion and stripping of the graphite layers, thus causing damage to LIB cycling stability. Those problems influence the service life span and make recycling and reuse more difficult.

3.2 Why do we recycle graphite?

The energy supply revolution in transportation is moving forward with the spread of electric vehicles (EVs) and hybrid EVs. It is predicted that the number of EVs will reach 125 million worldwide by 2030. The content of graphite in commercial batteries is normally between 12 and 21 wt% [3]. In practice, the power battery pack for a hybrid EV contains 10 kg graphite, and this content is increased to 50 kg in an EV [4]. As a result, the booming growth in LIB development has driven greater requirements for graphite. More importantly, only large flake graphite is suitable for preparing anode material in LIBs.

According to 2016 prices, battery-grade graphite costs $5000–20,000 per ton [5]. Therefore, all the evidence indicates that the economic benefit of recycling waste graphite from spent LIBs should be improved by this megatrend.

For sustainable resource development, the fact that graphite exploited from mines cannot be used without purification should not be neglected. Battery-grade graphite requires 99.5% purity with a spherical shape, whereas graphite mines can yield flake graphite with only 90%–98% purity. Inevitably, a purification process like high temperature and acid-alkaline leaching is employed, which accounts for a substantial proportion of the cost of battery-grade graphite production. Properly applying waste graphite could avoid the tedious selection and purification procedures of graphite mines. As proposed above, MCMBs synthesized through relatively high temperatures possess a stable structure that can reduce damage from long-term cycling. It is obvious that recycling waste MCMBs is an easier way to obtain battery-grade graphite than exploiting graphite from mines. Thus, waste graphite recovery is tremendously important for supply expansion and sustainable resource development.

Finally, the environmental impacts of improper waste graphite disposal must be fully recognized. As discussed above, some side reactions and electrochemical processes introduce harmful chemicals into graphite. SEI formation is necessary for steady long-term cycling performance but conversely may increase impedance and affect performance [6]. As an organic compound, the SEI adversely affects recycling and must be removed during the purification process. At the same time, some lithium is prone to deposit on graphite surfaces. Lithium recovery is profitable and necessary to avoid wasting nonrenewable resources. Those potential substances are both detrimental and profitable. With improper treatment or disposal, they could be released and irreversibly contaminate soil, water, and the atmosphere. But these potential harms can be transformed into economic benefits depending on how we handle them.

In summary, rapid demand growth, sustainable resource development, and potentially hazardous substances urge us to heighten the importance of waste graphite recycling.

4. Recycling methods for anode materials

4.1 Impurities removal

As discussed above, the impurities mingled in waste graphite can inevitably be adverse factors that impede waste graphite reuse. Thermal treatment is certainly a preferred method to deal with this problem. Graphite has naturally high melting and boiling points. The melting point is $3850 \pm 50°C$, and the boiling point is $4500°C$ [2]. This characteristic is the theoretical basis for graphite purification by the high-temperature method.

For common binders and additives, thermal treatment is the most effective method. Zhang et al. [7] discussed and clarified the details through corresponding thermal experiments. The main research objects consisted of acetylene black (AB), styrene-butadiene rubber (SBR), and carboxymethylcellulose sodium (CMC), which are frequently used as additives and binders. Through thermogravimetric analysis, graphite and AB were found to be relatively stable until 600 and 580°C, while the decomposition temperature of CMC and SBR started at 250 and 350°C and completely finished at 380 and 530°C, respectively. Finally, they considered 600°C as a suitable reaction temperature for impurities removal from these materials.

In practice, impurities include not only binders and additives but also byproducts produced with the electrochemical process and scrapped current collectors. Based on the thermal treatment, acid leaching is a solution to solve those metallic impurities [8]. Current collectors (copper and aluminum foil), AB, and binders (polyvinylidene fluoride, aka PVDF, or polytetrafluoroethylene, aka PTFE) can be mixed into the waste graphite during the dismantling and separation process. To use the pyrolysis temperature difference between the binder and current copper, the binder can be removed without damaging the copper at a relatively low temperature. In experiments, the condition of the first thermal treatment is settled at 400°C under the protection of inert gas. In that way, waste graphite can be successfully separated from the copper current. To assist with the selective leaching of the remaining metallic substances, aluminum and copper, a second thermal treatment is introduced to oxidize metal copper at 500°C in an air atmosphere. After that, the remaining metals in spent graphite can be easily leached in an acid solution (e.g., hydrochloric). It is worth mentioning that lithium, as a nonrenewable resource, receives a great deal of attention during the recycling process. So, further collection of lithium from the leaching solution proceeds through a series of beneficiation, purification, and extraction.

In addition to thermal treatment and acid leaching, the extraction method is a conductive way to remove the remaining electrolyte in the waste graphite, although it is presently limited to small-scale recycling. Sergej et al. [9] proposed that the main challenge in graphite recycling was the removal of the electrolyte and SEI. For this reason, they made a comprehensive comparison and investigation of three different electrolyte extraction concepts. The first involved only thermal treatment without an extraction process for electrolyte recovery. The second approach was based on the first. An extraction process with subcritical carbon dioxide was adopted before thermal treatment. The last method was similar to the second, except it replaced subcritical carbon dioxide with supercritical carbon dioxide as an extractant. For visual analysis of the different methodological results, scanning electron microscopy (SEM) was applied to demonstrate morphological changes. As shown in Fig. 14.1A, the morphology of graphite separated from the electrode with 100% SOH demonstrates a uniform distribution of graphite particles from 10 to 30 μm in size. After thermal treatment at 1000°C, the particles are prone to agglomerate and are covered by crystal-shaped compounds derived from conductive salt decomposition during the thermal treatment, as shown in Fig. 14.1B. By contrast, the

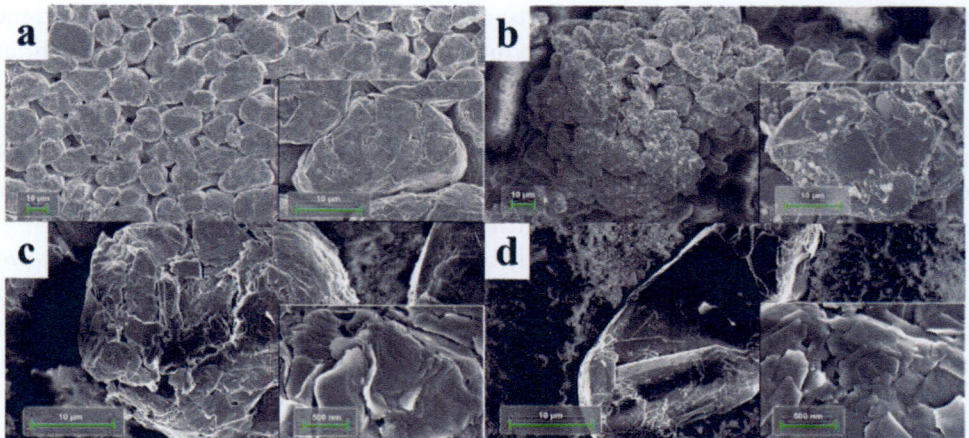

Figure 14.1 Scanning electron microscopy images of (A) graphite collected from the electrode (100% SOH) and waste graphite treated by (B) the first method—thermal treatment, (C) the second method—subcritical carbon dioxide as extractant and thermal treatment, (D) the third method—supercritical carbon dioxide as extractant and subsequent thermal treatment.

graphite shown in Fig. 14.1C is processed with a subcritical carbon dioxide-assisted electrolyte extraction with coaddition of acetonitrile and subsequent thermal treatment, which presents a glazed surface without a compound. Similarly, the application of supercritical carbon dioxide as the extraction agent to assist with thermal treatment can also effectively remove those impurities, as shown in Fig. 14.1D. It should be noted, however, that a high-pressurized carbon dioxide extraction method is unfavorable for the crystallinity size of the graphite particles, which may cause further negative effects on electrochemical performance. In conclusion, although the impurity removal results are the same, for further reuse of recycled graphite in LIBs, the subcritical carbon dioxide of the second method is considered the optimized method. According to the research paper, the recycled graphite employed in the second method exhibits satisfying electrochemical performance and a 90% recovery rate for the electrolyte, including the conductive salt.

4.2 Structural changes

Apart from impurities, structural changes can be beneficial or harmful depending on the chosen recycling method. It has been reported that the structural changes to waste graphite caused by long-term cycling can be repaired by thermal treatment in a graphitization furnace at 2600–3300°C [2]. Nothing but graphite can withstand such an extremely high-temperature ambience, while nearly all the impurities and remaining metallic substances are gasified. As expected, the structure of waste graphite can be simultaneously repaired. But this method consumes considerable energy, which increases the recycling cost.

Regardless of the ultra-high-temperature method, waste graphite reconstruction can be reasonably performed. As reported, waste graphite can be reconstructed by two steps, acid-treated and carbon-thermal reduction structural repair [10]. Firstly, waste graphite was treated with concentrated H_2SO_4 solution at 80°C for 5 h and further heated to 750°C for 8 h under an N_2 atmosphere to obtain acid-leached graphite (AG). This process aimed to remove the impurities, thus improving the purity of the final product, which is closely related to the electrochemical performance. Then, AG, $FeCl_2 \cdot 4H_2O$, and glucose were mixed and dried for the next procedure, calcination. In the roasted process, the composite was heated to 800°C, which was maintained for 8 h under an N_2 atmosphere to obtain the $AG/Fe_3O_4@Fe$ material. The obtained compound was treated with an acid solution for 2 h to remove the Fe element. The final recycled graphite product was obtained after washing and filtration. To confirm the structural changes from the recovery process, X-ray diffraction (XRD) was performed to observe the difference in the diffraction peak at $2\theta = 26.5$, as shown in Fig. 14.2 with an enlarged view. It is apparent that the peak of waste graphite has shifted to the left compared with that of commercial graphite, directly evidencing the increased layer spacing [7]. After the acid-leaching process, the peak is shifted more to the left, indicating additional layer spacing related to acid oxidizability. Subsequently and expectedly, the peak shifts back to the right after the reduction process, indicating the reconstruction of recycled graphite. The recycled graphite can be reused as an anode active material in LIBs due to the restored structure reflected in purity, reduced interlayer spacing, and reinforced bonding force.

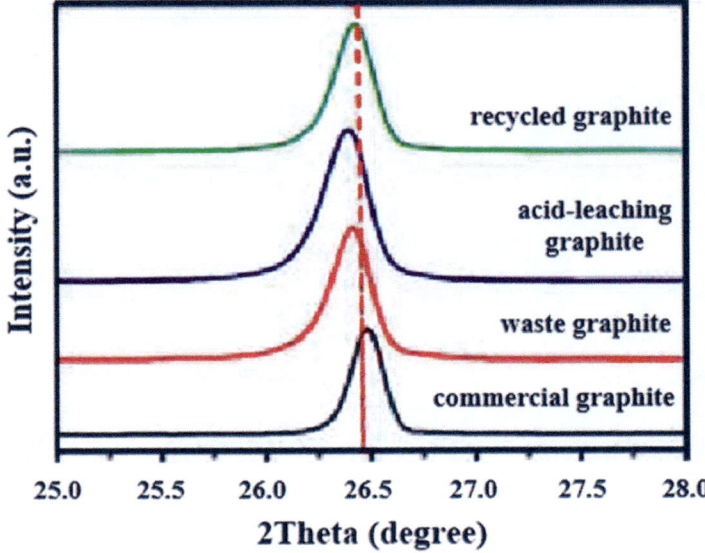

Figure 14.2 The (002) peak reflection of commercial graphite, waste material, acid-leaching graphite, and recycled graphite.

Considering this in reverse, structural damage can be an advantage in some special material preparation processes. The structural damage discussed here includes impurities and stretched d-spacing due to repeated de/intercalation of Li^+ in the long-term cycling process [11]. The loosened graphene layers in the cycled waste graphite can be seen in the SEM images of Fig. 14.3A. This structural change is an advantage in the preparation of graphene. Zhao et al. [12] adopted a modified Hummers' method to prepare graphene oxide (GO) from the waste graphite of spent LIBs. To testify to the positive effect of structural changes in waste graphite, the research' group also introduced natural graphite taken from the same treatment process as the contrasting sample. The capacity of exfoliation is directly illustrated by comparing two bottles of solution, as shown in Fig. 14.3B. The left one, prepared by waste graphite, is thicker and blacker than the right one that resulted from natural graphite. This difference arises because waste graphite possesses a thinner-layered structure associated with structural damage, confirming better exfoliation

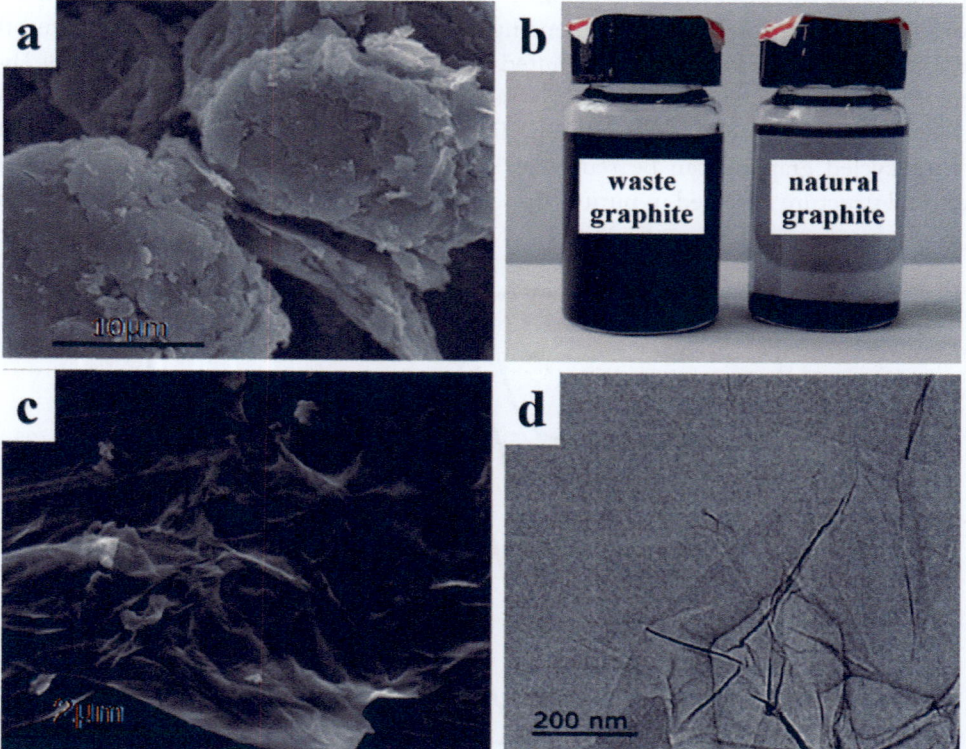

Figure 14.3 Scanning electron microscopy images of (A) waste graphite, (C) graphene oxide (GO) prepared by waste graphite; (B) two bottles of solution prepared by waste graphite and natural graphite left to stand for 1 h following ultrasonic dispersion in N-methyl-2-pyrrolidone for 15 min; (D) transmission electron microscopy image of GO.

capacity. As shown in Fig. 14.3C, GO is successfully obtained with the traditional morphology. At the 200 nm scale presented in Fig. 14.3D, GO exhibits an ultrathin structure. During preparation, the GO solution is observed as fairly stable without precipitation, even overnight, proving the effective exfoliation and production of few-layer graphene. The yield of GO is also positively correlated with the reduced layer bonding force. Hence, structural changes in waste graphite are a notable factor in the reuse process.

4.3 Application after recycling

Based on the previous discussion, the choice of application for regenerated graphite material is multitudinous, including alkaline metal-ion batteries, dual-ion batteries, capacitors, catalysts, electrochemical sensors, absorbents, and so on.

The first choice is to reuse various alkaline metal-ion batteries as anode materials. This kind of conception is beneficial to the establishment of a closed-loop battery system. Through the above research, waste graphite reuse is possible after impurities removal or structural repair.

Gao et al. [2] treated waste graphite with acid and thermal treatment. The regenerated graphite exhibited good electrochemical performance in charge capacity and cycling. The initial charge capacity and retention rate are 349 mAh/g and 98.8%, comparable to commercial graphite and confirming the successful regeneration process.

Liu et al. [10] used the reconstructed graphite as anode material for LIBs, and the material exhibited outstanding electrochemical performance. As a result, the reconstructed graphite delivers a high capacity of 427.9 mAh/g after 200 cycles at 0.5 C and also displays an outstanding rate capability (a capacity of 114.9 mAh/g is achieved at 3 C). Furthermore, the intermediate product of acid-treated graphite can also be directly used as anode material for sodium-ion batteries (SIBs), exhibiting high reversible capacity of 127 mAh/g at 50 mA/g and an excellent long cycle life (a reversible capacity of 106.8 mAh/g is achieved after 500 cycles at 2000 mA/g with high capacity retention of 90.98%).

Liang et al. [13] studied the recycling of waste graphite from spent LIBs through a simple pyrolysis process. The recycled graphite was used as anode material and tested in SIBs and potassium-ion batteries (KIBs). The recycled graphite performed well with a reversible capacity of 162 mAh/g at 0.2 A/g in SIBs. In a test of high-rate and long cycling performance, the graphite maintained 80% capacity at 5 A/g and retained 94.6% capacity for 1000 cycles at 2 A/g. More important, it was the first time for graphite used in KIBs, and the corresponding reaction mechanism and electrochemical performance were carefully analyzed by operando XRD. The recycled graphite exhibited a satisfying reversible capacity of 320 mA h/g at 0.05 A/g and 74% capacity retention at 0.02 A/g over 200 cycles.

In another study, the possibility of recycled graphite for a Li-ion-based all-carbon dual-ion battery (ACDIB) application was posited by the Kayakool' group [14]. They searched for the best temperature for the thermal treatment of waste graphite to obtain recycled graphite with the best electrochemical activity. Three different cells were assembled to illustrate electrochemical performance, including anode half-cells, cathode half-cells, and ACDIBs. For the anode half-cell, the best treatment temperature was 800°C, and the related product exhibited the highest specific capacity of 250 mAh/g at 372 mA/g over 300 cycles. As for the cathode half-cell, the ideal temperature was 650°C. The corresponding discharge capacity of recycled graphite reached 65 mAh/g at 100 mA/g in the first cycling, and it stably ran over 100 cycles. Finally, the above recycled graphite materials for the anode and cathode were coupled in an ACDIB. They delivered an initial discharge capacity of 58 mAh/g with a high average discharge voltage of ∼4.4 V, which equals an energy density of 255 Wh/kg.

Moreover, Schiavi et al. [15] proposed the use of recycled graphite in the application of an asymmetric supercapacitor. In the asymmetric supercapacitor, the waste graphite was reused as the negative electrode material, delivering a specific capacitance of 42 F/g and providing a maximum energy density of ∼9 Wh/kg and power density of 416 W/kg at 2.5 mA/cm^2.

In addition, graphite material with new functions can be an interesting recycling approach. Ruan et al. [16] recycled waste graphite to prepare an ORR electrocatalyst for use in fuel cells. Specifically, waste graphite was regarded as a carbon carrier and reacted with polyaniline andiron salt for the doping of N and Fe. The recovery catalyst displayed excellent catalytic activity, methanol resistance, and durability even better than the commercial Pt/C catalyst.

Jessielem et al. [17] provided another reuse approach for waste graphite—electrochemical sensors. The waste graphite was treated with nitric acid, and the counterpart cathode material was leached with citric acid. Then, a modified Hummers' method was adopted as the most practical preparation process for GO in dealing with the graphite following acid treatment. Specifically, they combined CoO from the cathode material with reduced GO (rGO) to synthesize the final product, rGO/CoO, through the sol-gel method and thermal treatment at 450°C for 2 h. As an electrochemical sensor, the performance of rGO/CoO was tested by cyclic voltammetry. The results verified the practical possibility and promising prospect of using these kinds of hybrid materials recovered from spent batteries as enzyme-free electrochemical sensors.

Zhao et al. [18] conceived the synthesis of MnO_2-modified graphite sorbents from waste graphite for treating lead-, cadmium-, and silver-contaminated water. As expected, this material exhibited enhanced selective separation capacity toward Pb(II), Cd(II), and Ag(I), with respective removal rates of 99.9%, 79.7%, and 99.8% for these three elements.

Finally, reusing waste graphite in an electro-Fenton process was proposed by the Cao' group [19]. They tested three materials from waste graphite distinguished by treatment method. RP (anode raw powder, waste graphite), AL (treated by acid leaching), and AAL (treated by acid and alkali leaching) were the investigation objects. The results indicated that functional groups of graphite adjusted by different processes were the main factor influencing material performance in the electro-Fenton system. According to electrochemical results, AAL has more advantages in the selection and yield of hydrogen peroxide (H_2O_2). AL material in the electro-Fenton system exhibited 100% bisphenol A (BPA) removal in 70 min and 87.4% COD removal in 240 min. The outstanding performance may be related to the higher carboxylic group content (35.83%) of AL material than of AAL (7%). Furthermore, compared with other carbon cathodes, the AL material maintained 100% BPA removal efficiency after 10 reuse cycles.

5. Defects in industry recycling

Large flake graphite qualified as battery grade is originally produced in a graphite mine and must be selected and purified. Those methods normally require extreme conditions such as strong acids and high temperatures that are environmentally unfriendly. Regardless of the complex preparation process, the excavation of existing resources may not keep pace with demand growth. Thus, future graphite production must include the recycling of waste graphite from spent batteries.

However, recycling companies, as well as industrial manufacturers like Umicore, currently choose to apply pyrometallurgical and a few hydrometallurgical processes. The recycled materials are judged mostly by their extracted values, meaning that the metals in cathode materials, such as Li, Co, and Ni, are the top priority in practical recycling [20]. Given this priority, the industry prefers pyrometallurgical processes to deal with all of the components at one time during smelting rather than introducing a preparation procedure for recovering waste graphite. As a result, waste graphite plays only a subordinate role in the recycling process as a heating material by burning or a reductant in metals extraction [21].

Some recycling companies, like Accurec, are achieving true preseparation of spent LIB components, like plastics, current collectors, and metallic cases, by various physical screening measures. Nevertheless, little importance has been attached to recovering waste graphite, resulting in graphite residue in the slag at the end of the recycling process. In addition, recently developed hydrometallurgical processes have paid greater attention to achieving a higher recovery rate for valuable metals [22].

There is a way to recycle waste graphite for structural repair by high temperature, as discussed above, at 2600–3300°C in the furnace. But this technique requires a large amount of energy, which means unfavorable economics. Additionally, chlorine gas (Cl_2) has been employed to convert metals and oxides into metal chlorides with much lower melting points to reduce energy consumption for heating. The drawback of this method is that Cl_2 is dangerous and toxic to the environment and human beings [2,23].

For now, the abundance of graphite resources assures a relatively low price and adequate supply for the replacement of waste graphite. But we should be aware of the importance of recycling waste graphite to guarantee sustainable resource development before it becomes too late to implement.

6. Prospects and challenges

As discussed, although valuable metals in cathode active materials are more attractive for research and industries, the waste graphite anode material should not be neglected going forward. LIB development peak in the coming years. It is crucial to find a proper and systematic recycling method to fully use spent LIBs, especially anode materials. In fact, no recycling industries have thus far targeted the recovery/reuse of graphite. Few industries have been involved in the graphite reduction process of metals and cathode materials, and they are often burned during the pyrometallurgical process. It should be recognized that graphite material has a huge potential in various fields depending on how we view it.

Fortunately, research groups have started to turn their attention to waste graphite recycling. Challenges in the recycling process are being uncovered, and the solutions do not confine graphite reuse only to batteries. An increasing number of eye-catching recycling processes have been proposed that provide practical examples for future industrial-scale recycling. Hence, the concerns for waste graphite recycling should be emphasized. Issues associated with the current graphite recycling/reuse process are as follows:

- Distinguishing different battery types. A series of industry-recognized labels, like colors or specialized symbols, is helpful to quickly sort the different battery types. With an effective sorted collection, object complexity and separation difficulty will be reduced, which will provide powerful help for recycling with high efficiency.
- The remaining Li element. As verified by previous research, the remaining Li element originates from the formation of SEI and other side reactions with electrolytes in long-term cycling. Therefore, the advantage of this potential high-value Li product should be considered in the recycling of waste graphite for increased economic benefits.
- Structural changes. Long-term service in LIBs changes the graphite structure by de/intercalation of Li^+. The enlarged interlayer spacing and reduced interlayer bonding force are beneficial for regenerating new graphite material, especially graphene. This strategy can be further exploited along with different applications of, for example, graphene.
- Impurities. Impurities, including SEI, lithium residue, cointercalation of electrolytes, and other byproducts, are harmful to the secondary life of recycled materials. Thus, purification techniques are critical to better use of waste graphite. Furthermore, it is meaningful to investigate the influence of impurities on graphite. According to various

research studies, some elements remaining in graphite may be beneficial, while others may have opposite effects. Illustrating the underlying impurities mechanism will provide good assistance in deciding on a specific and practical recycling method.
- Graphite-derived material. Development and research on graphite material and its derivatives are not constrained to the LIB sector. Various energy-storage devices, such as alkaline-ion batteries, dual-ion batteries, fuel batteries, and supercapacitors, are applications for graphite materials. Catalysts, absorbents, sensors, transistors, and diversified functional graphite materials provide feasible options for waste graphite recycling.
- Finally, waste graphite recycling is in line with the concept of sustainable development, which can effectively alleviate the pressure of growing demand for graphite and reduce dependence on original mines. Furthermore, reusing waste graphite avoids potential contamination to the environment.

References

[1] L. Chen, X. Fan, Z. Jiang, T. Ouyang, Y. Fei, Observation of "in-contact" characteristics of Brooks—Taylor mesophase spheres obtained by high-temperature centrifugation, Carbon 103 (2016) 421—424.
[2] Y. Gao, C. Wang, J. Zhang, Q. Jing, B. Ma, Y. Chen, W. Zhang, Graphite recycling from the spent lithium-ion batteries by sulfuric acid curing—leaching combined with high-temperature calcination, ACS Sustain. Chem. Eng. 8 (2020) 9447—9455.
[3] B. Moradi, G.G. Botte, Recycling of graphite anodes for the next generation of lithium ion batteries, J. Appl. Electrochem. 46 (2015) 123—148.
[4] S. Natarajan, V. Aravindan, An urgent call to spent LIB recycling: whys and wherefores for graphite recovery, Adv. Energy Mater. 10 (2020) 2002238.
[5] S.M. Badawy, Synthesis of high-quality graphene oxide from spent mobile phone batteries, Environ. Prog. Sustain. Energy 35 (2016) 1485—1491.
[6] X. Zheng, Q. Shi, Y. Wang, V.S. Battaglia, Y. Huang, H. Zheng, The role of carbon bond types on the formation of solid electrolyte interphase on graphite surfaces, Carbon 148 (2019) 105—114.
[7] J. Zhang, X. Li, D. Song, Y. Miao, J. Song, L. Zhang, Effective regeneration of anode material recycled from scrapped Li-ion batteries, J. Power Sources 390 (2018) 38—44.
[8] Y. Yang, S. Song, S. Lei, W. Sun, H. Hou, F. Jiang, X. Ji, W. Zhao, Y. Hu, A process for combination of recycling lithium and regenerating graphite from spent lithium-ion battery, Waste Manag. 85 (2019) 529—537.
[9] S. Rothermel, M. Evertz, J. Kasnatscheew, X. Qi, M. Grutzke, M. Winter, S. Nowak, Graphite recycling from spent lithium-ion batteries, ChemSusChem 9 (2016) 3473—3484.
[10] K. Liu, S. Yang, L. Luo, Q. Pan, P. Zhang, Y. Huang, F. Zheng, H. Wang, Q. Li, From spent graphite to recycle graphite anode for high-performance lithium ion batteries and sodium ion batteries, Electrochim. Acta 356 (2020) 136856.
[11] Y. Matsuo, K. Fumita, T. Fukutsuka, Y. Sugie, H. Koyama, K. Inoue, Butyrolactone derivatives as electrolyte additives for lithium-ion batteries with graphite anodes, J. Power Sources 119—121 (2003) 373—377.
[12] L. Zhao, X. Liu, C. Wan, X. Ye, F. Wu, Soluble graphene nanosheets from recycled graphite of spent lithium ion batteries, J. Mater. Eng. Perform. 27 (2018) 875—880.
[13] H.-J. Liang, B.-H. Hou, W.-H. Li, Q.-L. Ning, X. Yang, Z.-Y. Gu, X.-J. Nie, G. Wang, X.-L. Wu, Staging Na/K-ion de-/intercalation of graphite retrieved from spent Li-ion batteries: in operando X-ray diffraction studies and an advanced anode material for Na/K-ion batteries, Energy Environ. Sci. 12 (2019) 3575—3584.
[14] F.A. Kayakool, B. Gangaja, S. Nair, D. Santhanagopalan, Li-based all-carbon dual-ion batteries using graphite recycled from spent Li-ion batteries, Sustain. Mater. Technol. 28 (2021) e00262.

[15] P.G. Schiavi, P. Altimari, R. Zanoni, F. Pagnanelli, Full recycling of spent lithium ion batteries with production of core-shell nanowires//exfoliated graphite asymmetric supercapacitor, J. Energy Chem. 58 (2021) 336–344.
[16] D. Ruan, K. Zou, K. Du, F. Wang, L. Wu, Z. Zhang, X. Wu, G. Hu, Recycling of graphite anode from spent lithium-ion batteries for preparing Fe-N-doped carbon ORR catalyst, ChemCatChem 13 (2021) 2025–2033.
[17] J.S. Ribeiro, M.B.J.G. Freitas, J.C.C. Freitas, Recycling of graphite and metals from spent Li-ion batteries aiming the production of graphene/CoO-based electrochemical sensors, J. Environ. Chem. Eng. 9 (2021) 104689.
[18] T. Zhao, Y. Yao, M. Wang, R. Chen, Y. Yu, F. Wu, C. Zhang, Preparation of MnO_2-modified graphite sorbents from spent Li-ion batteries for the treatment of water contaminated by lead, cadmium, and silver, ACS Appl. Mater. Interfaces 9 (2017) 25369–25376.
[19] Z. Cao, X. Zheng, H. Cao, H. Zhao, Z. Sun, Z. Guo, K. Wang, B. Zhou, Efficient reuse of anode scrap from lithium-ion batteries as cathode for pollutant degradation in electro-Fenton process: role of different recovery processes, Chem. Eng. J. 337 (2018) 256–264.
[20] S. Natarajan, V. Aravindan, Burgeoning prospects of spent lithium-ion batteries in multifarious applications, Adv. Energy Mater. 8 (2018) 1802303.
[21] S. Natarajan, V. Aravindan, Recycling strategies for spent Li-ion battery mixed cathodes, ACS Energy Lett. 3 (2018) 2101–2103.
[22] X. Zhang, L. Li, E. Fan, Q. Xue, Y. Bian, F. Wu, R. Chen, Toward sustainable and systematic recycling of spent rechargeable batteries, Chem. Soc. Rev. 47 (2018) 7239–7302.
[23] A.D. Jara, A. Betemariam, G. Woldetinsae, J.Y. Kim, Purification, application and current market trend of natural graphite: a review, Int. J. Min. Sci. Technol. 29 (2019) 671–689.

CHAPTER 15

Recycling battery cathode materials

Xing Ou and Wei Wang

National Engineering Laboratory for High Efficiency Recovery of Refractory Nonferrous Metals, School of Metallurgy and Environment, Central South University, Changsha, Hunan, China

1. Introduction

Since Sony introduced the first generation of commercially available lithium-ion batteries (LIBs) in 1990, LIBs have been rapidly applied in mobile electronic devices because of their high energy density, long cycle life, and environmental friendliness [1]. With global energy and environmental issues becoming increasingly severe in recent years, electric vehicles are gradually gaining share around the world and are certain to further increase the demand for LIBs [2]. Meanwhile, due to the long-term charging and discharging process, LIBs suffer from deformation of the cathode material lattice, electrolyte decomposition, separator aging, and other problems that lead to capacity decay, so the overall life of LIBs is relatively short at only 3–5 years. As a result, large volumes of spent LIBs are currently available on international markets, and the number of spent LIBs should continue to rise in the next decade. The global quantity and weight of spent LIBs was forecast to exceed 25 billion units and 500,000 tons in 2020, and the size of the global LIB recycling market is expected to reach 641,595 tons by 2025 [3,4]. The toxic components in such large quantities of spent LIBs can be extremely harmful to the environment if not safely and efficiently recycled [5].

Generally, LIBs consist of five parts: cathode, anode, electrolyte, separator, and cell shell. Among these, the cathode, which can directly affect the LIB development space, plays the most important role in LIBs. In the process of cathode preparation, cathode materials are normally homogeneously mixed with a binder and conductive agent in N-methyl-2-pyrrolidone (NMP) solvent at a particular ratio. Then, the prepared electrode slurry is coated on Al foil as a current collector and finally dried to fabricate the cathode. A schematic of the cathode preparation process is shown in Fig. 15.1 [6].

LIB performance, including energy density, cycling life, and security, is mainly determined by cathode materials. $LiCoO_2$ (LCO) is a first-generation cathode material; to meet the needs of capacity delivery and stability, a large number of cathode materials have been developed, such as $LiMn_2O_4$ (LMO), $LiNiO_2$, $LiFePO_4$ (LFP), and $LiNi_xCo_yMn_{1-x-y}O_2$ (NCM). However, the heavy metal elements such as Ni, Co, and Mn contained in cathode materials can seriously harm the natural environment. Furthermore, Ni and Co indirectly enrich in the human body, affect human health, and may result in cancer [7]. At the same time, the scarcity of global reserves of Li and Co has contributed to their rising prices. Efficiently recycling of these elements would save resources while generating economic benefits [8,9].

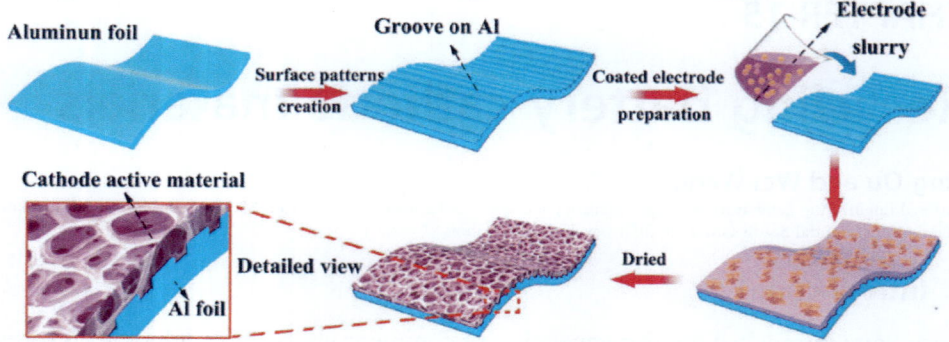

Figure 15.1 Schematic of the cathode preparation process [6].

2. Approaches for recycling battery cathode materials

With continuous performance improvements and expansions of their application areas, LIBs continue to experience growth in their market scale. The proper recycling of retired LIBs (especially cathode materials) has attracted growing attention from researchers. The first necessary step in LIB recycling is pretreatment. The purpose of this step is to separate different components safely and efficiently, which can greatly reduce the subsequent workload and energy consumption [1,10]. After pretreatment is completed, the separated components can be classified and recovered. The recycling approaches of cathode materials are mainly divided into pyrometallurgy, hydrometallurgy, and regeneration according to reaction characteristics [11].

2.1 Pretreatment

The structures of spent LIBs are very complicated. Therefore, to separate the components of spent LIBs and obtain cathode materials for subsequent leaching, the pretreatment process is critical and typically includes discharging, dismantling, and separating the cathode materials.

2.1.1 Discharging

Although spent LIBs are scrapped, a certain amount of energy remains in the battery. If the remaining energy is not fully discharged or discharged below a safe voltage before disassembly, there is a risk of short-circuiting and self-ignition during the posttreatment process [12]. LIB discharging methods can be either physical or chemical. Physical discharging usually refers to short-circuiting the battery's positive and negative poles or connecting a resistor to consume the battery's energy. Ku et al. [13] connected spent LIBs to a discharging device to bring the battery voltage below 0.1 V. This method effectively discharged the battery to a safe voltage, but obviously it was not suitable for processing large quantities of spent LIBs. Chemical discharging normally means immersing spent LIBs in a solution containing a certain concentration of an electrolyte, such as Na_2SO_4 [14] or NaCl [15], for a certain period to consume the residual energy. Li et al. [10] explored the influence of NaCl concentration and time on the discharging efficiency of spent LIBs. The results indicated that an ideal discharging efficiency of 71.96% could be obtained by discharging with 10 wt.% NaCl for 358 min.

2.1.2 Dismantling

After spent LIBs are completely discharged, dismantling is typically carried out manually or mechanically with the aim to separate various components from the spent LIBs [6,16]. Manual dismantling is always applied in the laboratory to obtain the spent cathode, anode, separator, and cell shell. When a large number of spent LIBs need to be recycled, industrial-scale mechanical dismantling reflects unique advantages over manual dismantling, such as large processing capacity and high efficiency. Because the components of spent LIBs have different physical properties, the cathode, anode, separator, Al foil, and Cu foil are simply separated by crushing [17], sieving [18], floating [19], magnetic separation [20], and gravity separation [21].

2.1.3 Separation of cathode materials

Cathode materials contain rare metals such as Li and Co with very high recovery values. However, cathode materials are normally firmly adhered to the Al foil by a binder, and therefore, a separation process is required to individually obtain them. Depending on the reaction mechanism, common separation methods include alkaline leaching, organic solvent dissolution, and heat treatment (Fig. 15.2) [22].

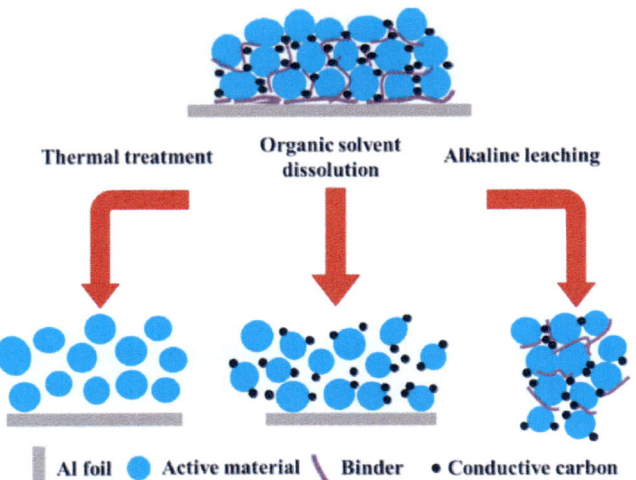

Figure 15.2 Methods of separating cathode materials from Al foil [22].

2.1.3.1 Alkaline leaching

Due to their different chemical properties, cathode materials can be selectively separated in the form of a solid, while Al foil will dissolve in alkaline solution, such as an NaOH solution. Ferreira et al. [23] found that 3 M NaOH could obtain a preferable separation effect under conditions of 50°C and a solid/liquid ratio of 100 g/L. The specific reaction is described by Eq. (15.1):

$$2Al + 2NaOH + 2H_2O \rightarrow 2NaAlO_2 + 3H_2 \uparrow \quad (15.1)$$

After leaching, the solid residues from filtration must be further calcinated to remove the binder and conductive agent. This method has the advantages of high separation efficiency and low cost. However, a certain volume of alkaline waste liquids and H_2 will be produced during the reaction, which may cause safety risks [24].

2.1.3.2 Organic solvent dissolution

According to the similarity-intermiscibility theory, polyvinylidene fluoride (PVDF) is widely used as a binder and can be dissolved by specific organic solvents, such as NMP [25], N,N-dimethylacetamide (DMAC) [26], and N,N-dimethylformamide (DMF) [27], to separate the cathode materials from the current collector. Among them, NMP is the most effective at dissolving PVDF owing to its high solubility. Li et al. [28] reported that PVDF could be completely removed by immersing the cathode in the organic solvent NMP at 100°C for 1 h. Based on the mechanism of cavitation by ultrasound, He et al. [29] developed ultrasonic cleaning in parallel with an NMP dissolution process to increase the peel-off efficiency of cathode materials to 99% under optimized conditions of 70°C for 90 min at an ultrasonic power of 240 W. The separated cathode materials displayed a low degree of agglomeration that was beneficial in the subsequent leaching process. Nonetheless, the toxicity, volatility, and high cost of NMP extremely limit its application. By contrast, Song et al. [26] considered DMAC and DMF more applicable to PVDF dissolution because of their lower prices. They found that the separation efficiency between the cathode materials and current collector could reach 98% with a 30°C reaction temperature, 30 min dissolving time, and 1:20 g/mL solid-to-liquid ratio.

2.1.3.3 Thermal treatment

Based on the different melting points and decomposition temperatures of cathode components, cathode materials can be separated by thermal treatment with the aim to burn off the binder, electrolyte, and conductive agent. The temperature of thermal treatment is required to increase above 400°C to discompose PVDF with stable properties. Meanwhile, the temperature should not be too high because Al will be oxidized to produce alumina, which is adverse to the subsequent leaching process [30]. In addition, the reaction atmosphere has a certain effect on thermal treatment. For example, in the process of recycling $LiFePO_4$, the air atmosphere can oxidize Fe^{2+} to Fe^{3+}, making it more likely to be leached by acid [22,31]. Conversely, the valence states of transition metal ions can be reduced by calcinating in the reducing atmosphere, which plays an important role in the subsequent leaching process. Sun et al. [32] proposed a method of vacuum pyrolysis in which the cathode materials of LCO were completely separated under the following conditions: reaction temperature of 600°C, vacuum evaporation time of 0.5 h, and residual gas pressure of 1.0 kPa. The vacuum atmosphere could reduce Co^{3+} to Co^{2+}, which was easier to dissolve in acid. Thermal treatment has a simple operation and no need for chemical reagents but has high energy consumption and generates poisonous gases such as HF.

2.2 Pyrometallurgy

Pyrometallurgy can efficiently recycle the valuable elements of spent LIBs, such as Co^{2+} and Ni^{3+}, by reducing them to metals or alloys. In the process, the electrolyte, binder, and conductive agent are incinerated by calcination and provide energy for the reaction. The pretreatment for pyrometallurgy only requires dismantling of the battery packs to individual cells. Many current industrial recycling technologies (Batrec, Umicore, Inmetco, Sony/Sumitomo, etc. [3]) use this process to complete the recovery of spent LIBs. Among them, Umicore is fairly representative because [33] of the maturity of its technology. They proposed a unique ultra-high-temperature technology to recycle spent LIBs using a high-temperature pyrolysis furnace divided into a preheating zone, plastics pyrolyzing zone, and smelting zone. In the preheating zone, the electrolyte was removed as steam by heating at a temperature below 300°C. In the plastics pyrolyzing zone, the operating temperature was raised to 700°C to burn off the plastic ingredients, thus maintaining the temperature and reducing energy loss. The smelting zone reduced Co^{2+} and Ni^{3+} to metals or alloys and obtained slugs containing Li, Al, Si, and Ca. This technology has the advantages of a short process time, low equipment requirements, and strong operability, but it can only receive semiproducts that need further hydrometallurgical processes to separate the metals of the alloys. Meanwhile, it is obvious that only Co, Cu, Ni, and Fe are recovered, while other metals such as Li, Al, Si, and Ca are lost in the slugs. With the gradual replacement of cobalt in current cathode materials and the scarcity of lithium resources, this technology no longer exhibits good economic benefits. Furthermore, pyrometallurgy has the defects of high energy consumption, toxic gas production, and materials loss. Therefore, recycling a new recycling process is needed for spent LIBs that is more environmentally friendly, is more efficient, and has lower energy consumption.

2.3 Hydrometallurgy

Hydrometallurgy uses chemical agents or acids generated by microorganisms to leach pretreated cathode materials to transform the valuable metals in spent LIBs to ion forms in solution. According to the type of leaching agent, the leaching process can be divided into four categories: organic acids, inorganic acids, bioleaching, and alkaline leaching.

Owing to its advantages of high sustainability, high exaction efficiency, low energy consumption, and environmental friendliness, hydrometallurgy is considered the most efficient recycling process for spent LIBs. Nevertheless, some challenges limit the development of hydrometallurgical processes, such as an intricate process flow, large quantities of acid or alkali waste, and the release of poisonous gases that are detrimental to the natural environmental and human health [34].

2.3.1 Inorganic acid leaching

The inorganic acids commonly used to leach cathode materials are HCl, H_2SO_4, and HNO_3. Early on, researchers found that HCl obtained better leaching efficiency than HNO_3 and H_2SO_4 when dealing with the same cathode materials. The phenomenon arises because the Cl^- of HCl has a certain reducibility. It can reduce metal ions in a high valence state to a low valence state, such as Co^{3+} to Co^{2+}, that can more easily dissolved by acid. Consequently, to solve the problem of metal ions with poor acid solubility, reducing acids or nonreducing acids together with reductant are always used to leach cathode materials. For instance, Wang et al. [35] utilized 4 M HCl to leach the hybrid cathode materials of LCO, LMO, and $LiCo_{1/3}Ni_{1/3}Mn_{1/3}O_2$. Most of the valuable metals could be leached with a high extraction efficiency, such as 99.5% of Co, 99.9% of Li, 99.8% of Ni, and 99.8% of Mn under the conditions of a temperature of 80°C, leaching time of 60 min, and solid/liquid ratio of 20 g/L. Rabia Sattar et al. [36] noted that the leaching efficiency of metals could be promoted by increasing the acid concentration, time, and temperature when $LiNi_xCo_yMn_zO_2$-type cathode materials were disposed by H_2SO_4. Under the optimal conditions, the maximal 92% Li and Ni could be leached, but the leaching efficiency of Co and Mn only reached 68% and 34.8%, respectively. Therefore, H_2O_2 was added to reduce the high valence of metal ions to lower valences such as Co^{3+} to Co^{2+} and Mn^{4+} to Mn^{2+}, which raised the efficiency of Co and Mn remarkably and had a positive influence on Li and Ni. With the addition of 4 vol.% H_2O_2, the leaching rate of all metals was up to 98% under conditions of 5% pulp density, 2 M H_2SO_4, 50°C, and 120 min leaching time. Lee et al. [37] discovered through reduction leaching with reductant H_2O_2 that the leaching rate of Co and Li increased by 45% and 10%, respectively, compared with nitric acid leaching alone. The experimental results indicated that the efficiency effect of Li and Co peaked at 95% when the H_2SO_4 concentration was 1 M and the amount of H_2O_2 added was 1.7 vol.%. H_3PO_4 [38], which is relatively mild, also proved capable of dissolving valuable metals from spent LIBs. Chen et al. [39] achieved extraction efficiencies for Ni, Co, Mn, and Li of 99.5%, 96.3%, 98.8%, and 100%, respectively, under optimum reaction conditions for the H_3PO_4 concentration (2 M), H_2O_2 dosage (4 vol.%), pulp density (20 mL/g), leaching temperature (60°C), and leaching time (60 min). Other common reducing agents include glucose [40], $NaHSO_3$ [41], and $Na_2S_2O_3$ [42], and the experimental results indicated that all could assist the acid solution to significantly enhance the

Table 15.1 Summary of inorganic acid leaching systems investigated in the literature.

Materials	Leaching agent	Conditions (T, t, S/L)	Efficiency (%)
LCO	4 M HCl [43]	80°C, 1 h, 10 g/L	Li = 99, Co = 99
LCO	3 M HCl + 10 vol.% H_2O_2 [44]	80°C, 1.5 h	Li = 99.4
LCO	2 M H_2SO_4 + 5 vol.% H_2O_2 [32]	80°C, 1 h, 50 g/L	Li = 99, Co = 99
LCO	1 M HNO_3 + 1.7 vol.% H_2O_2 [37]	75°C, 1 h, 20 g/L	Li = 95, Co = 95
LCO	2% H_3PO_4 + 2 vol.% H_2O_2 [45]	90°C, 0.5 h, 8 g/L	Li/Co > 95
LFP	0.6 M H_3PO_4 [46]	20 min, 50 g/L	Li = 94.29, Fe = 97.67
NCM	1 M H_2SO_4 [47]	95°C, 4 h, 50 g/L	Li = 93.4, Ni = 96.3, Co = 66.2, Mn = 50.2
NCM	1 M H_2SO_4 + 1 vol.% H_2O_2 [48]	40°C, 1 h, 40 g/L	Li/Co/Ni/Mn = 100
Mixed cathodes	4 M HCl [35]	80°C, 1 h	Ni, Co, Mn, Li > 99
Mixed cathodes	1 M H_2SO_4 + 0.075 M $NaHSO_3$ [41]	95°C, 4 h, 20 g/L	Li = 97, Ni = 96, Co = 92, Mn = 88

LCO, $LiCoO_2$; LFP, $LiFePO_4$; NCM, $LiNi_xCo_yMn_{1-x-y}O_2$.

leaching rate of metal ions. The detailed leaching parameters and efficiencies of various inorganic acid leaching systems are summarized in Table 15.1.

Inorganic acid leaching agents have the benefits of low costs and simple processes, but a number of shortcomings restrict their adoption, such as weak selective leaching for individual ions that increases the time and complexity of the subsequent separation and recovery processes, release of toxic gases during the process, like Cl_2, SO_x, NO_2, that require follow-up absorption treatment of the exhaust gas. For instance, Cl_2 is generated when LCO is treated by HCl. The specific reaction is described by Eq. (15.2):

$$2LiCoO_2 + 8HCl \rightarrow 2CoCl_2 + Cl_2\uparrow + 2LiCl + 4H_2O \quad (15.2)$$

2.3.2 Organic acid leaching

Considering the significant environmental hazards associated with inorganic acid leaching, including the emission of hazardous gases and the production of waste acid solutions, researchers have begun to pay more attention to organic acid leaching, which is more environmentally friendly. Although organic acids are relatively less acidic than inorganic acids, they are more easily decomposed; the main products of decomposition are H_2O and CO_2, which cause little pollution to the environment. Meanwhile, some organic acids inherently possess reducibility that considerably reduce the cost of chemical reagents and exhibit selective leaching characteristics, allowing the leaching and precipitation of metal ions to be separated in one step. These properties compensate for the weak acidity of organic acids and have led to extensive research in recent years.

Commonly available organic acid leaching agents are citric, malic, succinic, tartaric, formic, acetic, lactic, oxalic, and ascorbic acids. He et al. [49] adopted tartaric acid as the leaching agent and H_2O_2 as the reducing agent for the leaching of ternary cathode materials, and the recoveries of Ni, Co, Mn, and Li reached 99.31%, 98.64%, 99.31%, and 99.07%, respectively, under the optimum parameters. In consideration of the higher solubility of low-valent metal ions in acids, some organic acids with both reducing and acidic properties, such as ascorbic and oxalic acids, have been extensively investigated. Given that $C_2O_4^{2-}$ in $H_2C_2O_4$ forms a precipitate with Co^{2+}, and $H_2C_2O_4$ is reductive enough to convert Co^{3+} to Co^{2+}, Zeng et al. [50] suggested that oxalic acid could be employed for the selective leaching of LCO. The relevant reaction is given in Eq. (15.3):

$$5H_2C_2O_4 + 2LiCoO_2 = 2LiHC_2O_4 + 2CoC_2O_4\downarrow + 4H_2O + 2CO_2\uparrow \quad (15.3)$$

$$H_2C_2O_4 + CoC_2O_4 = Co(HC_2O_4)_2 \quad (15.4)$$

Eq. (15.3) reveals that the Co^{3+} and Li^+ in LCO are present in the residue and filtrate individually, which can be well separated. However, attention needs to be paid to the concentration of oxalic acid. When the concentration of oxalic acid is too high, the CoC_2O_4 precipitate may be dissolved, as in Eq. (15.4). The recycle efficiency of Li and Co achieved 98% and 97%, respectively, when optimum conditions were maintained at 2.5 h, 95°C, a solid-to-liquid ratio of 15 g/L, and a rotation rate of 400 rpm.

Since the low valence transition metal ions Ni^{2+}, Co^{2+} and Mn^{2+} can form oxalate complex of a five-membered ring structure with oxalic acid, Zhang et al. [51] considered that oxalic acid can also be employed for the efficient recovery of $LiNi_{1/3}Co_{1/3}Mn_{1/3}O_2$. The relevant reaction is described by Eq. (15.5):

$$4H_2C_2O_4 + 2LiNi_{1/3}Co_{1/3}Mn_{1/3}O_2 = Li_2C_2O_4 + 2(Ni_{1/3}Co_{1/3}Mn_{1/3})C_2O_4 + 4H_2O + 2CO_2 \quad (15.5)$$

During this reaction, Li^+ is dissolved in the solution, while Ni–Co–Mn ions are reduced to a lower valence state by oxalic acid to form a less soluble complex precipitation, thus achieving the separation between Li^+ and Ni–Co–Mn ions in one straightforward step. The detailed leaching parameters and efficiencies of the different organic acid leaching systems are summarized in Table 15.2.

Nonetheless, organic acids are typically expensive, which extremely limits their prospects for large-scale industrial applications. In addition, weak acidity contributes to the selection of lower-leaching solid-to-liquid ratios in the process parameters, which means that the processing capacity of the cathode materials during leaching is significantly reduced.

Table 15.2 Summary of organic acid leaching systems investigated in the literature.

Materials	Leaching agent	Conditions (T, t, S/L)	Efficiency (%)
LCO	1 M oxalic acid [52]	80°C, 2 h, 50 g/L	Li = 98, Co = 98
LCO	1.25 M ascorbic acid [53]	70°C, 20 min, 25 g/L	Li = 99, Co = 95
LCO	1.25 M citric acid + 1 vol.% H_2O_2 [54]	90°C, 0.5 h, 20 g/L	Li = 100, Co > 90
LCO	0.4 M tartaric acid + 0.02 M ascorbic acid [55]	80°C, 5 h, 2 g/L	Li = 100, Co = 97
LFP	0.8 M acetic acid + 6 vol.% H_2O_2 [56]	50°C, 0.5 h, 120 g/L	Li = 95.05
NCM	2 M citric acid + 2 vol.% H_2O_2 [57]	80°C, 1.5 h, 33.3 g/L	Li = 99, Ni = 97, Co = 95, Mn = 94
NCM	3 M trichloroacetic acid + 4 vol.% H_2O_2 [58]	60°C, 0.5 h, 50 g/L	Li = 100, Ni = 93, Co = 92, Mn = 90
NCM	2 M formic acid + 6 vol.% H_2O_2 [59]	60°C, 2 h, 50 g/L	Li = 100, Co/Ni/Mn = 85
Mixed cathodes	2 M tartaric acid + 4 vol.% H_2O_2 [49]	70°C, 0.5 h, 17 g/L	Li = 99, Co/Ni/Mn = 99

LCO, $LiCoO_2$; LFP, $LiFePO_4$; NCM, $LiNi_xCo_yMn_{1-x-y}O_2$.

2.3.3 Bioleaching

Bioleaching employs microorganisms that produce inorganic or organic acids to treat cathode materials from spent LIBs, mainly using S/Fe-oxidizing bacteria and heterotrophic fungi. In contrast to other leaching methods, bioleaching is more environmentally friendly, requires less energy, emits fewer greenhouse gases, and has lower operating costs. Jegan Roy et al. [60] used *Acidithiobacillus ferrooxidans* to treat spent LIBs and produce biogenic H_2SO_4 and Fe^{3+}. By controlling the reaction conditions at a temperature of 30°C, pH of 2.0, and pulp density of 10%, the recoveries achieved were 90% Ni, 92% Mn, 82% Co, and 89% Li. Heterotrophs including *Aspergillus niger*, *Penicillium simplicissimum*, and *Penicillium chrysogenum* fungi can leach metals by generating organic acids such as citric acid, gluconic acid, lactic acid, and malic acid. Horeh et al. [61] concluded that adapted *A. niger* could recycle 100% Li, 72% Mn, 94% Cu, 62% Al, 38% Co, and 45% Ni from spent LIBs at a pulp density of 1%, pH of 2.5, and temperature of 30°C. Although bioleaching has been extensively researched as an emerging approach in recent years, it is still struggling to be employed on a large scale in industry. There are two major reasons for this. First, the ability to process spent LIBs on a large scale is significantly limited because of the long bioleaching time, low bioleaching rate, and low pulp density required. In addition, the cultivation of bacteria and fungi is slow, and the conditions are demanding [62].

2.3.4 Alkaline leaching

The ammonia leaching method utilizes NH_4^+-containing systems, such as ammonia (NH_4OH), ammonium carbonate ((NH_4)$_2CO_3$), ammonium sulfate ((NH_4)$_2SO_4$), ammonium chloride (NH_4Cl), and other alkaline reagents, as leaching agents to recover the metal elements from cathode materials. During the reaction, NH_4^+ can selectively complex with metal ions such as Ni^{2+} and Co^{3+}, while Li^+ is also dissolved into the solution due to the destruction of cathode material's hierarchical structure, leaving most of Mn in the residue owing to their weak complexation with NH_4^+. As a result, the valuable metals Ni, Co, and Li in cathode materials are efficiently separated from the inexpensive metal Mn.

Zheng et al. [63] leached spent LIB cathode materials using NH_3-(NH_4)$_2SO_4$ as the leaching agent and Na_2SO_3 as the reducing agent and found that during the reaction, Ni, Co, and Li remained in solution as metal ions or amine complexes, while Mn was reduced by Mn^{4+} to Mn^{2+} to precipitate as (NH_4)$_2Mn(SO_3)_2 \cdot H_2O$. Meanwhile, they explored the relationship between various experimental factors and the leaching rate, finding that the leaching rate of the valuable metal ions Ni^{2+}, Co^{2+}, and Li^+ was enhanced with an increase in the concentration of (NH_4)$_2SO_4$, concentration of Na_2SO_3, time, and temperature. In the first step of the leaching procedure, 95.3% Li, 89.8% Ni, 80.7% Co, and only 4.3% Mn could be obtained by controlling the experimental parameters at 1.5 M (NH_4)$_2SO_4$, 4 M NH_3, 0.5 M Na_2SO_3, 353 K, 5 h, and 10 g/L pulp density. Under the same conditions after a two-step leaching procedure, the leaching efficiencies of Ni, Co, and Li obtained 94.8%, 88.4%, and 96.7% individually, while the leaching efficiency of Mn was merely 6.34%. Ma et al. [64] developed a novel three-stage process for the efficient recovery of valuable metals from spent LIBs. Initially, products consisting of Li_2CO_3, $(NiO)_m \cdot (MnO)_n$, Ni, and Co were obtained by reduction roasting at 650°C for 1 h using the spent anode powders of LIBs as the reducing agent. Afterward, Li^+ was recovered in the next step by water leaching. Finally, the water leaching residue was leached by 3 M $NH_3 \cdot H_2O$ associated with 1.5 M (NH_4)$_2SO_3$ at an optimized condition of 353 K, 120 min, and a 50 mL/g liquid–solid ratio, which achieved direct separation and recycling among Ni, Co, and Mn.

The ammonia leaching method has a considerable potential for application due to its unique selectivity, but it has a multitude of problems that must be addressed. For example, solutions containing the valuable elements Li, Ni, and Co after ammonia leaching require further treatment to separate the metals, which increases the process costs. It is also difficult to recover some elements enriched in the precipitate, such as Mn. The ammonia leaching step also produces some alkaline wastes that must be subsequently treated.

2.3.5 Separation and purification

After the leaching process, separation and purification are required to separate, purify, and recover the valuable metals Ni, Co, Mn, Li, and Al. The primary approaches involve solvent extraction, chemical precipitation, and electrodeposition.

2.3.5.1 Solvent extraction

The mechanism of solvent extraction accords with the inhomogeneous distribution of metal ions in a two-phase system. Extractants commonly available for the disposal of spent LIB leachates are Cyanex 272, D2EHPA, PC-88A, P507, and Acorga M5640.

T. Suzuki et al. [65] proposed a three-stage extraction process for the effective separation of Ni, Co, Mn, and Li from acidic leaching solutions. They carried out the extraction of Cu, Al, and Co one by one by adjusting the pH and using a variety of extractants while leaving Li in the leachate. This process can yield high metal recovery rates. But given the wide variety of metal ions contained in the leachate from spent LIBs, there are many obstacles in separating the various ions using only simple solvent extraction. Thus, a strategy that combines solvent extraction with chemical precipitation can be considered. Chen et al. [66] developed a solvent extraction in conjunction with a chemical precipitation method to recover valuable metals from spent LIB leachates. Initially, Ni^{2+} was precipitated using a dimethylglyoxime reagent wherein the pH was adjusted to 5. Co^{2+} was then precipitated by the addition of ammonium oxalate under an equilibrium pH of 6. Subsequently, Mn^{2+} was recovered by the extractant D2EHPA at a pH of 5. Ultimately, 0.5 M Na_3PO_4 was added to form a precipitate Li_3PO_4 with Li^+. By this method, over 98% of Ni, Co, and Mn and over 89% of Li were separated and recovered.

Solvent extraction, a well-established method of separation and purification, has the merits of excellent separation, less energy consumption, and superior product purity. But some extractants are particularly expensive and may be detrimental to the environment and human health.

2.3.5.2 Chemical precipitation

The chemical precipitation process separates and recovers valuable metal ions from spent LIBs by adding specific anions that combine with the target cations to form low-solubility precipitates in the leaching solution. In the recycling process of spent LIBs, the frequently applied precipitants are NaOH, Na_2CO_3, and $H_2C_2O_4$.

Pant et al. [67] developed a stepwise precipitation method using $H_2C_2O_4$ with saturated Na_2CO_3 as a precipitant to separate and recover Li, Ni, Co, and Mn from spent LIB leachates at various pH values. Similarly, Castillo et al. [68] utilized NaOH to precipitate Fe, Mn, and Li in turn by adjusting the pH of the solution to achieve the recycling of valuable metals. In contrast to other methods, precipitation is simpler to operate and has better recovery rates, but the consumption of chemical reagents remains considerable, and product purity is susceptible to impurities.

2.3.5.3 Electrodeposition

Electrodeposition can efficiently recover valuable metals from spent LIB leachates as pure metal or metal hydroxide. The mechanism is based on the redox reaction of metal ions in the leachates with additional energy provided by the two electrodes. Prabaharan et al. [69] developed an environmentally friendly and commercially feasible electrodeposition approach to separate Co^{2+} and Mn^{2+} in the leaching solution. They controlled the experimental conditions at a temperature of 90°C, a leaching solution pH of 2—2.5, and a current density of 200 A/m^2 and discovered that Co^{2+} in the leaching solution was deposited as pure metal at the cathode, and Mn^{2+} was produced as MnO_2 at the anode. The recovery rate of Co and Mn reached 96% and 99%, respectively. Electrodeposition significantly avoids the introduction of impurities and improves product purity, but the high consumption of electrical energy limits further application of this method.

2.4 Regeneration

Due to the complexity of the traditional recovery method of leaching-separation and purification, researchers have introduced a novel strategy of directly regenerating the cathode materials or precursors, which avoids the complicated separation process and yields a high-value-added product for economic benefit. The major regeneration methods available are coprecipitation, solid-state reaction, and the sol-gel method. The whole flow of regeneration and reutilization of the cathode electrode is shown in Fig. 15.3 [6].

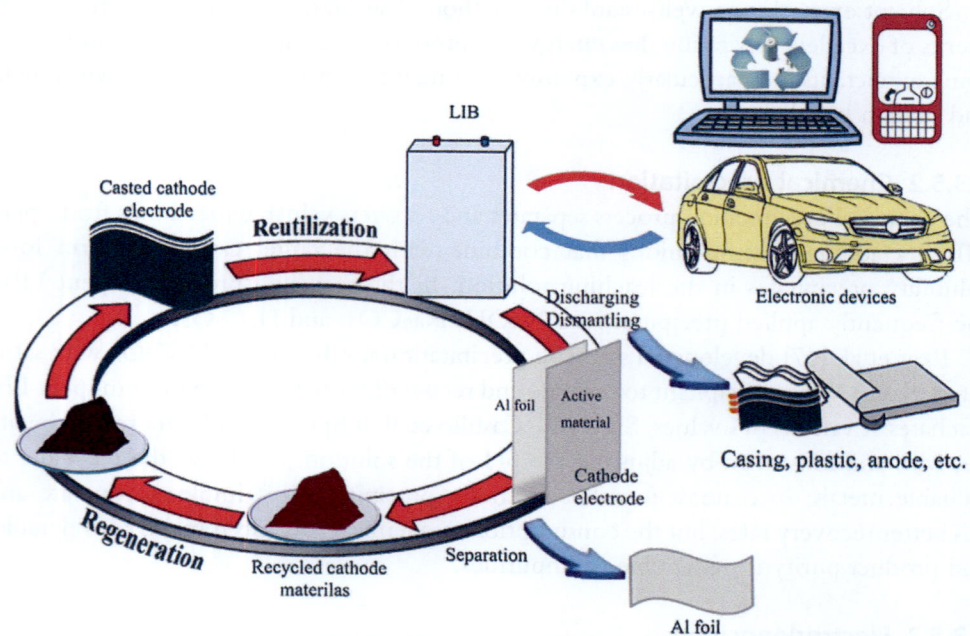

Figure 15.3 Whole flow of cathode electrode regeneration and reutilization [6].

2.4.1 Solid-state method

Regeneration of cathode materials by the solid-state method is a two-step process: spent cathode materials such as LCO, LFO, and Li sources are mechanically mixed in a specific molar ratio, and the mixture is calcined at a temperature of 600–950°C to obtain the regenerated materials.

Chen et al. [70] fabricated a regenerated LCO cathode material using this method by mixing spent cathode materials with $LiCO_3$ and calcinating it at 850°C. This kind of material exhibited excellent electrochemical properties: the initial discharge capacity was 150.3 mAh/g at a current of 0.1 C, and the capacity remained at 140.1 mAh/g in the potential range of 3.0–4.3 V after 100 cycles. They also explored the effect of impurities in spent LIBs on the regenerated material and concluded that small amounts of residual impurities Al (<0.4 wt.%) and Cu (<0.6 wt.%) could improve its electrochemical performance. Zhang et al. [71] pretreated the spent LFP to remove impurities such as $FePO_4$, Li_3PO_4, Fe_2O_3, and P_2O_5 first, then directly calcinated the spent LFP cathode powder at 650°C for 1 h under an atmosphere of Ar/H_2 to synthesize a new cathode showing nearly identical electrochemical properties to the LFP cathode material at high discharge currents.

The simple generation process and fewer steps contribute to the suitability of the solid-state method for large-scale industrial production. Nevertheless, impurities in spent LIBs are especially difficult to eliminate, which is detrimental to the properties of cathode materials.

2.4.2 Coprecipitation method

Coprecipitation is the most frequently used regeneration method, allowing the preparation of homogeneous cathode materials and the prospect of industrial application. The regeneration of cathode materials by coprecipitation can normally be divided into five procedures:

(1) leaching of pretreated cathode materials using chemical reagents
(2) purification of the leachate with the aid of a precipitant or extractant
(3) according to the target cathode materials to be synthesized, adjusting the ionic ratio of the leaching solution by adding a certain amount of metal salt solution
(4) producing the precursor by the coprecipitation method by adding a precipitant to the leachate
(5) mixing the precursor with the lithium sources to produce the target cathode by calcination

Following these steps, Sa et al. [72] regenerated NCM-111 cathode material by coprecipitation using 5 M NaOH as precipitant and obtained good electrochemical properties, with a discharge capacity of 158 mAh/g at 0.1 C and capacity retention of 80% after 100 cycles. The coprecipitation method can achieve high recycling efficiency and product purity for valuable metals. However, some problems remain, such as complicated processes and high costs.

2.4.3 Sol-gel method

The sol-gel method is mainly used to regenerate cathode materials by adding a complexing agent to the leaching solution to form a gel and then roasting to obtain the precursor. This method is always applied in combination with organic acid leaching, as the organic acid can replace the complexing agent to form a gel network with the metal ions. For example, Yao et al. [25] leached the spent cathode materials by a citric acid-hydrogen peroxide system, then supplemented the leachate with corresponding metal salts and adjusted the pH to make sol-gel. Finally, heat treatment was carried out to prepare the regenerated $LiCo_{1/3}Ni_{1/3}Mn_{1/3}O_2$ cathode, exhibiting a specific capacity of 152 mAh/g and a cycling performance of 95.6% after 100 cycles. The sol-gel method makes full use of valuable metals and obtains cathode materials with excellent performance. Nevertheless, this method is more complicated than the solid-state method, and the introduction of a complexing agent results in higher costs.

3. Challenges and perspectives

The number of spent LIBs has grown continuously in recent years, bringing about increasingly serious problems of environmental pollution and wasted resources. Therefore, it is urgent to explore reasonable methods of recycling spent LIBs and reusing their valuable metal resources. Nowadays, spent LIB recycling has been extensively investigated around the world. However, the various recycling methods still have shortcomings. To address some typical issues and promote the establishment of a more environmentally friendly, efficient, and economical spent LIB recycling process, the following perspectives are proposed:

(1) Most existing discharge methods immerse spent LIBs in NaCl; as a result, Cl^- may corrode the cell shell, leading to electrolyte leakage. Therefore, a safe and efficient discharging method must be developed.

(2) During the pretreatment process, efficient mechanical dismantling procedures must be established for large-scale recycling of spent LIBs. At the same time, pollution sources such as dust, hazardous gases, and organic solvents generated during dismantling need to be strictly controlled.

(3) Currently, cathode materials recovery by H_2SO_4 acid leaching has been broadly used on an industrial scale, but the acid utilization rate is not sufficient, and the residual acid concentration in the leachate is relatively high during the reaction. Hence, enhancing the acid utilization rate, increasing the treatment capacity, and avoiding or reducing possible secondary pollution in the leaching process should be topics of future research.

(4) Coprecipitation can regenerate cathode materials efficiently, but the presence of impurity ions in the leachate may affect the electrochemical performance of the regenerated electrode. So the removal of impurities from the leaching solution before coprecipitation needs further investigation.

(5) The reuse of spent LIBs need not be restricted to producing new electrode materials and can be used to fabricate other materials directly.

References

[1] B. Huang, et al., Recycling of lithium-ion batteries: recent advances and perspectives, J. Power Sources 399 (2018) 274–286.

[2] Q.X. Wu, Z.F. Pan, L. An, Recent advances in alkali-doped polybenzimidazole membranes for fuel cell applications, Renew. Sustain. Energy Rev. 89 (2018) 168–183.

[3] L. Li, et al., The recycling of spent lithium-ion batteries: a review of current processes and technologies, Electrochem. Energy R. 1 (4) (2018) 461–482.

[4] X.L. Zeng, J.H. Li, N. Singh, Recycling of spent lithium-ion battery: a critical review, Crit. Rev. Environ. Sci. Technol. 44 (10) (2014) 1129–1165.

[5] X.L. Li, et al., Direct regeneration of recycled cathode material mixture from scrapped Lifepo4 batteries, J. Power Sources 345 (2017) 78–84.

[6] Y. Zhao, et al., Regeneration and reutilization of cathode materials from spent lithium-ion batteries, Chem. Eng. J. (2020) 383.

[7] W.G. Lv, et al., A critical review and analysis on the recycling of spent lithium-ion batteries, ACS Sustain. Chem. Eng. 6 (2) (2018) 1504–1521.

[8] J.T. Hu, et al., A promising approach for the recovery of high value-added metals from spent lithium-ion batteries, J. Power Sources 351 (2017) 192–199.

[9] P. Liu, Recycling waste batteries: recovery of valuable resources or reutilization as functional materials, ACS Sustain. Chem. Eng. 6 (9) (2018) 11176–11185.

[10] J. Li, G.X. Wang, Z.M. Xu, Generation and detection of metal ions and volatile organic compounds (VOCs) emissions from the pretreatment processes for recycling spent lithium-ion batteries, Waste Manag. 52 (2016) 221–227.

[11] V. Innocenzi, et al., A review of the processes and lab-scale techniques for the treatment of spent rechargeable nimh batteries, J. Power Sources 362 (2017) 202–218.

[12] L.Q. Zhuang, et al., Recovery of valuable metals from $LiNi_{0.5}Co_{0.2}Mn_{0.3}O_2$ cathode materials of spent Li-ion batteries using mild mixed acid as leachant, Waste Manag. 85 (2019) 175–185.

[13] H. Ku, et al., Recycling of spent lithium-ion battery cathode materials by ammoniacal leaching, J. Hazard Mater. 313 (2016) 138–146.

[14] H.H. Nie, et al., $LiCoO_2$: recycling from spent batteries and regeneration with solid state synthesis, Green Chem. 17 (2) (2015) 1276–1280.

[15] M. Lu, et al., The Re-synthesis of $LiCoO_2$ from spent lithium ion batteries separated by vacuum-assisted heat-treating method, Int. J. Electrochem. Sci. 8 (6) (2013) 8201–8209.

[16] E.S. Fan, et al., Sustainable recycling technology for Li-ion batteries and beyond: challenges and future prospects, Chem. Rev. 120 (14) (2020) 7020–7063.

[17] J. Ordonez, E.J. Gago, A. Girard, Processes and technologies for the recycling and recovery of spent lithium-ion batteries, Renew. Sustain. Energy Rev. 60 (2016) 195–205.

[18] X. Wang, G. Gaustad, C.W. Babbitt, Targeting high value metals in lithium-ion battery recycling via shredding and size-based separation, Waste Manag. 51 (2016) 204–213.

[19] Y.Q. He, et al., Recovery of $LiCoO_2$ and graphite from spent lithium-ion batteries by fenton reagent-assisted flotation, J. Clean. Prod. 143 (2017) 319–325.

[20] E. Gratz, et al., A closed loop process for recycling spent lithium ion batteries, J. Power Sources 262 (2014) 255–262.

[21] D.A. Bertuol, et al., Application of spouted bed elutriation in the recycling of lithium ion batteries, J. Power Sources 275 (2015) 627–632.

[22] Y.Q. Wang, et al., Recent progress on the recycling technology of Li-ion batteries, J. Energy Chem. 55 (2021) 391–419.

[23] D.A. Ferreira, et al., Hydrometallurgical separation of aluminium, cobalt, copper and lithium from spent Li-ion batteries, J. Power Sources 187 (1) (2009) 238–246.

[24] J.P. Chen, et al., Environmentally friendly recycling and effective repairing of cathode powders from spent $LiFePO_4$ batteries, Green Chem. 18 (8) (2016) 2500–2506.

[25] L. Yao, Y. Feng, G.X. Xi, A new method for the synthesis of $LiNi_{1/3}Co_{1/3}Mn_{1/3}O_2$ from waste lithium ion batteries, RSC Adv. 5 (55) (2015) 44107–44114.

[26] X. Song, et al., Direct regeneration of cathode materials from spent lithium iron phosphate batteries using a solid phase sintering method, RSC Adv. 7 (8) (2017) 4783–4790.
[27] D.W. Song, et al., Heat treatment of $LiCoO_2$ recovered from cathode scraps with solvent method, J. Power Sources 249 (2014) 137–141.
[28] L. Li, et al., Recovery of metals from spent lithium-ion batteries with organic acids as leaching reagents and environmental assessment, J. Power Sources 233 (2013) 180–189.
[29] L.P. He, et al., Recovery of cathode materials and Al from spent lithium-ion batteries by ultrasonic cleaning, Waste Manag. 46 (2015) 523–528.
[30] X.X. Zhang, et al., Sustainable recycling and regeneration of cathode scraps from industrial production of lithium-ion batteries, ACS Sustain. Chem. Eng. 4 (12) (2016) 7041–7049.
[31] R.J. Zheng, et al., Optimized Li and Fe recovery from spent lithium-ion batteries via a solution-precipitation method, RSC Adv. 6 (49) (2016) 43613–43625.
[32] L. Sun, K.Q. Qiu, Vacuum pyrolysis and hydrometallurgical process for the recovery of valuable metals from spent lithium-ion batteries, J. Hazard Mater. 194 (2011) 378–384.
[33] G. Alvial-Hein, H. Mahandra, A. Ghahreman, Separation and recovery of cobalt and nickel from end of life products via solvent extraction technique: a review, J. Clean. Prod. (2021) 297.
[34] C.W. Liu, et al., Recycling of spent lithium-ion batteries in view of lithium recovery: a critical review, J. Clean. Prod. 228 (2019) 801–813.
[35] R.C. Wang, Y.C. Lin, S.H. Wu, A novel recovery process of metal values from the cathode active materials of the lithium-ion secondary batteries, Hydrometallurgy 99 (3–4) (2009) 194–201.
[36] R. Sattar, et al., Resource recovery of critically-rare metals by hydrometallurgical recycling of spent lithium ion batteries, Separ. Purif. Technol. 209 (2019) 725–733.
[37] C.K. Lee, K.I. Rhee, Reductive leaching of cathodic active materials from lithium ion battery wastes, Hydrometallurgy 68 (1–3) (2003) 5–10.
[38] X.P. Chen, et al., Recovery of valuable metals from waste cathode materials of spent lithium-ion batteries using mild phosphoric acid, J. Hazard Mater. 326 (2017) 77–86.
[39] X.P. Chen, et al., A novel closed-loop process for the simultaneous recovery of valuable metals and iron from a mixed type of spent lithium-ion batteries, Green Chem. 21 (23) (2019) 6342–6352.
[40] Q. Meng, Y.J. Zhang, P. Dong, Use of glucose as reductant to recover Co from spent lithium ions batteries, Waste Manag. 64 (2017) 214–218.
[41] P. Meshram, B.D. Pandey, T.R. Mankhand, Hydrometallurgical processing of spent lithium ion batteries (LIBs) in the presence of a reducing agent with emphasis on kinetics of leaching, Chem. Eng. J. 281 (2015) 418–427.
[42] K. Tanong, et al., Study of the factors influencing the metals solubilisation from a mixture of waste batteries by response surface methodology, Environ. Technol. 38 (24) (2017) 3167–3179.
[43] P. Zhang, et al., Hydrometallurgical process for recovery of metal values from spent lithium-ion secondary batteries, Hydrometallurgy 47 (2) (1998) 259–271.
[44] Y. Guo, et al., Leaching lithium from the anode electrode materials of spent lithium-ion batteries by hydrochloric acid (HCL), Waste Manag. 51 (2016) 227–233.
[45] E.G. Pinna, et al., Cathodes of spent Li-ion batteries: dissolution with phosphoric acid and recovery of lithium and cobalt from leach liquors, Hydrometallurgy 167 (2017) 66–71.
[46] Y. Yang, et al., A closed-loop process for selective metal recovery from spent lithium iron phosphate batteries through mechanochemical activation, ACS Sustain. Chem. Eng. (2017) acssuschemeng.7b01914.
[47] P. Meshram, B.D. Pandey, T.R. Mankhand, Recovery of valuable metals from cathodic active material of spent lithium ion batteries: leaching and kinetic aspects, Waste Manag. 45 (2015) 306–313.
[48] L.-P. He, et al., Leaching process for recovering valuable metals from the $LiNi_{1/3}CO_{1/3}Mn_{1/3}O_2$ cathode of lithium-ion batteries, Waste Manag. 64 (2017) 171–181.
[49] L.-P. He, et al., Recovery of lithium, nickel, cobalt, and manganese from spent lithium-ion batteries using L-tartaric acid as a leachant, ACS Sustain. Chem. Eng. 5 (1) (2017) 714–721.
[50] X. Zeng, J. Li, B. Shen, Novel approach to recover cobalt and lithium from spent lithium-ion battery using oxalic acid, J. Hazard Mater. 295 (2015) 112–118.

[51] X. Zhang, et al., Innovative application of acid leaching to regenerate Li(Ni$_{1/3}$CO$_{1/3}$Mn$_{1/3}$)O$_2$ cathodes from spent lithium-ion batteries, ACS Sustain. Chem. Eng. 6 (5) (2018) 5959–5968.
[52] L. Sun, K. Qiu, Organic oxalate as leachant and precipitant for the recovery of valuable metals from spent lithium-ion batteries, Waste Manag. 32 (8) (2012) 1575–1582.
[53] L. Li, et al., Ascorbic-acid-assisted recovery of cobalt and lithium from spent Li-ion batteries, J. Power Sources 218 (2012) 21–27.
[54] L. Li, et al., Recovery of cobalt and lithium from spent lithium ion batteries using organic citric acid as leachant, J. Hazard Mater. 176 (1) (2010) 288–293.
[55] G.P. Nayaka, et al., Dissolution of cathode active material of spent Li-ion batteries using tartaric acid and ascorbic acid mixture to recover Co, Hydrometallurgy 161 (2016) 54–57.
[56] Y. Yang, et al., Selective recovery of lithium from spent lithium iron phosphate batteries: a sustainable process, Green Chem. 20 (13) (2018) 3121–3133.
[57] X. Chen, T. Zhou, Hydrometallurgical process for the recovery of metal values from spent lithium-ion batteries in citric acid media, Waste Manag. Res. 32 (11) (2014) 1083–1093.
[58] X. Zhang, et al., A closed-loop process for recycling LiNi$_{1/3}$CO$_{1/3}$Mn$_{1/3}$O$_2$ from the cathode scraps of lithium-ion batteries: process optimization and kinetics analysis, Separ. Purif. Technol. 150 (2015) 186–195.
[59] W. Gao, et al., Lithium carbonate recovery from cathode scrap of spent lithium-ion battery: a closed-loop process, Environ. Sci. Technol. 51 (3) (2017) 1662–1669.
[60] J. Jegan Roy, M. Srinivasan, B. Cao, Bioleaching as an eco-friendly approach for metal recovery from spent nmc-based lithium-ion batteries at a high pulp density, ACS Sustain. Chem. Eng. 9 (8) (2021) 3060–3069.
[61] N. Bahaloo-Horeh, S.M. Mousavi, M. Baniasadi, Use of adapted metal tolerant *Aspergillus niger* to enhance bioleaching efficiency of valuable metals from spent lithium-ion mobile phone batteries, J. Clean. Prod. 197 (2018) 1546–1557.
[62] J.J. Roy, B. Cao, S. Madhavi, A review on the recycling of spent lithium-ion batteries (LIBs) by the bioleaching approach, Chemosphere 282 (2021) 130944.
[63] X. Zheng, et al., Spent lithium-ion battery recycling – reductive ammonia leaching of metals from cathode scrap by sodium sulphite, Waste Manag. 60 (2017) 680–688.
[64] Y. Ma, et al., A promising selective recovery process of valuable metals from spent lithium ion batteries via reduction roasting and ammonia leaching, J. Hazard Mater. 402 (2021) 123491.
[65] T. Suzuki, et al., A hydrometallurgical process for the separation of aluminum, cobalt, copper and lithium in acidic sulfate media, Separ. Purif. Technol. 98 (2012) 396–401.
[66] X. Chen, et al., Separation and recovery of metal values from leach liquor of waste lithium nickel cobalt manganese oxide based cathodes, Separ. Purif. Technol. 141 (2015) 76–83.
[67] D. Pant, T. Dolker, Green and facile method for the recovery of spent lithium nickel manganese cobalt oxide (NMC) based lithium ion batteries, Waste Manag. 60 (2017) 689–695.
[68] S. Castillo, et al., Advances in the recovering of spent lithium battery compounds, J. Power Sources 112 (1) (2002) 247–254.
[69] G. Prabaharan, et al., Electrochemical process for electrode material of spent lithium ion batteries, Waste Manag. 68 (2017) 527–533.
[70] S. Chen, et al., Renovation of LiCOO$_2$ with outstanding cycling stability by thermal treatment with Li$_2$CO$_3$ from spent Li-ion batteries, J. Energy Storage 8 (2016) 262–273.
[71] J. Chen, et al., Environmentally friendly recycling and effective repairing of cathode powders from spent LiFePO$_4$ batteries, Green Chem. 18 (2016).
[72] Q. Sa, et al., Synthesis of high performance LiNi$_{1/3}$Mn$_{1/3}$CO$_{1/3}$O$_2$ from lithium ion battery recovery stream, J. Power Sources 282 (2015) 140–145.

CHAPTER 16

Recycling battery metallic materials

Ziwei Zhao[1], Gurleen Kaur Walia[2], Ge Li[1] and Tian Tang[1]
[1]Department of Mechanical Engineering, University of Alberta, Edmonton, AB, Canada; [2]School of Electronics and Electrical Engineering, Lovely Professional University, Phagwara, Punjab, India

1. Commercial batteries and recyclability of metallic materials

Batteries, a connection of multiple electrochemical cells, aim to convert chemical energy into electrical energy owing to the redox reactions occurring inside them and the electron flow between two electrodes via an external circuit. Depending on the form of the electrolyte it contains, a battery can be a "wet cell" or a "dry cell." The wet cell is the original battery design, with liquid electrolyte as the conducting fluid. In 1886, Carl Gassner, a German scientist, introduced the dry cell as a modified form of the Leclanché cell. A completely dry battery does not work at room temperature. The term "dry" here means that the contents of the battery are unspillable irrespective of its orientation because of the immobilized electrolytic paste used. Another common classification is according to batteries' rechargeability, making them primary or secondary.

1.1 Primary batteries

Primary batteries, which utilize their chemicals only once with a single discharge, cannot be recharged. A few commonly used primary batteries are described below.

1.1.1 Zinc–carbon battery

The zinc–carbon (Zn–C) battery is one of the earliest primary cells developed and uses zinc and carbon electrodes. A Zn–C battery has a zinc anode, which is typically a solid can, housing the components of the battery as shown in Fig. 16.1A. Along with 99.9% pure zinc, this container is made of 0.03% cadmium and 0.3% lead for mechanical strength and ductility [1]. A mixture of aqueous powdered NH_4Cl (10%), manganese dioxide (MnO_2) (60%), carbon black (20%), and water (10%) form the cathode material [1]. A carbon rod, which acts as a current collector from the cathode, is placed inside this mixture to provide better conductivity and retain moisture. The carbon rod's porosity allows accumulated gas to escape while preventing electrolyte leakage. Two paper separators, a disc at the bottom of the zinc can and a cylindrical separator inserted into the can, are used to accommodate the positive cathode materials, keeping them from contacting the negative anode. The electrolyte constitutes a paste of zinc chloride for heavy-duty applications or a mixture of zinc chloride ($ZnCl_2$) and ammonium chloride

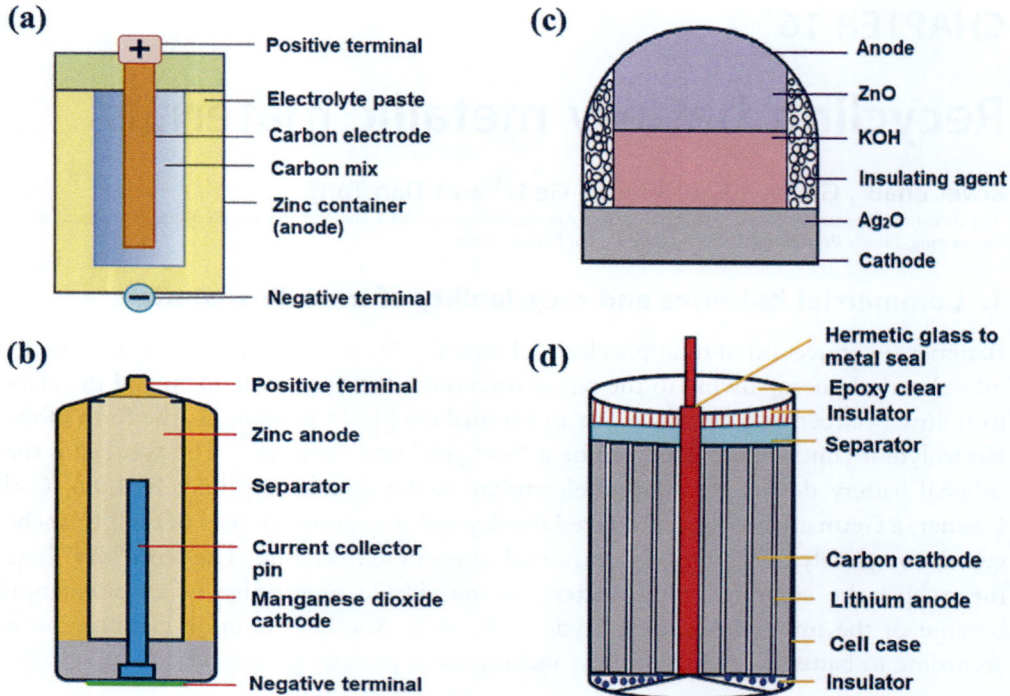

Figure 16.1 Structures of (A) Zinc—carbon battery, (B) Alkaline battery, (C) Zn/Ag$_2$O battery, and (D) Li—SO$_2$ battery.

(NH$_4$Cl) for general-purpose applications. This battery offers a voltage of 1.5 V. Zinc had a recyclability rate of 25% in 2019 [2], with China being the major mining region.

These batteries are generally used in low-drain devices, such as flashlights, remote controls, clocks, battery-operated toys, etc. They usually cannot operate for an extended period (with a shelf life, or how long a battery can be stored without losing its capacity, of 1—2 years) but are cost-effective, practical, and reliable. The Zn—C industry had been rising globally for over a century before 2009. But over the past few decades, Zn—C battery use has been declining worldwide despite its continual growth in Asia, Eastern Europe, and South America [3]. The global market for Zn—C batteries is more than 30 billion units per year [3]. A recent article reported that in 2018, the global Zn—C battery market accounted for USD 1721 million and is estimated to reach USD 1848 million by 2024 [4]. Zn—C batteries are predicted to have a compound annual growth rate (CAGR) of 1.1% over the period 2019—24 [4].

1.1.2 Alkaline zinc/manganese dioxide battery

Although alkaline and Zn—C batteries look alike from the outside, they are completely different from within in terms of material composition. Compared with Zn—C batteries, alkaline batteries have longer shelf and service (the total life in use from sale to discard) lives, higher energy density (measurement of energy in proportion to weight), and lower leakage (the escape of electrolytes or gas from a battery). Hence, they are preferred in toothbrushes, MP3 players, CD players, digital cameras, pagers, radios, toys, and game controllers, to name a few. The first primary alkaline battery was patented by a Canadian pioneer, Lewis Urry, in 1959, and used a zinc anode and MnO_2 cathode, as shown in Fig. 16.1B. Alkaline batteries received their name because of the use of an alkaline electrolyte and potassium hydroxide (KOH), compared with the acidic electrolytes used in the Zn—C battery. The current flows due to the redox reactions between Zn and MnO_2, as the standard redox potentials of Zn and MnO_2 are highly negative (−0.76 V) and highly positive (1.23 V), respectively [5]. The open-circuit voltage of this battery is around 1.6 V. In 2020, Mn was reported to have an insignificant recycling rate [6], with South Africa being the major mining region.

Alkaline batteries account for 80% of the manufactured batteries in the United States, 46% of all primary battery sales in Japan, and over 10 billion individual units produced worldwide [7]. In addition, in Switzerland, the United Kingdom, and the European Union, alkaline batteries account for 68%, 60%, and 47% of all battery sales, respectively [7—11]. In 2005, the production cost of alkaline batteries was USD 5.8 billion worldwide, which increased to USD 7.58 billion in 2020. Production is expected to reach USD 10.86 billion in 2028, with a CAGR of 4.9% in the 2021—28 period [5,12].

1.1.3 Zinc/silver-oxide battery

The zinc/silver-oxide battery uses a powdered amalgamated zinc anode and a silver oxide cathode, as shown in Fig. 16.1C. These flat, circular batteries are generally available in the form of button cells, offering a nominal cell voltage of 1.55 V. The electrolyte used is alkaline, usually sodium hydroxide (NaOH) or KOH. Although these batteries have used a certain amount of mercury to inhibit corrosion, Sony came out with the first mercury-free silver-oxide batteries in 2004 [13]. Zinc/silver-oxide batteries offer numerous advantages, such as long service life, long shelf-life, high capacity per unit weight, high energy per unit volume, and flat discharge characteristics (preferred because this ensures that the voltage remains constant as the battery drains). They are quite useful as miniature power sources and find applications in hearing aids, analog and digital watches, film cameras, medical instruments, small sensors, low-power devices, measuring instruments, calculators, etc. The recycling rate of silver was recorded to be around 8% in 2020, with Mexico being the major mining region. However, sodium and potassium have insignificant recycling rates [6].

In 2020, the sales of zinc/silver-oxide batteries accounted for 30% of all primary batteries in Japan [14]. The growing market of hearing aids and electrical wearables is fueling the growth of the zinc/silver-oxide battery market worldwide. In 2017, the market was valued at USD 15.55 billion and is predicted to record a CAGR of 3% in terms of revenue over the forecast period (2017—2025) to reach USD 19.70 billion by 2025 [15].

1.1.4 Lithium battery

Also known as lithium-metal batteries, lithium (Li) batteries use an anode of metallic lithium and a carbon cathode, as shown in Fig. 16.1D. As lithium is highly reactive in aqueous environments, nonaqueous electrolytes are required, e.g., polar organic liquids [16–18]. There are many variants of lithium batteries available in the market, such as lithium carbon fluoride (Li–CFx), lithium manganese dioxide (Li–MnO_2), lithium iron disulfide (Li–FeS_2), lithium thionyl chloride (Li–$SOCl_2$), and lithium-sulfur dioxide (Li–SO_2), to name a few. They are lightweight and have excellent durability, high cell voltage (3–4 V), high energy density, low self-discharge, wide operating temperature range, superior antileakage properties, and long shelf and service lives. Owing to these properties and the flexibility of shapes in which they can be constructed, these batteries find applications in cameras, pacemakers, defibrillators, and other implantable electronic medical devices. The recycling rate of Li was less than 1% in 2011 [19], and for iron, this rate was insignificant as recorded in 2020 [6]. Australia is the major mining region for both these metals.

Unlike other primary batteries, whose performance is reduced remarkably at low temperatures, Li–SO_2 and Li–$SOCl_2$ batteries can function at temperatures as low as −55°C. In addition, Li–$SOCl_2$ batteries can withstand high temperatures exceeding 150°C [20]. The global market for primary lithium batteries is predicted to reach USD 2850 million in 2024 from USD 2200 million in 2019 for a CAGR of around 5.3% [21].

1.2 Secondary batteries

Secondary batteries, also known as storage cells or charge accumulators, are rechargeable via an external source of electric power. In other words, the electrochemical reactions in rechargeable batteries are reversible. As shown in Fig. 16.2, when the battery discharges, the power is supplied to the load owing to the electrochemical reaction between the active materials in the electrodes and the electrolyte. During the recharging process, the roles of the cathode and anode are interchanged, as the electric current is now forced to flow in the opposite direction, as shown by the dashed line in Fig. 16.2, and the load is replaced by a direct current power source. A few commonly used storage cells are described below.

1.2.1 Lead–acid battery

The first secondary battery, the lead–acid battery, was designed by the French physicist Gaston Planté in 1859 and has been used commercially since 1890 [22]. It consists of a spongy metallic lead anode, lead dioxide (PbO_2) cathode, and an electrolyte of a diluted mixture of aqueous sulfuric acid (H_2SO_4) with a voltage range of 1.8–2.2 V. Lead–acid batteries are shock-resistant, reliable, durable, cheap, and capable of withstanding extreme temperatures [1]. They are commonly used as engine batteries in cars and trucks and provide motive power in forklifts and other construction equipment. During

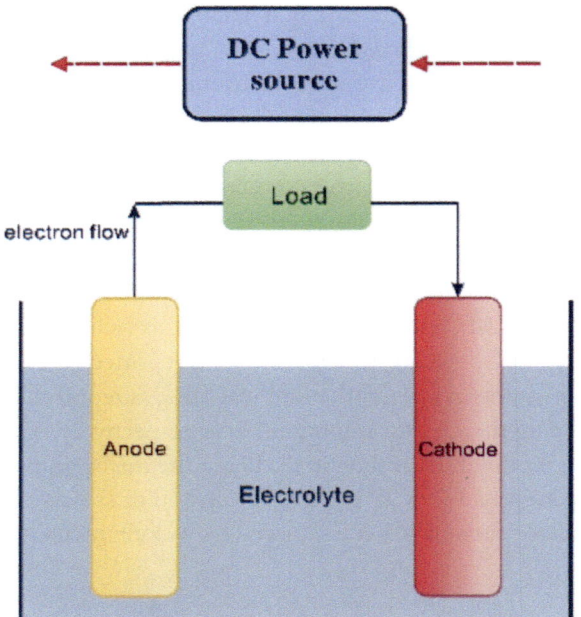

Figure 16.2 Phases of a secondary battery.

discharge, both electrodes become lead sulfate ($PbSO_4$), and sulfuric acid is highly consumed, leaving behind water, whereas in the recharging phase, the lead sulfate is again converted to metallic lead and lead oxide. In contrast to other metals studied, lead has a high recycling rate of 73% as recorded in 2020, with China being the major mining region [6].

Lead—acid batteries suffer from a few drawbacks, such as low life cycle, low power density, low energy density, slow charging times, and high self-discharge rates [23]. These batteries have been in use for over 130 years [24] in numerous applications and are the most extensively used rechargeable batteries to date, exceeding more than 60% of total market sales despite the rapid development of other secondary batteries [25]. In 2020, the market size of these batteries exceeded USD 50 billion, compared with USD 41.6 billion in 2019 [26]. It is predicted that annual installation may surpass one billion units by 2027 [27], reaching a USD 80 billion market in 2027 with a CAGR of 4.5% [27].

1.2.2 Nickel—cadmium battery

The nickel—cadmium (Ni—Cd) battery consists of an anode made from a mixture of cadmium and iron, a nickel-hydroxide ($Ni(OH)_2$) cathode, and an alkaline electrolyte of aqueous KOH. Ni—Cd batteries have an operating voltage of 1.2 V and are used in digital cameras, laptops, calculators, medical devices, space applications, etc. [1]. They perform well in high-discharge and low-temperature applications along with having excellent reliability, low maintenance, and long shelf and service lives [28]. Their power density, energy density, and number of cycles are higher than those of lead—acid batteries [22]. The main issue with the Ni—Cd battery is cadmium, an active material of the anode, which is very toxic, and hence the batteries must be properly sealed so that Cd is not exposed. Additionally, Ni—Cd batteries suffer from the "memory effect," also known as the temporary loss of cell capacity, where the cells tend to discharge to the same level they are used to being repeatedly discharged to, as described in Ref. [3]. To overcome this effect, Ni—Cd batteries should be discharged completely once after a few cycles. Much research was also done to overcome this issue by introducing additives into the cadmium electrode, such as iron, polyvinyl alcohol, $Ni(NO_3)_2$ (0.25—0.1 M), and $FeSO_4$ (0.1—0.4 M), as well as TiO_2 (0.01—0.03 M) and Na_2S (0.01—0.03 M) in $Cd(NO_3)_2$ [29—32]. Ni was reported to have a recycling rate of 50% in 2020 [6] whereas for cadmium, the rate was 15%, as reported in 2011 [19], with Indonesia and China being their major mining regions, respectively. The recycling rate of titanium is insignificant, with China and Australia being the major mining regions for ilmenite and rutile ores, respectively. Ni—Cd batteries dominated the market in secondary batteries until the 1990s and since then have been replaced by nickel—metal hydride and lithium-ion batteries [28].

1.2.3 Nickel—metal hydride battery

Nickel—metal hydride batteries were introduced in 1990 by Sanyo, a Japanese company [33]. They use a nickel hydroxide cathode and an anode of rare-earth metal (such as Ce, La, or Nd) or nickel alloy with many metals (such as V, Ti, Zr, Ni, Cr, Co, or Fe [1]). An alkaline electrolyte of about 30% KOH, having a very high ion conductivity (10^{-3} S/cm), is used [28]. Ni—MH batteries are similar to Ni—Cd batteries in construction, except that Ni—MH batteries have a hydrogen-absorbing negative electrode [22]. Both battery types have a voltage of 1.2 V and hence are often used interchangeably in many applications. Compared with Ni—Cd cells, Ni—MH cells are relatively expensive and have half the service life. But their advantages are larger capacity (30% more), higher power density (25% more) [1], reduced "memory effect," and being more eco-friendly [22]. There is no formation of dendrites, unlike in lithium-ion (Li-ion) batteries, leading to minimal chances of short circuits and overheating [34]. Ni—MH cells have several applications, such as in mobile phones, laptops, camcorders, emergency lights, uninterruptible power sources (UPSs), portable and electric vehicles. In 2020, chromium and cobalt had a recycling rates of 25% and 29% [6], with South Africa and Congo being the major mining regions, respectively. In 2007, worldwide production was over one billion Ni—MH batteries per year [35]. In 2017, Ni—MH cells were reported to account for 85% of the batteries in hybrid electric vehicles [36]. Unfortunately, the future for Ni—MH batteries is dark owing to the growing demand for Li-ion batteries. Global market share is estimated to decline from USD 2120 million in 2019 to USD 1810 million in 2024 with a CAGR of -2.6% [37].

1.2.4 Lithium-ion battery

Because using lithium batteries as rechargeable cells could be a safety hazard due to the presence of lithium metal, they were redesigned to use lithium compounds in place of lithium metal [1]. Li-ion batteries were introduced in 1991 by Sony, a Japanese company [33], although they were first proposed in the 1960s [22]. In the past few decades, Li-ion batteries have gained huge popularity owing to their remarkable characteristics, such as high voltage, low self-discharge, high energy density, almost 100% efficiency, high charge and discharge time constants, and applicability under a wide temperature range [20]. The nominal cell voltage of Li-ion batteries is 3.7 V. The function of these batteries is primarily based on the movement of lithium ions between the cathode and anode. A graphite anode, with a theoretical specific capacity of 372 mAh/g, serves as a substitute for lithium metal in lithium primary batteries [38]. The cathode is made of lithium-containing transition metal oxide, namely, lithium-cobalt oxide ($LiCoO_2$), lithium-nickel-cobalt-manganese oxide ($LiNiCoMnO_2$), lithium-manganese oxide ($LiMn_2O_4$), lithium-nickel oxide ($LiNiO_2$), lithium-nickel-cobalt-aluminium oxide ($LiNiCoAlO_2$), lithium titanate ($Li_4Ti_5O_{12}$), or lithium-iron-phosphate ($LiFePO_4$) [22,28,39,40]. Many more materials are being experimented with to find chemical compounds that offer better performance. A nonaqueous electrolyte of a lithium salt solvated in an organic solvent [28] is used because an aqueous electrolyte would completely ionize under such a high voltage [33]. The redox reactions in these batteries use an intercalation process, which makes them unique compared with other batteries. Li-ion batteries are generally used in notebooks, mobile phones, medical equipment, power backups (UPSs), and electric vehicles. The global market for Li-ion cells is anticipated to reach USD 94.4 billion in 2025 from USD 42.2 billion in 2020 with a CAGR of 16.4% [41].

Table 16.1 provides a comparison of metallic materials typically used in batteries based on their cost, major mining regions, reserves, production, and recyclability rate (the percentage of products that are recycled or reused). Recycling metallic materials prevents them from contributing to landfills and from being exhausted, as reserves are limited. Metals like aluminum, lead, and nickel have a high recyclability rate ($\geq 50\%$), while chromium, cobalt, copper, tin, and zinc have recyclability rates in the range of 20%—40%. Silver, a precious metal, has a very low recyclability rate (8%) because such metals are rarely discarded or disowned. Low recyclability (8%) is also reported for cadmium because the cadmium recycling industry is still in its infancy. Apart from these, some metals are hard to recycle because of modern complex designs and their existence in alloy form, making it increasingly difficult to separate all metals. It may require significant time, cost, and technical maturity to extract, sort, and reuse them. In addition, radioactive or toxic metals, such as lead and mercury, cannot be recycled. Some metals also get contaminated when they come in contact with harmful toxins and it cannot be guaranteed that these chemicals are completely removed even after a thorough cleaning of the metals.

Table 16.1 Comparison of metallic materials used in batteries.[a]

Metals	Cost (USD/metric tons (mt))	Major mining regions	Reserves (1000 mt)	Production (1000 mt)	Recyclability (%)
Aluminum	1962	China, India/Russia, Canada, UAE	75,000,000	65,200	51
Cadmium	2300	China, Korea, Japan/Canada, Kazakhstan, Mexico	Data NA	23	15 (2011) [19]
Chromium	7900	South Africa, Kazakhstan, Turkey, India, Finland	570,000	40,000	25
Cobalt	35,274	Congo, Russia, Australia, Philippines, Cuba	7100	140	29
Copper	6171	Chile, Peru, China, Congo, US	870,000	20,000	38
Iron (ore)	108	Australia, Brazil, China, India, Russia	180,000,000	2,400,000	Insignificant
Lead	1979	China, Australia, US, Peru/Mexico, Russia	88,000	4400	73
Lithium	8000	Australia, Chile, China, Argentina, Zimbabwe	21,000	82	<1 (2011) [19]
Manganese	4.72	South Africa, Australia, Gabon, China, Brazil	1,300,000	18,500	Insignificant
Mercury	75,000 [2] (2019)	China, Tajikistan, Mexico, Argentina, Peru	Data NA	3.7	1-10 (2011) [19]

Table 16.1 Comparison of metallic materials used in batteries.[a]—cont'd

Metals	Cost (USD/metric tons (mt))	Major mining regions	Reserves (1000 mt)	Production (1000 mt)	Recyclability (%)
Nickel	14,000	Indonesia, Philippines, Russia, New Caledonia, Australia	94,000	2500	50
Potash	830	Canada, Russia, Belarus, China, Germany	>3,700,000	43,000	Insignificant
Silicon	2117	China, Russia, Brazil, Norway, US	Data NA	8000	Insignificant
Silver	643,014	Mexico, Peru, China, Russia, Australia/Chile/Poland	500	25	8
Soda ash	154	US, Turkey, Kenya, Botswana, Ethiopia	26,000,000	52,000	Insignificant
Tin	17,416	China, Indonesia, Burma, Peru, Bolivia	4300	270	24
Titanium (Ilmenite)	NA	China, South Africa, Australia, Canada, Mozambique	700,000	7600	Insignificant
Titanium (Rutile)	1200	Australia, Sierra Leone, South Africa, Ukraine, Kenya	740,000	8200	Insignificant
Zinc	2403	China, Australia, Peru, India, US	250,000	12,000	25 (2019) [2]

[a] Data from Ref. [6] in 2020, unless otherwise specified.

2. Technologies for metal recycling

2.1 Conventional recycling techniques

Battery technologies occupy large market shares, such as those of alkaline zinc, Ni–Cd, lead-acid, and Li-ion batteries. Though they have different working principles, the recycling processes are similar. Typically, recycling consists of physical pretreatment and further processes, including pyrometallurgical or hydrometallurgical. Pretreatment is necessary to improve recycling efficiency. It usually includes dismantling, sorting, crushing, grinding, and sieving. After the dismantling and physical separation process (i.e., magnetic and gravitational method), iron shells and plastic parts can easily be separated from the active materials which are subjected to further processing. Significantly, to ensure operation safety, a discharging process is also required in pretreatment [42].

Pyrometallurgical and hydrometallurgical processes are the main methods to recycle valuable metal elements. For the pyrometallurgical technique, the general process is to reduce the metal salt or metal oxide to a metallic state with a reductant in the furnace at a high temperature. Based on their different volatility, metals can be separated at different temperatures [43]. To ensure the purity of the metal, further refinery process, such as electrowinning, is necessary. Besides pure metal, alloys are also common products of the pyrometallurgical method [44]. In a typical hydrometallurgical process, leaching is the first step. The scraps are leached by chemical agents or microorganisms to concentrate the valuable elements in the solution. To further separate the elements from the leachate, precipitation, sol-gel, and solvent extraction are used [45–47].

Generally, pyrometallurgical or thermal processes can be easily applied in large-scale production with relatively small consumption of chemical agents. However, it can consume a large amount of energy, and the selectivity of metal is unsatisfactory [48]. In comparison, the hydrometallurgical process is highly selective and energy-saving, yet it involves the use of many chemical agents. In this part, recycling technologies are explained for the types of batteries with high market shares.

2.1.1 Alkaline and zinc–carbon battery
2.1.1.1 Pyrometallurgical process

The pyrometallurgical process has been widely adopted for alkaline and zinc-carbon batteries. For example, Batrec et al. [49] in Switzerland can process 2000 tons of batteries and produce 780 tons of ferromanganese, 400 tons of zinc alloy, and 3 tons of Hg. According to Belardi et al. [49], at relatively low temperatures, HgO can be decomposed into Hg and oxygen. 400°C is adequate for the volatilization of metallic Hg, which could be further collected by chemical agents such as HCl-KI or via condensation at low temperatures. When the temperature reaches 1200°C for 30 min, 99% of Zinc can be collected by volatilization, and the residue material can be used to produce ferromanganese in further processing. To reduce the temperature during processing, vacuum pyrometallurgical processes can also be applied to these types of batteries, where the volatilization of the metal is enhanced under a low-pressure condition [50].

2.1.1.2 Hydrometallurgical process
Due to the existence of alkaline electrolyte (KOH), neutral leaching is necessary to decrease the usage of acid leaching agents. Many pieces of research have demonstrated the high efficiency of sulfuric acid for Zn leaching. For example, more than 99.0% of Zn can be leached at the temperature of 80°C with 1.1 M sulfuric acid, while only 23.1% of manganese can be transferred into aqueous liquor [43]. Mn_2O_3 and Mn_3O_4 in the battery scraps cannot be leached out, as the produced MnO_2 is insoluble. Reducing tetravalent to the soluble form of divalent is critical to improving the efficiency of manganese leaching. Because of this, a reducing agent is needed before leaching. ZnS, FeS_2, Fe, organic acid, and H_2O_2 have been studied to reduce Mn, and H_2O_2 is considered the most effective one. ZnS and FeS_2 introduce sulfides, which may lead to SO_2 generation. To avoid pollution, the facilities to collect or eliminate SO_2 should be designed, which increases the complexity of the recycling process.

Solvent extraction and chemical precipitation are the most studied separation techniques. Single cation extraction systems, such as D2EHPA, Cyanex 272, and PC88A, cannot effectively separate Zn and Mn. In comparison, the selectivity of neutral extractants or the combination of neutral and cation extractants, such as Cyanex 923, Cyanex 302 15% + D2EHPA 5%, is higher when separating these two elements. Lee et al. reported that a combination of Ionquest 801 30% and TBP 5% can reach an extraction rate of 96% and 1.5% for Zn and Mn respectively, with a separation factor of 1578 [51]. Chemical precipitation separation relies on the change in solubility of Zn and Mn under different pH to achieve high separation efficiency. The coprecipitation effect for these two elements is severe when the pH is in the range of 8–10, but $Zn(OH)_2$ can dissolve again in the form of $Zn(OH)_4^{2-}$ when the pH is above 13. Based on this phenomenon, Chen et al. reported that the total recovery rates of Zn and Mn were about 91% and 94%, respectively [52].

2.1.2 Lead–acid battery
2.1.2.1 Pyrometallurgical process
As shown in Fig. 16.3, Pb paste is sent to a furnace operating at 1100–1300°C to be smelted with a reductant agent, typically C, CO, and Fe. The furnace can be rotary, reverberatory, or blast, and Pb bullion (85–92 wt%) is produced at the end of the smelting process. The slag containing Pb (about 8–10 wt%) and other elements such as Sb, As, and Sn can be used as the raw materials to produce alloy after further refinery [53]. Due to the existence of $PbSO_4$, direct thermal treatment can generate SO_2 emissions. To avoid this, a desulfurization process is necessary. Soda ash is widely used to eliminate sulfur from the Pb Paste [54].

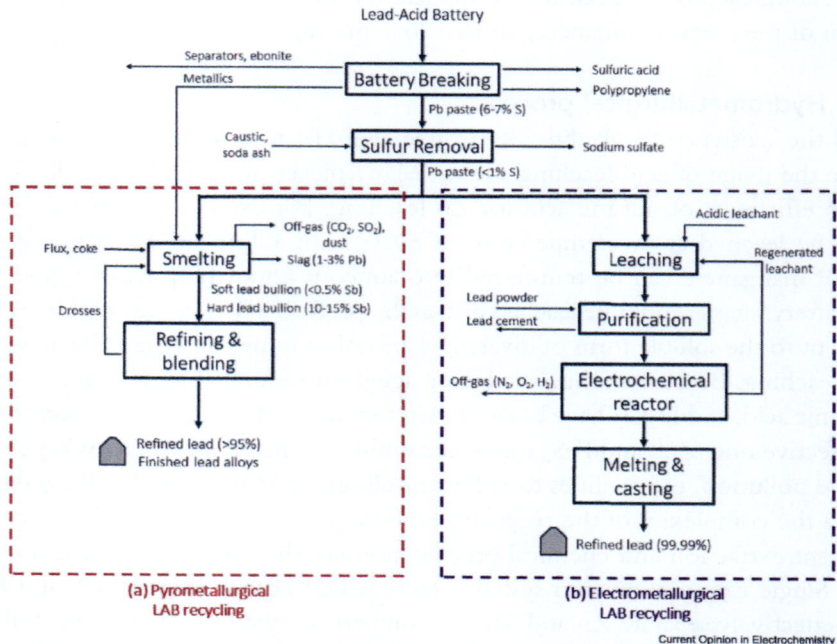

Figure 16.3 General recycling process for lead–acid battery (A) pyrometallurgical (B) electrometallurgical or hydrometallurgical. *(Reproduced with permission from Ref. [53] under license no. 5192671266069, Copyright 2019 Elsevier.)*

Although this technique has already been adopted in large-scale recycling, sulfur-containing and metallic element-containing wastes are still hard to eliminate. For example, when a soda solution is introduced to desulfurize the Pb paste, the lean sulfur elements can still lead to small amounts of SO_2 and wastewater [55]. Complete Pb elimination from the final slag is impossible, too, which leaves landfills as the last resort.

2.1.2.2 Hydrometallurgical process

In the hydrometallurgical process, many chemical agents are used to react with Pb-containing scraps in an aqueous environment, which can effectively eliminate the high-temperature condition and hazardous material emission in the pyrometallurgical process. From the perspective of purity and recycling efficiency, 90%–95% Pb can be recovered with more than 99.9% purity.

A typical hydrometallurgical process has four steps: desulfurization, reduction, leaching, and electrowinning. Usually, Na_2CO_3 and $(NH4)_2CO_3$ are used to realize the desulfurization with the product paste of PbO, PbO_2, and $PbCO_3$ [56]. Since PbO_2 is hard to be dissolved in acid, reducing Pb from +4 to +2 is a critical process to improve the leaching efficiency. Na_2SO_3 and H_2O_2 are the prevalent reduction agents in this process, and HBF_4 and H_2SiF_6 are the typical leaching acids [57,58]. However, the involvement of toxic leaching acid poses a challenge for the design of the recycling plant.

2.1.3 Ni−Cd battery
2.1.3.1 Pyrometallurgical process
Based on the different volatilities of Ni, Cd, Co, and Fe, vaporization of Cd at high temperature is the primary strategy for Ni recycling in large-scale production (Fig. 16.4). CdO can react with reductants (typically coal and hydrogen) followed by vaporization at 900°C with an extraction efficiency of 99.2% [59,60]. In the residue, Ni becomes the main element, and Ni−Co alloy is produced in the further refining process. For example, under the vacuum pressure of 1−3.1 Pa at 1073 K for 2.5 h, the recycling efficiency is more than 99.89% for Ni with more than 99.99% purity [61]. This process has been adopted at the industry level by some companies [62].

2.1.3.2 Hydrometallurgical process
Acid leaching and bioleaching are the main approaches for Ni−Cd leaching. For acid leaching, in H_2SO_4, a typical leaching agent, Cd can be effectively dissolved with a relatively low acid concentration. However, Ni only dissolves in heated sulfuric acid due to the formation of the passivation layer [63]. For example, when treated with 10% sulfuric acid at 298 K for 5 h, 100% Cd and only 20% Ni can be leached. When the temperature is raised to 325 K, the leaching efficiency of Ni in 10% sulfuric acid can be improved to 96% with the addition of H_2O_2 [64]. Bioleaching technique as an alternative method has also been widely investigated. For instance, Bajestani et al. reported that *Acidithiobacillus ferrooxidans* could recover 87% Ni, 67% Cd, and 93.7% Co, respectively [65]. Low emission and few chemical agent involvements are the main advantages of bioleaching. However, culturing the bacteria is costly, which limits the large-scale application in the industry.

Precipitation and solvent extraction are subsequently used for metal separation from the leachate. Based on the different solubility of metal salts, precipitation is feasible to separate these valuable elements with simple requirements on the pH. Xue and coworkers [66] proposed that by changing the pH of the sulfuric acid-leached solution to 4 to 6 with MnO_2, Fe can be precipitated first. Then by adding ammonium sulfate, Ni is precipitated in the form of $(NH_4)_2Ni(SO_4)_2·H_2O$. Finally, $CdCO_3$ can be recycled with ammonium bicarbonate under the pH of 6−6.5. By combining with the electrowinning method, the purity of Cd can be further improved, and the overall recovery rate of Cd is more than 99%. As for solvent extraction, proper organic solvent extractants such as DEHPA, Cyanex 272, TOPS99, and PC88A can effectively separate Ni, Cd, Co under different pH [67]. By introducing the combination of DEHPA and Cyanex 272, the extraction rate can reach 99.7% and 99.5%, respectively, for Cd and Co [68].

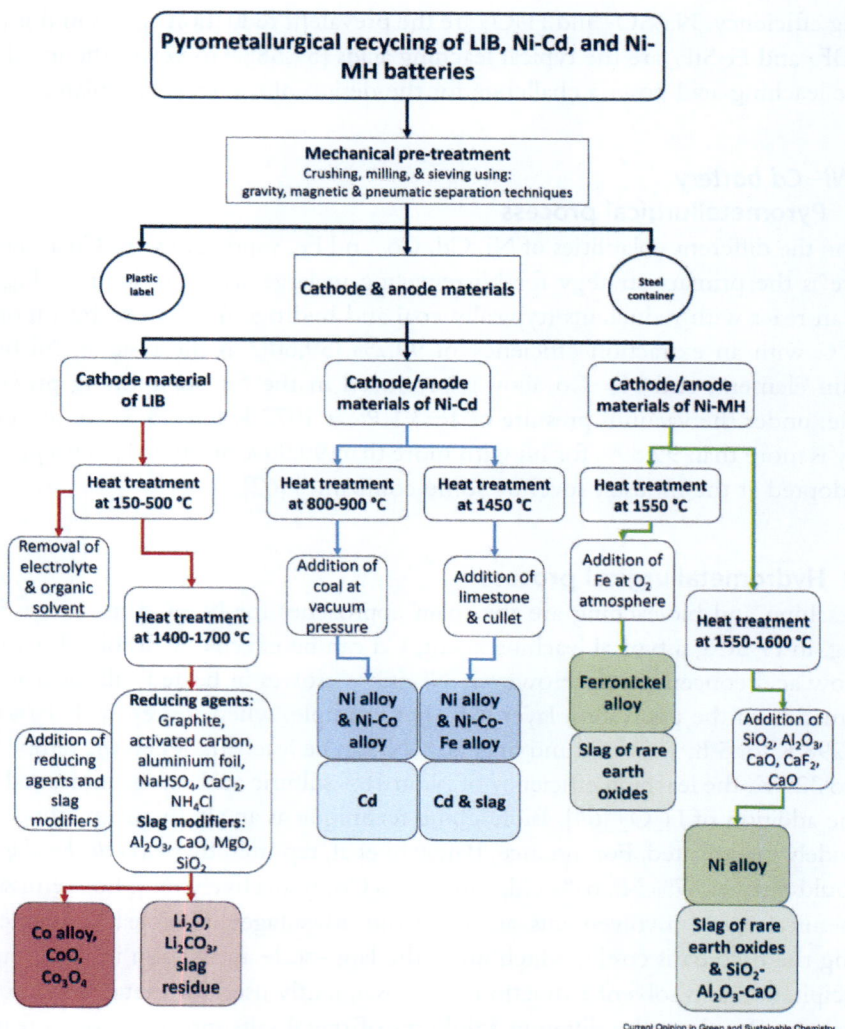

Figure 16.4 General pyrometallurgical processes for Li-ion, Ni–Cd, and Ni–MH battery. *(Reproduced with permission from Ref. [48] under license no. 5192680131142 Copyright 2020 Elsevier.)*

2.1.4 Ni–MH battery
2.1.4.1 Pyrometallurgical process

The pyrometallurgical process for the Ni–MH battery recycling is usually related to alloy production, such as Ni–Co alloy and ferronickel alloy. Therefore, the products are relatively limited compared to the hydrometallurgical process. For example, the scraps can be heated to 1000°C to be oxidized in air and then reduced with Fe at

1550°C to produce alloy. The rare earth elements (REEs) are usually transferred into the slag phase for further refining, including electrowinning and hydrometallurgical processes. During the treatment, fluxing agents, such as $CaO-CaF_2$, $CaO-SiO_2$, and $SiO_2-Al_2O_3$, are usually introduced into the furnace to improve separation efficiency between REE slags and alloys [44]. A subsequent hydrometallurgical process can covert the REE in the slags to REE chloride. Further electrowinning of the smelting REE salt can produce pure REE metals or REE mixture, which can be used in the electronic industry [69].

2.1.4.2 Hydrometallurgical process

An example of the hydrometallurgical process to recover cathode elements in Ni−MH batteries is shown in Fig. 16.5. For both electrodes, the typical leaching agent is acid, including HCl and H_2SO_4. According to Zhang et al., under the condition of 3 M HCl at 95°C for 3 h with the solid: liquid ratio of 1:9, over 96% of nickel, 99% of REEs and 100% of cobalt could be leached [71]. Sulfuric acid cannot reach such a high leaching level, but its cost is much lower than HCl, making it the most popular leaching agent for Ni−MH battery recycling. For instance, 2 M sulfuric acid can recover over 96% REEs, 97% Ni, and 100% Co at 95°C in 4 h [72].

2.1.4.2.1 Cathode recycling Due to the complexity of the metal content of anode material, it is often a good choice to separately recycle the metallic parts of the electrodes of Ni−MH batteries. Nickel and cobalt contents are usually up to 70% in the cathode material. The chemical characters of these two elements are similar since both belong to group VIII, and their radii are close to each other. Solvent extraction and precipitation are widely used in the separation of Ni and Co. Most precipitation techniques are based on the low solubility of Co_2O_3. By adding oxidation agents such as $(NH_4)_2S_2O_8$, Co^{2+} can be converted to Co^{3+} and precipitate in the form of Co_2O_3 by regulating the pH in the range of 6−8 [70]. Ni in the leachate can be further precipitated by dimethylglyoxime. Solvent extraction is more popular than precipitation. Cyanex 272 has been demonstrated as an effective extraction agent by many researchers [73,74]. For instance, at pH 5.7 with 0.6−0.7 M Cyanex 272 in aliphatic kerosene Exxsol D-80, the selectivity factor for Co and Ni can be more than 1000 [75]. To avoid the complicated separation process, regeneration of cathode active material has been reported. Changing the ratio of Ni and Co by adding other metal sources, β-$Ni(OH)_2$ containing Co can be directly used as the electrode material in Ni−MH batteries. However, due to the impurity content in the regenerated electrode, the quality is hard to control and its performance cannot reach the commercial cathode material. These disadvantages have limited the application of the regeneration technique.

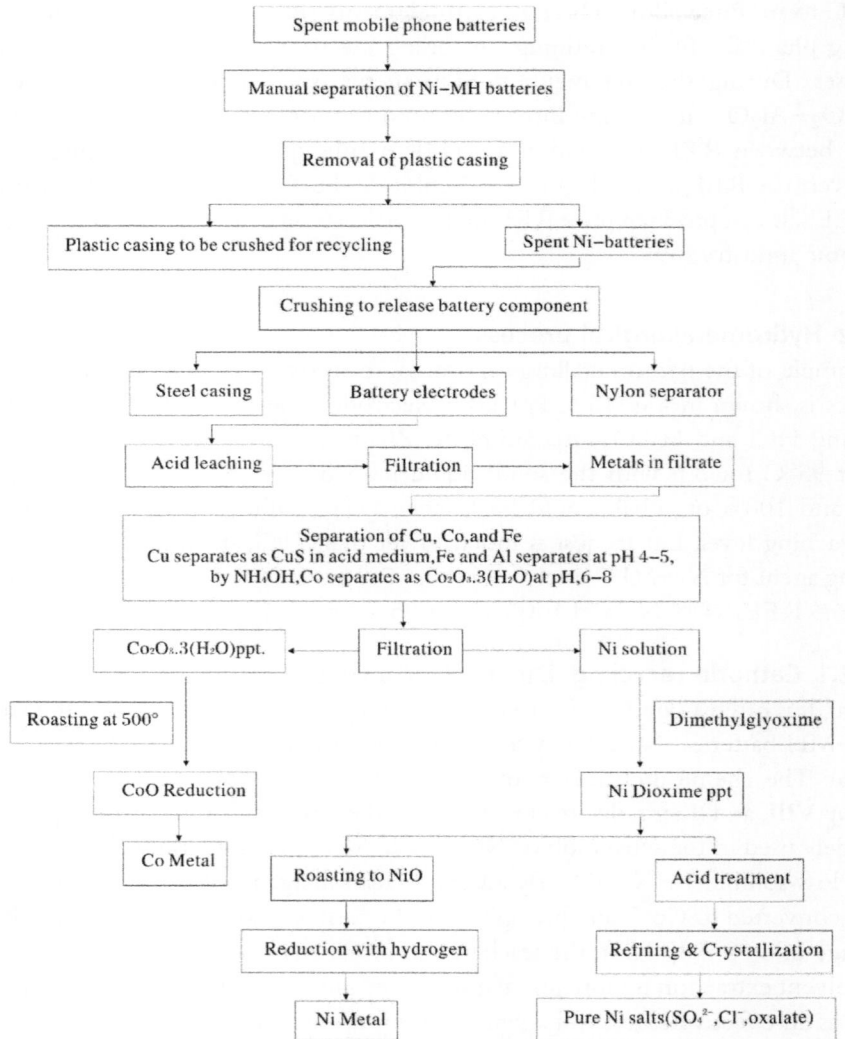

Figure 16.5 A hydrometallurgical process for the recovery of cathode elements (Ni and Co) from Ni—MH battery. *(Reproduced with permission from Ref. [70] under license no. 5192680468973 Copyright 2008 Elsevier.)*

2.1.4.2.2 Anode recycling Compared with the positive electrodes, the element composition of negative electrodes is complex, including REE which is valuable to be recycled. Based on the precipitation of sodium rare earth (RE) double sulfate, REE can be separated from Ni and Co, which can be further recycled using the same techniques described for the positive electrodes. The leaching efficiency is generally in the range of 90%—95% for acid-leached REE. For instance, Pietrelli et al. leached the anode material with 2 M sulfuric acid, and the leaching efficiency was 77% for Ni, 98% for Co, 93% for both La and Ce, 92% for Pr, and 96% for Nd. Afterward, NaOH was added to the leach liquors, and sodium RE double sulfate was precipitated with a recovery ratio of 82%–88% for the REE [76]. However, the quality of the REE precipitation is unsatisfactory. An oxalic acid-assisted technique proposed by Yang et al. [76] can generate high-purity REE products. In this approach, REE oxalates are obtained by adding oxalic acid to the HCl leachate, and the final product can reach 99% purity.

2.1.5 Li-ion battery
2.1.5.1 Pyrometallurgical process

The current collector and cathode are valuable parts for Li-ion battery recycling, but only cathodes containing Co or Ni are worth recycling from the economic stand. Just like the pyrometallurgical processes described for Ni–Cd and Ni–MH batteries, the main product is alloy. To take advantage of the organic materials such as electrolytes and binders, these materials are sent to the furnace to burn, which can provide considerable energy to lower the cost of heating in the next step. The anode material (graphite) and current collector (Al) can serve as reductants to convert active cathode materials, $LiCoO_2$(LCO) or $LiCoNiO_2$(LNO), to metallic state to form an Ni–Co alloy, while Li can be converted to lithium carbonate. During this process, fluxing agents, typically CaO and SiO_2, are usually added to the furnace to form a slag system with aluminum oxide, which can decrease the temperature for the smelting of Co and Ni alloy and concentrate other elements such as Li, Cu, and Mn to the slag.

The above process has been implemented in the industry on a large scale. For instance, Inmetco can process steel waste along with Li-ion batteries as the secondary feed in the same facility, and their main product is nickel-cobalt-iron alloy [77]. The reduction reaction occurs at 1260°C and the product is then sent to the electric arc furnace to be melted. Through the pyrometallurgical process, techniques to recycle lithium, for instance, the EcoBatRec method [78], are also feasible. In this process, active materials are reduced at 1400°C to obtain metallic lithium, and through vacuum distillation or entraining gas evaporation, the final product is lithium oxide. Besides the common shortcomings of pyrometallurgical processes such as high energy consumption, cathodes without Ni and Co such as $LiFePO_4$(LFP) and $LiMn_2O_4$(LMO) are not profitable to be recycled at the industry level.

2.1.5.2 Hydrometallurgical process
2.1.5.2.1 Leaching The main leaching approaches for Li-ion batteries include inorganic acid leaching, organic acid leaching, alkali leaching, and bioleaching. As described for the Ni–MH batteries, to maximize the leaching efficiency of Co, Co^{3+} should be converted into Co^{2+} with a reducing agent. For inorganic acid leaching, sulfuric acid and hydrochloric acid are the most prevalent for LFP, LCO, and $Li(NiCoMn)O_2$(NCM)

batteries. For example, He et al. proposed that by using 1 M H_2SO_4 and 1 vol% H_2O_2 at the temperature of 40°C, the leaching the efficiency of NCM111 (the ratio of Ni, Co, and Mn is 1:1:1 in the cathode material) can reach 99.7% for Li, Ni, Mn and Co [79]. In the HCl leaching system, 100% of valuable elements can be transferred into leachate without a reducing agent. However, the volatility and Cl_2 generation are the main drawbacks for HCl. Besides H_2SO_4 and HCl, other acids can also be used in the leaching process for Li-ion batteries, such as nitric acid and phosphoric acid, which can leach almost 100% Li and 98% Co with the assistance of glucose and H_2O_2 [80].

As for organic acid leaching, citric acid is the most studied owing to its high efficiency, which is more than 95% for Li, Ni, Co, and Mn with H_2O_2 being the reducing agent, as demonstrated by Li and coworkers [81]. In addition to the above leaching technique, bioleaching techniques are also proposed such as *A. ferrooxidans* [65] used to leach Co and Li from LCO. However, the application is limited for its immaturity and the high cost of the bacteria culture.

2.1.5.2.2 Metal separation Solvent extraction, precipitation, and the sol-gel process can be further applied to the leachate to obtain products with high purity. Solvent extraction, similar to what has been explained before, is based on the solubility of metal ions in different organic solvents. D2EHPA, Acorga M5640, Cyanex 272, and PC-88A are the commonly used agents [82–84]. For instance, Nguyen et al. proposed that 99.8% Co is recycled with 60% Na-0.56 M PC-88A, and over 99.0% nickel is recycled by 5.0% (v/v) PC-88A at the pH of 6 [85]. Their process is illustrated in Fig. 16.6.

During precipitation, lithium is always precipitated as $LiCO_3$, which can be used as the precursor for Li-ion battery production. Other valuable elements, such as Ni and Co, are precipitated with OH^-, $C_2O_4^{2-}$, CO_3^{2-}, S^{2-} and PO_4^{3-}. However, coprecipitation can lead to the impurity of the final product. In some research, by regulating the molar ratio of metal elements in the leachate, the precipitate can be used to synthesize new cathode materials directly to circumvent the impurity problem. He et al. [86] used $NiSO_4$, $CoSO_4$, and $MnSO_4$ to regulate the leachate molar ratio and then obtained metal hydroxide by adding aqueous NaOH and a specific amount of $NH_3 \cdot H_2O$ as the precipitation agent. After a series of solid-state reactions with $LiCO_3$ (the lithium source in the regenerated cathode) precipitated by Na_2CO_3, the cathode material (NCM111) can be obtained, whose electrochemical performance is comparable to new cathode material. The sol-gel process is quite similar to the coprecipitation process, and the difference lies in the transferring of the valuable element to the gel state by hydrolyzing [87].

2.1.5.3 Direct recycling process
Besides the conventional pyrometallurgical and hydrometallurgical processes, direct recycling techniques have been developed that are unique for Li-ion batteries. Lithium plating and phase change of the cathode materials lead to capacity losses for used

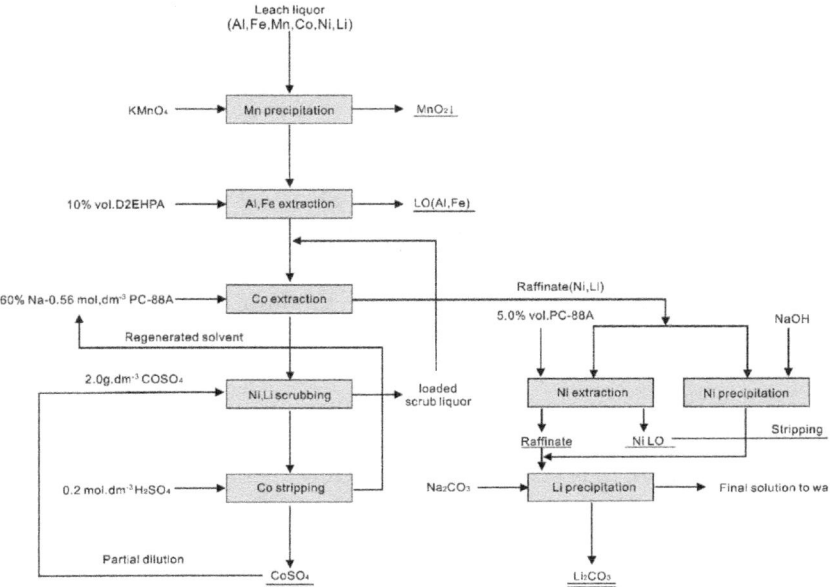

Figure 16.6 PC-88A assisted solvent extraction process with high selectivity for recycling Ni and Co from Li-ion battery. *(Reproduced with permission from Ref. [85] under license no. 5192680718574 Copyright 2014 Springer Nature.)*

Li-ion batteries. Based on relithiation and structural restoration through a hydrothermal process or solid-state sintering, capacity can be recovered without decomposition reactions of the cathode materials. The advantage of direct recycling is apparent: it eliminates many complex processes to regenerate cathode materials and reduces the emission and pollution by using a lower level of chemical agents. For solid-state sintering, cathode materials and lithium source agents are heated to achieve relithiation. Liu et al. reported that LFP cathode can be directly recycled by heating the spent cathode material with new LFP materials, which serve as the lithium source, at 700°C for 8 h. Such a regenerated LFP cathode performs with a capacity of 144 mAh/g, which could be kept at 93.75% after 100 cycles at 0.1C [88]. Sloop et al. proposed a hydrothermal technique in which used cathode materials are harvested with concentrated 4 M LiOH and 30% hydrogen peroxide aqueous solution. The solution was first heated to 125°C and maintained for 30 min to relithiate, followed by heating to 300–900°C to achieve structural restoration. The result showed that healed NCM 523 can be subjected to 2000 charge-discharge cycles at the rate of 0.5C with 80% capacity retention [89]. The direct recycling process also has some disadvantages: (1) regenerated cathode materials cannot perform as well as the newly produced ones; and (2) there is a high requirement for sorting since the different compositions of the cathode materials are hard to be processed simultaneously. Because of these, most direct recycling processes are still limited in the laboratory.

2.2 Recent green recycling techniques

In addition to the recycling methods discussed above, new recycling technologies are urgent and evolving, especially for the Li-ion batteries (LIBs).

2.2.1 Donnan dialysis

A Donnan dialysis (DD)-based technology [90] utilizes cation exchange membranes to separate and retrieve several metals in the course of the hydrometallurgy process for recycling cathodes in LIBs. This method promises to recover a higher amount of Li as compared to the traditional methods. In step 1, lithium is recovered from leachate, whereas the transition metals are recovered in step 2, as shown in Fig. 16.7. Step 3 involves the neutralization precipitation with barium hydroxide and in step 4, lithium bicarbonate is obtained, which is not a stable solid and precipitates as lithium carbonate in step 5. In step 6, the stripping solution is recycled and could be reused in its entirety. In contrast to electrodialysis, the DD technology is highly energy-saving as it does not use an electric current and can be operated continuously without any need for regeneration.

2.2.2 Hierarchical mesosponge γ-Al_2O_3 monolith extraction

This technology along with the optical adsorption process is used to recover cobalt ions (Co^{2+}) from the spent LIBs (SLIBs), which are present with a mass fraction of 5%–15% [91]. Grafting techniques are used in the fabrication of chemical optical extractors for

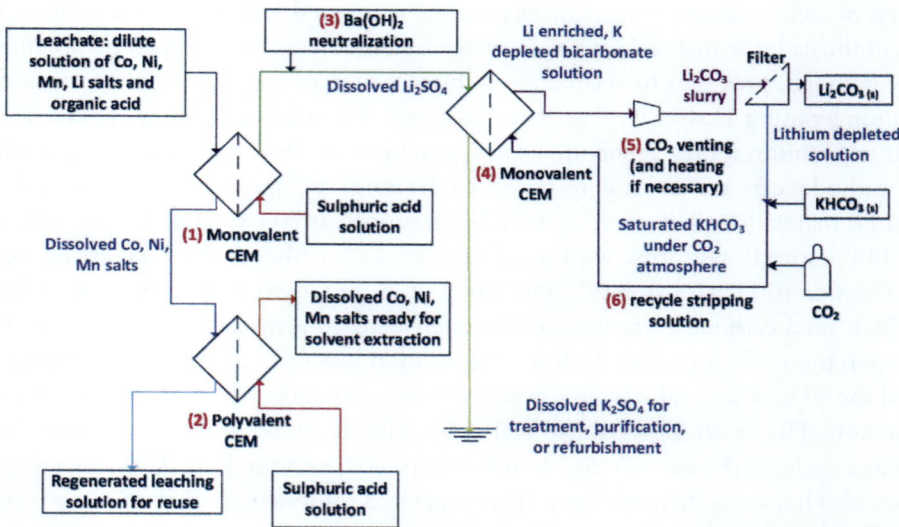

Figure 16.7 A proposed process for recovering lithium carbonate from organic acid LIB cathode powder leachate. *(Reproduced with permission from Ref. [90] under license no. 5103480928524. Copyright 2018 Elsevier.)*

colorimetric and visual extraction, detection, and recovery of cobalt ions at low concentrations. Optical extractors are made of two main components: a solid carrier/substrate/platform material and an organic chromophore that acts as a selective chelating agent (indicator/ligand/probe). A selective binding with the synthesized chelating agents, such as (E)-5-((1,3,4-thiadiazol-2-yl) diazenyl) benzene-1,3-diol (TDDB) and (E)-4-((2-mercaptophenyl)diazenyl)-2-nitrosonaphthalen-1-ol (MPDN), is done to achieve Co^{2+} ion-capture system. The construction of a solid/sponge Co^{2+} ion extractor from SLIB leach fluid is made possible by the dense dressing assembling of TDDB and MPDN into tiny, mesosponge γ-Al_2O_3 monoliths. This technology offers multiple benefits, such as the cost-effective, selective, and sensitive recovery of cobalt ions in a one-step process which in turn, leads to efficient management and exploitation of accumulated e-waste and enhanced environmental protection. The optical adsorption method surpasses the other physical and chemical methods, namely, ion exchange, chemical precipitation, reverse osmosis, and electrochemical treatment, in extracting the cobalt ions from its solutions, owing to its numerous advantages, the most significant one being the ability to observe, with naked eyes, color changes caused by the change in concentration.

2.2.3 Electrochemical cathode reduction method
This method deals with the leaching of cobalt from $LiCoO_2$ cathodes of SLIBs [92]. Various reductants were attempted, such as hydrogen peroxide [93], ascorbic acid [94], sodium hydrogen sulfite, glucose, and sodium sulfite [95]. However, in 2018 Meng et al. came up with glucose as a reductant and achieved 90% leaching efficiency for Co and 94% for Li using 1.25 mol/L of malic acid, a green organic acid, at 70°C for 180 min at a voltage of 8 V [80].

2.2.4 Prerecycling processes
Before chemical processes and purification, several separation methods can be applied, such as magnetic and eddy current separation, screening, gravity separation, flotation, electrolyte separation, etc. The process of shredding offers many advantages, for example, the liberation of the materials, an increase of the surface area, material segregation, and increased efficiency for the rest of the recycling processes. Li, in 2020 [96], introduced an automated semidestructive disassembly system that dismantles and separates the cell case, electrodes, metal tab, and separators from an LIB pouch cell, requiring minimal human intervention. Furthermore, the organic solvent method was used for reclaiming cathode coatings from electrodes. As opposed to the destructive crushing method used widely in industry for ages, this proposed technology has a significant potential for attaining greater coating material recovery rates and yielding purer cathode powder for mechanical pretreatment operations in mass production. Wang et al. [97] proposed a prerecycling process that uses shredding and sorting technologies to effectively separate the target LIB materials into size fractions, allowing recyclers to choose upgrading

technologies and material recovery hierarchies that maximize the amount of mass or value recovered. It was hypothesized that the uncertainty of the input materials would decrease and the overall purity of the output would improve if sorting was done before pretreatment. Five size fractions were obtained from the shredded LIB using sorting sieves of various sizes, which were tested on a Vibration Machine Test System. The XRF (X-ray fluorescence) was then used to analyze the metallic composition of the material that had already been separated into each size fraction. According to the findings, presorting by cathode type has the potential to improve the segregation efficiency of battery waste streams. The effectiveness of the prerecycling system could be improved by implementing a battery labeling system that may provide relevant information such as electrode chemistry, casing materials, manufacturer, etc.

Many new projects have been initiated worldwide to recycle LIB. For instance, the American Battery Technology Company (formerly known as the American Battery Metals Corporation) in Nevada, USA, plans to process 20,000 metric tons of recycled material per year and works on recycling, extraction, and resource production [98]. Battery Resourcers in Massachusetts, USA, an LIB recycling and manufacturing start-up, engineers closed-loop ultra-high-quality cathode materials with 97% metal recovery and produces new materials as per the specifications of their customers [98]. Presently, it offers NMC hydroxide, lithium carbonate (Li_2CO_3), and sintered cathode materials, with single-crystal cathodes and battery-grade graphite under development. There are many other projects—for instance, the Green Li-ion start-up in Singapore, Blue-Green Solution by Rice University, Texas, Lithium-ion Battery Recycling Solution, Fortum's plant in Ikaalinen, Finland, SMCC Recycling by South Korea, ReLIB by England, Primobius by the joint effort of Australia and Germany, Li-Cycle by Canada, and Brunp Recycling Technology Co. by China, to name a few [98].

3. Conclusion and future perspectives

The high demands for portable electronics and electric vehicles have boosted the rapid growth of various batteries. Based on their diverse advantages, different batteries have their own market positions. For example, the major advantages of Lead-acid, Ni–Cd, Ni–MH, and zinc alkaline batteries are their low price and high safety. In contrast, the incomparably high energy density consolidates the status of the lithium-ion battery in the industry field. At the same time, the increasing concerns about environmental problems and raw material consumption provide a great opportunity for the development of the battery recycling industry. Compared to the direct recycling process, the pyrometallurgical and hydrometallurgical recycling processes are relatively mature technologies, and both of them have been used at the industry level. Although battery recycling contributes to protecting the environment, high recycling cost is the main challenge for its broad development at the industrial level. The attraction brought by battery recycling can be overshadowed by the lack of economic advantage compared to battery production, especially when the transportation cost and safety issues associated with used batteries are taken into account.

For the pyrometallurgical methods, huge energy consumption hinders its further application, and better energy control systems are required. For the hydrometallurgical processes, complex chemical processes and massive chemical agent consumption are the major problems. Although the bioleaching technique can reduce the use of chemical agents, the cultivating time and cost for the microorganism limit its industrialization. The promising direct recycling or regeneration technologies are suffering from reduced energy capacity, which is a key disadvantage for their commercialization. As raw metallic materials continue to be mined, their limited supply will likely cause the market price to rise, providing a stronger driving force for their recycling. Further advancements in technology are required to overcome the existing limitations and make the recycling of battery metallic materials safer, more efficient, and more economical.

References

[1] B. Viswanathan, Batteries, in: B. Viswanathan (Ed.), Energy Sources, Elsevier, 2017, pp. 263–313.
[2] U.S. Geological Survey, Mineral Commodity Summaries 2020, US Geological Survey, 2020, Available from: http://doi.org/10.3133/mcs2020.
[3] J.-Y. Huot, Chemistry, electrochemistry, and electrochemical applications | zinc, in: J. Garche (Ed.), Encyclopedia of Electrochemical Power Sources, Elsevier, 2009, pp. 883–892.
[4] K.D. Market Insights, Zinc Carbon Battery Market Size and Forecast, 2021 [cited 2021 July 20]. Available from: Kdmarketinsights.com https://www.kdmarketinsights.com/pr/420/zinc-carbon-battery-market.
[5] T. Takamura, Primary batteries—aqueous systems | alkaline manganese–zinc, in: J. Garche (Ed.), Encyclopedia of Electrochemical Power Sources, Elsevier, 2009, pp. 28–42.
[6] U.S. Geological Survey, Mineral Commodity Summaries 2021, US Geological Survey, 2021, Available from: http://doi.org/10.3133/mcs2021.
[7] E. Olivetti, J. Gregory, R. Kirchain, Life Cycle Impacts of Alkaline Batteries with a Focus on End-Of-Life, 2011 [cited 2021 July 20]. Available from: Epbaeurope.net https://www.epbaeurope.net/techdoc/life-cycle-impacts-alkaline-batteries-focus-end-life/.
[8] Monthly Battery Sales Statistics, 2010 [cited 2021 July 20]. Available from: http://www.baj.or.jp/e/statistics/02.php.
[9] Inobat.ch, Absatzzahlen, 2008, 2008 [cited 2021 July 20]. Available from: http://www.inobat.ch/fileadmin/user_upload/pdf_09/Absatz_Statistik_2008.pdf.
[10] K. Fisher, E. Wallén, P.P. Laenen, M. Collins, Battery Waste Management Life Cycle Assessment Final Report for Publication, 2006 [cited 2021 July 20]. Available from: Epbaeurope.net http://www.epbaeurope.net/090607_2006_Oct.pdf.
[11] EPBA Battery Statistics, European Portable Battery Association, 2000 [cited 2021 July 5]. Available from: Epbaeurope.net http://www.epbaeurope.net/statistics.html.
[12] Alkaline Battery Market Size, Share and COVID-19 Impact Analysis, 2021 [cited 2021 July 20]. Available from: Fortunebusinessinsights.com https://www.fortunebusinessinsights.com/alkaline-battery-market-103298.
[13] World's First Environmentally Friendly Mercury Free Solver Oxide Battery Commercialised by Sony, 2004 [cited 2021 July 20]. Available from: Azom.com https://www.azom.com/article.aspx?ArticleID=2651.

[14] Machinery statistics (Ministry of Economy, Trade and Industry). Baj.or.jp. [cited 2021 July 20]. Available from: https://www.baj.or.jp/statistics/mechanical/index.html.
[15] Silver Oxide Battery Market Analysis, 2018 [cited 2021 July 20]. Available from: Coherentmarketinsights.com https://www.coherentmarketinsights.com/market-insight/silver-oxide-battery-market-1942.
[16] J.T. Nelson, C.F. Green, Organic Electrolyte Battery Systems, 1972 [cited 2021 July 20]. Available from: Dtic.mil https://apps.dtic.mil/sti/pdfs/AD0741786.pdf.
[17] J.O. Besenhard, G. Eichinger, High energy density lithium cells, J. Electroanal. Chem. Interfacial Electrochem. 68 (1) (1976) 1–18.
[18] G. Eichinger, J.O. Besenhard, High energy density lithium cells, J. Electroanal. Chem. Interfacial Electrochem. 72 (1) (1976) 1–31.
[19] United Nations Environment Programme, International Resource Panel, Recycling Rates of Metals: A Status Report, 2011.
[20] G. Pistoia, Applications—portable | portable devices: batteries, in: J. Garche (Ed.), Encyclopedia of Electrochemical Power Sources, Elsevier, 2009, pp. 29–38.
[21] Global Primary Lithium Battery Market Industry Scenario, Key Drivers, Challenges and Trends Forecast 2024, 2021 [cited 2021 July 20]. Available from: Marketwatch.com https://www.marketwatch.com/press-release/global-primary-lithium-battery-market-industry-scenario-key-drivers-challenges-and-trends-forecast-2024-2021-04-19.
[22] S.T. Revankar, Chemical energy storage, in: H. Bindra, S. Revankar (Eds.), Storage and Hybridization of Nuclear Energy, Elsevier, 2019, pp. 177–227.
[23] F. Hussain, M.Z. Rahman, A.N. Sivasengaran, M. Hasanuzzaman, Energy storage technologies, in: Energy for Sustainable Development, Elsevier, 2020, pp. 125–165.
[24] C. Sulzberger, Pearl street in miniature: models of the electric generating station [history], IEEE Comput. Appl. Power 11 (2) (2013) 76–85.
[25] J. Liu, Addressing the grand challenges in energy storage, Adv. Funct. Mater. 23 (8) (2013) 924–928.
[26] Lead Acid Battery Market, 2009 [cited 2021 July 20]. Available from: Marketsandmarkets.com https://www.marketsandmarkets.com/Market-Reports/lead-acid-battery-market-161171997.html.
[27] A. Gupta, N. Paranjape, Lead Acid Battery Market Size by Application (Stationary {telecommunications, UPS, Control and Switchgear,}, Motive, SLI {automobiles, Motorcycles}), by Construction (Flooded, VRLA), by Sales Channel (OEM, Aftermarket), Industry Analysis Report, Regional Outlook, Application Potential, Price Trend, Competitive Market Share & Forecast, 2021 – 2027, Global Market Insights, Inc., 2021. Mar [cited 2021 July 20]. Available from: https://www.gminsights.com/industry-analysis/lead-acid-battery-market.
[28] B. Sundén, Battery technologies, in: B. Sundén (Ed.), Hydrogen, Batteries and Fuel Cells, Elsevier, 2019, pp. 57–79.
[29] S. Tamil Selvan, S. Nathira Begum, V. Chidambaram, R. Sabapathi, K.I. Vasu, Effect of iron addition to the cadmium electrode, J. Power Sources 32 (1) (1990) 55–62.
[30] M.Z.A. Munshi, A.C.C. Tseung, D. Misale, The behaviour of polyvinyl alcohol at the planar Cd/Cd(OH)2 electrode interface, J. Power Sources 23 (4) (1988) 341–350.
[31] M.Z.A. Munshi, A.C.C. Tseung, J. Parker, J.L. Dawson, Effect of an organic additive on the impedance of cadmium in alkaline solution, J. Appl. Electrochem. 15 (5) (1985) 737–744.
[32] G.P. Kalaignan, C. Umaprakatheeswaran, B. Muralidharan, A. Gopalan, T. Vasudevan, Electrochemical behaviour of addition agents impregnated in cadmium hydroxide electrodes for alkaline batteries, J. Power Sources 58 (1) (1996) 29–34.
[33] F. Torabi, P. Ahmadi, Battery technologies, in: P.A. Farschad Torabi (Ed.), Simulation of Battery Systems, Elsevier, 2020, pp. 1–54.
[34] P.-J. Tsais, L.I. Chan, Nickel-based batteries: materials and chemistry, in: Electricity Transmission, Distribution and Storage Systems, Elsevier, 2013, pp. 309–397.
[35] M.A. Fetcenko, S.R. Ovshinsky, B. Reichman, K. Young, C. Fierro, J. Koch, et al., Recent advances in NiMH battery technology, J. Power Sources 165 (2) (2007) 544–551.

[36] Z. Abdin, K.R. Khalilpour, Single and polystorage technologies for renewable-based hybrid energy systems, in: Polygeneration with Polystorage for Chemical and Energy Hubs, Elsevier, 2019, pp. 77—131.

[37] Global Ni-MH Battery Market Growth 2019-2024 - Fior Markets, 2019 [cited 2021 July 20]. Available from: Fiormarkets.com https://www.fiormarkets.com/report/global-ni-mh-battery-market-growth-2019-2024-371189.html.

[38] J.-M. Tarascon, M. Armand, Issues and challenges facing rechargeable lithium batteries, in: Materials for Sustainable Energy, Co-Published with Macmillan Publishers Ltd, UK, 2010, pp. 171—179.

[39] G.M. Koenig Jr., I. Belharouak, H. Deng, Y.-K. Sun, K. Amine, Composition-tailored synthesis of gradient transition metal precursor particles for lithium-ion battery cathode materials, Chem. Mater. 23 (7) (2011) 1954—1963.

[40] J. Xu, G. Chen, H. Zhang, W. Zheng, Y. Li, Electrochemical performance of Zr-doped Li3V2(PO4)3/C composite cathode materials for lithium ion batteries, J. Appl. Electrochem. 45 (2) (2015) 123—130.

[41] Lithium-Ion Battery Market, Marketsandmarkets.com, 2021 [cited 2021 July 20]. Available from: https://www.marketsandmarkets.com/Market-Reports/lithium-ion-battery-market-49714593.html.

[42] G. Dorella, M.B. Mansur, A study of the separation of cobalt from spent Li-ion battery residues, J. Power Sources 170 (1) (2007) 210—215.

[43] I. De Michelis, F. Ferella, E. Karakaya, F. Beolchini, F. Vegliò, Recovery of zinc and manganese from alkaline and zinc-carbon spent batteries, J. Power Sources 172 (2) (2007) 975—983.

[44] T. Müller, B. Friedrich, Development of a recycling process for nickel-metal hydride batteries, J. Power Sources 158 (2) (2006) 1498—1509.

[45] G. Xi, L. Yang, M. Lu, Study on preparation of nanocrystalline ferrites using spent alkaline Zn—Mn batteries, Mater. Lett. 60 (29—30) (2006) 3582—3585.

[46] M. Sadegh Safarzadeh, M.S. Bafghi, D. Moradkhani, M. Ojaghi Ilkhchi, A review on hydrometallurgical extraction and recovery of cadmium from various resources, Miner. Eng. 20 (3) (2007) 211—220.

[47] A. Sobianowska-Turek, W. Szczepaniak, P. Maciejewski, M. Gawlik-Kobylińska, Recovery of zinc and manganese, and other metals (Fe, Cu, Ni, Co, Cd, Cr, Na, K) from Zn-MnO2 and Zn-C waste batteries: hydroxyl and carbonate co-precipitation from solution after reducing acidic leaching with use of oxalic acid, J. Power Sources 325 (2016) 220—228.

[48] M. Assefi, S. Maroufi, Y. Yamauchi, V. Sahajwalla, Pyrometallurgical recycling of Li-ion, Ni—Cd and Ni—MH batteries: a minireview, Curr. Opin. Green Sustainable Chem. 24 (2020) 26—31.

[49] G. Belardi, R. Lavecchia, F. Medici, L. Piga, Thermal treatment for recovery of manganese and zinc from zinc-carbon and alkaline spent batteries, Waste Manag. 32 (10) (2012) 1945—1951.

[50] X. Xiang, F. Xia, L. Zhan, B. Xie, Preparation of zinc nano structured particles from spent zinc manganese batteries by vacuum separation and inert gas condensation, Separ. Purif. Technol. 142 (2015) 227—233.

[51] J.Y. Lee, Y. Pranolo, W. Zhang, C.Y. Cheng, The recovery of zinc and manganese from synthetic spent-battery leach solutions by solvent extraction, Solvent Extr. Ion Exch. 28 (1) (2010) 73—84.

[52] W.-S. Chen, C.-T. Liao, K.-Y. Lin, Recovery zinc and manganese from spent battery powder by hydrometallurgical route, Energy Proc. 107 (2017) 167—174.

[53] S.-Y. Tan, D.J. Payne, J.P. Hallett, G.H. Kelsall, Developments in electrochemical processes for recycling lead—acid batteries, Curr. Opin. Electrochem. 16 (2019) 83—89.

[54] S. Maruthamuthu, T. Dhanibabu, A. Veluchamy, S. Palanichamy, P. Subramanian, N. Palaniswamy, Elecrokinetic separation of sulphate and lead from sludge of spent lead acid battery, J. Hazard Mater. 193 (2011) 188—193.

[55] Z. Sun, H. Cao, X. Zhang, X. Lin, W. Zheng, G. Cao, et al., Spent lead-acid battery recycling in China—a review and sustainable analyses on mass flow of lead, Waste Manag. 64 (2017) 190—201.

[56] M. Olper, M. Maccagni, Pb Battery Recycling New Frontiers in Paste Desulphurization and Lead Production, Lead and Zinc, 2008, pp. 237—246.

[57] Y. Ma, J. Zhang, Y. Huang, J. Cao, A novel process combined with flue-gas desulfurization technology to reduce lead dioxide from spent lead-acid batteries, Hydrometallurgy 178 (2018) 146—150.

[58] R.D. Prengaman, Recovering lead from batteries, JOM (J. Occup. Med.) 47 (1) (1995) 31—33.

[59] D.C.R. Espinosa, J.A.S. Tenório, Recycling of nickel–cadmium batteries using coal as reducing agent, J. Power Sources 157 (1) (2006) 600–604.

[60] D.C.R. Espinosa, J.A.S. Tenório, Fundamental aspects of recycling of nickel–cadmium batteries through vacuum distillation, J. Power Sources 135 (1–2) (2004) 320–326.

[61] K. Huang, J. Li, Z. Xu, Characterization and recycling of cadmium from waste nickel-cadmium batteries, Waste Manag. 30 (11) (2010) 2292–2298.

[62] D.C.R. Espinosa, A.M. Bernardes, J.A.S. Tenório, An overview on the current processes for the recycling of batteries, J. Power Sources 135 (1–2) (2004) 311–319.

[63] T. Du, A. Vijayakumar, K.B. Sundaram, V. Desai, Chemical mechanical polishing of nickel for applications in MEMS devices, Microelectron. Eng. 75 (2) (2004) 234–241.

[64] N.S. Randhawa, K. Gharami, M. Kumar, Leaching kinetics of spent nickel–cadmium battery in sulphuric acid, Hydrometallurgy 165 (2016) 191–198.

[65] M. Ijadi Bajestani, S.M. Mousavi, S.A. Shojaosadati, Bioleaching of heavy metals from spent household batteries using Acidithiobacillus ferrooxidans: statistical evaluation and optimization, Separ. Purif. Technol. 132 (2014) 309–316.

[66] Z. Xue, Z. Hua, N. Yao, S. Chen, Separation and recovery of nickel and cadmium from spent Cd–Ni storage batteries and their process wastes, Separ. Sci. Technol. 27 (2) (1992) 213–221.

[67] E. Vahidi, A. Babakhani, F. Rashchi, A. Zakeri, Modeling of synergistic effect of Cyanex 302 and D2EHPA on separation of nickel and cadmium from sulfate leach liquors of spent Ni-Cd batteries, REWAS (2013) 262–271.

[68] C. Nogueira, F. Delmas, New flowsheet for the recovery of cadmium, cobalt and nickel from spent Ni–Cd batteries by solvent extraction, Hydrometallurgy 52 (3) (1999) 267–287.

[69] Closed-loop recycling of nickel, cobalt and rare earth metals from spent nickel-metal hydride-batteries, in: H. Heegn, B. Friedrich, T. Müller, R. Weyhe (Eds.), Proceedings: XXII International Mineral Processing Congress, 2003. Cape Town, South Africa.

[70] M.A. Rabah, F.E. Farghaly, M.A. Abd-El Motaleb, Recovery of nickel, cobalt and some salts from spent Ni–MH batteries, Waste Manag. 28 (7) (2008) 1159–1167.

[71] P. Zhang, T. Yokoyama, O. Itabashi, Y. Wakui, T.M. Suzuki, K. Inoue, Hydrometallurgical process for recovery of metal values from spent nickel-metal hydride secondary batteries, Hydrometallurgy 50 (1) (1998) 61–75.

[72] V. Innocenzi, N.M. Ippolito, I. De Michelis, M. Prisciandaro, F. Medici, F. Vegliò, A review of the processes and lab-scale techniques for the treatment of spent rechargeable NiMH batteries, J. Power Sources 362 (2017) 202–218.

[73] L. Pietrelli, B. Bellomo, D. Fontana, M. Montereali, Characterization and leaching of NiCd and NiMH spent batteries for the recovery of metals, Waste Manag. 25 (2) (2005) 221–226.

[74] P. Zhang, T. Yokoyama, O. Itabashi, Y. Wakui, T.M. Suzuki, K. Inoue, Recovery of metal values from spent nickel–metal hydride rechargeable batteries, J. Power Sources 77 (2) (1999) 116–122.

[75] L.E.O.C. Rodrigues, M.B. Mansur, Hydrometallurgical separation of rare earth elements, cobalt and nickel from spent nickel–metal–hydride batteries, J. Power Sources 195 (11) (2010) 3735–3741.

[76] X. Yang, J. Zhang, X. Fang, Rare earth element recycling from waste nickel-metal hydride batteries, J. Hazard Mater. 279 (2014) 384–388.

[77] F. Saloojee, J. Lloyd, Lithium Battery Recycling Process, Department of Environmental Affairs Development Bank of South Africa, 2015 (Project No DB-074 (RW1/1016)).

[78] T. Träger, B. Friedrich, R. Weyhe, Recovery concept of value metals from automotive lithium-ion batteries, Chem. Ing. Tech. 87 (11) (2015) 1550–1557.

[79] L.P. He, S.Y. Sun, X.F. Song, J.G. Yu, Leaching process for recovering valuable metals from the LiNi1/3Co1/3Mn1/3O2 cathode of lithium-ion batteries, Waste Manag. 64 (2017) 171–181.

[80] Q. Meng, Y. Zhang, P. Dong, Use of glucose as reductant to recover Co from spent lithium ions batteries, Waste Manag. 64 (2017) 214–218.

[81] L. Li, J. Ge, F. Wu, R. Chen, S. Chen, B. Wu, Recovery of cobalt and lithium from spent lithium ion batteries using organic citric acid as leachant, J. Hazard Mater. 176 (1–3) (2010) 288–293.

[82] J. Nan, D. Han, X. Zuo, Recovery of metal values from spent lithium-ion batteries with chemical deposition and solvent extraction, J. Power Sources 152 (2005) 278–284.

[83] J. Kang, G. Senanayake, J. Sohn, S.M. Shin, Recovery of cobalt sulfate from spent lithium ion batteries by reductive leaching and solvent extraction with Cyanex 272, Hydrometallurgy 100 (3–4) (2010) 168–171.

[84] S.-H. Joo, S.M. Shin, D. Shin, C. Oh, J.-P. Wang, Extractive separation studies of manganese from spent lithium battery leachate using mixture of PC88A and Versatic 10 acid in kerosene, Hydrometallurgy 156 (2015) 136–141.

[85] V.T. Nguyen, J.-C. Lee, J. Jeong, B.-S. Kim, B.D. Pandey, Selective recovery of cobalt, nickel and lithium from sulfate leachate of cathode scrap of Li-ion batteries using liquid-liquid extraction, Met. Mater. Int. 20 (2) (2014) 357–365.

[86] Y. Yang, S. Xu, Y. He, Lithium recycling and cathode material regeneration from acid leach liquor of spent lithium-ion battery via facile co-extraction and co-precipitation processes, Waste Manag. 64 (2017) 219–227.

[87] L. Li, E. Fan, Y. Guan, X. Zhang, Q. Xue, L. Wei, et al., Sustainable recovery of cathode materials from spent lithium-ion batteries using lactic acid leaching system, ACS Sustain. Chem. Eng. 5 (6) (2017) 5224–5233.

[88] X. Song, T. Hu, C. Liang, H.L. Long, L. Zhou, W. Song, et al., Direct regeneration of cathode materials from spent lithium iron phosphate batteries using a solid phase sintering method, RSC Adv. 7 (8) (2017) 4783–4790.

[89] S.E. Sloop, L. Crandon, M. Allen, M.M. Lerner, H. Zhang, W. Sirisaksoontorn, et al., Cathode healing methods for recycling of lithium-ion batteries, Sustainable Mater. Technol. 22 (2019).

[90] A.C. Sonoc, J. Jeswiet, N. Murayama, J. Shibata, A study of the application of Donnan dialysis to the recycling of lithium ion batteries, Hydrometallurgy 175 (2018) 133–143.

[91] H. Gomaa, M. Shenashen, H. Yamaguchi, A. Alamoudi, S. El-Safty, Extraction and recovery of Co 2+ ions from spent lithium-ion batteries using hierarchical mesosponge γ-Al2O3 monolith extractors, Green Chem. 20 (8) (2018) 1841–1857.

[92] Q. Meng, Y. Zhang, P. Dong, Use of electrochemical cathode-reduction method for leaching of cobalt from spent lithium-ion batteries, J. Clean. Prod. 180 (2018) 64–70.

[93] X. Chen, H. Ma, C. Luo, T. Zhou, Recovery of valuable metals from waste cathode materials of spent lithium-ion batteries using mild phosphoric acid, J. Hazard Mater. 326 (2017) 77–86.

[94] G. Nayaka, K. Pai, G. Santhosh, J. Manjanna, Dissolution of cathode active material of spent Li-ion batteries using tartaric acid and ascorbic acid mixture to recover Co, Hydrometallurgy 161 (2016) 54–57.

[95] X. Zheng, W. Gao, X. Zhang, M. He, X. Lin, H. Cao, et al., Spent lithium-ion battery recycling– Reductive ammonia leaching of metals from cathode scrap by sodium sulphite, Waste Manag. 60 (2017) 680–688.

[96] L. Li, The Material Separation Process for Recycling End-Of-Life Li-Ion Batteries, Virginia Tech, 2020.

[97] X. Wang, G. Gaustad, C.W. Babbitt, Targeting high value metals in lithium-ion battery recycling via shredding and size-based separation, Waste Manag. 51 (2016) 204–213.

[98] J. Kumagai, Lithium-Ion Battery Recycling Finally Takes off in North America and Europe, Ieee.org, 2021 [cited 2021 July 20]. Available from: https://spectrum.ieee.org/energy/batteries-storage/lithiumion-battery-recycling-finally-takes-off-in-north-america-and-europe.

CHAPTER 17

Recycling battery casing materials

Tony Lyon[1], Malena T.L. Staudacher[1], Thomas Mütze[2] and Urs A. Peuker[1]
[1]Institute of Mechanical Process Engineering and Mineral Processing, TU Bergakademie Freiberg, Freiberg, Germany;
[2]Helmholtz Institute Freiberg for Resource Technology, Helmholtz-Zentrum Dresden-Rossendorf, Freiberg, Germany

1. Introduction

Lithium-ion batteries (LIBs) have become an indispensable part of modern life. Whether in tools, mobile phones, scooters, or electric cars, the demand for LIBs is increasing in many areas of everyday life. Various scenarios predict an almost exponential growth in the demand for electrochemically stored energy and thus also in the demand for key elements for LIBs such as lithium, cobalt, and nickel [1,2]. Many of the current recycling processes are limited to recovering valuable metals such as cobalt and nickel due to economic viability [3]. Base metals such as aluminum are usually not recovered and are thus lost to the raw material cycle.

An essential component of LIBs is the cell housing. It protects against the ingress of water and the leakage of electrolytes. Both cases would not only render the battery inoperable but also entail safety risks. Depending on the type of battery, the battery housing differs in shape and material. Round cells, which are often found in consumer electronics, usually have a steel casing. Prismatic and pouch cells, as often used in the automotive industry, usually have a housing made of an aluminum alloy. Depending on the design of the battery, the mass share of the housing in relation to the total cell weight can be up to 24% [4]. For this reason, housing should not be neglected and must rather be considered as an important secondary source of raw materials for metals such as aluminum and steel. Furthermore, housing components can make an important contribution to achieving current and future recycling rates.

2. Target

Pyrometallurgical recycling processes, which use the complete LIB cell as input, transfer the base metals such as aluminum, lithium, or manganese to the slag phase [5]. This is usually only used later as a building material [3]. Applying mechanical processing upstream, a large part of the cell housing can be separated and is thus not lost in the pyrometallurgical process. The mechanically separated metal fraction can in turn be fed into existing recycling processes, such as for steel or aluminum [3,5].

However, a high scrap quality is also required for these processes. For example, almost all other metals can dissolve in an aluminum melt, but these can only be removed with difficulty or not at all due to the high oxygen affinity of aluminum [6]. The separated metals can either be sold at a low profit to specialized recyclers or can be further concentrated directly to increase the potential profit. The aim is therefore not only to separate the components of the cell housings from the other components through mechanical processing but also to discharge all metals in pure concentrates.

3. Structure of cell housings

The external appearance of LIB cells can be divided into three main categories, prismatic, cylindrical, and pouch cells (see Fig. 17.1).

The housing of LIB cells is always metal-based, as a pure plastic housing cannot prevent the ingress of water or the outward diffusion of organic solvents to 100% [7]. In most cases, stainless steel or aluminum is used for the housing. For the thin-walled pouch cells, highly refined aluminum composite foils are also used, which consist of several layers of plastic and aluminum [7]. The wall thickness of prismatic cell housings is typically between 0.5 and 1.5 mm [8].

The housing parts are deep-drawn or extruded and the cell covers are attached by laser or ultrasonic welding. This process is comparatively complex and requires many intermediate steps and machines. Therefore, changes in the format of the cells are difficult to realize. This in turn has the advantage that a high number of cells with low tolerances and a robust housing allows automated handling [9].

The electrical connection points (battery poles) are also integrated into the cell housing. Their task is to contact the electrodes inside the cell. To do this, they must conduct the electrical current through the housing to the outside, but must not endanger the hermetic seal of the cell and, like the housing itself, must be inert to the electrolyte inside the cell. Especially in the case of the negative pole of prismatic cells made of aluminum, these can consist of a material composite of copper, aluminum, plastic, and in some cases steel, to fulfill all requirements. Fig. 17.2 shows the light microscope images of microstructural

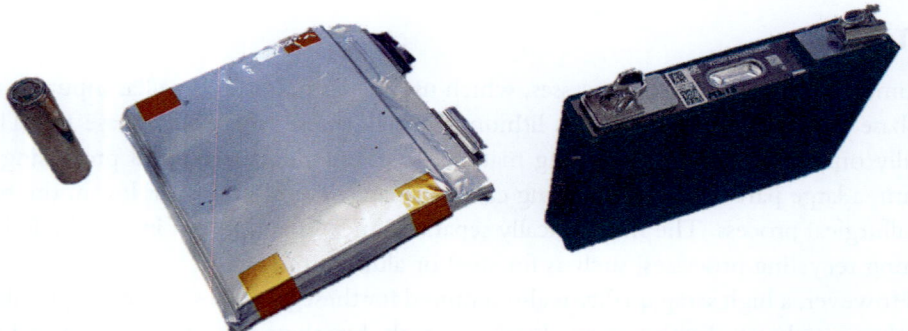

Figure 17.1 Types of lithium-ion battery cells: Cylindrical (*left*), pouch cell (*middle*), prismatic (*right*).

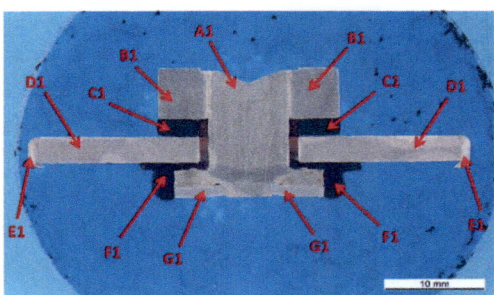 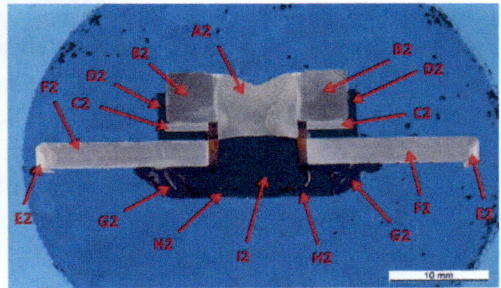

Figure 17.2 Light microscope image of a microstructure section through the positive pole (*left*) and the negative pole (*right*).

Table 17.1 Materials list for battery poles.

Symbol	Material
A1, A2, B1, B2, D1, E1, E2, F2, G1	Aluminum
C1, C2, D2, F1, G2	Plastics
H2, I2	Copper

sections of a positive and a negative pole of a prismatic battery cell from e-mobility. The materials used are listed in Table 17.1. Various aluminum alloys can be used.

Furthermore, for electrical insulation among them, cells are usually covered with a nonconductive thin plastic foil. In some cells, additional components made of plastic or metal serve to contact and fix the electrode windings inside the cell. Normally, these are directly connected to the housing and are similar to it in structure and material thickness, which is why they can be counted as part of it.

4. Housing fraction and case disassembly

Battery systems, especially for automotive applications, are usually too large to be processed in their entirety regardless of the recycling method.

Depending on the process and type of battery, the systems are dismantled down to the module or cell level. Currently, this process is carried out manually, but for the increased mass flow of end-of-life (EoL) batteries expected in the future, this process must be automated. Research is currently being conducted into options for automating disassembly as well as alternative design options for the modules [10,11]. With the number of EoL batteries expected in the future, automation is necessary to achieve the required throughput. This in turn requires uniform system designs [12].

When the separated cells are available, the case can be separated from the rest in one of two ways. One method is manual disassembly of the battery cell itself. For this, the housing must be opened mechanically at a suitable point and the electrode windings and electrolyte are removed and separated. However, this method requires a substantial amount of manual labor. For an industrial application, this process step must be automated [13]. Prototypes for this step have been developed for pouch cells [14].

The second possibility is the mechanical processing of LIB cells. In this process, a combination of different comminution and sorting processes can produce various material concentrates that can be purified in a subsequent, for example, metallurgical, step. The process of mechanical processing is explained in more detail in the following chapter.

5. Generating the housing fraction by mechanical processing

Mechanical processing aims to break down all material compounds through suitable comminution processes to archive the liberation of material compounds and connections. The liberated materials can then be sorted according to various physical property differences, the so-called separation characteristics.

When processing LIBs, various safety precautions must first be taken against the thermal and chemical hazards. In most cases, this involves discharging all cells to a 0% state of charge. In addition, the machines of the first comminution stage are encapsulated and filled with an inert atmosphere (for example, nitrogen) [15].

Gellner et al. [4] describe a process in which the cells are first comminuted and liberated in a high-speed uniaxial rotor gap shear. During this process, mainly shearing forces act on the material. In that study, the cells are precrushed in a two-axis rotor shear before they enter the liberation step. On an industrial scale, a one-step crushing is the state of the technology using a one-shaft or multishaft shredder with sieve baskets. Subsequently, the active material is delaminated off the electrodes and separated by classification, e.g., sieving. The screen overflow is then separated into a plastic concentrate, an electrode foil mixture, and the casing material by airflow sorting according to the separation characteristic of the quasi-stationary settling velocity e.g., in a zigzag air classifier. A similar procedure was also developed in the LithoRec project [15,16]. The main difference here lies in the sequence of the processes. For example, in the LithoRec process, the casing is separated directly by airflow sorting after liberation. Figs. 17.3 and 17.4 show the flowcharts of the two processes. In the case of cells with a ferromagnetic housing, this can also be separated directly from the remaining comminution products via a magnetic separator [18]. The Duesenfeld process is also based on a very similar combination of preparation steps [5].

For the precrushing of prismatic LIB cells in a double-shaft rotor shear, Wuschke [8] found a linear relationship between the specific mechanical stress energy, $w_{B,mech}$, and the tensile strength of the housing material, R_m.

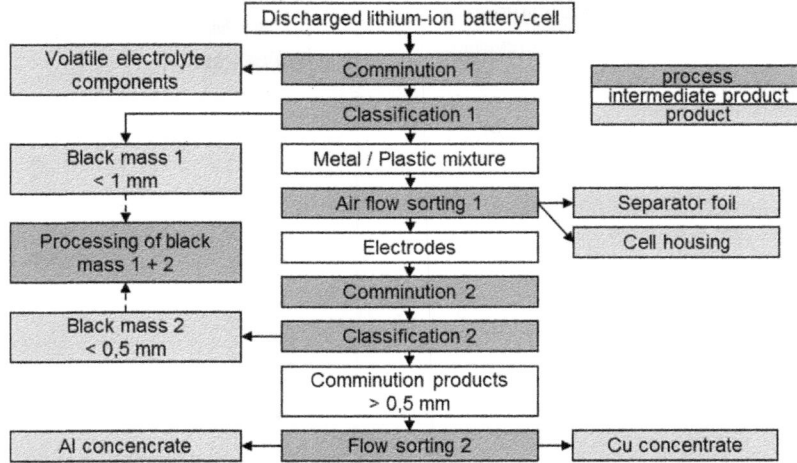

Figure 17.3 Process flowchart according to Wuschke et al. [17].

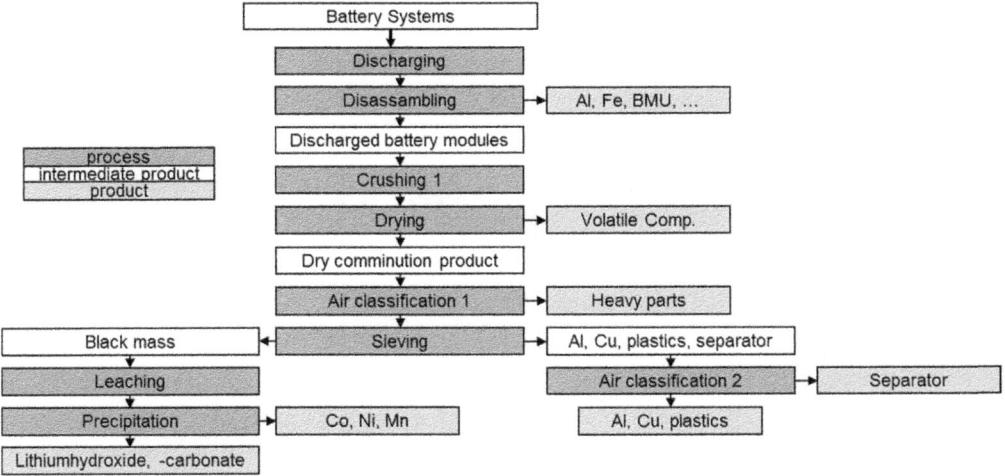

Figure 17.4 LithoRec process flowchart according to [15].

From this correlation, it can be seen that cells with steel housings (examples A and B, Fig. 17.5) specifically require more energy to be crushed than cells with aluminum housings (examples C and D). This can be used for an initial estimation of the energy required for preshredding, e.g., for dimensioning a plant.

Furthermore, Wuschke [8] showed a correlation between the specific mechanical stress energy and the resulting piece size. Thus, the comminution of LIB also follows the trend known from the fundamental works of Rumpf [19].

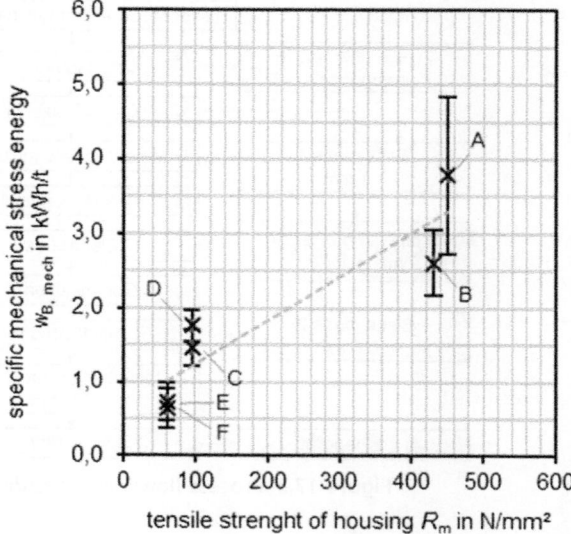

Figure 17.5 Estimation of the required comminution energy according to [8].

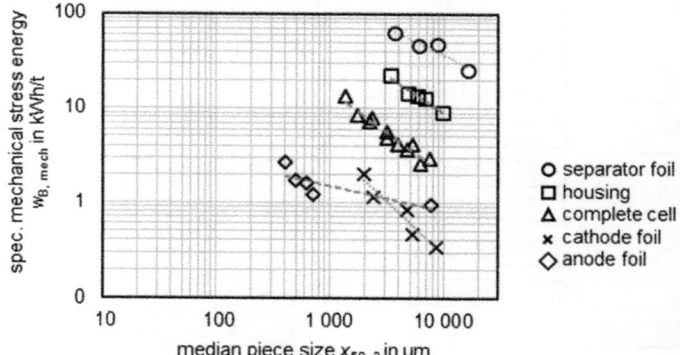

Figure 17.6 Relationship between energy and piece size according to [8].

Based on the graph shown in Fig. 17.6, it can also be concluded that the specific mechanical stress energy to be applied to liberate the components of the battery cell does not only depend on the tensile strength of the housing but rather represents the sum of the comminution energies from all components building up the cell. This means that the relationship derived from Fig. 17.5 cannot be used for the design of the liberation comminution alone. Here, the comminution result is causally related to all components building up the cell.

After crushing and separating the electrolyte, the already delaminated black mass is separated by classification according to the scheme shown in Fig. 17.3. The housing fraction is then produced in a two-stage air-classification process.

Fig. 17.7 shows the settling velocity distribution of different types of crushed automotive LIB cells for separation in a zigzag air classifier. In this binary separation step, all components except for the housing were combined into one class. It shows for various prismatic LIB cells the possibility to separate a large part of the housing in the range around an 8 m/s inflow velocity while discharging as few other components as possible into this fraction. This confirms that the mechanical process routes presented to this point can be used for different cells.

The housing fraction produced consists of the housing of the cells/modules. In addition, parts such as screws, cell connectors, current conductor clips, and residual electrode foils can be found in this fraction. Table 17.2 shows the composition of the housing fraction from different cells, processed according to the flowchart shown in Fig. 17.3. The aluminum content of the housing fraction for the three cell examples is between 61% and 77%. However, aluminum melts have the property that other metals such as copper can dissolve in them [6]. Due to the simultaneously high oxygen affinity of the aluminum, it is then no longer possible to refine it. This means that the actual housing fraction is not a useable product, yet. By separating it at higher incident flow velocities at around 12 m/s, the impurities of electrodes and plastic could be reduced. However, as shown in Fig. 17.7, this would also result in the loss of between 25% and 70% of

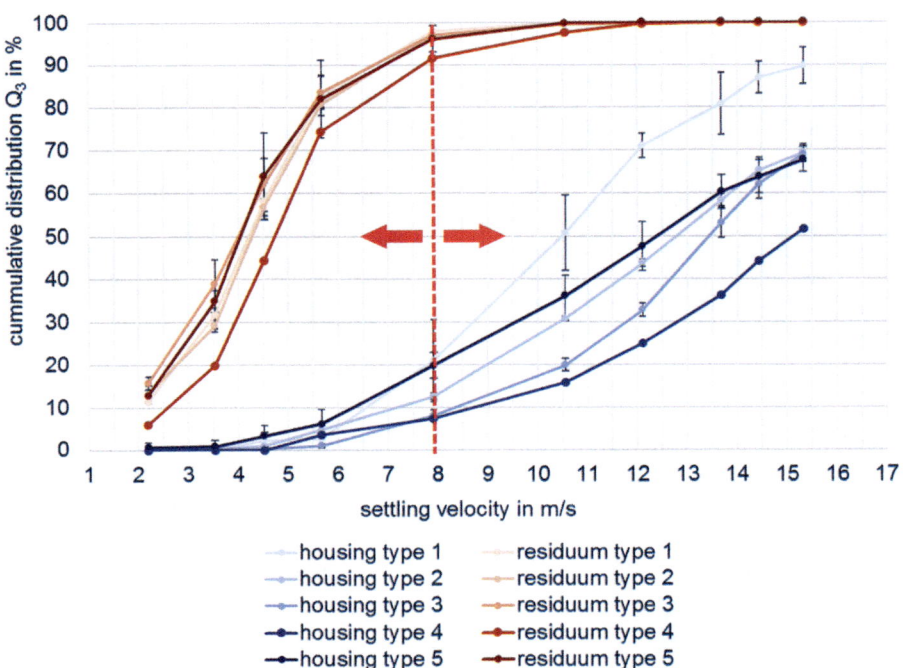

Figure 17.7 Settling velocity distribution of different cells after the same comminution.

Table 17.2 Composition of housing products.

Content c in %	Typ A	Typ B	Typ C
Aluminum	77.1	63.2	60.7
Copper	6.5	21.0	9.8
Steel	3.4	2.8	1.6
Plastics	5.8	7.3	25.1
Cathode foil	3.9	0.7	0.4
Anode foil	2.9	4.1	2.2
Black mass	0.4	0.4	0.0
Residual compounds	0.0	0.5	0.2

the mass of the housing. This would simultaneously increase the copper and steel content in the remaining aluminum fraction, as the pieces of these metals usually have a higher settling velocity than aluminum.

Cell types A, B, and C are prismatic cells with an aluminum housing, a mass between 665 and 1850 g, and a volume of 303—936 cm^3. So they differ in their mass, size, and housing thickness. The housing thickness is 0.8 mm for Type A, 1 mm for Type B, and between 0.9 and 1.4 mm for Type C depending on whether it is measured at the bottom, side, or top of the cell. The four main components within the housing fraction are shown in Fig. 17.8 using the LIB of type A. The pictures make it clear that the

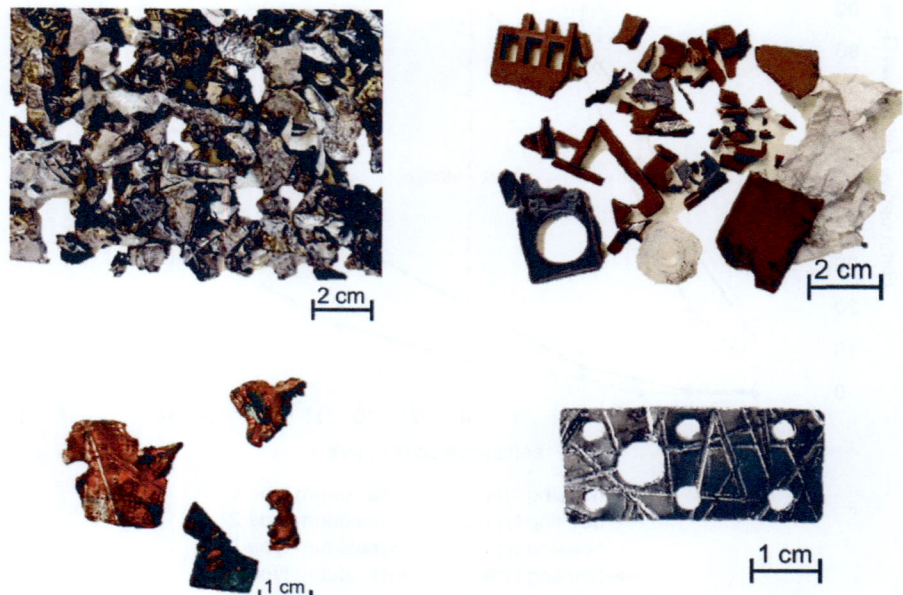

Figure 17.8 Housing fraction components: Aluminum (*top left*), Plastic (*top right*), Copper (*bottom left*), Steel (*bottom right*).

particles have a very heterogeneous particle shape. Furthermore, it can be seen that different plastics are used in the LIB.

For the reasons mentioned, further purification of the produced fraction is necessary to return the aluminum and other materials used in the cell housings to the raw material cycle. The possibilities of achieving this goal are discussed in the next chapter.

6. Processing the housing fraction

6.1 Liberation comminution

A basic prerequisite for sorting the present housing fraction into its individual components is their complete liberation. The degree of liberation in Eq. (17.1) is defined by the quotient of the mass of the particles that are fully liberated m_{free} and the total mass of the fraction/cell m_{total}.

$$A = \frac{m_{free}}{m_{total}} \cdot 100\% \tag{17.1}$$

Fig. 17.9 shows the degree of liberation of the crushed housing with all its components as a function of the specific mechanical stress energy introduced into the cell at the liberation comminution. The values shown in yellow are from Wuschke [8] and refer to the digestion of the main component electrodes, the housing, and the separator foil. The other data points show examples of the housing fractions of various cells concerning

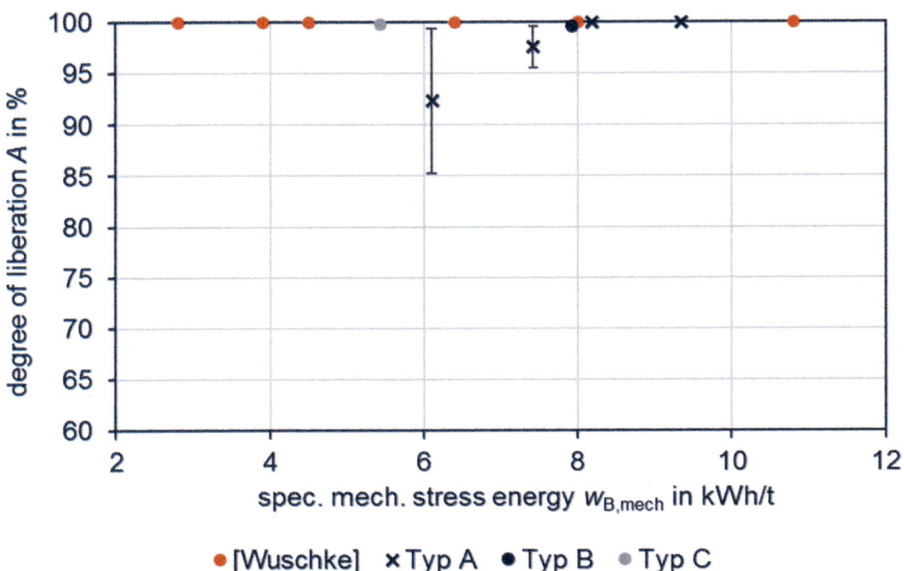

Figure 17.9 Degree of liberation—reference Wuschke [8].

metals and coarser plastic parts. By comparing the data sets, two important points can be highlighted. On the one hand, different cells nevertheless behave differently under the same crushing conditions. On the other hand, liberation is strongly determined by the reference system. For example, the separation of the housing from the electrode and separator foils can be chosen as the target. If only the housing is considered, it consists of many material composites. For example, the cells are usually covered with a thin plastic film. If this combination is considered strictly as a composite, the housing can be separated from the electrodes but would still have a degree of liberation of 0%.

Nevertheless, the data show that the pieces of the housing fraction have a complete liberation of the essential components at mechanical stress energy from about 8 kWh/t for the whole cell for different cell types.

6.2 Eddy current separation

One possibility to further process the housing fraction and to enrich different material concentrates is using an eddy current separator. This separator is often used in the processing of secondary raw materials. The particles are sorted into conductors and nonconductors.

A rotating magnet system induces eddy currents in the electrically conductive particles. This leads to a field in the opposite direction to the inducing magnetic field, resulting in a repulsive force. In addition to the particle shape and size, the conductivity/density ratio of the particles is decisive for the repulsive effect. Fig. 17.10 shows a schematic drawing of an eddy current separator. The example shown can also separate ferromagnetic

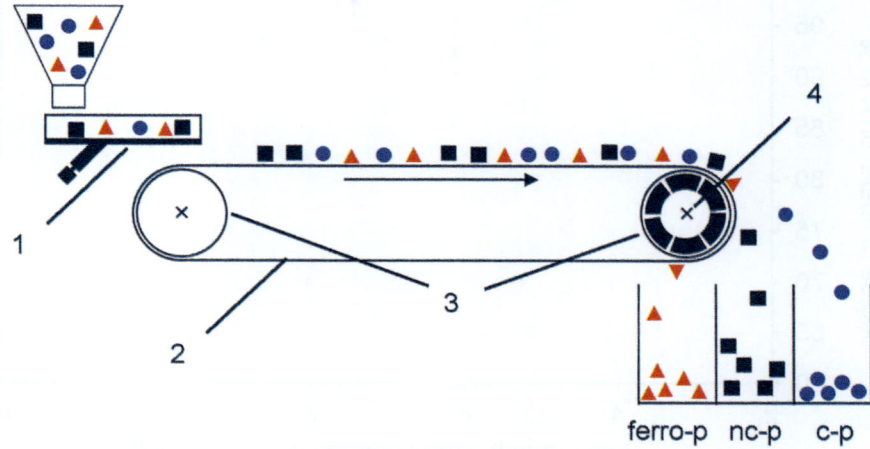

1 - vibrating feeder, 2 - conveyor, 3 - belt drum, 4 - centric pole wheel
ferro-p — ferromagnetic product, nc-p — non-conductive product, c-p — conductive product

Figure 17.10 Schematic of eddy current separator.

materials. The only danger is that these particles heat up too much and burn through the belt. For this reason, eddy current separation should always be preceded by magnetic separation.

$$B = \frac{A_T}{A_{T,\text{ideal}}} \quad (17.2)$$

To evaluate and compare sorting results, the ratio of separation B is often calculated. For this purpose, the mass yield of a sorting process is shown in a diagram on the ordinate axis above the recovery of the valuable material in a product concentrated on the abscissa axis. Now the enclosed area of the graph of the real sorting in the diagram is determined by integration and put in relation to the enclosed area of the graph of the ideal sorting (cf. Eq. 17.2). It can accept values up to 1, which corresponds to an ideal separation of valuable material and tailings.

Fig. 17.11 shows the ratio of separation B over different specific mechanical stress energies for the products iron (ferromagnetic material) and plastic (nonconductive material) out of the eddy current separation. To improve the sorting results, the housing fraction has been mechanically stressed again before. By choosing a hammer mill and thus an impact stress, the particles should be compacted and thus their shape is unified and made more spherical. The principle of the deformation of ductile particles is shown in Fig. 17.12. With increasing stress, the particles become more compacted [20].

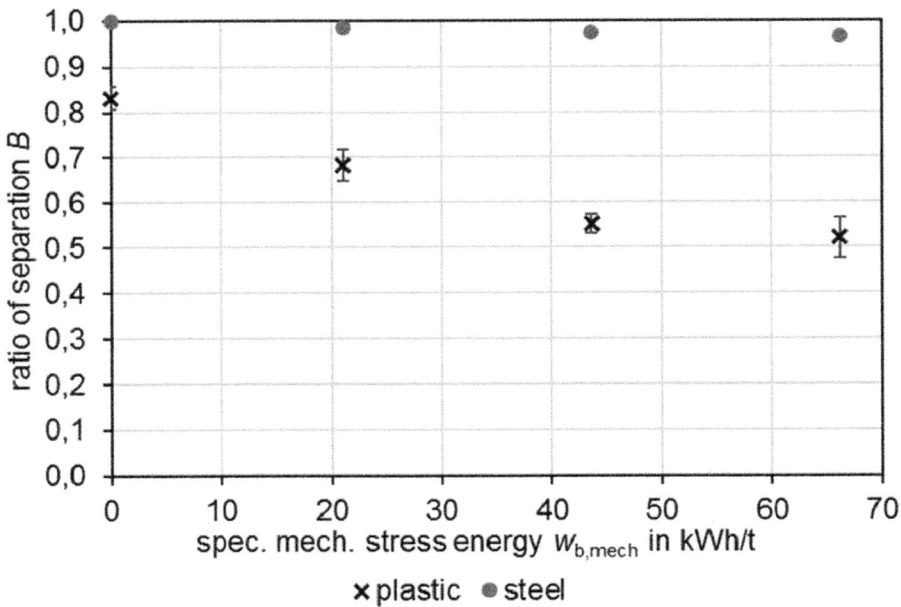

Figure 17.11 Ratio of separation of eddy current separator.

Figure 17.12 Deformation of ductile particles through mechanical stress.

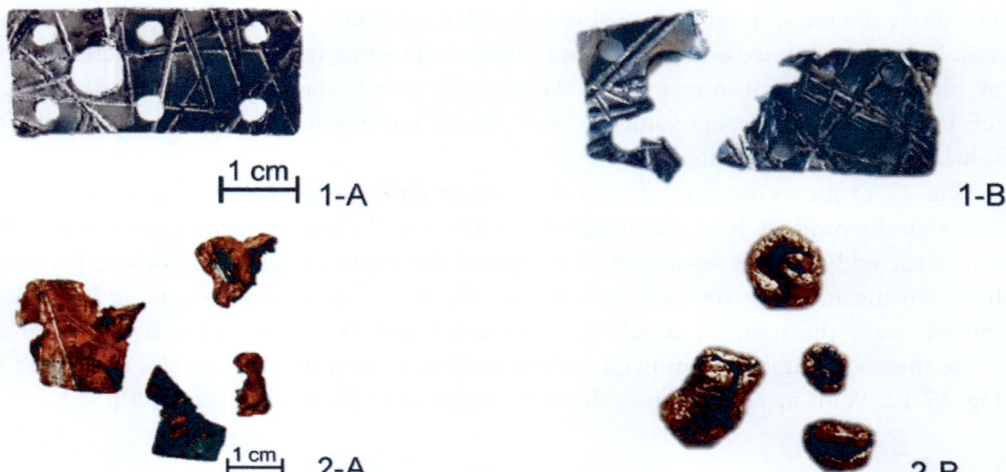

Figure 17.13 Examples for deformation through impact stress steel before (1-A), steel after (1-B), copper before (2-A), copper after (2-B).

The results show that the ferromagnetic steel can be separated almost completely and in a single type, regardless of the stress energy. Nevertheless, the results show a downward trend in separation success with increasing stress. The values for the separation success of the plastic fraction show a clear decrease with increasing stress.

The results can be explained by the fact that steel deforms comparatively little because of impact stress [21] and thus the sorting properties change only insignificantly. This can be seen very well in Fig. 17.13. Picture 1-A and 1-B show the steel before and after the treatment. The piece was cut once but not deformed. On the other hand, examples 2-A and 2-B show the clear deformation of the copper pieces.

The next goal is to separate the plastic to increase the quality of the recyclable concentrate, consisting of aluminum and copper. As far as possible, no valuable material should be lost in the plastic product. This means that the higher the ratio of separation B for plastic, the better the recyclable metal concentrate. The deterioration of the sorting results of the plastic fraction is not directly related to the stress on the plastic parts. It is more due to the crushing and reshaping of the aluminum pieces. Compared to larger pieces, smaller pieces also form smaller eddy currents and therefore have a less intensive repulsion. In contrast, platy/flat pieces result in greater throwing distances [22]. The conclusion is, that compacting the particles by additional mechanical stress has no disadvantages for the separation of the ferromagnetic components, but more aluminum is discharged into the plastic fraction and thus represents a loss of valuable material. Expressed in figures, this means that in the unstressed sample, around 6 wt.% of the solid aluminum and copper are discharged into the plastic fraction, which together with the metal foils corresponds to a metal content of 50 wt.% in this fraction. After stressing with about 65 kWh/t, about 30 wt.% of the aluminum and 65 wt.% of the copper are discharged into the fraction, which together with the metal foils, which are discharged with 75 wt.% (cathode) and 90 wt.% (anode) respectively, corresponds to a metal content of over wt.70%.

6.3 Selective crushing and sieve classification

Another possibility of processing is the selective comminution and subsequent classification of the comminuted product. In this process, the different materials differ in their properties and can be comminuted to different degrees under the same stress. As already described in the previous chapter, the housing fraction can, for example, be stressed by a hammer mill. However, most of the metallic particles are not ground but the particle size decreases due to compaction, change of particle shape respectively. Fig. 17.14 shows the particle size distribution of the individual components before and after an impact stress with a specific mechanical energy of approximately 21 kWh/t.

It can be seen that especially the electrode foil residues remaining in the housing fraction can significantly be reduced in size due to compaction, since those foils have a low wall thickness (green graphs in Fig. 17.14). This means that these could be separated by sieve classification after stressing. Fig. 17.15 shows the ratio of separation B as a function of the stress energy for this sieving step. For the unstressed sample, B is very small at only 0.07 and thus shows no possibility of enrichment through sieving. After stressing the sample with a specific energy of approximately 21 kWh/t, B is highest for this example with 0.84. At higher energy levels, B decreases again. It can be assumed that the actual optimum can already be achieved with a less intensive load. This step makes it possible to remove the electrode foils from the actual housing fraction with the main components aluminum, steel, copper and plastic. Although the electrode foils themselves are made of aluminum and copper as well, they could probably not be processed together with the other aluminum and copper components due to their different handling properties caused by the low wall thickness. Furthermore, the active materials remaining on the foils, which could also contain the metals lithium, cobalt, manganese and nickel in

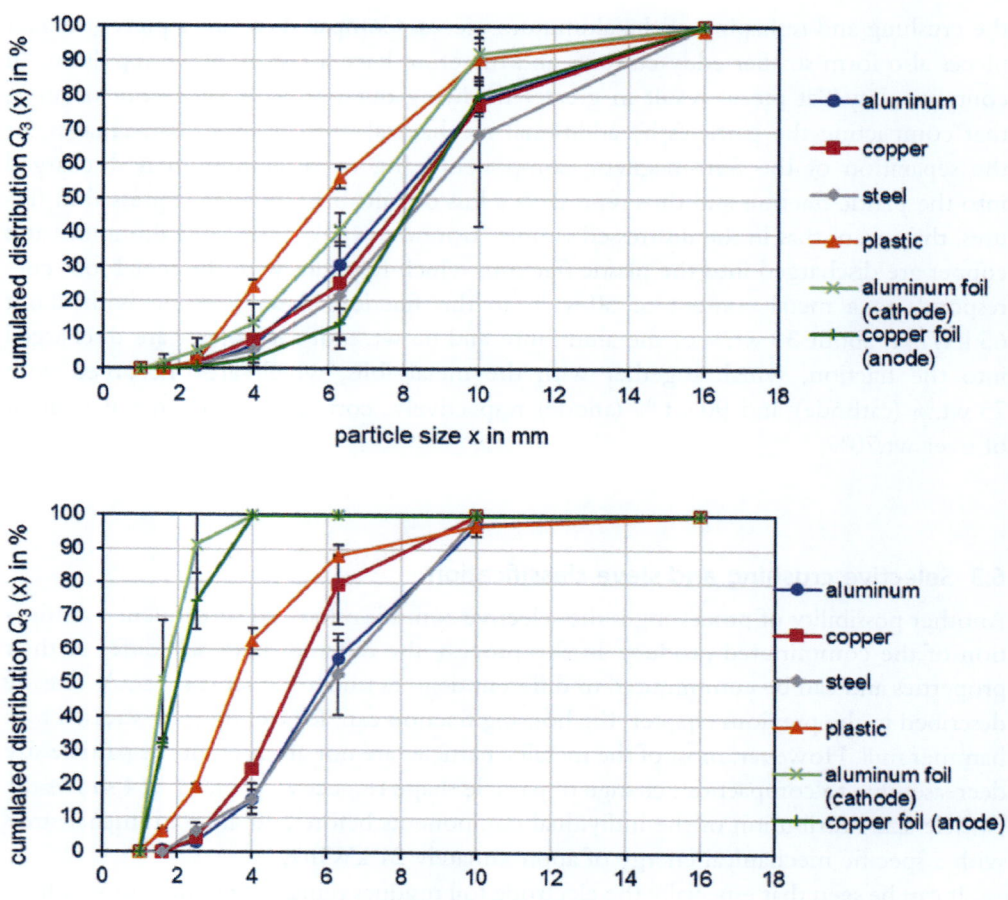

Figure 17.14 Particle size distribution of the components before comminution (*top*), after comminution (*below*) determined by sieving.

addition to graphite (depending on the battery type), can be separated. In the end, concentrates with fewer impurities are produced.

It can be said that by applying minimal stress, selective crushing of the electrode foils is possible, allowing them to be separated by sieving and thus producing a cleaner housing concentrate.

6.4 Air classification

A very common sorting device used in recycling is an upstream air classifier, e.g., a zigzag air classifier. It separates particles according to their quasi-stationary settling velocity. The

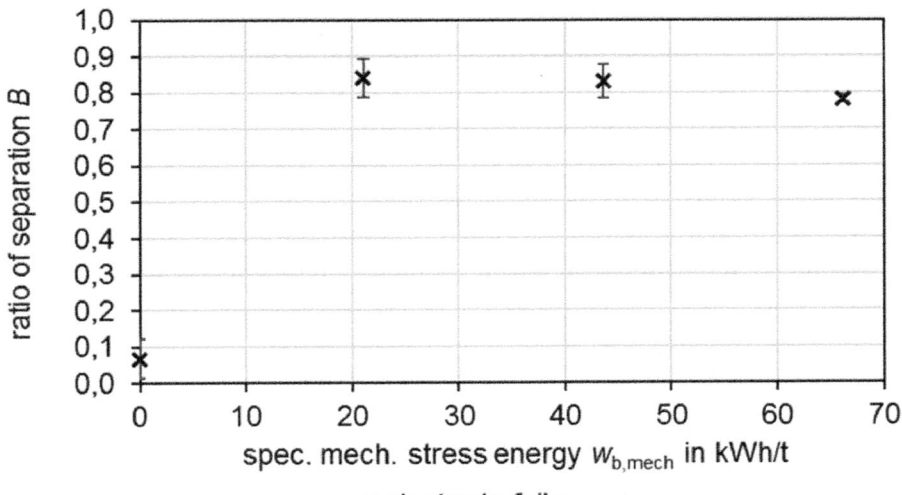
x electrode foils

Figure 17.15 Ratio of separation for sieve classification.

shape, size, thickness, and density of the particles influence the settling velocity. According to Böhme [23], besides density, thickness is the greatest influencing factor for platy particles.

The upper diagram from Fig. 17.16 shows the settling velocity distribution of the components of the housing fraction, determined by the experimental procedure. According to the results, it is possible to separate a large part of the electrode foils by airflow sorting without losing much of the actual housing fraction. It is therefore also possible to separate impurities from the fraction using this method. Furthermore, the fraction could be separated into two products by separation at approximately 16 m/s. The first one contains aluminum and plastic and the second one mainly steel and copper, but both products still have a certain high level of contamination. For this sorting method, too, it is possible to improve the sorting properties by applying a selected stress to the particles prior to the classification.

Eq. (17.3) is the quasi-stationary settling velocity v_m for nonspherical particles according to Böhme [23]. In the equation, ρ_s is the density of the solids, and ρ_f is the density of the fluid. The last term is the mass of the particle m divided by the surface area A, which is exposed to the airflow. This surface is usually the largest cross-sectional area of the particle in a stable position:

$$v_m = \sqrt{2g \cdot \frac{\rho_s - \rho_f}{\rho_s \rho_f} \cdot \frac{1}{c_{wT}} \cdot \frac{m}{A}} \quad (17.3)$$

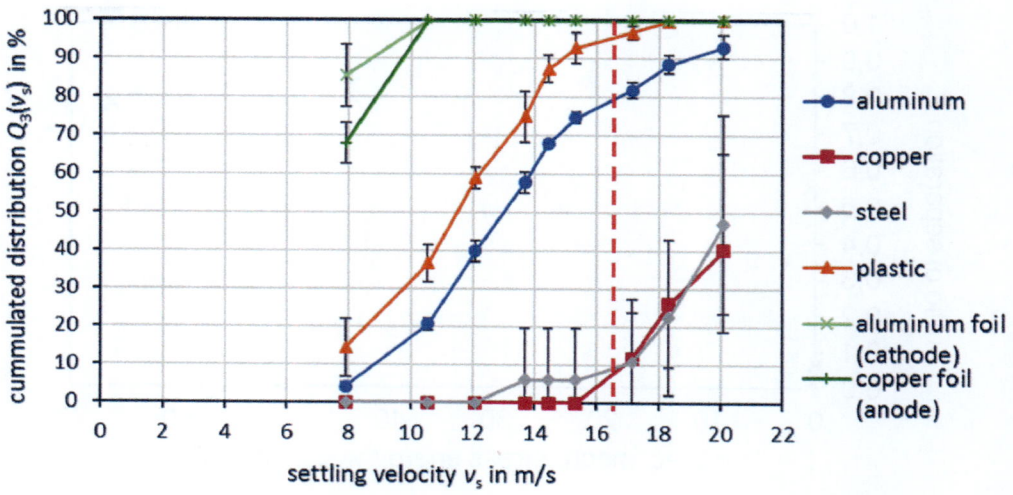

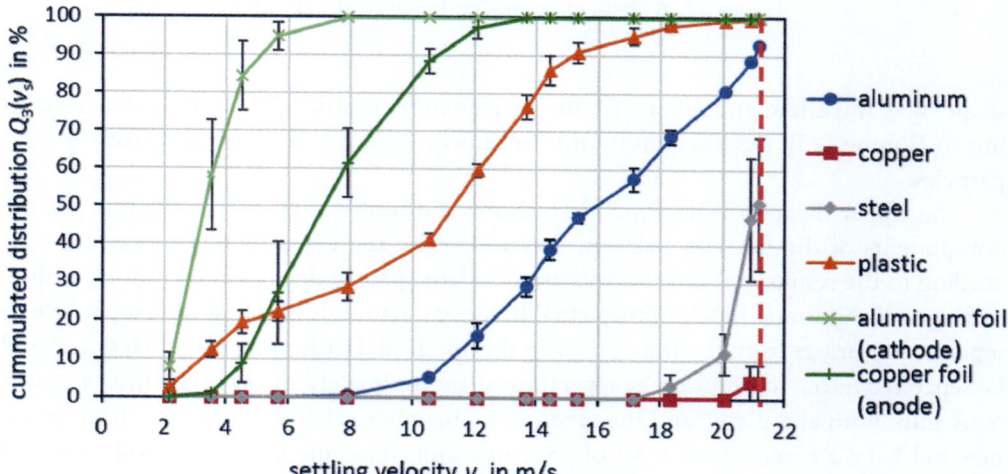

Figure 17.16 Settling velocity distribution of the components before comminution (*top*), after comminution (*below*).

$$v_\mathrm{m} = \sqrt{3g \cdot x_\mathrm{s} \cdot \frac{(\rho_\mathrm{s} - \rho_\mathrm{f})}{\rho_\mathrm{f}}} \qquad (17.4)$$

Due to the transformation of the particles into spheres, area *A* reduces its influence on the settling velocity, and the influence of the mass *m* of the particles increases. For particles of approximately the same size, the mass depends on their density. Therefore, the

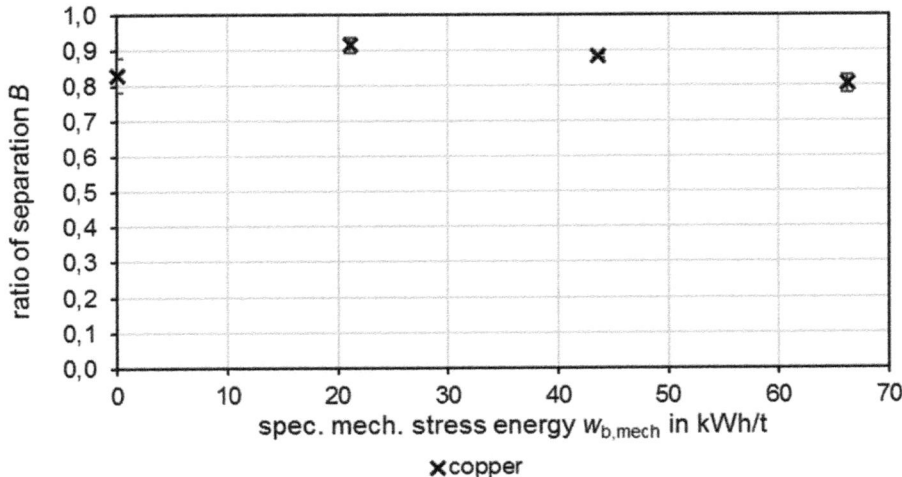

Figure 17.17 Ratio of separation for air classification.

differences in settling velocity between copper and aluminum particles due to their different densities should increase after stressing (Fig. 17.17).

The ratio of separation also shows an improvement for the separation of copper. Here, too, the optimum is at the lowest energy input (21 kWh/t) as for the sieve classification. The reason why the sorting success deteriorates again is the following. With increasing stress, the particles become more and more deformed, and this also applies to aluminum. This also increases the settling velocity, which at the maximum inflow velocity of around 21 m/s (cf. Fig. 17.16 below) leads to more aluminum being discharged into the sinking material and thus into the copper fraction. According to Eq. (17.4), the settling velocity of a spherical aluminum particle with a diameter of 10 mm is approximately 28 m/s. This shows that the inflow velocity for separating spherical copper and aluminum particles of this size is too low.

7. Process flowchart

The flow diagram shown in Fig. 17.18 could result from the sorting procedures explained above. The first step is to separate the casing fraction itself from the remaining components of the battery after comminution. Then the ferromagnetic components are separated from the material mixture by a magnetic separator. Normally, this fraction should consist entirely of iron and steel. The material is then fed into an eddy current separator. Here, the nonmagnetic and nonconductive components are separated. These are the plastics. What remains is a mixture of nonmagnetic but conductive materials. Essentially, this fraction consists of aluminum and copper (cf. "metal mix" in Fig. 17.18).

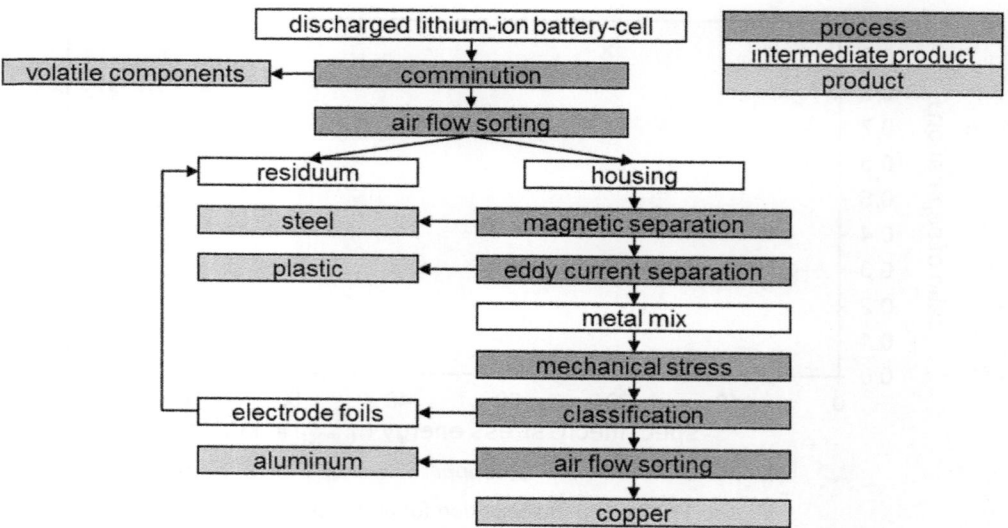

Figure 17.18 Possible process flowchart based on [17].

Table 17.3 Content of valuable material in corresponding products.

	Valuable material content c_c in %		
	Type A	Type B	Type C
Aluminum	90	81	85
Copper	48	79	33
Plastic	55	64	88
Steel	97	93	90

The mixture of the two metals can be further processed in various ways. One possibility is further air classification. Fig. 17.16 shows, among other things, the settling velocity distribution of both materials. After the separation of all other components, which are shown in the diagram, enrichment for copper and aluminum can take place in different concentrates. It is also possible to choose other processes or process combinations. For example, Chapter 23 "Mechanical and physical processes of battery recycling" describes the sensor-based sorting of copper and aluminum components from housing fractions.

Table 17.3 shows the recyclable material contents of the corresponding concentrates for the three considered cell types A, B, and C that can be achieved according to the procedure shown in Fig. 17.18. Compared with Table 17.2, it can be seen that the content can be increased for all recyclables.

Figs. 17.19–17.21 show the whole composition of the different products for the three cell types. The respective components are shown here with their respective masses.

The colored markings show the proportion of concentration for the various component materials. For example, for aluminum cell type A (Fig. 17.19), this means that approximately 86% of the 91.3 g of aluminum is discharged in the aluminum concentrate. Another 9% end up in the copper concentrate and about 5% end up in the plastic concentrate.

For all cell types, the ferromagnetic steel can be completely discharged into a product. Nearly no other material enters this fraction. For all of them, more than 80% of the plastic in the starting material is discharged into the plastic product. Most of the remaining electrode foils end up in the aluminum fraction. Due to the comparatively low mass, these represent only minimal contamination. Furthermore, they could be separated by additional classification (cf. the chapter "Selective Crushing and Sieve Classification"). The solid aluminum of the housing is enriched with more than 80% in the aluminum concentrate for cell types A and B. For cell type C, a value of over 70% is achieved. In this case, the process could be further optimized for a higher yield. Depending on the cell type, the copper is discharged between 69% … 87% into the copper concentrate. Due to the different masses of the materials used, no pure metal concentrates are produced, despite the comparatively high output. Thus, between 33 … 79% copper content is achieved for the copper concentrates of the example cells. Nevertheless, this means an enrichment of the copper content according to Eq. (17.5) between 3.4 … 7.1. Here, the mass fraction of the valuable material (e.g., copper) in the starting material c_a is related to that in the concentrate c_c. The enrichment of aluminum (i on average 1.3) is only minimal. This means that the quality as an aluminum raw material after processing is only

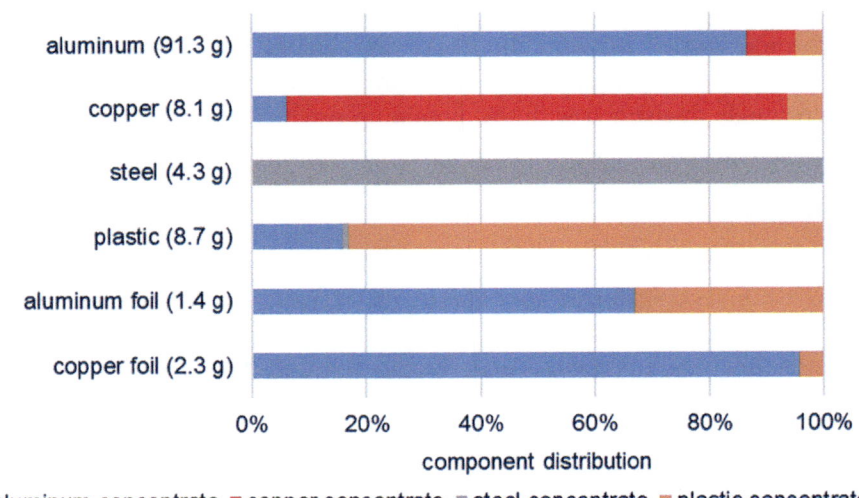

Figure 17.19 Composition of the products of cell type A.

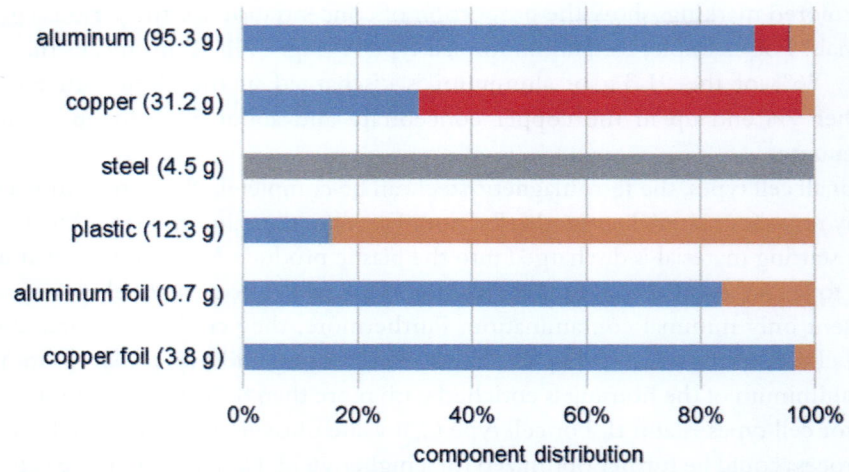

Figure 17.20 Composition of the products of cell type B.

marginally better than in the starting material. In contrast, a value of up to 9 is achieved for i for plastics and up to 58 for steel.

$$i = \frac{c_c}{c_a} \qquad (17.5)$$

The main differences here are due to the different cell structures. The thickness of the cell housing is between 0.8 and 1.4 mm, which results in significant differences in settling velocity. Furthermore, a nail safety device is installed in some cells, e.g., cell type C. This is a copper foil located between the electrode winding and cell housing. This foil is thicker than the anode foil but thinner than the copper from the pole. As a result, this foil is increasingly discharged with the aluminum product (Fig. 17.21).

8. Summary

The case fraction of LIBs obtained by mechanical processes is an important source of raw materials. It can account for up to a quarter of the total mass of the cell. Many of the commercially available prismatic cells have a housing made of aluminum, which thus also accounts for the largest mass share of this fraction. The other main components are copper and plastic.

A large part of this fraction can be separated from the functional battery components by mechanical processes such as air classifying. This fraction can be separated into different potential product concentrates by further mechanical processes. These include magnetic separation for ferromagnetic components, eddy current separation for

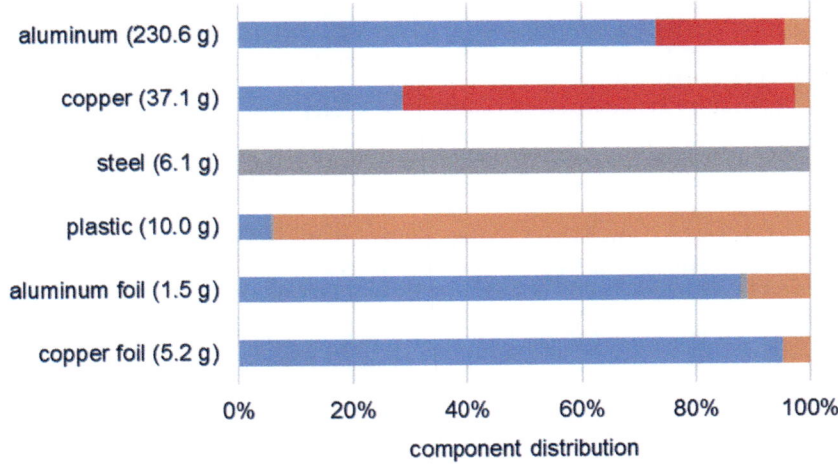

Figure 17.21 Composition of the products cell type C.

nonconductive materials such as plastic, and another air classification for sorting aluminum and copper. Other processes such as sensor-based sorting, air-separation tables, or setting machines are common in scrap processing and could be used here as well.

The properties of individual materials can be influenced by targeted impact stress and thus the sorting result can also be changed/improved. This also makes it possible to selectively comminute the remaining electrode foils, whereby these can be separated from the housing materials through classification.

The mass fraction of aluminum, which is the main component in the enclosures considered, is only marginally improved by the processes when comparing the product and the output material. However, all other materials can be enriched in various concentrates and thus made accessible again as raw materials.

In addition, there may also be batteries with steel cases [24]. If this is a type of steel with a ferritic or martensitic structure, it can be separated with the help of a magnetic separator and the procedure described in this chapter can be used without major adjustments. In this process, the casing pieces are discharged with the other steel pieces in the steel concentrate. If the steel is nonmagnetic austenitic steel, it must be investigated at which sorting step a steel concentrate can be produced. If necessary, a further process step must be added.

References

[1] M. Buchert, P. Dolega, S. Degreif, Gigafactories für Lithium-Ionen-Zellen — Rohstoffbedarfe für die globale Elektromobilität bis 2050, 2019.

[2] D.L. Thompson, et al., The importance of design in lithium ion battery recycling—a critical review, Green Chem. 22 (22) (2020) 7585−7603.
[3] B. Friedrich, L. Schwich, New science based concepts for increased efficiency in battery recycling, Metals 11 (4) (2021) 533.
[4] M. Gellner, et al., Akkus mechanisch aufbereiten, Recycl. Mag. 16 (2015) 26−29.
[5] T. Elwert, F. Frank, Auf dem Weg zu einem geschlossenen Stoffkreislauf für Lithium-Ionen-Batterien. Recycling und Sekundärrohstoffe, Thomé-Kozmiensky Verlag GmbH, 2020, pp. 525−530. Band 13.
[6] H. Martens, D. Goldmann, Recyclingtechnik - Fachbuch für Lehre und Praxis, Springer, Wiesbaden, 2016.
[7] T. Woehrle, Lithium-ion cell, in: R. Korthauer (Ed.), Lithium-Ion Batteries: Basics and Applications, Springer Berlin Heidelberg, Berlin, Heidelberg, 2018, pp. 101−111.
[8] L. Wuschke, Mechanische Aufbereitung von Lithium-Ionen-Batteriezellen. 1. Auflage ed. Freiberger Forschungshefte/A 927, TU Bergakademie Freiberg, Freiberg, 2018.
[9] R. Korthauer, Handbuch Lithium-Ionen-Batterien. SpringerLink Bücher, Springer Vieweg, Berlin, Heidelberg, 2013.
[10] J. Schäfer, et al., Challenges and solutions of automated disassembly and condition-based remanufacturing of lithium-ion battery modules for a circular economy, Procedia Manuf. 43 (2020) 614−619.
[11] K. Wegener, et al., Disassembly of electric vehicle batteries using the example of the audi Q5 hybrid system, Procedia CIRP 23 (2014) 155−160.
[12] C. Herrmann, et al., Scenario-based development of disassembly systems for automotive lithium ion battery systems, Adv. Mater. Res. 907 (2014) 391−401.
[13] J. Marshall, et al., Disassembly of Li ion cells—characterization and safety considerations of a recycling scheme, Metals 10 (6) (2020) 773.
[14] L. Li, et al., Disassembly automation for recycling end-of-life lithium-ion pouch cells, JOM 71 (12) (2019) 4457−4464.
[15] J. Diekmann, et al., Ecological recycling of lithium-ion batteries from electric vehicles with focus on mechanical processes, J. Electrochem. Soc. 164 (1) (2017) A6184−A6191.
[16] A. Kwade, J. Diekmann, Recycling of Lithium-Ion Batteries: The LithoRec Way, Springer International Publishing, 2017.
[17] L. Wuschke, et al., Zur mechanischen aufbereitung von Li-Ionen-Batterien, BHM Berg-und Hüttenmännische Monatshefte 161 (6) (2016) 267−276.
[18] C. Hanisch, et al., Recycling of Lithium-Ion Batteries, Handbook of Clean Energy Systems, 2015.
[19] H. Rumpf, Wirtschaftlichkeit und ökonomische Bedeutung des Zerkleinerns, in: 4th Eur. Symp. Comminution. Aus: Dechema-Monographien Nr. 1549-1575/Band 79, Teil A/1. 1975. Nürnberg. S 19-41. Weinheim/Bergstr.: Verlag Chemie GmbH. in Schubert, H. Aufbereitung fester mineralischer Rohstoffe. 1: Kennzeichnung von Körnerkollektiven, Kennzeichnung von Aufbereitungs- und Trennerfolg, Zerkleinerung, Klassierung. 4., stark überarb. Auflage, Dt. Verl. für Grundstoffindustrie, Leipzig, 1989, 363 S.
[20] S. Sander, G. Schubert, G. Timmel, Characterisation of fragments produced by the comminution of metals especially considering the fragment shape, Powder Technol. 122 (2) (2002) 177−187.
[21] D. Schubert, Zerkleinerung der Metalle in einer diskontinuierlich arbeitenden Hammermühle unter dem Gesichtspunkt der Stückformveränderung und die daraus entstehenden neuen Möglichkeiten der Sortierung, Technische Universität Bergakademie Freiberg, 1998.
[22] H. Schubert, Handbuch der mechanischen Verfahrenstechnik, vol. 1, Wiley-VCH, Weinheim, 2003.
[23] S. Böhme, Zur Stromtrennung Zerkleinerter Metallischer Sekundärrohstoffe. 1. Aufl. Ed. Freiberger Forschungshefte/A, Deutscher Verlag für Grundstoffindustrie, Leipzig, 1989.
[24] L. Wuschke, et al., Crushing of large Li-ion battery cells, Waste Manag. 85 (2019) 317−326.

SECTION 4

Battery remanufacturing and reusing

CHAPTER 18

Regeneration technologies for electrode nanomaterials of recycled batteries

Xing Ou and Haiqiang Gong
National Engineering Laboratory for High Efficiency Recovery of Refractory Nonferrous Metals, School of Metallurgy and Environment, Central South University, Changsha, Hunan, China

1. Introduction

Lithium-ion batteries (LIBs) are used widely in portable electronic equipment, electric vehicles, and various energy-storage systems [1]. As the future substitute for conventional batteries, LIBs have several distinctive characteristics, especially a rational structure with a light weight and small size but without sacrificing energy storage capacity. The application of new-generation batteries has made electric vehicles a reality [2]. But the fledgling electric vehicles are still immature and fault-prone. Over the past 6 years, the number of commercial LIBs has increased exponentially. Electrode materials have sustained serious damage from hundreds of charging and discharging cycles. The service life of most LIBs has been 1–3 years based on production technology and operating frequency. However, the stock of lithium resources is not sufficient to sustainably support the developing trade. Recently, there has been a trend to create higher energy density, higher capacity, more stable cycles, and higher security for new batteries in developing new electric equipment and recycling spent LIBs [3]. As a research direction with substantial development potential, systematic and deep study of nanomaterials has been ongoing for many years. Fig. 18.1 shows the structure and principle of operation of an LIB.

Attention has recently been paid to nanomaterials due to their excellent traits, including their mechanical, electrochemical, and optical properties. Nanomaterials are the materials that at least one dimension in the size of nanometers(1-100 nm) in three dimensions, which is a new generation of materials composed of nano-particles between the size of atoms, molecules and macroscopic system [4]. When its size is close to the coherence length of electrons, its properties change greatly because of the self-organization brought on by strong coherence. Moreover, its scale is close to the wavelength of light, and it has the special effect of a large surface. Therefore, its properties, such as melting point, magnetism, optics, heat conduction, and conductivity, are often different from those of the material in the overall state. Nanomaterial structures include

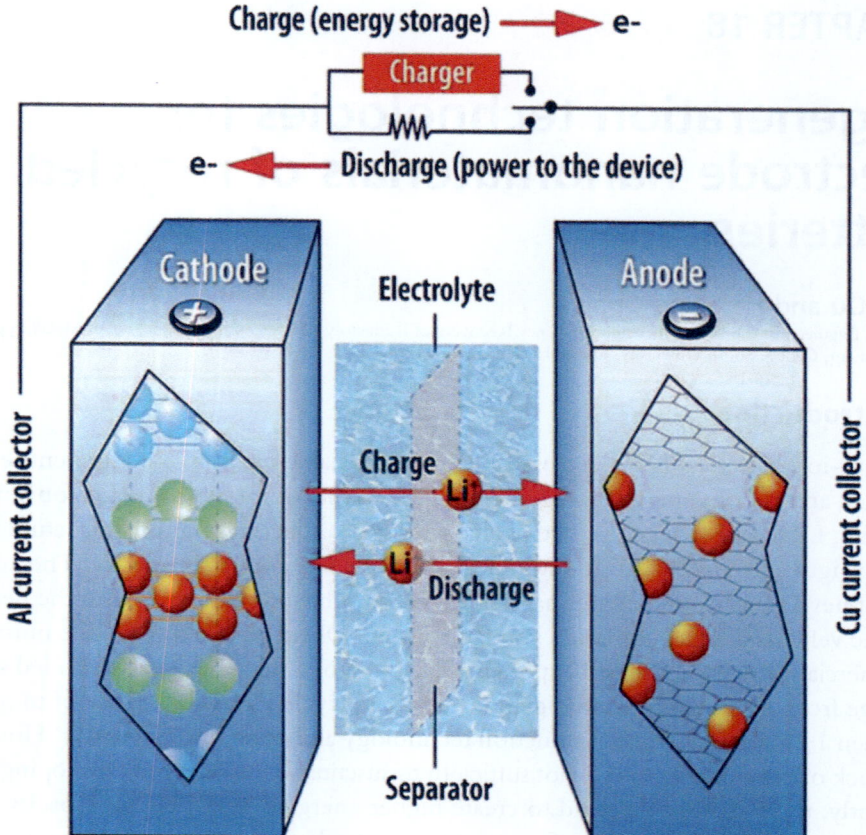

Figure 18.1 Structure and principle of operation of lithium-ion battery [3].

nanoparticles, nanorods, nanowires, nanotubes, and nanosheets [5]. Since the appearance of nanoparticle materials in the 1970s, research has occurred in three phases:

1. Before 1990, the main research area was exploring the synthetic methods of various materials and evaluating performance that differed from that of conventional materials. Worldwide research was focused on a single material, nanocrystalline.
2. Between 1990 and 1994, the focus changed to designing nanocomposites by utilizing the physical and chemical properties of nanomaterials and explored the performance and synthetic methods of nanocomposites.
3. After 1994, nanostructure assembling systems and artificially assembled nano-structured materials systems became new centers of nanomaterials research. Its basic connotation is that nano particles and nanowires and tubes composed of them are assembled and arranged in one-dimensional, two-dimensional and three-dimensional space into nanostructured systems.

For LIBs, functional materials with hierarchically nanostructured frameworks are much better than traditional electrode materials because of the larger contact area and the shorter path length for transportation. The advantages in microstructure lead to higher charge—discharge rates and easier accommodation of lithium insertion/extraction strain by covering the surface with conductive additives rather than mixing the additives randomly [6]. As a result, nanomaterials have been applied in energy-storage devices, especially LIBs, which are the key to the success of next-generation electric vehicles [7]. The amount of energy stored in a given LIB mass or volume is typically expected to be as high as possible. For hybrid electric vehicles, it is thought that the specific energy density and specific power density should be more than 50 Wh/kg and 3 kW/kg, which are much more for electric vehicles. For now, exploring nanostructure electrodes is the most promising path to reach these points. Consumer products with nanomaterials have developed quite rapidly, resulting in increasing amounts of waste nanomaterials. However, some nanomaterials rely on resource-limited metals, rare earth metals, sophisticated techniques, and so on. In addition, the potential health risks of nanomaterials waste are still unclear. Nanomaterials waste may also cause environmental problems due to the difficulty of detecting nanomaterials in nature without an appropriate solution. Because nanomaterials reactions in nature are not easy to predict, even small changes in structure may lead to complete changes in properties, making it difficult to confirm their environmental impact. Finally, their easy reaction characteristics make them difficult to trace and remove [8]. Considering the production and environmental risks posed by nanomaterials, their recovery is essential for sustainable development of the industry [9]. Fig. 18.2 shows the closed-loop materials cycle of batteries.

Traditional LIB recycling procedures include

1. mechanical separation and crushing to prepare the raw material for recycling
2. thermal treatment or dissolution by organic solvent to remove carbon and binders from electrode materials

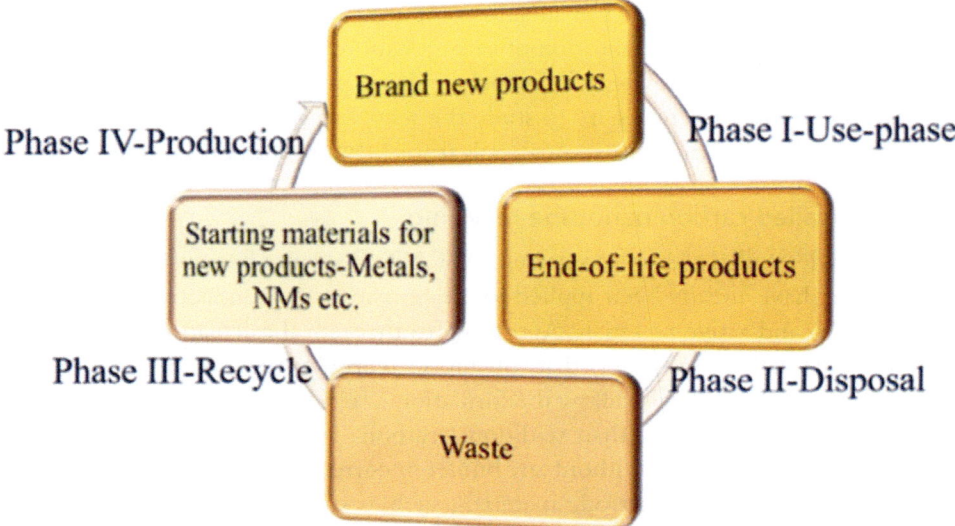

Figure 18.2 Closed-loop materials cycle [9].

3. chemical leaching to enrich nanomaterials, eliminate impurities, and separate other valuable materials
4. selection of appropriate collection methods according to the enrichment form

When we attempt to recycle nanomaterials from spent LIBs, the procedure above is applied, with its basic logic being separation, purification, and regeneration. Nanomaterials recycling falls into two categories: one is to directly extract nanomaterials from spent LIBs for regeneration, and the other is to synthesize other valuable materials with the recycled nanomaterials. Conventional recycling methods [10] (solvent extraction [11], heat treatment, mechanical treatment, plasma technology, etc.) provide ideas for recovering high-value nanomaterials. Nevertheless, the variety and complexity of nanowaste make it nearly impossible to develop a universal treatment for nanowaste recycling. The recovery method must be based on the properties of the nanowaste. This chapter introduces several recycling methods for various kinds of nanomaterials.

2. Carbonaceous nanomaterials recycling

Carbon nanomaterials refer to carbon materials with at least one dimension of the dispersed phase size less than 100 nm [12]. At present, research on carbon nanomaterials is quite active, with the continuous emergence of various carbon nanomaterials structures—carbon nanotubes, carbon fibers, and so on [13]. The related synthesis methods include laser evaporation synthesis, plasma jet deposition, condensed phase electrolysis, graphite arc, and chemical vapor deposition. As the most promising anode material, carbon nanomaterials can greatly improve the electrochemical properties and energy-storage capacity of LIBs. Carbon nanotubes have excellent stiffness and toughness, while graphene has excellent electrical and thermal conductivity [14]. At present, carbon anode materials are used only in low-power devices such as computers and smartphones. Further development brought on by the application of carbon nanomaterials in anodes may promote the development of high-power devices such as electric vehicles and energy-storage power grids. Therefore, it is necessary to study the recycling of related materials. Methods for partially recovering carbon nanomaterials from LIBs are introduced below.

2.1 Single-walled carbon nanotube recycling

Although single-walled carbon nanotubes (SWCNTs) have not been used in commercial batteries, they have already been applied to create many new conductive additives, current collectors, and active materials. According to theoretical calculations, the reversible capacity of SWCNTs outstrips that of LiC_2, whose reversible capacity is about 1116 mAh g^{-1}. SWCNTs also allow the formation of electroosmotic networks at significantly lower weight-loading than traditional carbon. Furthermore, SWCNTs can be used as a separated electrode (without any binder or current collector) or provide physical support for high-capacity electrode materials, such as silicon or germanium. Although there some research has been conducted about the recovery of anode graphite and the applications of graphite recycling, nanocarbon recycling has not been considered in any depth. As SWCNTs are incorporated into LIBs, advance development of a recycling method is necessary to avoid potential negative impacts on the environment.

Schauerman et al. found a way to recycle SWCNTs, which were synthesized by laser vaporization from fully discharged LIBs, without destroying the SWCNT nanostructure while reducing the SWCNT energy consumption per mass produced [14].

The recycling process consisted of four steps:
1. The end-of-life battery was fully discharged, and the SWCNTs were separated from the batteries. The SWCNT electrode material was stirred on a magnetic stirring plate at room temperature for 3 h and treated in an ultrasonic bath at 40°C for 1 h. The homogeneous material was recovered on filter paper by vacuum filtration and dried in a vacuum oven at 100°C for 2 h.
2. Acid reflux or hydrochloric acid treatment was performed on freestanding SWCNT paper for purifying.
3. The SWCNTs after the acid reflux were concentrated on 1 μm PTFE filter paper by vacuum filtration, then the acid that remained in the SWCNT filter paper was washed away with deionized water. The processed SWVNTs were dried in the drying oven.
4. Thermal oxidation of SWCNTs in air was performed to remove impurities and apply the refunctionalized SWCNTs to new graphite electrodes in the LIBs.

Fig. 18.3 provides a schematic of the recycling process.

New electrodes are made with SWCNTs synthesized by laser vaporization in a 1150°C three-zone flowing argon tube furnace and purified by a series of acid and thermal treatments, such as refluxing at 125°C in 3 M nitric acid for 16 h and thermally oxidizing in air at 525°C. The acid leaching step is divided into two parts, mixed acid leaching and hydrochloric acid leaching. The acid used for mixed acid leaching consists of 6 M nitric acid and 3 M hydrochloric acid. The material is recovered by reflux leaching for 16 h, and the material obtained after leaching is dried in a vacuum oven at 100°C for 2 h. Then, after calcination heat treatment at 620°C, the recovered SWCNTs are obtained. Concentrated hydrochloric acid with a concentration of 37% is used for hydrochloric acid leaching. The recovered material is stirred and leached for 1 h and then immersed ultrasonically for 15 min. The leached SWCNTs are filtered on filter paper by vacuum filtration and dried at 100°C for 2 h. Then, the recovered SWCNTs are prepared after calcination heat treatment at 575°C.

The recycling method makes it possible to recycle and refunctionalize SWCNTs without destroying their nanoscale structure and function. It should be noted that the recycled SWCNTs are mixed with the newly synthesized SWCNTs to make a new electrode material. There is nearly no difference between the electrodes made by the mixture of SWCNTs. Moreover, the recycled SWCNTs show better coulombic efficiencies than virgin SWCNTs. Compared with synthesizing and purifying virgin SWCNTs, there is an

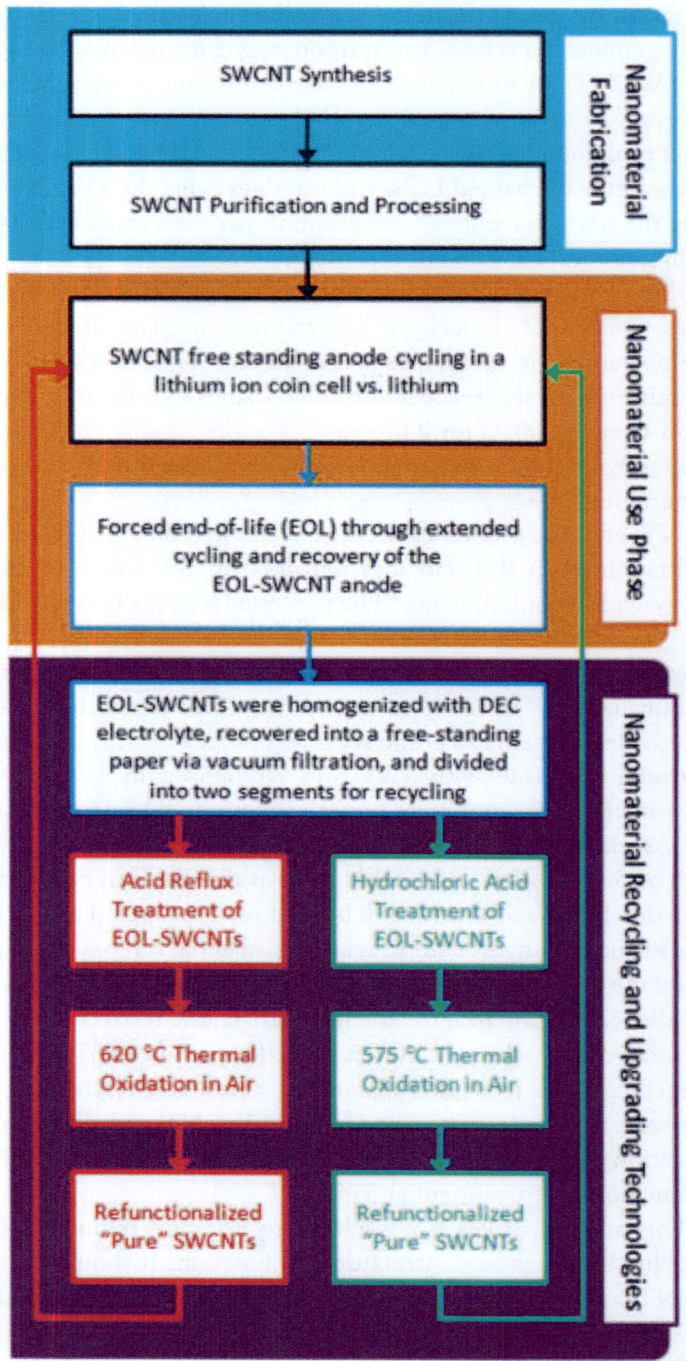

Figure 18.3 Schematic of the single-walled carbon nanotube recycling process from end-of-life material [14].

electricity reduction of about 50% with the acid reflux treatment and about 75% with the hydrochloric acid treatment. Further research is needed to deal with the more complex battery waste, which requires additional separation procedures for other contaminants.

2.2 Soluble graphene nanosheet recycling

Graphene is considered a revolutionary material due to its unique structure, a single-layer two-dimensional honeycomb lattice. Graphene has excellent optical, electrical, and mechanical properties, which means it has application prospects in materials science, micro-nano processing, energy, biomedicine, and drug delivery. The main synthesis methods include mechanical stripping, redox, oriented epigenetic, silicon carbide epitaxy, hummer, and chemical vapor deposition. For graphene to be recovered, it must be dissolved with organic or inorganic solvents. However, due to the strong dispersion between graphene plates, graphene shows low solubility in most solvents, especially in water. Although some high-boiling solvents can help graphene disperse or dissolve at high temperatures, these solvents also make it difficult and expensive to manufacture dissolved graphene into thin sheets or other forms. Even if water-soluble molecules or polymers can dissolve graphene, the entry of these reagents will inevitably deteriorate the properties of graphene. Therefore, reducing graphene oxide is one of the most commonly used processes. Following oxidation, graphene has good solubility, but most reducing agents are toxic and corrosive. Some convincing studies indicate that graphene oxide can be heated under alkaline conditions to prepare water-soluble graphene suspension, with the reduction rate of graphene oxide under alkaline conditions depending on the pH value. The higher the pH value is, the faster the reaction rate. With that in mind, Zhao et al. developed a method for treating graphene oxide recycled from spent LIBs with potassium hydroxide and sodium hydroxide solution to obtain soluble graphene flakes [15].

The recycling process consists of four steps:
1. 10 g P_2O_5 and 10 g $K_2S_2O_8$ is put into 80 mL of concentrated sulfuric acid to form a solution. The graphite recovered from the spent LIB is added to the solution and reacted at 90°C for 4.5 h with stirring. After leaching, the suspension is cooled naturally, diluted slowly, and filtered, and the filter residue is dried at 500°C to obtain graphene oxide.
2. The obtained graphene oxide is dissolved in 150 mL concentrated sulfuric acid, adding 10 g potassium permanganate under vigorous stirring. After standing at 35°C–38°C for 2 h, the suspension is stirred in an ice bath with 100 mL deionized water for 6–7 h. Then, 30% hydrogen peroxide is added to the solution until it turns bright yellow; the solution is then stirred for 2–3 h.
3. The obtained precipitate is mixed with 250 mL dilute hydrochloric acid, stirred for 30 min, and then filtered and separated. The above steps are repeated three to four times.

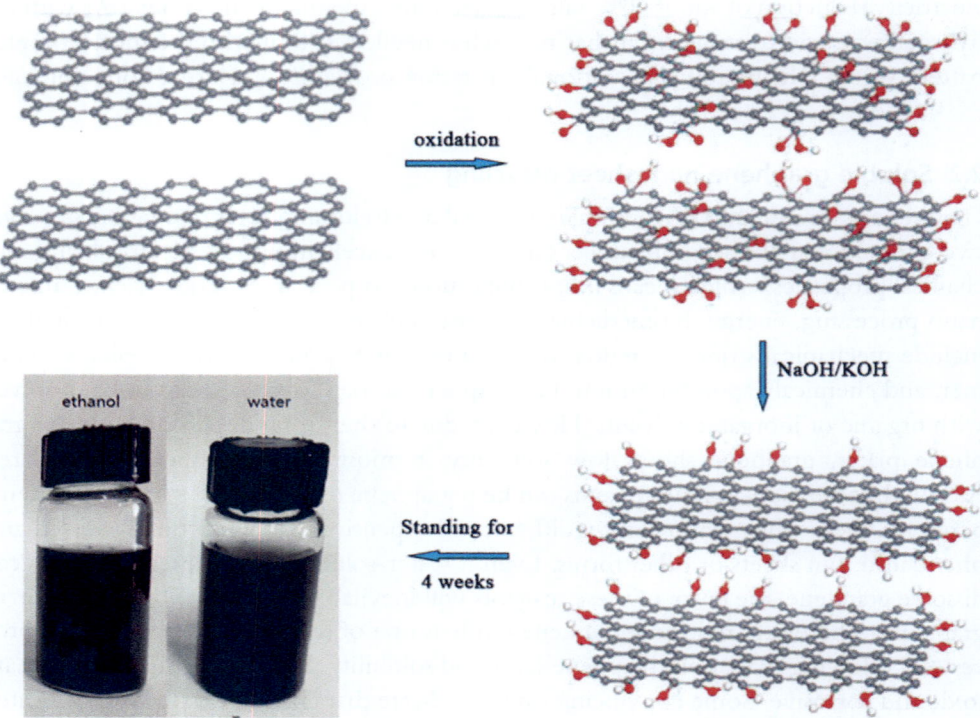

Figure 18.4 Schematic diagram for the formation of soluble graphene [15].

4. The obtained solid graphite oxide is dispersed into deionized water by ultrasound and then filtered and dried to obtain graphene oxide sheets.
5. The obtained graphene oxide sheets are mixed with hydroxide and stirred at about 220°C for 10 h. After the reaction, the reduced graphene sheets are cooled naturally to room temperature, washed with pure water, and dried.

Fig. 18.4 provides a schematic of soluble graphene formation.

The method is characterized by treating graphene oxide with hydroxide, which greatly reduces the unsaturated part and produces more hydroxyl functional groups for good dispersion in aqueous solution and ethanol, which is conducive to the subsequent recovery and regeneration of graphene.

3. Metal nanomaterials recycling

Metal nanomaterials are metals and alloys that form nanograins. They have a characteristic grain boundary ratio, a specific surface energy, and a large proportion of surface atoms. For a decrease in particle size from 100 to 5 nm, the ratio of surface energy to total energy increased from 0.8% to 14%, the proportion of grain boundary increased from 3% to 50%, and the proportion of surface atoms increased to 40% (and increased to 80% at 2 nm). The atomic arrangement of nanosolids is different from that not only of long-range ordered crystals but also of long-range disordered and short-range ordered solid structures. It is a metastable intermediate material between solids and molecules [16]. The special structure of nanomaterials produces four effects, namely small-size, quantum, surface effect, and interface effects. Therefore, nanomaterials have physical and chemical properties that traditional materials do not and show unique optical, electrical, magnetic, catalytic, chemical, and superconducting properties [17].

Metal nanomaterials in LIBs are usually alloy nanomaterials mixed with transition metals. Its LIB action mechanism is different from the conventional lithium-ion insertion and de/intercalation mechanisms. It involves the oxidation-reduction of metal nanoparticles and the formation and decomposition of metal lithium alloys [18]. The overall mechanism is as follows:

$$MX + zLi^+ + ze^- = Li_zX + M \qquad (18.1)$$

where M represents the transition metals Co, Ni, Fe, etc., and X represents O, S, F, and N [19].

The recovery of LIB nanomaterials is different from their synthesis. It is necessary to consider the gap between the costs of nanomaterials recovery and nanomaterials synthesis and how to separate and recover nanomaterials without damaging their structures. Or the valuable components recovered from the battery can be used to synthesize nanomaterials. What is the difference between the synthesis costs and materials properties of nanomaterials? The regeneration and recovery methods of some metal nanomaterials are introduced below.

3.1 Cobalt ferrite nanoparticle recycling

Cobalt ferrite nanoparticles have unique physical, chemical, catalytic and magnetic properties which have been widely used in kinds of electronic and magnetic applications, such as high-density magnetic recording media, ferromagnetic fluid, magnetic resonance imaging, etc [20]. The electrode containing cobalt is generally provided with excellent properties but can also pollute the environment, increase the production cost of the battery, and decrease the inventory of cobalt resources because of widespread application of the materials. So it is necessary to recycle the cobalt electrode to reduce mining pressure on cobalt resources and environmental pollution.

Considering the unique magnetic and mechanical properties of cobalt ferrite nanoparticles, several methods are available for their fabrication, including coprecipitation, self-propagating combustion, and sol-gel. Yang et al. found a way to recycle spent LIBs by combining the sol-gel and combustion processes to obtain cobalt ferrite nanoparticles of high value. Compared with other synthetic methods, sol-gel operates at lower temperatures, its derived materials are homogeneous with accurate element composition, and self-propagating combustion makes full use of the energy inside the materials to react. The advantages of this method are simple—low energy consumption and high purity [20].

The specific procedure can be divided into seven steps:

1. The spent LIBs were discharged completely and disassembled mechanically to separate the electrode to be recovered.
2. The active electrode materials were recycled by thermal treatment, crushing, and calcination.
3. The active materials were dissolved in 4 M HNO_3 blended with 2.5 wt% H_2O_2 stirred continuously at 80°C, and the cobalt solution was precipitated with 40 wt% NaOH into cobaltous hydroxide until the pH of 9–10. The mixed solution was centrifuged for 12 min at 2500 rpm to obtain the cobaltous hydroxide precipitation.
4. The cobaltous hydroxide precipitation was dissolved in 2 M HNO_3, blended with H_2O_2 again, and the cobalt concentration was measured by ICP.
5. Appropriate concentrations of $Co(NO_3)_2$ and $Fe(NO_3)_2$ were added to adjust the cobalt/iron ratio according to the cobalt ferrite nanoparticles ($Co_{0.8}Fe_{2.2}O_4$). (The amounts of $Co(NO_3)_2$ and $Fe(NO_3)_2$ depended on the concentrations of Co^{2+} and Fe^{3+} analyzed by ICP.)
6. Citric acid was added to the solution to adjust the molar ratio of the citric acid and metal ions, and the pH was increased to 7.0 by aqua ammonia. Then the gels were made by stirring continuously, dried in an oven at 135°C, and ignited to initiate the self-propagating combustion process.
7. The gels were calcinated at 900°C for 2 h to prepare the cobalt ferrite ($Co_{0.8}Fe_{2.2}O_4$).

The self-propagating combustion is fundamentally the thermally induced autocatalytic anionic redox reaction. In this method, the nitrate ion makes a great difference in providing the oxidation environment, reducing the decomposition temperature of organic components, and improving the oxidation reaction rate, which leads to self-propagating combustion. Self-propagating combustion realizes the reaction between powders in a self-propagating way. Compared with the traditional process for preparing materials, the process is reduced, shortened, and simpler. Once the ignition start-up process is complete, it does not need to provide further energy. Due to the high temperature when the combustion wave passes through the sample, volatile impurities can be eliminated, and the product purity is high. At the same time, there is a large thermal gradient and fast condensation rate in the combustion process, which may form a complex phase and make it easy to directly change raw materials into another product. It is also possible to mechanize and automate the process. In addition, it is possible to produce another high value-added product with cheap raw material, thus resulting in low costs and good economic benefits.

3.2 Fabrication of nano-Co_3S_4

Cobalt sulfide has been widely used in a new series of batteries bearing outstanding performance. Replacing the oxygen in cobalt oxide crystals with sulfur always leads to better electrochemical performance and higher lithium capacity. The higher reactivity of cobalt sulfide creates high crystalline flexibility for battery cycling capacity with high performance. Cobalt sulfide performs better than cobalt oxide because the Co-S bond length is longer than that of Co-O. In addition, the chemical stability, redox conductivity, and reversibility of cobalt-based crystalline compounds with nonmetal group (VI) are made much higher by the increased ion radius from oxygen to sulfur.

Vakylabad et al. discovered a method to synthesize Co_3O_4 nanopowders from spent LIB leaching solution [21]. Cobalt and nickel ions are selectively extracted from the leaching solution with xanthine complex at normal temperature, and the cobalt complex is obtained by washing the complex with ammonia. The composite is transformed into uniform and pure Co_3O_4 nanospheres through mild heat treatment. The key aspect of the method is the difference in the solubility product constant of xanthate complexes with metal ions when using potassium amyl xanthate as the chelating agent. The lower the constant is, the more stable the metal-xanthate complexes. In this system, the cobalt ion concentration can only reach 1.0×10^{-13} mol/L, while manganese and lithium ionic complexes with xanthate can be almost completely dissolved in the solution as a sulfate phase [21].

The specific procedure can be described in four steps:

1. The spent LIBs were soaked in NaCl solution overnight to fully discharge the batteries and dried at 60°C for 12 h. Active cathode materials were separate from the batteries and cut up by hand. Pyrolyzation was performed in a muffle furnace at 300°C for 30 min to remove the binder and other organic materials. The pyrolyzation removed the organic chemical, making it simple and easy to pull the graphite from the electrode materials.
2. The active materials were dissolved in H_2SO_4 blended with H_2O_2. The leaching procedure operated at 80°C for 120 min. The concentration of H_2SO_4 was 2M and of H_2O_2 was 6%. The leaching solution containing the active materials was diluted to a pH of 1.8.
3. The solution was dropped with 30% potassium amyl xanthate until the final pH of 4.5, and then the pure xanthate complex was obtained from the complexation reaction. Considering that xanthate complex species are associated with decreased precipitation of the xanthate complex, the pH of the leaching solution should first be adjusted to 4.5 using sodium hydroxide solution, which also prevents the decomposition of xanthate in an acidic medium.
4. The precipitation was washed out with ammonia for 2 h and calcinated at 250°C for 1 h to obtain Co_3O_4 nanoparticles. The coprecipitated nickel xanthate was completely separated from the cobalt xanthate through the ammonia washing process.

Sulfuric acid leaching was selected because the properties of sulfuric acid conform with other chemicals in the procedure, especially the xanthates. The materials recycled are difficult to be dissolved by the sulfuric acid alone for a strong bond in the crystalline of active materials and a cobalt capacity of three. To enhance dissolution, hydrogen peroxide is added to play a reducing role by participating in the leaching reaction, and cobalt is reduced to two capacities. Reducing the capacity of cobalt is conducive to the dissolution of cobalt and increases the leaching capacity. The conditions of the cathode active materials should be properly reduced for optimal leaching.

The key points of the method are its wide applicability and relatively loose recovery conditions. Compared with other methods, this method of controlling precipitation dissolution through the difference of the complex stability constant has a better separation effect and shorter process.

3.3 Regeneration of $LiMn_xFe_{1-x}PO_4/C$ cathode materials

Lithium iron phosphate has attracted extensive attention in the field of LIB cathode materials because of its low cost, friendly environment, good thermal stability, and excellent electrochemical performance. However, the application of $LiFePO_4$ in batteries is seriously affected by its poor conductivity, lithium-ion diffusion rate, energy density, and capacitance. Three choices are available to overcome these problems. One is to refine the grain, reduce the diffusion path of lithium ions, and improve the magnification of the material. Another is surface coating material to improve performance. Yet another is doping elements to improve conductivity. $LiMn_xFe_{1-x}PO_4/C$ nanomaterials are lithium iron phosphate cathode materials coated with carbon and doped with manganese ions at the iron site. The doping of manganese ions at the iron site can increase the cell parameters, change the diffusion of lithium ions from two-phase interface control in lithium iron phosphate to single-phase diffusion control, further improve the conductivity of the material, and improve the comprehensive electrochemical properties of the material. Doping at the lithium site can also improve conductivity but will affect the lithium-ion diffusion channel and hinder lithium-ion diffusion. The purpose of carbon coating is to improve the surface conductivity and refine the grain.

Shi et al. developed a method to regenerate high-performance nano-$LiMn_xFe_{1-x}PO_4/C$ with spent $LiFePO_4$ and $LiMn_2O_4$ electrodes as raw materials by a mechanochemical activation method inspired by the removal of aluminum foil fragments by alkaline leaching. In this method, electrode additives, such as polyvinylidene fluoride, are recycled without any sintering, leaching, or washing steps. The only wastes produced in the recycling process are water, carbon dioxide, and ammonia [22].

The specific procedure can be divided into several steps:
1. The spent LIBs to be recovered were discharged completely and disassembled. The waste $LiFePO_4$ and $LiMn_2O_4$ were to be separated, screened out, and then evenly mixed.

2. The crushed and mixed electrode material powders were immersed in a 20 wt.% sodium hydroxide solution and stirred at 80°C for 12 h. After leaching, the materials were washed with 5 wt.% sodium hydroxide solution at 80°C five times and then with deionized water three times.
3. Using ethanol as a dispersant, lithium carbonate, ammonium dihydrogen phosphate, polyvinyl alcohol, and the treated recycled material were mixed and milled in a ball mill for 10 h.
4. The milled slurry was calcined at 450°C for 4 h in air, then calcined at 650°C for 6 h, and naturally cooled to room temperature to obtain the nano-$LiMn_xFe_{1-x}PO_4/C$ material (Fig. 18.5).

High-efficiency ball milling leads to changes in the original structure, grain refinement, and uniform mixing. Combined with calcination, the amorphous carbon coating can be regenerated. Relatively small main lattice parameters can alleviate the shrinkage and expansion effects of repeated ion insertion and disembedding to improve material reversibility. The regenerated $LiMn_{0.5}Fe_{0.5}PO_4$ and $LiMn_{0.8}Fe_{0.2}PO_4$ materials show good electrochemical properties. The discharge capacity and cycle performance of the electrode meet the expectations.

This recovery process adopts the method of mechanical ball milling activation, so the recovered materials are mixed evenly, and the particles size sufficiently small, which is conducive to the preparation of nano-$LiMn_xFe_{1-x}PO_4/C$ material. There are no sintering, leaching, or washing steps anywhere in the process, and the spent electrode materials

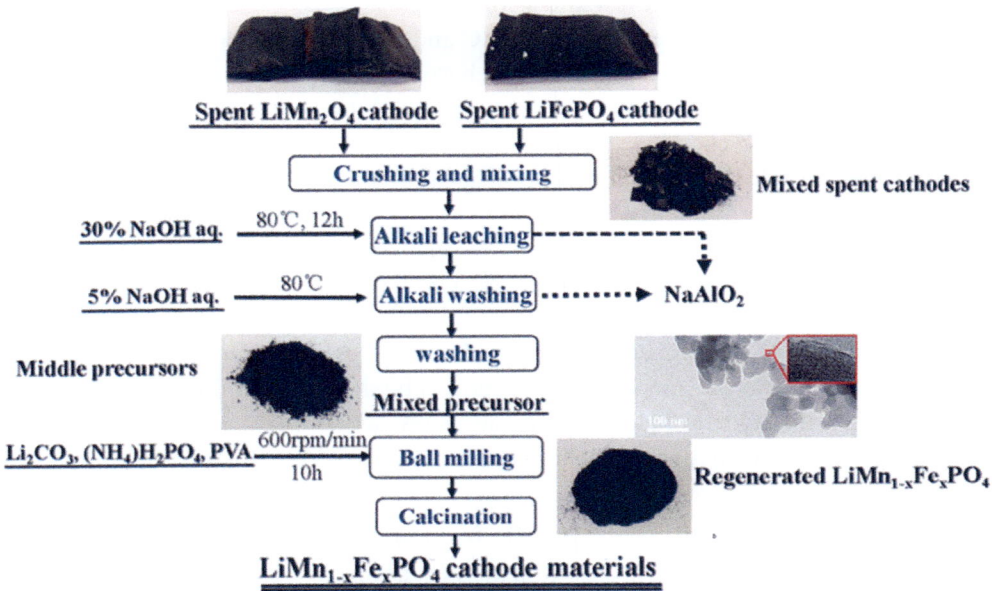

Figure 18.5 Schematic of the recycling process [22].

are effectively recycled with a high value. This method overcomes the shortcomings of long traditional metallurgical processes, complex processes, and high costs and can handle relatively complex mixed electrode wastes.

3.4 Regeneration of $Co_{1+x}Al_zZn_xZr_xFe_{2-3x}O_4$ from spent lithium-ion batteries

$CoFe_2O_4$ material has the advantages of good mechanical properties, corrosion resistance, and strong tensile strength, especially good magnetostrictive performance, which leads to the wide use of $CoFe_2O_4$ in various sensors. To further improve its magnetostrictive properties, the preparation conditions can be optimized to change its particle size and morphology, or the crystal structure can be changed by doping metal ions to realize ion rearrangement. Xi et al. used a sol-gel self-propagating method to mix nickel, cobalt, and zirconium into $CoFe_2O_4$ to prepare magnetostrictive nanomaterials with various properties.

The specific procedure can be divided into four steps:

1. The spent LIB was fully discharged and disassembled. The obtained cathode material was crushed and dissolved in 3.5 mol/L sulfuric acid and 10% hydrogen peroxide. After the cathode was dissolved, a solution containing metal ions was obtained by filtration and separation.
2. $Co(NO_3)_2$, $Al(NO_3)_3$, $Fe(NO_3)_3$, and $Zr(NO_3)_4$ were added to the spent LIB solution, and the corresponding molar ratio of metal ions was n (Co): n (Al): n (Zr): n (Fe) = (1 + x): x: x: (2-x). Citric acid was added continuously to the mixed solution so that n (total metal ions): n (citric acid) = 1:1.
3. All the reagents were mixed evenly at 60°C and stirred in a water bath for heating for 30 min. Ammonia was added to adjust the pH to 6.9, and then the solution was stirred at 80°C in a constant-temperature water bath to form a gel. The gel was transferred to the drying chamber at 110°C to obtain the precursor of xerogel. The dry gel was added to ethanol to ignite self-propagation, and the powder was ground in a mortar.
4. The obtained powder was compressed into small cylinders by adding PVA and then calcined at 1450°C in air atmosphere for 6 h to obtain the target material $Co_{1+x}Al_zZn_xZr_xFe_{2-3x}O_4$.

The grain size of the prepared nanomaterials changes, resulting in lattice distortion and changes in magnetism and magnetostriction. The derivative of the sample increases, and the required magnetic field intensity decreases, which plays an important role in applying the magnetostrictive sensor in a low magnetic field.

3.5 Regeneration of $LiNi_{0.6}Co_{0.2}Mn_{0.2}O_2$ from spent lithium-ion batteries

Nickel cobalt lithium manganate electrode material is widely used in various LIB cathode materials because of its outstanding advantages, such as high energy density, good cycle performance, high voltage platform, good thermal stability, long cycle life, ideal crystal structure, small self-discharge, and no memory effect [23]. With the rapid development of LIBs, the need to improve the energy density of batteries is becoming more and more urgent. Therefore, commonly used lithium nickel cobalt manganate electrode materials are gradually being eliminated. $LiNi_{0.6}Co_{0.2}Mn_{0.2}O_2$, with higher energy density and better stability, has become one of the main research directions. The main reasons for the failure of $LiNi_{0.6}Co_{0.2}Mn_{0.2}O_2$ have been severe side phase formation and lithium nickel disorder caused by repeated charge and discharge cycles. Zhang et al. developed a method to directly recycle spent $LiNi_{0.6}Co_{0.2}Mn_{0.2}O_2$, mainly through the reconstruction of lithium ions, conducting Mn-rich-over-lithiated oxide in the outer layer of spent $LiNi_{0.6}Co_{0.2}Mn_{0.2}O_2$. To further improve the electrochemical performance of the recovered $LiNi_{0.6}Co_{0.2}Mn_{0.2}O_2$ material, $Li_{1.20}Mn_{0.54}Ni_{0.13}Co_{0.13}O_2$ was selected as the outer coating material [24].

The specific procedure can be divided into five steps:

1. The spent LIBs were soaked in saturated saline solution for several days, fully discharged, and then manually dismantled to separate the electrode materials.
2. The obtained cathode material was immersed in sodium hydroxide solution at 60°C to dissolve the aluminum foil, washed with 5% sodium solution three times to fully remove aluminum, and then washed with pure water three times. After washing, the obtained material was dried at 90°C for 12 h.
3. The treated material was calcined in a muffle furnace at 600°C for 5 h to decompose organics such as polyvinylidene fluoride and carbon. After calcination, the material was cooled naturally to room temperature. After grinding for half an hour, the powder material was obtained through a 200-mesh sieve.
4. $LiOH \cdot H_2O$ was added to the recovered powder material to obtain the precursor material. Polyurethane balls were added to the precursor material. After mixing for 30 min, the polyurethane balls were separated. The precursor mixture was calcined at 500°C for 5 h and then at 830°C for 10 h to obtain the recovered product $LiNi_{0.6}Co_{0.2}Mn_{0.2}O_2$.
5. The recovered $LiNi_{0.6}Co_{0.2}Mn_{0.2}O_2$ was dissolved in ethanol solution. $Ni(CH_3COO)_2 \cdot 4H_2O$, $Co(CH_3COO)_2 \cdot 4H_2O$, and $C_4H_6MnO_4 \cdot 4H_2O$ were dissolved in ethanol solution as raw materials and the molar ratio was controlled according to the coating material. Then, the metal ion solution was dropped into $LiNi_{0.6}Co_{0.2}Mn_{0.2}O_2$ ethanol solution at a rate of 5 mL/min at 40°C, and the pH was maintained at about 10.5 with LiOH ethanol solution. After the coating was completed, it was washed and filtered with ethanol three times and then dried in a vacuum oven at 80°C for 12 h to obtain the NCM precursor material with improved properties. Finally, it was mixed with $LiOH \cdot H_2O$ and sintered at 850°C for 2 h.

This method regenerates the $LiNi_{0.6}Co_{0.2}Mn_{0.2}O_2$ material in the spent LIB and uses a coating of appropriate materials to improve the recovered $LiNi_{0.6}Co_{0.2}Mn_{0.2}O_2$

discharge capacity, cycle stability, and rate capacity. The coating adopts $Li_{1.20}Mn_{0.54}Ni_{0.13}Co_{0.13}O_2$ material, and its nanostructure provides a high ion-conductivity diffusion channel for lithium ions that can limit the occurrence of side reactions during the cycle process.

4. Conclusions

The application of nanomaterials in LIBs is one of the hot research directions for realizing further development of LIBs. At present, some nanomaterials have been applied in LIBs to improve battery performance. However, many studies have focused on the synthesis of nanomaterials and the performance improvement brought about by the application of nanomaterials in LIBs. Few studies have given attention to the recovery of nanomaterials, and the impact of nanomaterials on the environment is difficult to evaluate. Their high reaction activity and the great impact of their structural properties make it difficult to predict property changes in nature.

Considering the large-scale application of LIBs, it is urgent to recycle the valuable components of spent LIBs. This chapter summarizes and introduces the existing technologies of regenerating nanomaterials from spent LIBs. There is no obvious commonality between related technologies. But according to recovery objectives and their own properties, most relevant recovery methods rely on their preparation methods or other methods combined with traditional recovery processes.

The recovery of carbonaceous nanomaterials tends to carry out acid leaching and separation without damaging the structure of nanomaterials and then uses appropriate solvents to disperse the nanomaterials as evenly as possible and recover them, thus obtaining nanomaterials with good performance.

Metal nanomaterials tend to dissolve the recovered materials, adjust the proportion of metal ions, synthesize nanomaterials by the sol-gel method, or further coat a layer of material on the outer layer to improve performance.

At present, the recovery of nanomaterials in LIBs only includes the recovery of carbon anode nanomaterials and some metal cathode nanomaterials. Applied research on other LIB nanomaterials such as silicon and titanium is insufficient, and a huge development gap remains. There is also a lack of systematic and detailed research on the recovery of applied nanomaterials. Many recovery methods remain in the laboratory stage and will require some time for complete industrialization.

References

[1] Y. Sun, N. Liu, Y. Cui, Promises and challenges of nanomaterials for lithium-based rechargeable batteries, Nat. Energy 1 (7) (2016).
[2] S. Goriparti, E. Miele, F. De Angelis, et al., Review on recent progress of nanostructured anode materials for Li-ion batteries, J. Power Sources 257 (2014) 421–443.

[3] M. Osiak, H. Geaney, E. Armstrong, et al., Structuring materials for lithium-ion batteries: advancements in nanomaterial structure, composition, and defined assembly on cell performance, J. Mater. Chem. 2 (25) (2014).

[4] Y. Tang, X. Rui, Y. Zhang, et al., Vanadium pentoxide cathode materials for high-performance lithium-ion batteries enabled by a hierarchical nanoflower structure via an electrochemical process, J. Mater. Chem. 1 (1) (2013) 82–88.

[5] H.X. Zong, X. Xia, Y.R. Liang, et al., Designing function-oriented artificial nanomaterials and membranes via electrospinning and electrospraying techniques, Mat. Sci. Eng. C-Mater. 92 (2018) 1075–1091.

[6] A. Mishra, A. Mehta, S. Basu, et al., Electrode materials for lithium-ion batteries, Mater. Sci. Energy Technol. 1 (2) (2018) 182–187.

[7] R. Marom, S.F. Amalraj, N. Leifer, et al., A review of advanced and practical lithium battery materials, J. Mater. Chem. 21 (27) (2011).

[8] E. Kabir, V. Kumar, K.-H. Kim, et al., Environmental impacts of nanomaterials, J. Environ. Manag. 225 (2018) 261–271.

[9] T. Dutta, K.-H. Kim, A. Deep, et al., Recovery of nanomaterials from battery and electronic wastes: a new paradigm of environmental waste management, Renew. Sustain. Energy Rev. 82 (2018) 3694–3704.

[10] Z. Yao, Z. Huang, Q. Song, et al., Recovery nano-flake (100 nm thickness) of zero-valent manganese from spent lithium-ion batteries, J. Clean. Prod. (2021) 278.

[11] X. Ji, X. Huang, Q. Zhao, et al., Facile synthesis of carbon-coated Zn_2SnO_4 nanomaterials as anode materials for lithium-ion batteries, J. Nanomater. 2014 (2014) 1–6.

[12] X.J. Duan, J.X. Deng, X. Wang, et al., Manufacturing conductive polyanilineigraphite nanocomposites with spent battery powder (SBP) for energy storage: a potential approach for sustainable waste management, J. Hazard Mater. 312 (2016) 319–328.

[13] Y. Goto, Y. Kamebuchi, T. Hagio, et al., Electrodeposition of copper/carbonous nanomaterial composite coatings for heat-dissipation materials, Coatings 8 (1) (2017).

[14] C.M. Schauerman, M.J. Ganter, G. Gaustad, et al., Recycling single-wall carbon nanotube anodes from lithium ion batteries, J. Mater. Chem. 22 (24) (2012).

[15] L. Zhao, X. Liu, C. Wan, et al., Soluble graphene nanosheets from recycled graphite of spent lithium ion batteries, J. Mater. Eng. Perform. 27 (2) (2018) 875–880.

[16] R. Chen, T. Zhao, X. Zhang, et al., Advanced cathode materials for lithium-ion batteries using nanoarchitectonics, Nanoscale. Horiz. 1 (6) (2016) 423–444.

[17] P. Poizot, S. Laruelle, S. Grugeon, et al., Nano-sized transition-metal oxides as negative-electrode materials for lithium-ion batteries, Nature 407 (2000) 496–499.

[18] C. Jiang, E. Hosono, H. Zhou, Nanomaterials for lithium ion batteries, Nano Today 1 (4) (2006) 28–33.

[19] L. Yao, Y. Xi, G. Xi, et al., Synthesis of cobalt ferrite with enhanced magnetostriction properties by the Sol–gel–hydrothermal route using spent Li-ion battery, J. Alloys Compd. 680 (2016) 73–79.

[20] L. Yang, Q. Yan, G. Xi, et al., Preparation of cobalt ferrite nanoparticles by using spent Li-ion batteries, J. Mater. Sci. 46 (18) (2011) 6106–6110.

[21] A. Behrad Vakylabad, E. Darezereshki, A. Hassanzadeh, Selective recovery of cobalt and fabrication of nano-Co_3S_4 from pregnant leach solution of spent lithium-ion batteries, J. Sustain. Metall. 7 (2021).

[22] H. Shi, Y. Zhang, P. Dong, et al., A facile strategy for recovering spent $LiFePO_4$ and $LiMn_2O_4$ cathode materials to produce high performance $LiMnxFe_{1-x}PO_4/C$ cathode materials, Ceram. Int. 46 (8) (2020) 11698–11704.

[23] X.Q. Meng, J. Hao, H.B. Cao, et al., Recycling of $LiNi_{1/3}Co_{1/3}Mn_{1/3}O_2$ cathode materials from spent lithium-ion batteries using mechanochemical activation and solid-state sintering, Waste Manag. 84 (2019) 54–63.

[24] Y. Zhang, T. Hao, X. Huang, et al., Synthesis of high performance nano-over-lithiated oxide coated $LiNi_{0.6}Co_{0.2}Mn_{0.2}O_2$ from spent lithium ion batteries, Mater. Res. Express 6 (8) (2019).

CHAPTER 19

LIB industry waste valorization for battery production

Basudev Swain[1], Jae-chun Lee[2] and Chan-Gi Lee[1]

[1]Advanced Materials & Processing Center, Institute for Advanced Engineering (IAE), Yongin, Republic of Korea; [2]Mineral Resources Research Division, Korea Institute of Geoscience and Mineral Resources (KIGAM), Daejeon, Republic of Korea

1. Introduction

Cobalt is a critical metal for lithium-ion batteries (LIBs), and a large share of global cobalt production is used for battery manufacturing. LIBs are the dominant technology for electric vehicles (EVs), and their use is projected to increase exponentially [1]. Along with the established application of LIBs in mobile, laptop, and other electronic devices, exponential growth is projected for LIBs in EVs and alternative energy-storage devices. Concerns about the carbon footprint of internal-combustion-powered automobiles, projections of a future hydrocarbon shortage, and the quest for alternative energy sources are estimated to occupy the LIB market. High energy density and fast charging characteristics have accelerated the growth of power and transport applications, which drove a compound annual growth rate (CAGR) of 24% for LIBs from 2015–2018 [2]. LIB automotive applications were 43% of total shipments in 2015, which jumped to 70% in 2018 [2]. S&P Global Market Intelligence has estimated that demand for LIBs in automotive applications will grow at a CAGR of 26% to 2025 [3]. Greim et al. have reported that LIB demand will likely grow by more than 30% per annum to 2030 [2]. Massive demand signifies that extraordinary LIB waste is being produced during manufacturing and at end-of-life (EOL). Although cobalt is not a rare metal, it is considered critical because of its finite availability and geopolitical issues. Considering environmental and resource scarcity, a cradle-to-grave approach to material utilization is neither economically suitable nor environmentally sustainable. Hence, the cradle-to-cradle valorization of cobalt from EOL LIBs is inevitable and essential, for which urban mining and recycling play a vital role. EOL LIB recycling is occurring worldwide to recover cobalt from these secondary resources to maintain a stable supply and address issues like the environment, circular economy, and stringent environment directives. Although these EOL LIBs are being recycled, the recycling rate is not synchronized with growth in demand. Hence, EOL LIBs and related waste components must be recycled and circularized to material flows. LIB recycling benefits the environment and can be an important economic component. MarketsandMarkets Research Private Ltd. has estimated that LIB recycling is worth USD 1.5 billion and is projected to grow from USD 12.2 billion in 2025 to USD 18.1 billion by 2030, for a CAGR of 8.2% [4].

Several studies have been conducted to recover cobalt from waste cathodic active material ($LiCoO_2$) by pyrometallurgical and hydrometallurgical processes [5–10]. The methods reported by several researchers thus far commonly include crushing, physical separation, acid leaching, and precipitation or solvent extraction to recover cobalt and lithium from battery waste [11,12]. Contestabile et al. have developed a process for recovering cobalt as $Co(OH)_2$ from waste LIBs using crushing, physical separation, hydrochloric acid leaching, and neutralization precipitation [11]. Lee et al. have suggested a process for cobalt recovery as $Co(OH)_2$ from waste LIBs using a pyro-hydrometallurgical technique followed by roasting, crushing, physical separation, HNO_3 or H_2SO_4 leaching, and precipitation [13]. Zhang et al. used the leaching-solvent extraction process to recover cobalt from waste $LiCoO_2$ [14]. Waste $LiCoO_2$ was leached in hydrochloric acid, and $CoSO_4 \cdot 6H_2O$ and Li_2CO_3 were recovered from the leached liquor using solvent extraction by PC-88A and the precipitation technique. With this technique, 99.99% cobalt was recovered as $CoSO_4 \cdot 6H_2O$ from waste $LiCoO_2$. Wu proposed a recycling process in which waste $LiCoO_2$ was leached using sulfuric acid and hydrogen peroxide. The impurities in the leach liquor were selectively and completely extracted with D2EHPA. The cobalt in the leach liquor was extracted selectively with PC-88A in kerosene at an equilibrium pH of 5.5. The separation factor for the cobalt and lithium was 1×10^5. The lithium remaining in the raffinate was recovered as lithium carbonate precipitate at 95°C. Nan et al. proposed a process consisting of leaching with H_2SO_4, followed by both precipitation and solvent extraction. The cobalt was extracted from leach liquor as cobalt oxalate, and the remaining cobalt was extracted using Cyanex 272. But in this process, solvent extraction was mainly employed to extract copper [7]. Kim et al. reported the renovation of $LiCoO_2$ from spent LIBs by a hydrothermal route [15].

Among the reported processes, hydrometallurgy is superior, versatile, flexible, and industrially feasible for simultaneously addressing waste battery recycling, e-waste management, and environmental challenges for the circular economy of battery metals and sustainable waste battery recycling. We have developed a closed-loop industrial hydrometallurgical process for valorizing LIB industry waste for battery manufacturing, as discussed in this chapter. Through a sequential combination of hydrometallurgical techniques like leaching-solvent extract-scrubbing-(precipitation) stripping, a process for 5N pure cobalt recovery from LIB industry waste has been developed. First, the cobalt is extracted by acid leaching. Then, cobalt is separated/purified by solvent extraction from the leach liquor. Following solvent extraction, cobalt is recovered by precipitation stripping via two routes, (1) as cobalt sulfate and (2) as cobalt oxalate. Cobalt oxide, which can be used for battery production, is synthesized and characterized from the cobalt oxalate. In the developed process, LIB industry waste material is leached using 2000 mol/m^3 of sulfuric acid and 5 vol. % of H_2O_2 solution at a pulp density of 100 kg/m^3 under a leaching time of 30 min at 75°C. Upon optimization from the leach liquor, the cobalt is separated/

purified by 65% saponified Cyanex 272 using 1500 mol/m^3 Cyanex 272 at an organic to aqueous phase ratio of 2/3 and pH of 5. From cobalt-loaded Cyanex 272, the residual lithium is scrubbed by 100 mol/m^3 Na$_2$CO$_3$ solution (three times) at an organic-to-aqueous phase ratio of 0.5. From the purified cobalt-loaded Cyanex 272, cobalt oxalate is recovered using 1000 mol/m^3 of oxalic acid through precipitation stripping. Finally, the precipitate is calcined to synthesize Co$_3$O$_4$, which is a precursor for LIB manufacturing. TGA-DTA followed by XRD analysis confirmed that at 200°C, CoC$_2$O$_4\cdot$2H$_2$O can be converted to anhydrous CoC$_2$O$_4$; at 350°C, anhydrous CoC$_2$O$_4$ can be converted to Co$_3$O$_4$; and at 1100°C, Co$_3$O$_4$ can be converted to CoO. SEM analysis indicates that rod-shaped CoC$_2$O$_4$ and Co$_3$O$_4$ powder were synthesized and can add value to the cobalt recovery process. Waste LIBs can support LIB manufacturing through a versatile and flexible industrial approach through the reported route.

2. Waste characterization

Characterization of waste is a rudimentary and vital stage for understanding waste and the possibility for valorization and extraction of interest to the circular economy. Waste battery recycling and recovery of valued materials from cathode active material or battery manufacturing industry waste are characterized by (1) destructive analysis and (2) nondestructive analysis. Nondestructive analysis includes various material characterizations, such as XRF, XRD, and SEM, whereas destructive analysis includes digestion, strong acid leaching, pressure leaching, and other processes. We characterized LIB industry waste by XRD, SEM, and a particle size analyzer. The analyzed XRD patterns, particle size, and SEM image for waste LIB cathode material are presented in Fig. 19.1A−C. The XRD pattern for waste shown in Fig. 19.1A indicates that the waste is LiCoO$_2$. Fig. 19.1B shows an average particle size of about 8.26 μm. The SEM shown in Fig. 19.1C indicates that the waste powder is irregularly shaped with very limited agglomeration behavior. The limited agglomeration appearing in the SEM may be due to the presence of a binder in the cathode waste powder. The powder is a black, very light material. The structure and texture indicate that it can be leached safely without further treatment. Followed by characterization, the waste material was leached to bring the cobalt into the acid solution from solid LiCoO$_2$ powder.

3. Cobalt extraction by acid leaching

Characterization indicated that the waste contained mainly LiCoO$_2$ material, which is essential content for the modern revolution in electronic devices and EVs. For metals extraction by hydrometallurgy, leaching is the rudimentary stage, which various authors have reported on using mineral acids. The leaching chemistry of LiCoO$_2$ with mineral

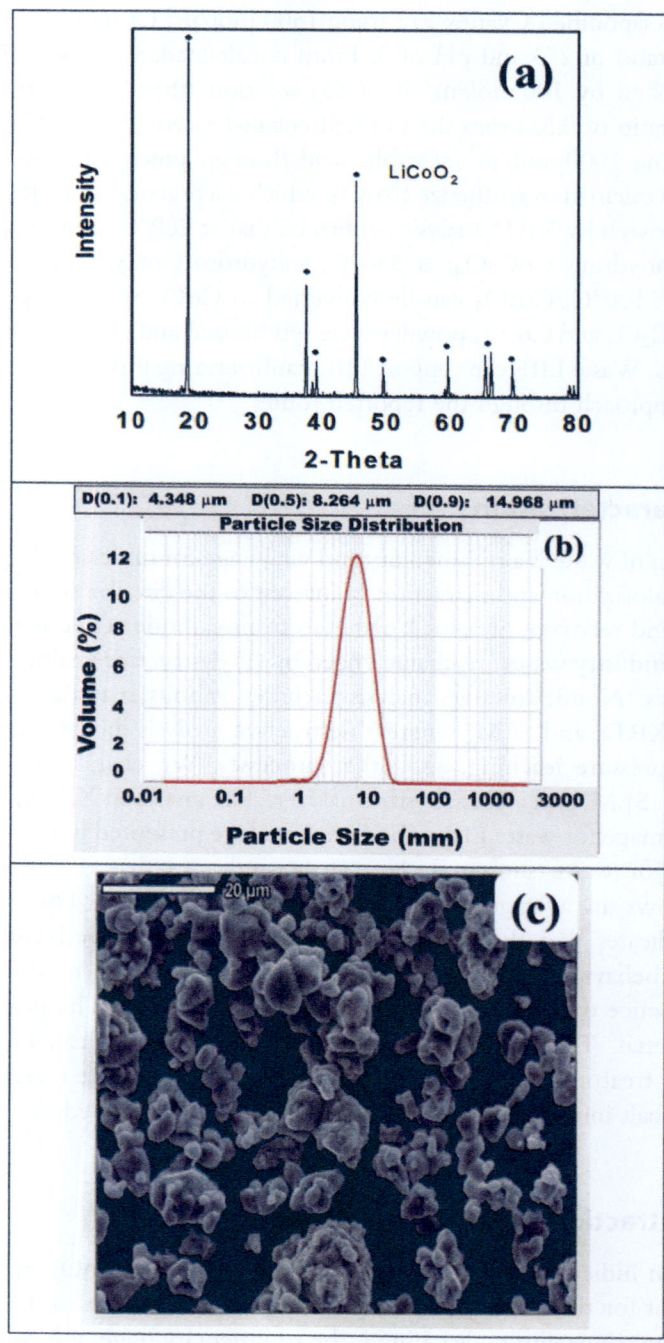

Figure 19.1 (A) XRD patterns, (B) particle sizes, and (C) SEM image for waste lithium-ion battery cathode material.

acids is provided below with chemical equations. The chemistry for leaching $LiCoO_2$ with HCl, HNO_3, and H_2SO_4 lixiviant can be explained using Eqs. (19.1)–(19.3):

$$LiCo(III)O_{2(s)} + 4HCl_{(aq)} \rightarrow Co(III)Cl_{3(aq)} + LiCl_{(aq)} + 2H_2O \quad (19.1)$$

$$LiCo(III)O_{2(s)} + 4HNO_{3(aq)} \rightarrow Co(III)(NO_3)_{3(aq)} + Li(NO)_{3(aq)} + 2H_2O \quad (19.2)$$

$$2LiCo(III)O_{2(s)} + 4H_2SO_{4(aq)} \rightarrow Co(III)_2(SO_4)_{3(aq)} + Li_2SO_{4(aq)} + 4H_2O \quad (19.3)$$

When oxidant H_2O_2 is used with a mineral acid lixiviant like HCl, HNO_3, or H_2SO_4, the chemistry for leaching $LiCoO_2$ can be explained differently using Eqs. (19.4)–(19.6):

$$2LiCo(III)O_{2(s)} + 6HCl_{(aq)} + H_2O_{2(aq)} \rightarrow 2Co(II)Cl_{2(aq)} + 2LiCl_{(aq)} + 4H_2O + O_{2(g)} \quad (19.4)$$

$$2LiCo(III)O_{2(s)} + 6HNO_{3(aq)} + H_2O_{2(aq)} \rightarrow 2Co(II)(NO_3)_{2(aq)} + 2Li(NO_3)_{aq} + 4H_2O + O_{2(g)} \quad (19.5)$$

$$2LiCo(III)O_{2(s)} + 3H_2SO_{4(aq)} + H_2O_{2(aq)} \rightarrow Co(II)SO_{4(aq)} + Li_2SO_{4(aq)} + 4H_2O + O_{2(g)} \quad (19.6)$$

We have selected H_2SO_4 as a lixiviant, considering advantages such as its industrial favorability and minimal effects on the environment and occupational safety. Various process parameters typically having significant effects on efficiencies are discussed sectionwise.

3.1 Effect of H_2O_2 on cobalt leaching from $LiCoO_2$

The effect of the reductant H_2O_2 solution concentration on the leaching efficiencies of waste cathode active materials has been studied. The waste material was leached with an H_2SO_4 solution of 2000 mol/m^3 at an initial pulp density of 100 kg/m^3. During leaching, the temperature was maintained at 75°C with a leaching time of 60 min for each set of experiments. Fig. 19.2 demonstrates that only 34.09% cobalt was leached without adding H_2O_2 solution. In contrast, cobalt recovery was increased from 34.09% to 99.99% by increasing the H_2O_2 concentration from zero to 25 vol. %. The results also indicate that the initial leaching rate of cobalt is high and in the same range the initial lithium leaching rate is relatively slower. It was also observed that the initial leaching rate for lithium is high compared with cobalt up to the addition of 10 vol. % of H_2O_2. The result showed

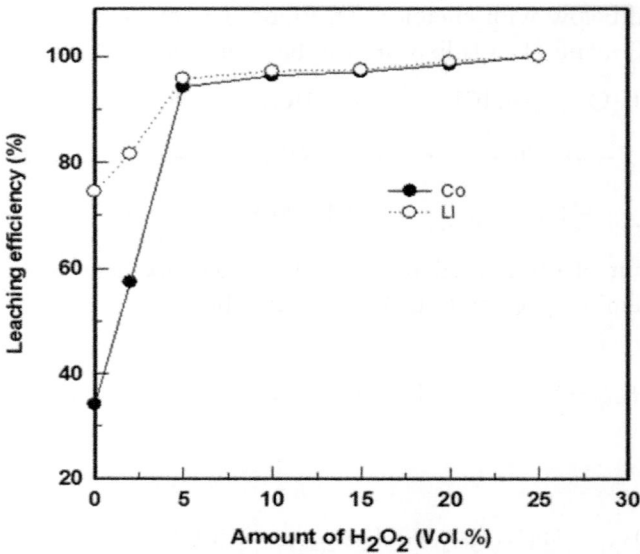

Figure 19.2 Effect of H_2O_2 vol. % on the leaching of waste $LiCoO_2$ with 2000 mol/m^3 H_2SO_4 solution at 75°C for 30 min (pulp density = 100 kg/m^3, agitation speed = 300 rpm).

the percentage leaching of cobalt and lithium was nearly the same for more than 5 vol. % H_2O_2 solution. The extraction of cobalt and lithium was found to be 94.15% and 95.72% using 5 vol. % H_2O_2 reductants. Recovery increased with increases in H_2O_2 until the concentration of H_2O_2 reached 20 vol. %. For further study, the material was directly leached in sulfuric acid using a minimum 5 vol. % H_2O_2. The leaching chemistry of using oxidant H_2O_2 can be explained using Eq. (19.6), above [16].

Aaltonen et al. have reported that among all mineral acids, H_2SO_4 is the most efficient lixiviant in the presence of H_2O_2 or the absence of H_2O_2 [17]. Fig. 19.3 compares the

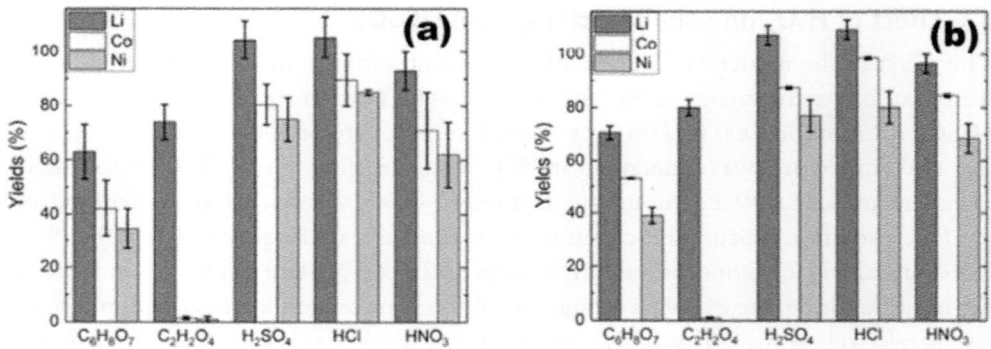

Figure 19.3 Battery metal leaching efficiency, (A) without H_2O_2 and (B) in the presence of 1% H_2O_2 using different acid solutions. Acid concentrations = 2 M $C_6H_8O_7$; 1 M $C_2H_2O_4$; 2 M H_2SO_4; 4 M HCl; 1 M HNO_3; slurry density = 5% (w/v); T = 25°C; t = 24 h [17].

efficiencies of metal leaching from LIB waste using mineral acid and organic reagents using lixiviant. The figure indicates H_2SO_4 with added of H_2O_2 can be the most efficient lixiviant and oxidant combination for efficient leaching cobalt [17].

3.2 Effect of leaching time on cobalt leaching from $LiCoO_2$

The relationship between the amount of metal leached for various time intervals and orders of reaction using a 2000 mol/m³ H_2SO_4 solution at 75°C (H_2O_2 = 5 vol. %, pulp density = 100 kg/m³, agitation speed = 300 rpm) is depicted in Fig. 19.4. Fig. 19.4A shows that the leaching efficiency of cobalt increased from 80.56% to 94.15% and for lithium increased from 80.67% to 95.72% as a function of a reaction time varying from 5 to 60 min. The results shown in Fig. 19.4A indicate that 80.56% cobalt and 80.67% lithium can be leached in 5 min, and 93.14% cobalt and 94.53% lithium were recovered with a leaching time of 30 min. After 30 min, no significant increase in leaching efficiency was observed. The saturation of cobalt and lithium leaching efficiencies comes from decreases in $LiCoO_2$ as a reactant for the leaching reaction. For industrial-scale leaching, a 30 min period can be quite efficient for industrial operation and management and reduce production time.

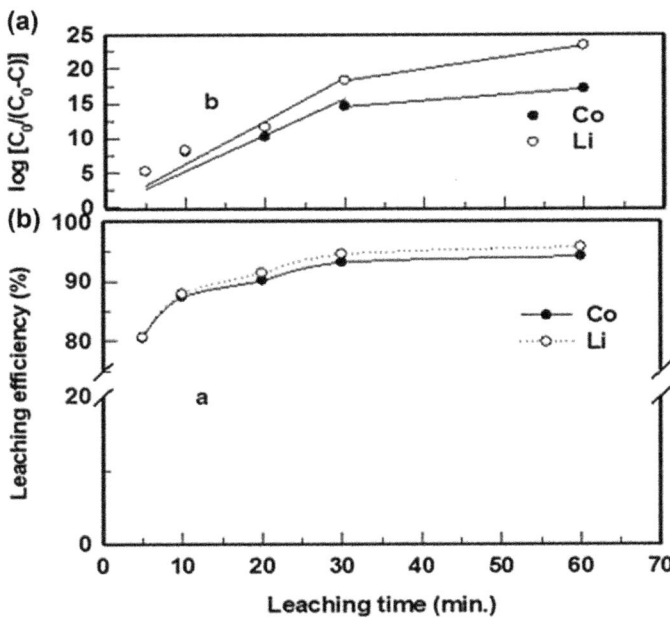

Figure 19.4 (A) Effect of leaching time on leaching efficiency and (B) the leaching rate (effect of leaching time on logarithm of initial over unleached number of metals) of waste $LiCoO_2$ with 2000 mol/m³ H_2SO_4 solution at 75°C (H_2O_2 = 5 vol. %, pulp density = 100 kg/m³, agitation speed = 300 rpm).

3.3 Effect of temperature on cobalt leaching from LiCoO₂

The effect of temperature on leaching has also been studied using H_2SO_4 (2000 mol/m³) as a lixiviant while maintaining a pulp density of 100 kg/m³, reductant concentration of 5 vol. %, and leaching time of 30 min. The result presented in Fig. 19.5 indicates that 66.86% of cobalt and 84.50% of Lithium were leached at 25°C. Cobalt and lithium recovery increased with increases in the leaching temperature. With an increase in temperature to 75°C, the leaching efficiency of cobalt and lithium increased to 93.13% and 94.53%, respectively. Further increase in temperature has not shown any significant increase in the recovery of cobalt and lithium.

3.4 Effect of H₂SO₄ concentration on cobalt leaching from LiCoO₂

The effect of H_2SO_4 concentration on the leaching of waste $LiCoO_2$ has been studied. The concentration of H_2SO_4 is varied from 1000 to 4000 mol/m³ at a pulp density of 100 kg/m³, temperature 75°C, reductant 5 vol. % H_2O_2 and time 30 min. The results presented in Fig. 19.6 indicate that with an increased concentration of H_2SO_4, increases in cobalt and lithium leaching efficiency predominate. The figure shows cobalt and lithium leaching efficiencies of 93.13% and 94.53%, respectively, with 2000 mol/m³ H_2SO_4 as a lixiviant. Cobalt recovery increased slowly from 95.17% to 97.34% with an increase in acid concentration from 3000 mol/m³ to 4000 mol/m³. Fig. 19.6 indicates that increases beyond 2000 mol/m³ in the H_2SO_4 concentration add no value

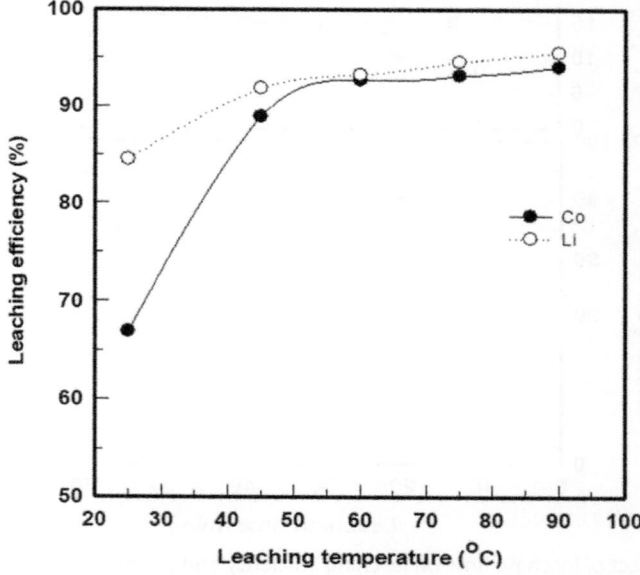

Figure 19.5 Effect of leaching temperature on the leaching of waste $LiCoO_2$, with 2000 mol/m³ H_2SO_4 for 30 min (H_2O_2 = 5 vol. %, pulp density = 100 kg/m³, agitation speed = 300 rpm).

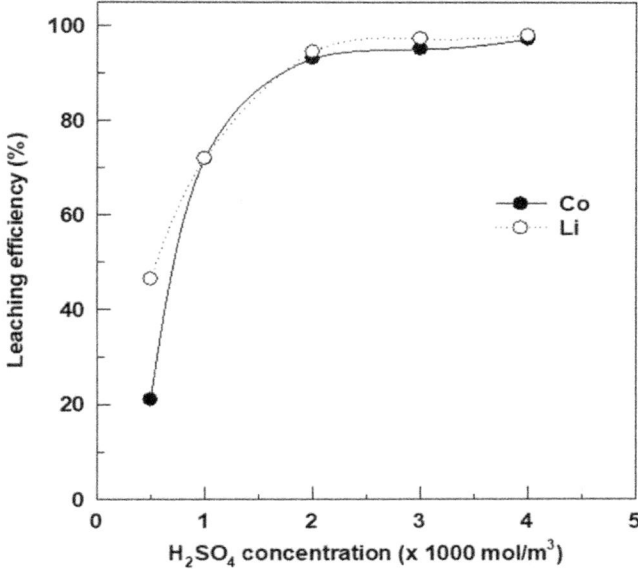

Figure 19.6 Effect of H_2SO_4 concentration on leaching of waste $LiCoO_2$ at 75°C for 30 min ($H_2O_2 = 5$ vol. %, pulp density = 100 kg/m³, agitation speed = 300 rpm).

to leaching efficiency. The figure also indicates that 2000 mol/m³ H_2SO_4 is enough to quantitatively leach out cobalt and lithium using a pulp density of 100 kg/m³. Hence, a higher concentration of lixiviant can be ruled out, as quantitative leaching is achieved in 2000 mol/m³ H_2SO_4. Hence, further studies have been conducted using 2000 mol/m³ H_2SO_4.

3.5 Effect of pulp density on cobalt leaching from $LiCoO_2$

The leaching efficiency of cobalt and lithium from waste $LiCoO_2$ as a function of pulp density keeping other parameters (2000 mol/m³ H_2SO_4 for 30 min ($H_2O_2 = 5$ vol. %, agitation speed 300 rpm) constant has been studied and is presented in Fig. 19.7. The experiments were conducted by varying the pulp density while maintaining the other process parameters as above. The results in Fig. 19.7 indicate total cobalt and lithium recovery at a pulp density of 10–20 kg/m³. Complete extraction of metal from waste $LiCoO_2$ at a pulp density of 10–20 kg/m³ is mainly contributed by a stoichiometric excess of H_2SO_4 lixiviant, where reactant $LiCoO_2$ is the limiting condition. Fig. 19.7 also indicates that despite increases in the quantitative extraction of cobalt and lithium, the cobalt and lithium recovery efficiency decreases as pulp density increases.

However, the experimental results presented in Fig. 19.7 indicate the 85.30% of cobalt and 90.50% of lithium can be achieved at a pulp density 125 kg/m³. When the pulp density decreases to 100 kg/m³, recovery increases to 93.0% and 94.5% for cobalt and

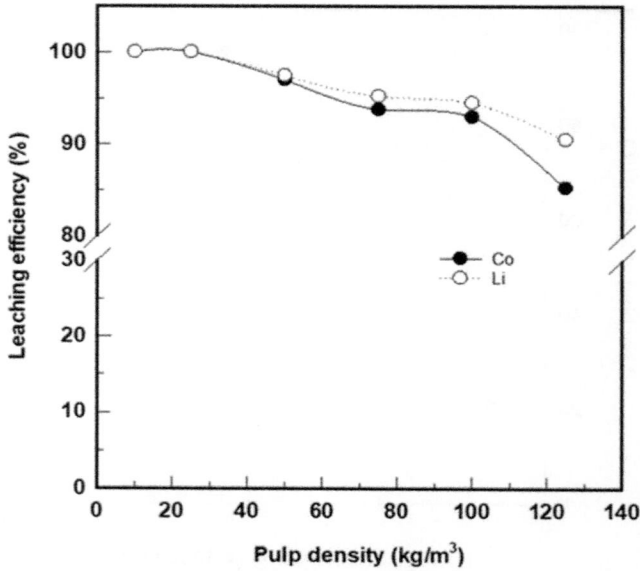

Figure 19.7 Effect of pulp density on leaching of waste $LiCoO_2$ with 2000 mol/m^3 H_2SO_4 for 30 min (H_2O_2 = 5 vol. %, agitation speed = 300 rpm).

lithium, respectively. At higher pulp densities, filtration problems were observed, so the pulp density of 100 kg/m^3 is suitable for leaching waste $LiCoO_2$. Leaching efficiency as a function of pulp density reflects the fact that leaching chemistry is mainly controlled by stoichiometry. From a thorough stochiometric calculation using Eq. (19.3), the theoretical pulp density for cobalt and lithium leaching can be estimated as 130 kg/m^3.

The effect of agitation has also been examined. Due to the fast chemical reaction, no significant effect has been found from agitation velocity, so the 300-rpm agitation velocity was fixed for each set of experiments. From the above studies, considering the best suitable conditions to reduce energy and chemical consumption, an optimum condition for leaching cobalt and lithium from waste $LiCoO_2$ was obtained. The result showed that 93% of cobalt and 94.50% of lithium could be leached under optimized experimental conditions, i.e., pulp density of 100 kg/m^3, lixiviant concentration of 2000 mol/m^3, reductant H_2O_2 concentration of 5 vol. %, leaching time of 30 min, and temperature of 75°C. The leach liquor generated under optimized conditions of 44.72 kg/m^3 cobalt and 5.43 kg/m^3 lithium was used for cobalt and lithium separation and recovery using a solvent extraction process.

4. Selective separation of cobalt by solvent extraction from LiCoO$_2$ leach liquor

For industrial-scale recovery of valuable metals from primary or secondary resources by solvent extraction, commonly available commercial solvents like Cyanex 272, PC-88A, and D2EPHA are used [5,18,19]. Solvent extraction is used because its processes are efficient, versatile, and flexible to variations in metal content and different metals in wastes compared with other processes [20,21]. Ref. [22] Extractants like Cyanex 272, PC-88A, and D2EPHA are used because they are commercially available, cheaper, and recyclable. Devi et al. have compared cobalt and nickel extraction behavior using Cyanex 272, PC-88A, and DEHPA in kerosene to extract 0.6 g/L of both metals from a sulfate medium and have indicated $pH_{1/2}$ Ni—Co values of approximately 3, 1.5, and 0.5, respectively. Devi et al. (1998) used a 0.6 M solution of Cyanex 272, PC-88A, and DEHPA in kerosene to extract cobalt and nickel (both 0.6 g/L) from a sulfate medium and obtained $pH_{0.5}$ Ni—Co values of approximately 3, 1.5, and 0.5, respectively. The same research revealed that separation factors for cobalt over nickel ranged between 4000 and 7000, 70 and 1700, and 6 and 10 for Cyanex 272, PC-88A, and DEHPA, respectively. From reported research, it is quite clear that Cyanex 272 is a better solvent than PC-88A or DEHPA for cobalt separation from leach liquor of waste LiCoO$_2$ [23]. Cyanex 272 is a weaker acid than its competitors (DEHPA, PC-88A), signifying that cobalt extracted by Cyanex 272 can easily be stripped by a relatively weaker acid; therefore, we have considered cobalt extraction from leach liquor of waste LiCoO$_2$. More than 50% of a cobalt-producing hydrometallurgy plant uses Cyanex 272, contributing to its efficiency. Most companies regard the use of Cyanex 272 as confidential, and thus, not much plant and operating data are available [24].

For cobalt extraction from leach liquor of waste LiCoO$_2$, are rationale for choosing Cyanex 272 is the following. Cyanex 272 (1) offers much higher separation factors than DEHPA or PC-88A, (2) is a reagent that can effectively extract cobalt or selectively separate cobalt from nickel even at a cobalt/nickel concentration ratio of 1:100, (3) is the weakest acid of all the organophosphorus acids, thus providing room for easy stripping. Extractant Cyanex 272 was partially saponified up to 65% by adding a stoichiometric amount of concentrated NaOH solution to the extractant in kerosene under constant stirring of the mixture. The kerosene was used as a diluent. Tri-n-butyl-phosphoric acid (TBP) 5 vol. % was used as a phase modifier to the Cyanex 272 extractant solution. The requisite pH of the aqueous leach liquor feed was adjusted with the addition of sodium hydroxide or sulfuric acid solution. Solvent extraction experiments were carried out using suitable volumes of aqueous leach liquor of waste cathode active material and Cyanex 272 solutions. The aqueous leach liquor was equilibrated with organic

Cyanex 272 for 10 min, and then the phase was separated by a separating funnel. The aqueous raffinate was analyzed to determine the amount of cobalt(II) and lithium(I) extracted by the organic extractant. This time, 10 min was found to be sufficient to attain equilibrium, as verified in preliminary tests. The preliminary tests suggest that about 3–4 min of shaking is sufficient to achieve equilibrium. Scrubbing and stripping tests were carried out using Na_2CO_3 solution and H_2SO_4 solution, respectively. All experiments were carried out at ambient temperature ($25 \pm 1°C$). The extraction was carried out in a crosscurrent solvent extraction manner. The same stock aqueous leach liquor was used in each set of experiments, but the raffinate from the first stage of extraction was used as the aqueous feed in the second stage for solvent extraction. In all the solvent extraction studies, the comparatively lower end of lithium contamination into the organic phase was preferred for subsequent study to avoid scrubbing difficulties and achieve better cobalt purity in the organic phase. Enough leach liquor was generated under the optimized condition. The leach liquor generated under the optimized condition contained 44.72 kg/m^3 cobalt and 5.43 kg/m^3 lithium at a pH of 5.00, which was used for cobalt and lithium separation and recovery using a solvent extraction process. To recover the cobalt from a sulfate leach liquor containing 44.72 kg/m^3 cobalt and 5.43 kg/m^3 lithium, solvent extraction studies were carried out using Cyanex 272 as extractant, 5 vol. % TBP as phase modifier, and kerosene for dilution.

4.1 pH-extraction isotherm for cobalt from $LiCoO_2$ leach liquor

The pH-extraction isotherm of cobalt and lithium was studied using 1000 mol/m^3 65% saponified Cyanex 272 at various equilibrium pH values with an aqueous/organic (A/O) volume ratio of 1. The extraction percentage, distribution coefficient, and separation factor at various equilibrium pH values were determined. The effect of equilibrium pH on the extraction and separation of cobalt and lithium is shown in Fig. 19.8. The pH-isotherm presented in Fig. 19.8 indicates that cobalt and lithium recovery increases with increases in equilibrium pH. For an equilibrium pH range of 4.46–5.35, which corresponds to an initial pH of 1.61–5.05, cobalt extraction ranged from 25.18% to 51.23%, and lithium extraction ranged from 5.62% to 8.23%. Lithium recovery was very low compared with cobalt extraction. No significant increase in lithium extraction occurred at an equilibrium pH higher than 4.75. Above an equilibrium pH of 5.35, extraction behavior could not be determined accurately, as cobalt precipitation starts. The separation factor of cobalt and lithium also increases with an increase in equilibrium pH.

As shown in Fig. 19.8, the separation factor also increases with increases in equilibrium pH. In an equilibrium pH range of 4.45–5.35, the separation factor ranged from 5.65 to 11.74. The separation factor increases significantly within the range of pH 4.90–5.35. The maximum separation factor for cobalt and lithium was 11.74, and the

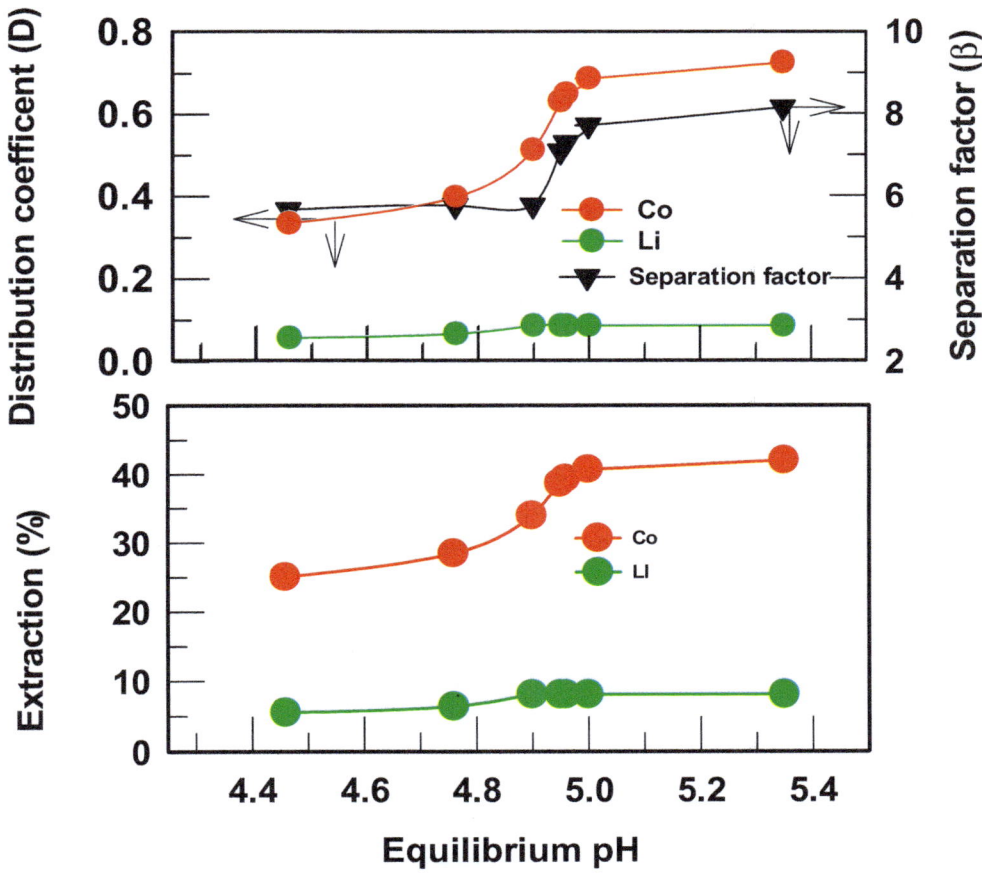

Figure 19.8 pH isotherm for leach liquor at 1000 mol/m^3, 65% Cyanex 272, and effect of pH on distribution coefficient and separation factor.

maximum cobalt extraction was 51.23% at equilibrium pH = 5.35 (initial feed pH = 5.00). For this equilibrium pH, only 8.23% lithium was coextracted. Hence, an initial pH of 5.00 was considered suitable for further studies. Saponified Cyanex 272 was used to better control the equilibrium pH, as it provides suitable buffering activity for that purpose. The saponification reaction of Cyanex 272 using concentrated NaOH can be explained using Eq. (19.7). Through saponification, the dimeric (Cyanex 272)$_2$ converts to monomeric Na–Cyanex 272 derived by inherent cobalt, and Cyanex 272 chemistry provides suitability for efficient extraction of cobalt from leach liquor of LiCoO$_2$. Similarly, the possible chemical reaction using Cyanex 272 as extractant (a mixture of dimeric and monomeric Cyanex 272) as a function of equilibrium pH from cobalt sulfate media can be explained using Eq. (19.8):

$$NaOH_{aq} + (Cyanex\ 272)_2 \rightarrow 2NaCyanex\ 272_{org} + H_2O \quad (19.7)$$

$$2Co(II)SO_{4(aq)} + 2NaCyanex\ 272_{org} + (Cyanex\ 272)_2 \rightarrow Co(II)(Cyanex\ 272)_{2org}$$
$$+ Co(II).2(Cyanex\ 272)_{org} + Na_2SO_{4(aq)} + H_2SO_{4(aq)} \quad (19.8)$$

4.2 Effect of Cyanex 272 concentration and volume of LiCoO$_2$ leach liquor

Optimization of the extraction condition was carried out in terms of extractant concentration and its volume ratio at an initial pH of 5.00. The concentration of the extractant Cyanex 272 was varied at 500 mol/m^3, 750 mol/m^3, 1000 mol/m^3, 1250 mol/m^3, 1500 mol/m^3, 1750 mol/m^3, and 2000 mol/m^3 for various volume ratios. The percentage extraction and separation factors for cobalt and lithium were calculated and are depicted in Tables 19.1 and 19.2, respectively.

Tables 19.1 and 19.2 show that the extraction of cobalt/lithium and the separation factor of cobalt increase as the Cyanex 272 concentration is steadily increased. The cobalt extraction percentage increases as a truncated S-shaped curve when the O/A volume ratio varies from 1.0 to 4.0. For the efficient use of Cyanex 272 and the best quantitative extraction of cobalt from leach liquor, a model is proposed. The proposed model is constructed on the assumption that the transfer of metal species into the organic phase occurs through a nonlinear sigmoid phenomenon. In the proposed model, the volume of extractant is the forcing variable, while the metal extraction percentage is the response.

4.3 Metal extractability behavior and modeling

The percentage of cobalt extracted can be considered a response to the active extractant percentage in the fresh solvent system. The extracted cobalt species percentage exhibits a truncated sigmoid-shaped trend with an insignificant lower part as a function of Cyanex 272 volume for a particular concentration of Cyanex 272. A nonlinear modified sigmoid function was proposed to predict cobalt extraction efficiency ($E_{\%(Co)}$) as a function of the volume of Cyanex 272 ($V_{Cyanex\ 272}$) at a constant concentration in fresh organic feed and presented in Eq. (19.4):

$$E_{\%(Co)} = 100 - Ae^{V_{Cyanex\ 272}/B} \quad (19.9)$$

where the unstructured parameters A, B were estimated by a multivariate nonlinear regression analysis using trust-region algorithms [25,26]. $V_{Cyanex\ 272}$ is the volume of Cyanex 272 at a constant concentration, and $E_{\%(Co)}$ is the percentage of extraction. For various Cyanex 272 organic concentration phases, i.e., at 500 mol/m^3, 750 mol/m^3, 1000 mol/m^3, 1250 mol/m^3, 1500 mol/m^3, 1750 mol/m^3, and 2000 mol/m^3, the percentage of cobalt extraction from aqueous phase was predicted and compared with observed values. The predicted and observed results were compared graphically by

Table 19.1 Effect of Cyanex 272 concentrations and phase ratio on metals extraction. Experimental conditions: aqueous leach liquor contains 44.72 kg/m³ of cobalt and 5.43 kg/m³ of lithium at pH of 5.00.

	(%) Cobalt at various concentrations of Cyanex 272							
Aq.: Org.	500 mol/m³	750 mol/m³	1000 mol/m³	1250 mol/m³	1500 mol/m³	1750 mol/m³	2000 mol/m³
1:1	20.747	27.210	42.156	49.779	55.782	70.566	77.187
1:2	43.75	66.24	78.75	90.3	97.928	99.95	99.89
1:2.5	62	77.451	99.60	99.911	99.923	>99.99	>99.99
1:3	70.3	90.719	99.8	99.949	99.952	>99.99	>99.99
1:4	77	>99.99	>99.99	>99.99	>99.99	>99.99	>99.99

	(%) Lithium coextracted at various concentrations of Cyanex 272							
Aq.: Org.	500 mol/m³	750 mol/m³	1000 mol/m³	1250 mol/m³	1500 mol/m³	1750 mol/m³	2000 mol/m³
1:1	2.3	5.635	8.217	9.303	9.55	11.900	15.052
1:2	2.3	10.034	13.81	26.505	39.52	43.830	45.398
1:2.5	2.3	12.284	25.122	43.91	56.48	54.886	56.480
1:3	2.3	14.533	42.449	51.78	66.142	65.935	66.142
1:4	2.3	48.615	52.249	58.57	86.989	75.380	86.989

Table 19.2 Effect of Cyanex 272 concentrations and phase ratio on separation factor of cobalt over lithium. Experimental conditions: aqueous leach liquor contains 44.72 kg/m^3 of cobalt and 5.43 kg/m^3 lithium at a pH of 5.00.

	Separation factor at various concentrations of Cyanex 272						
Aq.: Org.	500 mol/m^3	750 mol/m^3	1000 mol/m^3	1250 mol/m^3	1500 mol/m^3	1750 mol/m^3	2000 mol/m^3
1:1	12.12	6.26	8.14	9.66	11.94	17.75	19.09
1:2	36.00	14.39	23.12	25.81	72.37	2603	1096.40
1:2.5	43.45	24.54	754.29	1433	1012	4082	1012.57
1:3	109.59	57.49	672.25	1825	1089	5774	1089.44
1:4	154.99	11,814	406,104	7027	1671	3650	1671.82

plotting $E_{\%(Co)}$ versus $V_{Cyanex\ 272}$, as shown in Fig. 19.9. In this figure, symbols denote experimental points. The curves were generated using estimated parameters in the generic model equation. The MATLAB 6.5 Interactive Curve Fitting procedure was used for prediction and compared for each case, and the set of best-fit parameters was estimated using the nonlinear least-squares method, trust-region algorithms, and LAR robust method for performing minimization of the objective function given by the sum of squares due to error (SSE) $\sum_i e_i^2$. The error (e) is the difference between the observed and predicted responses for metal extraction ($E_{\%(Co)}$). The predicted responses

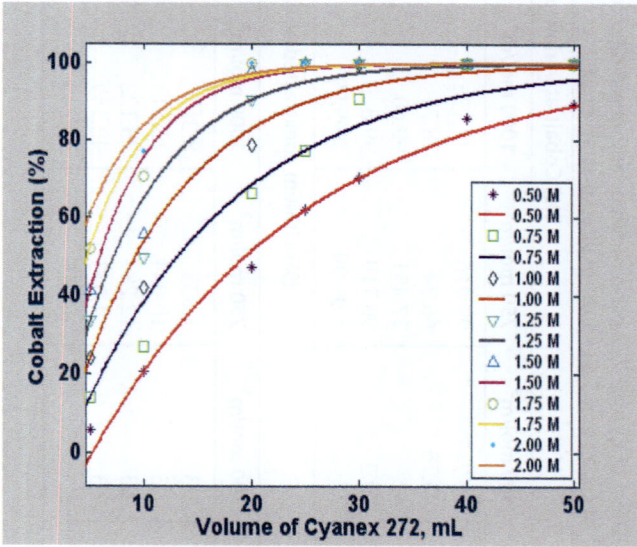

Figure 19.9 Extraction of cobalt at various concentrations of Cyanex 272 from leach liquor at pH 5. The lines are the best-fitted lines generated by the model equation, and the points are experimental results at various concentrations of Cyanex 272.

are the fitted values obtained by evaluating the best-fit equation at the observed values of the independent variable $V_{Cyanex\ 272}$. The asymptotic standard error values (SEs) were also computed for each best fit parameter. The 95% confidence interval value, giving a range within a parameter that is expected to lie with a certain probability of 95%, is calculated for each case.

The parameters estimated by the above method are given in Table 19.3, and the model is shown by the lines in Fig. 19.9. The random patterns of the residuals indicate that the model is free of systematic errors.

Similarly, the percentage of lithium coextracted was considered as a response to the active extractant percentage in the fresh solvent system. As our earlier fundamental investigation demonstrates, when there is competition between cobalt and lithium, the availability of extractants is small, and lithium shows poor extractability [27]. The coextracted lithium percentage is limited toward a truncated sigmoid-shaped trend with a significant lower part as a function of the Cyanex 272 volume for a particular concentration of Cyanex 272. For lithium, a nonlinear modified sigmoid function was proposed to predict $E_{\%(Li)}$ in terms of $V_{Cyanex\ 272}$ in Equation (19.10) [5], with $V_{Cyanex\ 272}$ denoting the volume of Cyanex 272 in fresh organic feed:

$$E_{\%(Li)} = 100 - A_1 e^{V_{Cyanex\ 272}/B_1} \qquad (19.10)$$

where the unstructured parameters A_1, B_1 were similarly estimated by a multivariate nonlinear regression analysis using the trust-region algorithm [25,26]. For all different Cyanex 272 organic concentration phases at 500 mol/m^3, 750 mol/m^3, 1000 mol/m^3, 1250 mol/m^3, 1500 mol/m^3, 1750 mol/m^3, and 2000 mol/m^3, the percentages of lithium extraction from the aqueous phase were predicted and compared with observed values. The predicted and observed results were compared graphically by plotting $E_{\%(Li)}$ versus $V_{Cyanex\ 272}$, as shown in Fig. 19.10. The parameters estimated by the above method are given in Table 19.4, and the model is shown by the lines in

Table 19.3 Estimated model parameters and goodness of fit for cobalt.

Cyanex 272, mol/m^3	A ±SE	B ±SE	Goodness of fit		
			SSE	R-square	Adjusted R-square	RMSE
500	129.30 ± 7.60	20.40 ± 1.73	21.22	0.9965	0.9958	2.06
750	119.70 ± 20.2	14.98 ± 3.35	98.03	0.9864	0.9837	4.428
1000	124.90 ± 47.8	10.06 ± 4.61	244.00	0.9601	0.9521	6.985
1250	125.9 ± 38.0	5.70 ± 2.57	75.01	0.9841	0.9810	3.873
1500	140.60 ± 67.7	10.06 ± 2.53	81.28	0.9787	0.9744	4.032
1750	115.40 ± 38.8	5.70 ± 1.77	26.70	0.9885	0.9861	2.311
2000	93.76 ± 28.29	5.70 ± 1.58	14.15	0.9905	0.9886	1.682

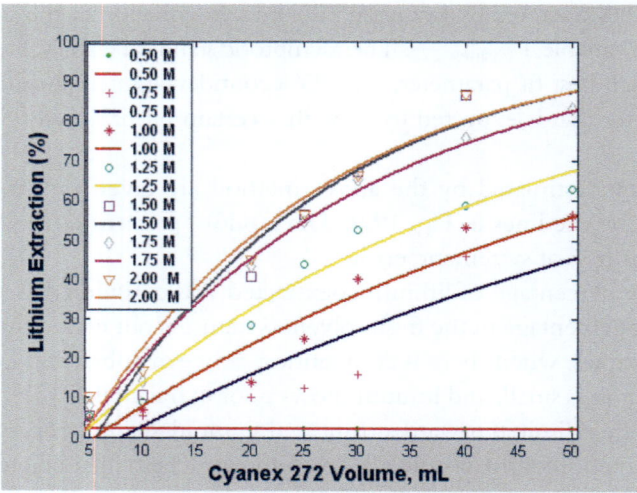

Figure 19.10 Extraction of lithium at various concentrations of Cyanex 272 from leach liquor at pH of 5. The lines are the best-fitted lines generated by the model equation, and points are experimental results at various concentrations of Cyanex 272.

Table 19.4 Estimated model parameters and goodness of fit for lithium.

Cyanex 272, mol/m³	$A_1 \pm SE$	$B_1 \pm SE$	Goodness of fit			
			SSE	R-square	Adjusted R-square	RMSE
500	97.7 ± 0	$-6.395e-014 \pm 7.365e-014$	Poor	Poor	Poor	Poor
750	112.2 ± 22.81	0.0144 ± 0.008	452.80	0.8184	0.7820	9.516
1000	110.8 ± 16.28	0.0185 ± 0.006	206.20	0.9259	0.9111	6.421
1250	110.1 ± 13.09	0.0244 ± 0.006	115.10	0.9653	0.9583	4.799
1500	130.1 ± 32.51	0.0466 ± 0.018	411.80	0.9470	0.9364	9.075
1750	116.9 ± 12.30	0.0380 ± 0.006	71.46	0.9860	0.9832	3.780
2000	121.4 ± 29.31	0.0452 ± 0.016	344.30	0.9486	0.9384	8.298

Fig. 19.10. The random patterns of the residuals indicate that the model is free of systematic errors. Efficient use of Cyanex 272 and the best quantitative extraction of cobalt can be predicted by a suitable compromise between the maximum cobalt extracted and the minimum consumption of extractant.

Fig. 19.11 shows cobalt loading kg/m³ and kg/m³ at various concentrations of Cyanex 272 and various Cyanex 272 volume ratios for aqueous and organic phases. Both cobalt loading kg/m³ and (kg/m³)/(mol/m³) decrease as the organic and aqueous volume ratio increases from 1.00 to 4.00. The decreasing behavior in cobalt loading was very

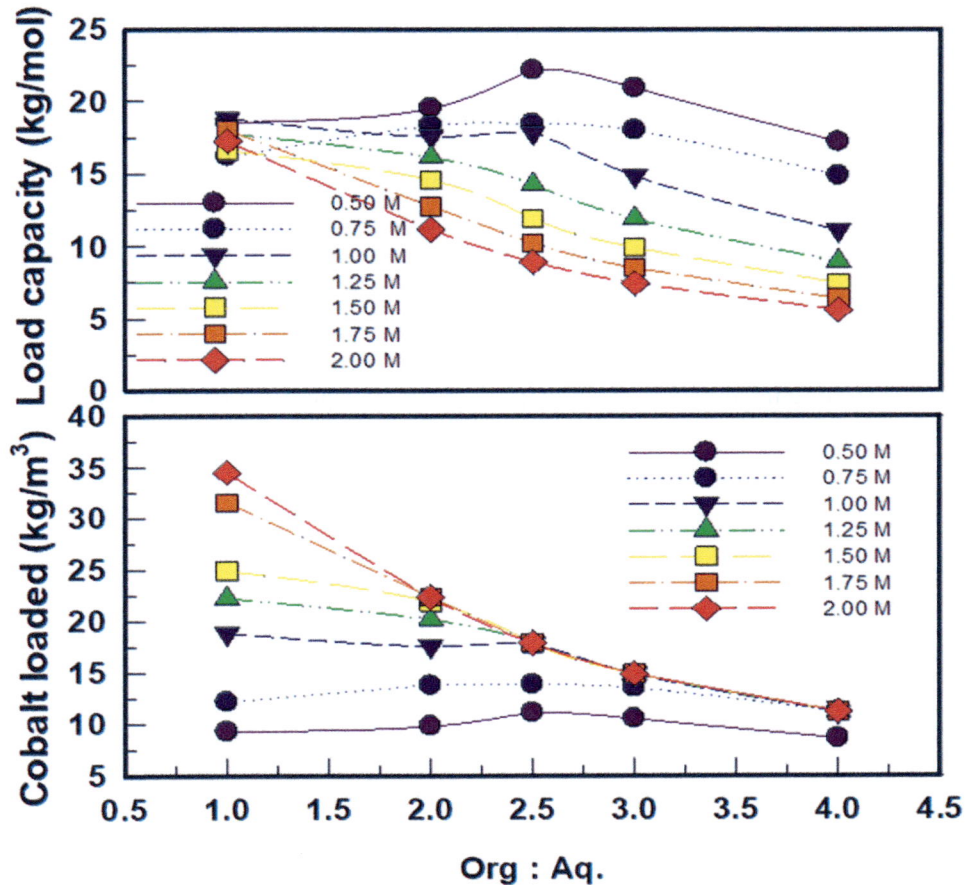

Figure 19.11 Loading of cobalt at different concentration of Cyanex 272. Experimental condition: fresh leach liquor at pH of 5.00%, 65% saponified Cyanex 272.

sharp at higher concentrations like 2000 mol/m^3, 1750 mol/m^3, and 1500 mol/m^3. The decreasing behavior in cobalt loading at lower concentrations like 1250 mol/m^3 and 1000 mol/m^3 was relatively slower. But in the case of Cyanex 272, concentrations of 750 mol/m^3 and 500 mol/m^3 remained constant. Although having a higher concentration and volume ratio is preferable from a quantitative extraction point of view, it has some inherent practical problems. The extractant investment per unit of a gram of cobalt extraction is high at a higher concentration of extractant, and on the other hand, the extractant investment per unit of a gram of cobalt extraction is lower at a lower concentration. This is because of the viscosity of the extractant.

The phase disengagement was slower, and Table 19.4 also shows that a higher coextraction of lithium occurs. This leads to purity concerns in cobalt production and subsequent stripping difficulties at higher concentrations, since pregnant organic extractant is viscous, and saponified Cyanex 272 at this concentration is very vulnerable to cake formation if the solution is kept overnight. From an industrial point of view, the viscous pregnant Cyanex 272 flow rate becomes very slow, and cake formation behavior adds another disadvantage to the problems. Theoretical prediction suggests that for quantitative extraction of cobalt at lower concentrations like 500 mol/m^3, Cyanex 272 requires a higher volume of the organic phase and tedious multistage extraction steps in the subsequent multistep scrubbing and stripping stage. Higher volume adds disadvantages from the points of view of flooding, total system handling, power, and investment. Considering various factors such as the separation factor, phase disengagement problem, the volume of solution, viscosity, scrubbing, purity of loaded organic, multistage extraction, and multistage scrubbing, the McCabe−Thiele extraction isotherm for two Cyanex 272 concentrations, i.e., 1000 mol/m^3 and 1500 mol/m^3, were studied.

4.4 McCabe−Thiele cobalt extraction isotherm from LiCoO$_2$ leach liquor

From our proposed model, cobalt concentrations in the organic phase, aqueous phase, and distribution were calculated for various phase ratios. Theoretical maximum loading of Cyanex 272 was also calculated for two selected concentrations, i.e., 1000 mol/m^3 and 1500 mol/m^3, based on the mechanism. For the construction of McCabe−Thiele, values for cobalt content in the organic and aqueous phases were calculated from the proposed model. Eq. (19.11) was used to calculate the McCabe−Thiele concentrations:

$$Y_{Org.} = \frac{V_{Aq.}}{V_{Org.}(X_0 - X_n)_{Aq.}} \qquad (19.11)$$

where Y and X represent the axis, V represents the volume of solution, subscript Org. represents an organic phase, subscript Aq. represents the aqueous phase, and subscripts 0 and n represent the number of stages.

The extraction isotherm is plotted in Figs. 19.12 and 19.13 to determine the number of stages required for quantitative cobalt extraction from leach liquor using 1000 mol/m^3 and 1500 mol/m^3 of Cyanex 272 at a chosen O/A ratio for observed and calculated values. The leach liquor pH was adjusted to 5.00. The O/A ratio was varied within 1/5 to 5/1, keeping the total volume of the organic and aqueous phases constant. In Fig. 19.12, the McCabe−Thiele diagram for 1000 mol/m^3 of Cyanex 272 indicated two stages are required for quantitative extraction of cobalt at a phase ratio O/A = 2/3 under these conditions both for calculated model and experimental results.

Fig. 19.13 shows the McCabe−Thiele diagram for 1500 mol/m^3 of Cyanex 272 and indicates that two stages are required for quantitative extraction of cobalt at a phase ratio of O/A = 2/3 under these conditions for experimental results, whereas the calculated

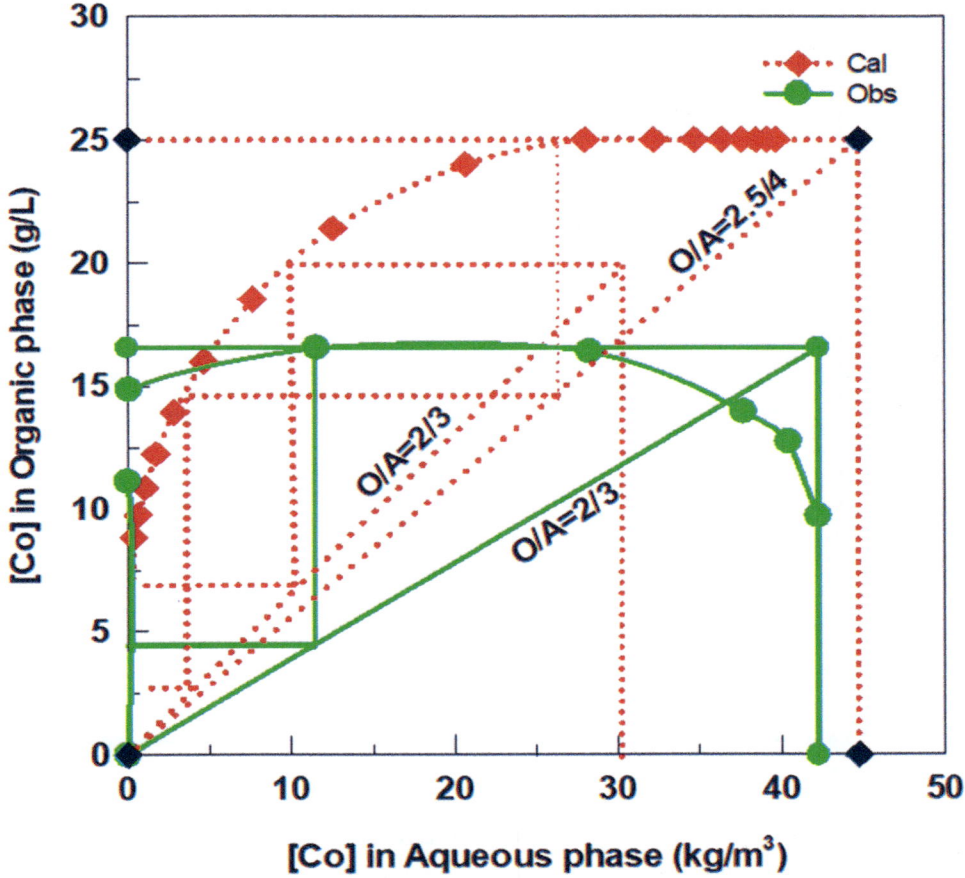

Figure 19.12 McCabe–Thiele isotherm for cobalt extraction from leach liquor using 1000 mol/m^3 Cyanex 272 at pH 5.00 (dotted line indicates calculated line, and dark line indicates the observed result).

model requires the same two stages based on the proposed model but at a different operating ratio of O/A = 0.85. Both studies of the McCabe–Thiele diagram indicate that two stages are required for quantitative extraction of cobalt at a phase ratio of O/A = 2/3 under these conditions for both concentrations of Cyanex 272, i.e., 1000 mol/m^3 and 1500 mol/m^3. Figs. 19.12 and 19.13 show that when the aqueous solution volume becomes higher, cobalt loading to the organic phase decreases after a plateau. From both McCabe–Thiele diagrams, it can be reasonably understood that a higher volume ratio of aqueous leads back to extraction or equilibrium shifts in a backward direction. The feasibility of quantitative cobalt extraction from the leach liquor by two or three stages was checked by a countercurrent simulation study.

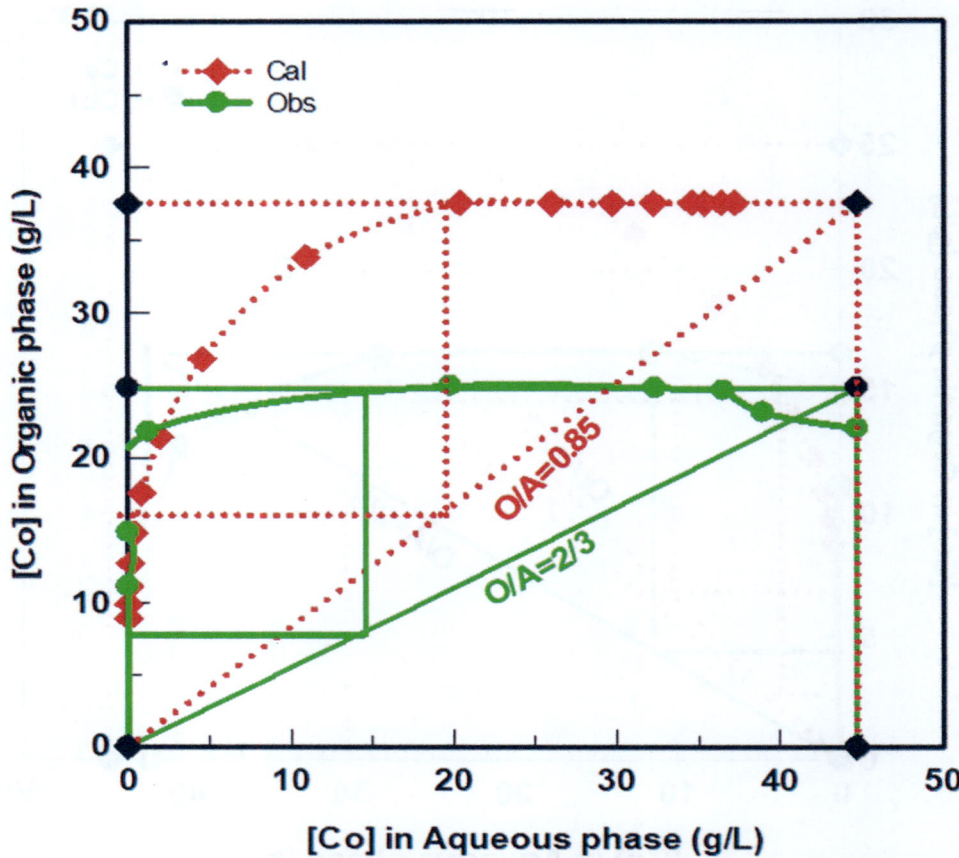

Figure 19.13 McCabe–Thiele isotherm for cobalt extraction from leach liquor using 1500 mol/m³ Cyanex 272 at pH 5.00 (dotted line indicates calculated line and dark line indicate the observed result).

4.5 Countercurrent batch simulation studies for cobalt and lithium separation

To achieve quantitative extraction of cobalt, a two-stage countercurrent batch simulation study was carried out using 1000 mol/m³ and 1500 mol/m³ saponified Cyanex 272 at an O/A phase ratio of 2/3 for each. The representative results obtained for loaded organic and raffinate had the compositions shown in Figs. 19.14 and 19.15. In the case of 1000 mol/m³ of Cyanex 272, only 60.83% of cobalt was extracted, along with 15.66% lithium contamination. So as far as the quantitative extraction of cobalt is concerned, the countercurrent simulation study is not suitable for purposes of this concentration.

On other hand, using 1500 mol/m³ of Cyanex 272, cobalt extraction of 99.99% was observed along with 35.55% of lithium contamination. Hence, this condition was used for our further study for scrubbing contaminated lithium.

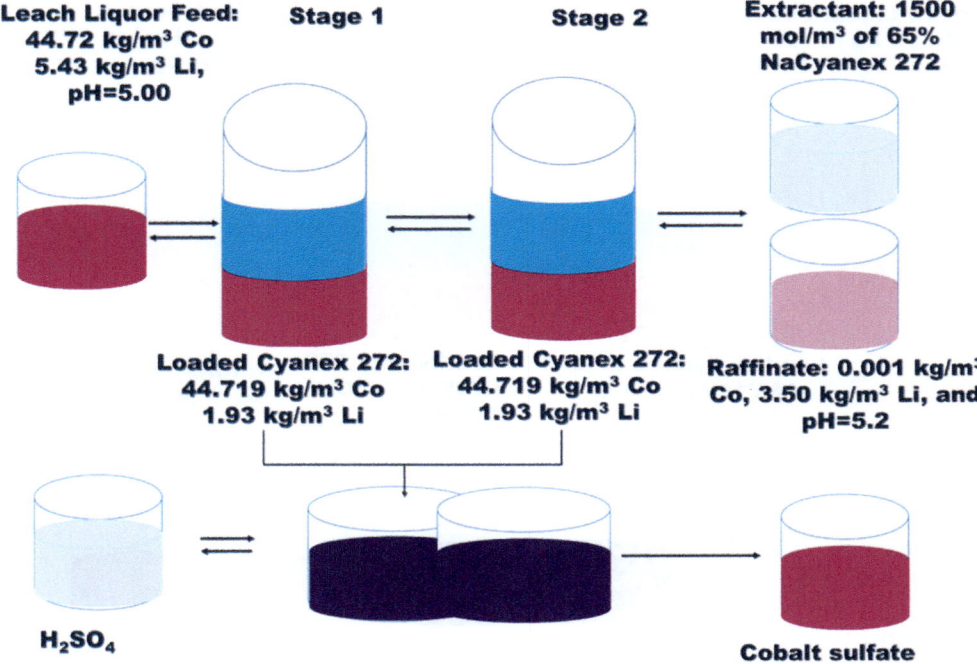

Figure 19.14 A countercurrent simulation study using 1000 mol/m³ Cyanex 272 (65% saponified).

4.6 Lithium scrubbing and cobalt purification from loaded Cyanex 272

The scrubbing studies were carried out as in our previous work [24] to remove lithium from the metal-loaded organic phase using an Na_2CO_3 scrub solution. The metal-loaded organic phase for the scrubbing test was loaded using 1500 mol/m³ of 65% saponified Cyanex 272 with the above countercurrent study. The loaded organic phase after the two-stage countercurrent contained 44.719 kg/m³ of cobalt and 1.93 kg/m³ of lithium. The scrubbing studies were carried out using various scrubbing solutions, such as double-distilled water, distilled water at various temperatures, oxalic acid, and sodium carbonate at various concentration levels. They follow the following order for scrubbed lithium under the same conditions, i.e., oxalic acid > double distilled water ≈ distilled water at various temperatures > sodium carbonate. Sodium carbonate was selected as the scrubbing reagent because others have drawbacks; for example, for oxalic acid, about 1 kg/m³ of cobalt get scrubbed, and in the case of distilled water, phase separation time becomes longer. Hence, the small amount of lithium coextracted in the organic phase can be easily removed with Na_2CO_3 solution. The scrubbing process using Na_2CO_3 can be represented by the displacement reaction given in Eq. (19.6):

$$2Li(Cyanex\ 272) + Na_2CO_3 \leftrightarrow Li_2CO_3 + 2Na(Cyanex\ 272) \qquad (19.12)$$

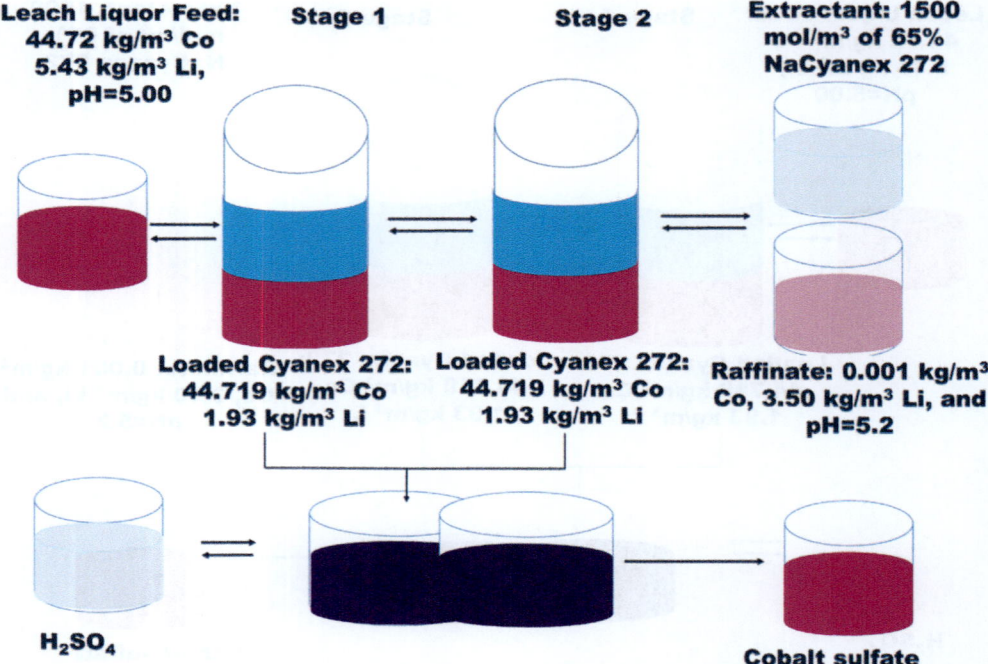

Figure 19.15 A countercurrent simulation study using 1500 mol/m^3 of Cyanex 272 (65% saponified).

Scrubbing studies were conducted using a loaded organic and scrubbing solution of 100 mol/m^3 Na$_2$CO$_3$ within a ratio range of 1/5 to 5/1. The scrubbing isotherm is plotted in Fig. 19.16. The McCabe–Thiele isotherm behavior for scrubbing shows that scrubbing lithium from loaded organic is very efficient, whereas cobalt removal is not easy under this condition.

Fig. 19.16 indicates that scrubbing is required about three times at an O/A ratio of 2.5 to obtain a quantitative scrubbing using 100 mol/m^3 of sodium carbonate. The result obtained from the above simulation was verified by three-stage scrubbing. About 99.99% pure cobalt was obtained in the organic phase. It was observed that as the phase ratio of organic to scrubbing solution increases, phase separation becomes an important problem. At an organic versus scrubbing solution ratio of 5/1, the phase disengagement time was more than 1 day, and after that, phase separation could not even be completed. So it is difficult to use a higher O/A ratio of more than 5/1 for scrubbing.

4.7 Cobalt stripping and cobalt sulfate recovery

Purified cobalt-loaded organic obtained from lithium scrubbing has been used for experimental stripping purposes. In all stripping study sets, the O/A ratio was maintained at 1.

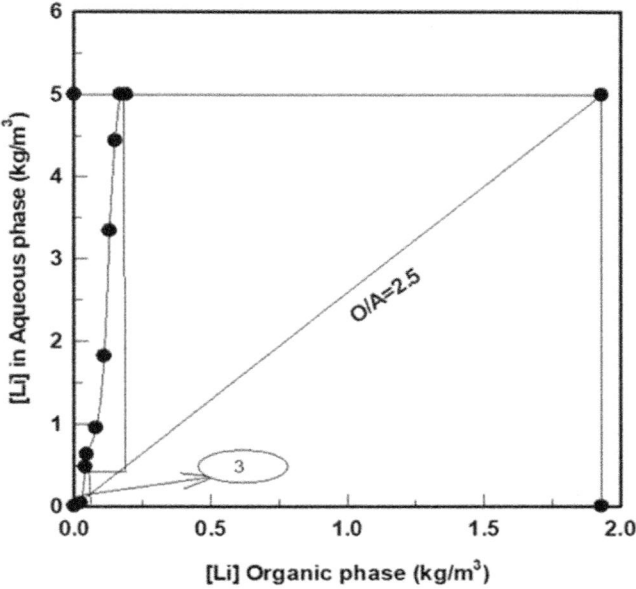

Figure 19.16 Equilibrium isotherms for lithium scrubbing from loaded 1500 mol/m³ Cyanex 272 from leach solution at O/A = 2/3. (Initial loaded organic solution 44.719 kg/m³ Co and 1.93 kg/m³ Li.)

The percentages of stripping and purity at various strip acid concentrations were determined. The results presented in Fig. 19.17 show that cobalt stripping depends on the H_2SO_4 concentration. Cobalt stripping was increased with increases in the concentration of sulfuric acid in the stripping solution. Cobalt stripping increased from 17.10% to 99.95% for an increase in the sulfuric acid concentration from 100 mol/m³ to 750 mol/m³. At a higher H_2SO_4 concentration range, there was no change in metal recovery, as cobalt stripping reaches about 5N purity. However, the lithium present as trace, i.e., 0.002 kg/m³, in the loaded organic was also completely stripped, even at lower sulfuric acid concentrations. The lithium concentration was negligible compared with the cobalt concentration in stripped solution. The change in purity concerning stripping is presented in Fig. 19.17. Fig. 19.17 indicates that the cobalt purity ranged from 99.951% to 99.992%, as the strip acid concentration varied from 100 mol/m³ to 200 mol/m³. From the purified cobalt-loaded organic, pure cobalt sulfate crystal was produced by stripping with 5000 mol/m³ H_2SO_4 contacted at an O/A ratio of 5. Cobalt stripping from loaded Cyanex 272 and the regeneration of Cyanex 272 chemistry using H_2SO_4 can be explained using Eq. (19.13):

$$Co(II)(Cyanex\ 272)_{2org} + Co(II).2(Cyanex\ 272)_{org} + 2H_2SO_{4(aq)} \rightarrow 2Co(II)SO_{4(aq)} + 2(Cyanex\ 272)_{2org} + 2H^+ \quad (19.13)$$

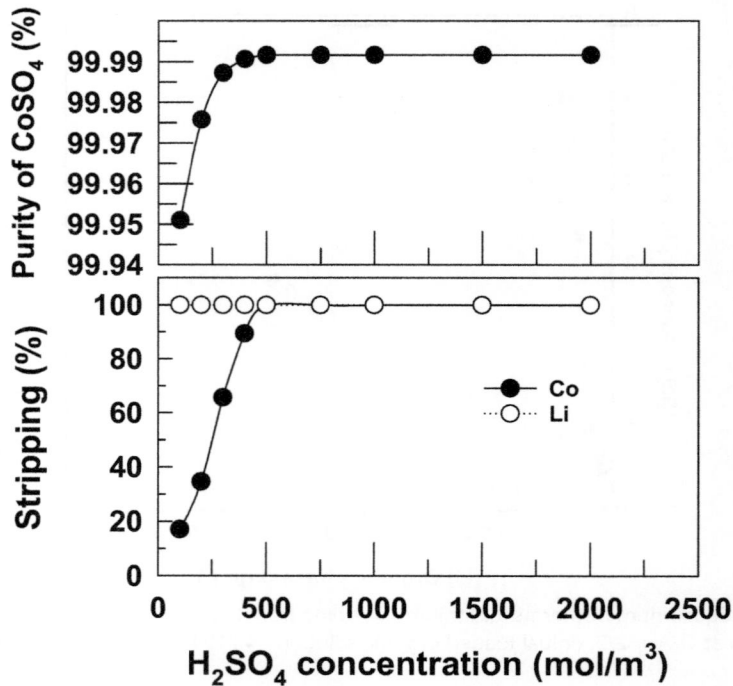

Figure 19.17 Effect of H$_2$SO$_4$ concentration on stripping of cobalt and lithium from loaded organic generated from 1.50 M Cyanex 272.

4.8 Solvent extraction recommendations

(i) More than 65% saponification of Cyanex 272 is neither possible nor suitable because of cake formation in the organic phase, depending on temperature and concentration.

(ii) For pH adjustment in the leach liquor, only concentrated NaOH is recommended because when concentrated NH$_4$OH is used for this purpose, some cobalt is precipitated as (NH$_4$)$_2$Co(SO$_4$)$_2 \cdot$2H2O. This phenomenon may be a disadvantage for the solvent extraction process but can be considered an advantage for obtaining (NH$_4$)$_2$Co(SO$_4$)$_2 \cdot$2H$_2$O salt in the simplest step.

(iii) Care must be taken if saponified Cyanex 272 is stored overnight, as cake formation of saponified Cyanex 272 is enhanced when the temperature drops below room temperature, depending on the environment.

5. Valorization through oxide synthesis

5.1 Synthesis of cobalt oxalate and cobalt oxide powder by precipitation stripping

Synthesis of cobalt oxalate and cobalt oxide powder was started from LIB industry waste and processed through leaching-solvent extract-precipitation stripping-heat treatment.

First, a total of 1×10^{-3} m^3 of leach liquor was produced by leaching waste LIB cathode material at the optimized condition. The leach liquor generated under this condition contains 44.72 kg/m^3 cobalt and 5.43 kg/m^3 lithium. This leach liquor was used for further purification of cobalt by solvent extraction using 65% saponified Cyanex 272 as an extractant. The extractant Cyanex 272 (1000 mol/m^3) was saponified up to 65% using concentrated NaOH solution to the extractant in kerosene under constant stirring of the mixture. A 5 vol. % of TBP was used as a phase modifier to the organic extractant solution. The 65% saponified Cyanex 272 was equilibrated with the leach liquor at pH 5.00 with an O/A ratio of two for 10 min. The requisite pH of the aqueous leach liquor feed was adjusted using sodium hydroxide. Following equilibrium, phase separation was achieved within 5 min, and both phases were separated. The aqueous raffinate was analyzed to determine the volume of cobalt and lithium extracted by the organic extractant. The analysis indicates 78.75% extraction of cobalt from leach liquor, which corresponds to 17.61 kg/m^3 cobalt loading in the organic phase. Similarly, the 13.81% coextraction of lithium corresponds to 0.38 kg/m^3 extraction in the organic phase. For further purification of cobalt, the lithium was scrubbed using 100 mol/m^3 Na$_2$CO$_3$ solution (3 times) at an O/A of 0.5. For this step, 0.1 kg/m^3 of cobalt and 0.09 kg/m^3 were scrubbed out in each stage. Purified and loaded organic was stripped using 1000 mol/m^3 of oxalic acid. In the presence of oxalic acid, cobalt oxalate was precipitated out. The precipitate was centrifuged at 5000 rpm for 10 min. After centrifuge, the precipitate was separated from the organic phase and acetone-washed to remove the organic extractant. Fig. 19.18 shows the XRD pattern and SEM hydrated CoC$_2$O$_4$ isolated/recovered/purified through precipitation stripping. The XRD pattern in Fig. 19.18A shows that the precipitated sample is CoC$_2$O$_4 \cdot$H$_2$O. Morphology analysis through the SEM image in Fig. 19.18B shows that the precipitated sample has a rodlike structure.

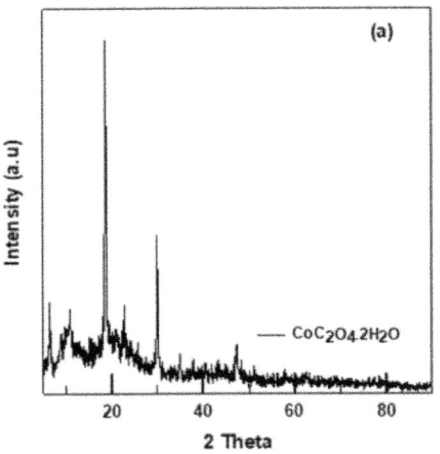 

Figure 19.18 (A) XRD pattern and (B) SEM image of precipitation stripped sample.

Hence, rod-shaped $CoC_2O_4 \cdot H_2O$ was synthesized through the proposed process. The precipitate was dried in an oven at 90°C overnight. The dried oxalate was characterized by TG-DTA, XRD, and SEM. Fig. 19.19 shows TG-DTA for the dry cobalt oxalate precipitate sample. TGA for the oxalate shows three different transitions at 220, 360, and 800°C corresponding to the conversion of $CoC_2O_4 \cdot 2H_2O$ to CoC_2O_4, CoC_2O_4 to Co_3O_4, and Co_3O_4 to CoO. In the same figure, DTA exhibits two different endothermic and exothermic peaks at 200 and 364°C, respectively. DTA indicates that the conversion of $CoC_2O_4 \cdot 2H_2O$ to CoC_2O_4 is an endothermic reaction, and conversion of CoC_2O_4 to the Co_3O_4 exothermic reaction is an exothermic reaction.

To better understand and characterize by TGA-DTA, the samples were synthesized by calcination at 100, 200, 250, 360, 450, 900, and 1100°C. The XRD pattern of all the synthesized samples was analyzed and is presented in Fig. 19.20. The XRD pattern indicates that up to 200°C, the synthesized cobalt oxalate remained as hydrated $CoC_2O_4 \cdot 2H_2O$, and at 200°C, the hydrated $CoC_2O_4 \cdot 2H_2O$ converted to anhydrous CoC_2O_4. At 350°C, the anhydrous CoC_2O_4 converted to Co_3O_4, and the Co_3O_4 converted to CoO at 1100°C.

Fig. 19.21 represents an SEM micrograph of two selective powders, i.e., CoC_2O_4 and Co_3O_4, synthesized at 250 and 450°C, respectively. The figure clearly shows that rod-shaped CoC_2O_4 powder can be conveniently synthesized from waste LIB. From the CoC_2O_4 powder, Co_3O_4 rod-shaped powder can be synthesized. The synthesis process can close the loop, and cobalt recovered from waste LIB cathode material can be valorized through rod-shaped Co_3O_4 powder synthesis and returned for battery manufacturing as a precursor for LIB manufacturing.

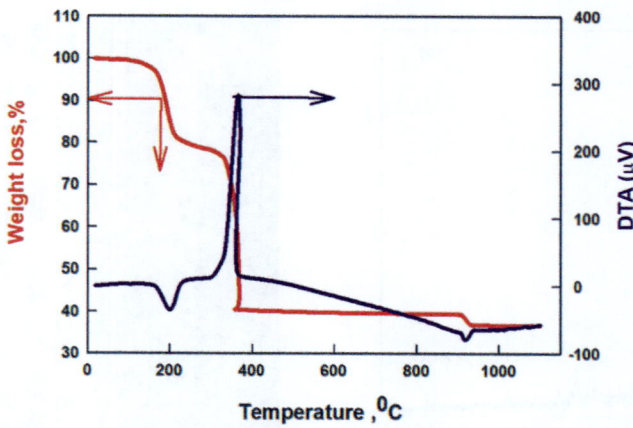

Figure 19.19 TG-DTA for dry cobalt oxalate precipitate sample.

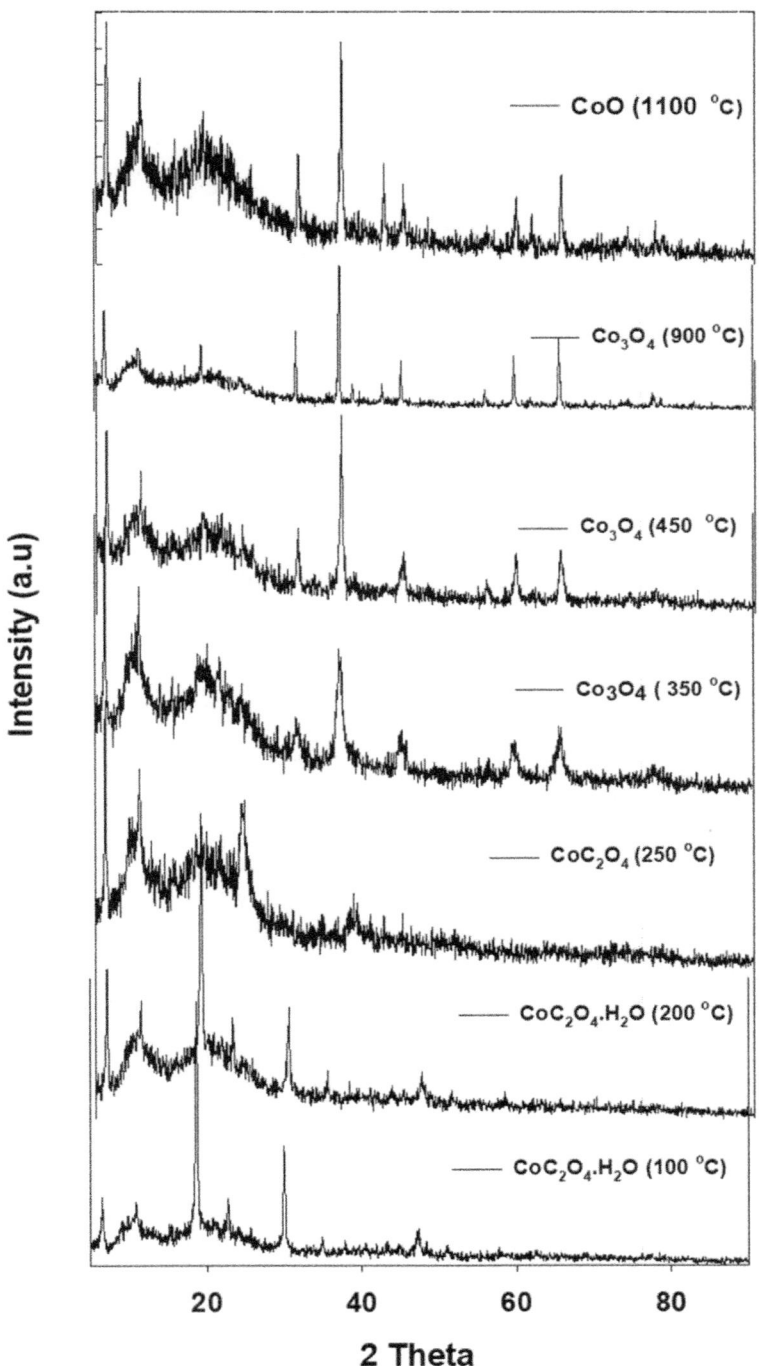

Figure 19.20 XRD pattern of pure CoC_2O_4 sample calcined at 100, 200, 250, 360, 450, 900, and 1100°C.

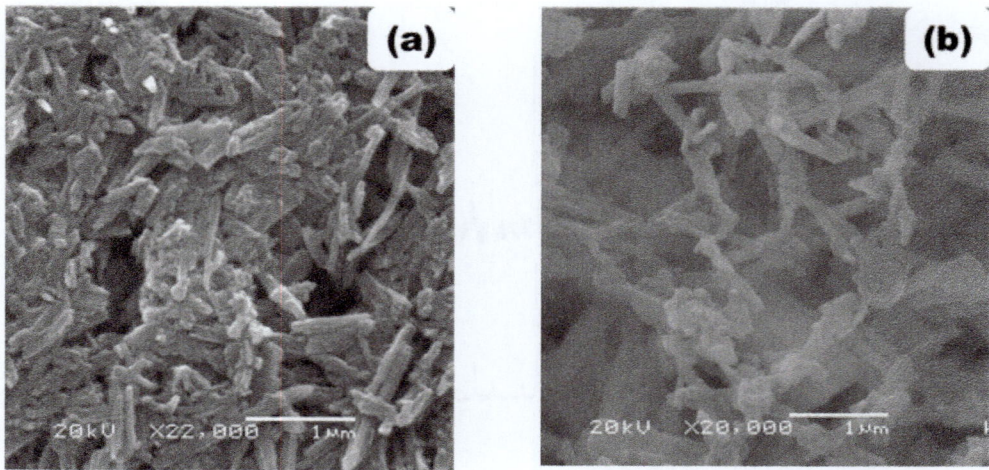

Figure 19.21 SEM image of (A) CoC_2O_4 powder and (B) Co_3O_4 powder synthesized at 250 and 450°C, respectively.

5.2 Chemistry of precipitation stripping and cobalt oxide powder synthesis

The possible chemical reaction mechanism for synthesizing cobalt oxalate from cobalt-loaded Cyanex 272 through stripping can be explained using Eqs. (19.14) and (19.15). In the solution phase, $C_2O_4H_2$ dissociates as $C_2O_4^{2-}$ through protonation. When equilibrated, charge transfer enters through the organic phase (cobalt-loaded Cyanex 272). In the organic phase, $C_2O_4^{2-}$ acts as a polydentate ligand. As shown in the equations [2], through displacement, $C_2O_4^{2-}$ forms a coordination complex with cobalt and dissociates the $[Cyanex\ 272]^{2-}$ from Co^{2+}. The Co^{2+} and $C_2O_4^{2-}$ form a hydrated $CoC_2O_4 \cdot 2H_2O$ complex. Through calcination at 220°C, the hydrated $CoC_2O_4 \cdot 2H_2O$ is converted to anhydrous CoC_2O_4. The Co_3O_4 can be synthesized from anhydrous CoC_2O_4 through calcination at around 350°C. The chemical Eqs. (19.16) and (19.17) represent the conversion of $CoC_2O_4 \cdot 2H_2O$ to CoC_2O_4, and the synthesis of CoC_2O_4 to Co_3O_4, respectively:

(19.14)

(19.15)

$$\text{Co(C}_2\text{O}_4\text{)(OH}_2\text{)}_2 \xrightarrow{220\ °C} \text{Co(C}_2\text{O}_4\text{)} + 2\text{H}_2\text{O} \qquad (19.16)$$

$$\text{Co(C}_2\text{O}_4\text{)} \xrightarrow{360\ °C} \text{Co}_3\text{O}_4 \qquad (19.17)$$

The Co_3O_4 powder synthesis chemistry explained in Eq. (19.17) is not obvious and has hardly been discussed in the literature. Hence, it is discussed further through Eqs. (19.18) to (19.26). As shown in Eqs. (19.18) to (19.26), several reaction processes occur. As shown in Eq. (19.19), some Co^{2+} undergoes an oxidation reaction to form Co^{3+}. Some of the oxalate ion undergoes a reduction to form a glycolic ion and generate O_2. Co^{3+} and Co^{2+} form Co_2O_3 and CoO, as shown in Eqs. (19.21) and (19.22), respectively. As shown in Eq. (19.23), the oxalate ion undergoes a pyrolysis reaction to form CO_2, and the glycolic ion undergoes a pyrolysis reaction to form CO and H_2O. Finally, Co_2O_3 and CoO undergo a solid-state reaction to form Co_3O_4:

$$3\,\text{Co(C}_2\text{O}_4\text{)} \xrightarrow{360\ °C} 3\,\text{C}_2\text{O}_4^{2-} + 3\,\text{Co}^{2+} \qquad (19.18)$$

$$2\,\text{Co}^{2+} \xrightarrow{360\ °C} 2\,\text{Co}^{3+} + 2e^- \qquad (19.19)$$

$$2\,\text{C}_2\text{O}_4^{2-} + 2e^- + 2\text{H}_2\text{O} \xrightarrow{360\ °C} 2\,\text{C}_2\text{H}_2\text{O}_3^{2-} + 2\text{O}_2 \qquad (19.20)$$

$$2\,\text{Co}^{3+} + 1.5\,\text{O}_2 \xrightarrow{360\ °C} \text{Co}_2\text{O}_3 \qquad (19.21)$$

$$\text{Co}^{2+} + 0.5\,\text{O}_2 \xrightarrow{360\ °C} \text{CoO} \qquad (19.22)$$

$$\text{C}_2\text{O}_4^{2-} \xrightarrow{360\ °C} 2\,\text{CO}_2 \qquad (19.23)$$

$$2 \; \text{(glyoxylate)} \xrightarrow{360\,°C} 4CO + 2H_2O \quad (19.24)$$

$$Co_2O_3 + CoO \xrightarrow{360\,°C} Co_3O_4 \quad (19.25)$$

$$3 \; \text{(Co oxalate)} \xrightarrow{360\,°C} Co_3O_4 + 2CO_2 + 4CO \quad (19.26)$$

Unlike most cobalt powder preparation reported in the literature, instead of a pure chemical precursor from a complex and impure LIB waste cathode, the value-added rod-shaped CoC_2O_4 and Co_3O_4 powder were synthesized, which is a novel, economical, and environmentally friendly process. Valorization of waste LIBs has been achieved through a technique of selective purification of cobalt by solvent extraction and the synthesis of rod-shaped CoC_2O_4 powder. From CoC_2O_4 powder, pure Co_3O_4 can be synthesized through simple calcination at 450°C. The developed cobalt powder recovery process is a techno-economically feasible process for commercial production of rod-shaped pure Co_3O_4 powder at about a 1 μm size from the reported one-pot synthesis through simple precipitation stripping. The technique is a versatile and flexible process for mass production of pure CoC_2O_4 and Co_3O_4 powder. As reported, the processes are simple and do not require any special industrial facilities, allowing the waste LIB recycling industry to produce pure CoC_2O_4 and Co_3O_4 powder. The reported technique can supplement the existing waste LIB cathode recycling industry for the valorization of waste LIB cathode material.

6. Process development

In this research, a hydrometallurgical process for the separation and recovery of cobalt and lithium from waste cathodic active material ($LiCoO_2$) generated during the manufacturing of LIBs from sulfate media was developed. The extraction/separation process is relatively simple. The process consists of (1) optimization of leaching parameters using H_2SO_4 as a lixiviant, (2) selective solvent extraction of cobalt over lithium using 65% saponified Cyanex 272, (3) scrubbing of extracted lithium by Na_2CO_3, and (4) recovery of cobalt as cobalt sulfate salt or synthesis of Co_3O_4 powder for battery manufacturing. The flowsheet is shown in Fig. 19.22. The process may be effective from waste management, recycling, and environmental safety points of view.

LIB industry waste valorization for battery production 423

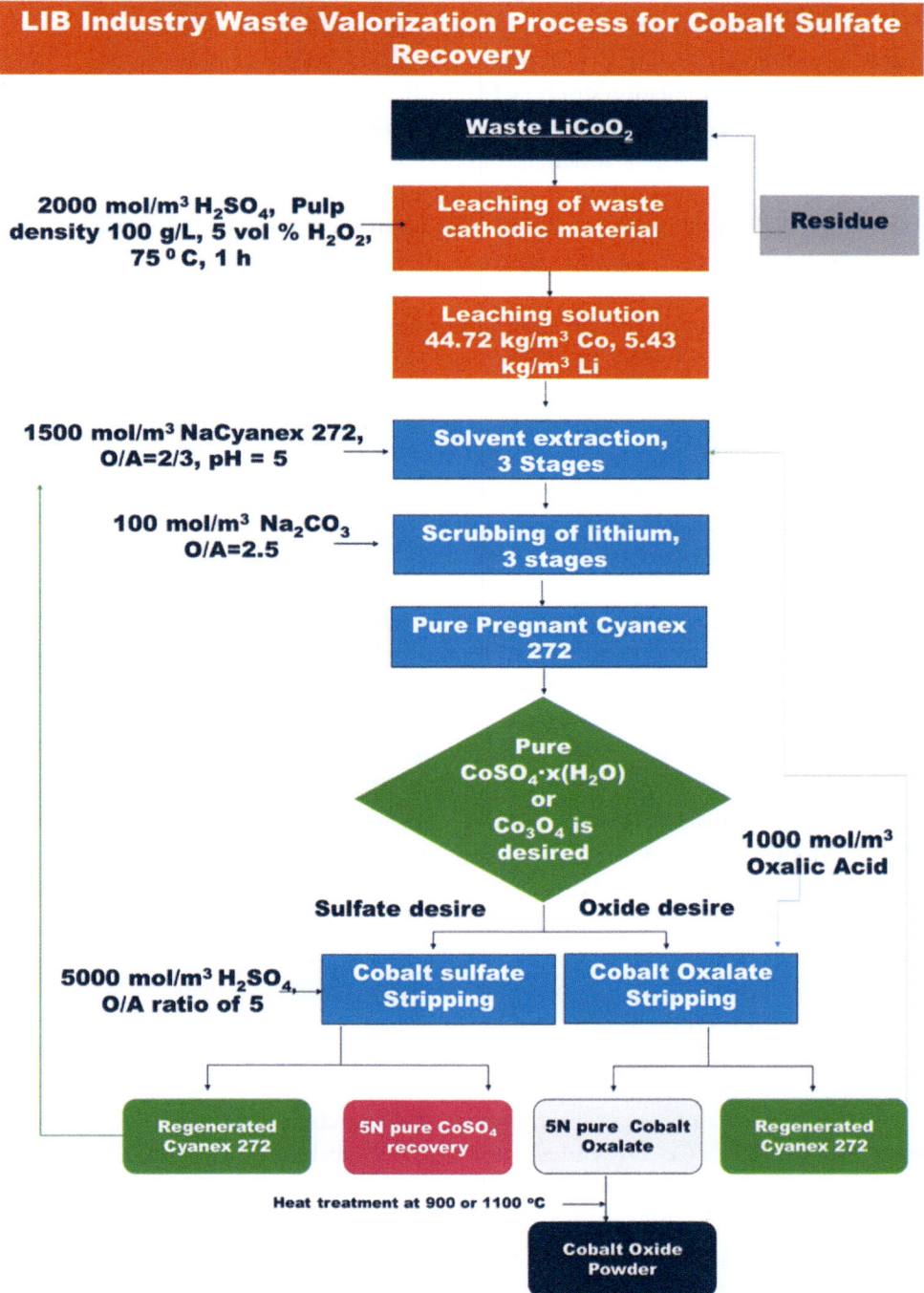

Figure 19.22 Flowsheet for recovery of high-purity cobalt from sulfate leach liquor of waste LiCoO$_2$. (The dotted line indicates that the organic phase is returned to the extraction step for reuse, and residue generated after leaching can be used for leaching again along with the waste.)

7. Conclusions

Report regarding valorization Co_3O_4 from waste LIB cathode material by leaching-solvent extract-scrubbing-precipitation stripping has hardly been reported elsewhere and discussed in this section [28]. Followed by the synthesis of Co_3O_4 through leaching-solvent extract-precipitation stripping-heat treatment from $LiCoO_2$ can be explained in the flowsheet presented in Fig. 19.22. Followed by the significance of the reported process for synthesis.

(i) In general, most Co_3O_4 used for $LiCoO_2$ reported in the literature is from a pure chemical precursor, whereas our currently reported process uses waste LIB cathode as the precursor material.

(ii) The best condition for leaching cobalt from waste $LiCoO_2$ in our system was found to be 2000 mol/m^3 of H_2SO_4 at a leaching temperature of 75°C, a pulp density of 100 kg/m^3, and an H_2O_2 addition amount of 5 vol. % for 30 min. Under these conditions, the leaching efficiencies of cobalt and lithium were 93% and 94.5%, respectively.

(iii) The leaching efficiency of cobalt from waste $LiCoO_2$ was very low without adding H_2O_2, but leaching efficiency greatly increased with an increase in H_2O_2. This is because H_2O_2 reduces Co(III) to Co(II).

(iv) Solvent extraction using Cyanex 272 was very effective for separating cobalt and lithium from the waste cathodic active material ($LiCoO_2$) leach liquor. The quantitative recovery of cobalt with minimum lithium coextraction was accomplished using a Cyanex 272 concentration of 1.50 M, an initial pH of 5.00, and an O/A ratio of 2/3 in two stages. Solvent extraction was followed by three-stage lithium scrubbing using a 0.1 M Na_2CO_3 solution with an O/A ratio of 3.8. Then it was stripped using 0.5 M sulfuric acid. About 99.99% of pure $CoSO_4$ solution can be obtained by managing the suitable volume and concentration of H_2SO_4 solution.

(v) From the above three-step optimization process, i.e., leaching, countercurrent solvent extraction, and scrubbing, a suitable flow circuit was developed for cobalt sulfate recovery (Fig. 19.22).

(vi) Through leaching-solvent extract-precipitation stripping, CoC_2O_4 was synthesized, followed by Co_3O_4 synthesized through simple calcination.

(vii) Most Co_3O_4 synthesis described in the literature is about a gram to subgram material synthesis process, but our process offers a versatile and flexible approach for mass production capability up to a kilogram scale. Unlike the reported process, valorization can be a continuous rather than batch process.

(viii) The developed Co_3O_4 synthesis techniques are techno-economically feasible processes for commercial production of Co_3O_4 powder through one-pot precipitation stripping without any complexity.

(ix) The process may be effective from waste management, waste recycling, and environmental safety points of view.

References

[1] C. Xu, Q. Dai, L. Gaines, M. Hu, A. Tukker, B. Steubing, Future material demand for automotive lithium-based batteries, Commun. Mater. 1 (1) (2020) 99.

[2] P. Greim, A.A. Solomon, C. Breyer, Assessment of lithium criticality in the global energy transition and addressing policy gaps in transportation, Nat. Commun. 11 (1) (2020) 4570.

[3] Sumangil AYM, Top Electric Vehicle Markets Dominate Lithium-Ion Battery Capacity Growth: S&P Global Market Intelligence, 2021 [Available from: https://www.spglobal.com/marketintelligence/en/news-insights/blog/top-electric-vehicle-markets-dominate-lithium-ion-battery-capacity-growth.

[4] Lithium-ion Battery Recycling Market by Battery Chemistry (Lithium-Nickel Manganese Cobalt, Lithium-Iron Phosphate, Lithium-Manganese Oxide, LTO, NCA, LCO), Industry (Automotive, Marine, Industrial, and Power), and Region - Global Forecast to 2030: Markets and Markets Research Private Ltd, 2021. https://www.marketsandmarkets.com/Market-Reports/lithium-ion-battery-recycling-market-153488928.html.

[5] B. Swain, J. Jeong, J.-C. Lee, G.-H. Lee, J.-S. Sohn. Hydrometallurgical process for recovery of cobalt from waste cathodic active material generated during manufacturing of lithium ion batteries. J. Power Sources. 167 (2) (2007) 536–544.

[6] N. Vieceli, R. Casasola, G. Lombardo, B. Ebin, M. Petranikova, Hydrometallurgical recycling of EV lithium-ion batteries: effects of incineration on the leaching efficiency of metals using sulfuric acid, Waste Manag. 125 (2021) 192–203.

[7] J. Nan, D. Han, X. Zuo, Recovery of metal values from spent lithium-ion batteries with chemical deposition and solvent extraction, J. Power Sources 152 (2005) 278–284.

[8] X. Zheng, Z. Zhu, X. Lin, Y. Zhang, Y. He, H. Cao, et al., A mini-review on metal recycling from spent lithium ion batteries, Engineering 4 (3) (2018) 361–370.

[9] P. Meshram, B.D. Pandey, T.R. Mankhand, Extraction of lithium from primary and secondary sources by pre-treatment, leaching and separation: a comprehensive review, Hydrometallurgy 150 (2014) 192–208.

[10] J. Lin, C. Liu, H. Cao, R. Chen, Y. Yang, L. Li, et al., Environmentally benign process for selective recovery of valuable metals from spent lithium-ion batteries by using conventional sulfation roasting, Green Chem. 21 (21) (2019) 5904–5913.

[11] M. Contestabile, S. Panero, B. Scrosati, A laboratory-scale lithium-ion battery recycling process, J. Power Sources 92 (1) (2001) 65–69.

[12] S.-G. Zhu, W.-Z. He, G.-M. Li, X. Zhou, X.-J. Zhang, J.-W. Huang, Recovery of Co and Li from spent lithium-ion batteries by combination method of acid leaching and chemical precipitation, Trans. Nonferrous Metals Soc. China 22 (9) (2012) 2274–2281.

[13] C.K. Lee, K.-I. Rhee, Reductive leaching of cathodic active materials from lithium ion battery wastes, Hydrometallurgy 68 (1–3) (2003) 5–10.

[14] P. Zhang, T. Yokoyama, O. Itabashi, T.M. Suzuki, K. Inoue, Hydrometallurgical process for recovery of metal values from spent lithium-ion secondary batteries, Hydrometallurgy 47 (2) (1998) 259–271.

[15] D.-S. Kim, J.-S. Sohn, C.-K. Lee, J.-H. Lee, K.-S. Han, Y.-I. Lee, Simultaneous separation and renovation of lithium cobalt oxide from the cathode of spent lithium ion rechargeable batteries, J. Power Sources 132 (1) (2004) 145–149.

[16] F. Larouche, F. Tedjar, K. Amouzegar, G. Houlachi, P. Bouchard, G.P. Demopoulos, et al., Progress and status of hydrometallurgical and direct recycling of Li-ion batteries and beyond, Materials 13 (3) (2020).

[17] M. Aaltonen, C. Peng, B. Wilson, M. Lundström, Leaching of metals from spent lithium-ion batteries, Recycling 2 (4) (2017).

[18] F. Wang, R. Sun, J. Xu, Z. Chen, M. Kang, Recovery of cobalt from spent lithium ion batteries using sulphuric acid leaching followed by solid–liquid separation and solvent extraction, RSC Adv. 6 (88) (2016) 85303–85311.

[19] X. Chen, T. Zhou, Hydrometallurgical process for the recovery of metal values from spent lithium-ion batteries in citric acid media, Waste Manag. Res. 32 (11) (2014) 1083–1093.

[20] B. Swain, C. Mishra, H.S. Hong, S.-S. Cho, S. Lee, Commercial process for the recovery of metals from ITO etching industry wastewater by liquid—liquid extraction: simulation, analysis of mechanism, and mathematical model to predict optimum operational conditions, Green Chem. 17 (7) (2015) 3979—3991.

[21] A.L. Mular, D.N. Halbe, D.J. Barratt, Mineral Processing Plant Design, Practice, and Control: Proceedings, Society for Mining, Metallurgy, and Exploration, 2002.

[22] X. Zhang, Y. Xie, X. Lin, H. Li, H. Cao, An overview on the processes and technologies for recycling cathodic active materials from spent lithium-ion batteries, J. Mater. Cycles Waste Manag. 15 (4) (2013) 420—430.

[23] N.B. Devi, K.C. Nathsarma, V. Chakravortty, Separation of divalent manganese and cobalt ions from sulphate solutions using sodium salts of D2EHPA, PC 88A and Cyanex 272, Hydrometallurgy 54 (2—3) (2000) 117—131.

[24] M.C. Olivier, Developing a Solvent Extraction Process for the Separation of Cobalt and Iron from Nickel Sulfate Solutions, Engineering at Stellenbosch University, 2011.

[25] W.S. Cleveland, Robust locally weighted regression and smoothing scatterplots, J. Am. Stat. Assoc. 74 (368) (1979) 829—836.

[26] W.S. Cleveland, S.J. Devlin, Locally weighted regression: an approach to regression analysis by local fitting, J. Am. Stat. Assoc. 83 (403) (1988) 596—610.

[27] B. Swain, J. Jeong, J. Lee, G.-H. Lee, Separation of cobalt and lithium from mixed sulphate solution using Na-Cyanex 272, Hydrometallurgy 84 (3) (2006) 130—138.

[28] B. Swain JCl, C.G. Lee, Valorization of cobalt from waste LIB cathode through cobalt oxalate and cobalt oxide synthesis by leaching-solvent extract-precipitation stripping, Arch. Metall. Mater. 63 (2) (2018) 1037—1042.

CHAPTER 20

Recycling lithium, cobalt, and nickel for return to the battery production cycle

Yue Yang[1,2], Shuya Lei[1] and Rui Xu[1]

[1]School of Minerals Processing and Bioengineering, Central South University, Changsha, Hunan, China; [2]Key Laboratory of Hunan Province for Clean and Efficient Utilization of Strategic Calcium-containing Mineral Resources, Central South University, Changsha, Hunan, China

1. Direct repair

Direct repair is a regeneration technology that supplements deficient ions to the crystal structure of the spent cathode material and restores its electrochemical performance without complex smelting processes. Based on repair characteristics, it can be divided into solid-phase calcination and the hydrothermal method.

1.1 Solid-phase calcination

Solid-phase calcination is a process that first mixes spent cathode materials with lithium salt or transition metal salt and then calcinates the mixture at a high temperature of about 600–900°C (Fig. 20.1A). During this process, lithium salt or transition metal salt is added according to the stoichiometric ratio of metal elements lacking in the cathode material. With the help of calcination, the binder and organic components decompose, and the lithium or transitional metals are fixed in the crystal structure, resulting in restoration of the crystal structure and electrochemical performance [1]. For example, Zhou et al. [2] successfully repaired $LiNi_{0.5}Co_{0.2}Mn_{0.3}O_2$ (NCM 523) cathode material, with the deficiency of lithium in the material lattice replenished by lithium acetate at a mole ratio of (Ni + Co + Mn):Li = 1:1.05 (Fig. 20.1B–G). The discharge capacities of the repaired cathode material were 164.6 mAh/g at 0.1 C and 147 mAh/g at 1 C, respectively. The capacity retention reached 89.12% after 100 cycles at 1 C. Similarly, Zhang et al. directly regenerated $Li(Ni_{1/3}Mn_{1/3}Co_{1/3})O_2$ (NCM 111) materials by adjusting the mole ratio of Ni, Co, and Mn at 1:1:1 with Li_2CO_3, $Ni(NO_3)_2 \cdot 6H_2O$, $Co(NO_3)_2 \cdot 6H_2O$, and $Mn(NO_3)_2 \cdot 4H_2O$. The mixture is first calcinated at 450°C for 5 h, and then calcined at 900°C for 20 h, the regenerated cathode materials show good spherical morphology, and the initial discharge capacity reaches 155.4 mAh/g at 0.1 C [3].

1.2 Hydrothermal method

The hydrothermal method first puts spent cathode materials into a lithium-containing solution and then heats the mixture in a reactor to obtain the regenerated cathode materials (Fig. 20.2A). The key step in hydrothermal repair is lithiation. In lithiation, Li^+ in solution

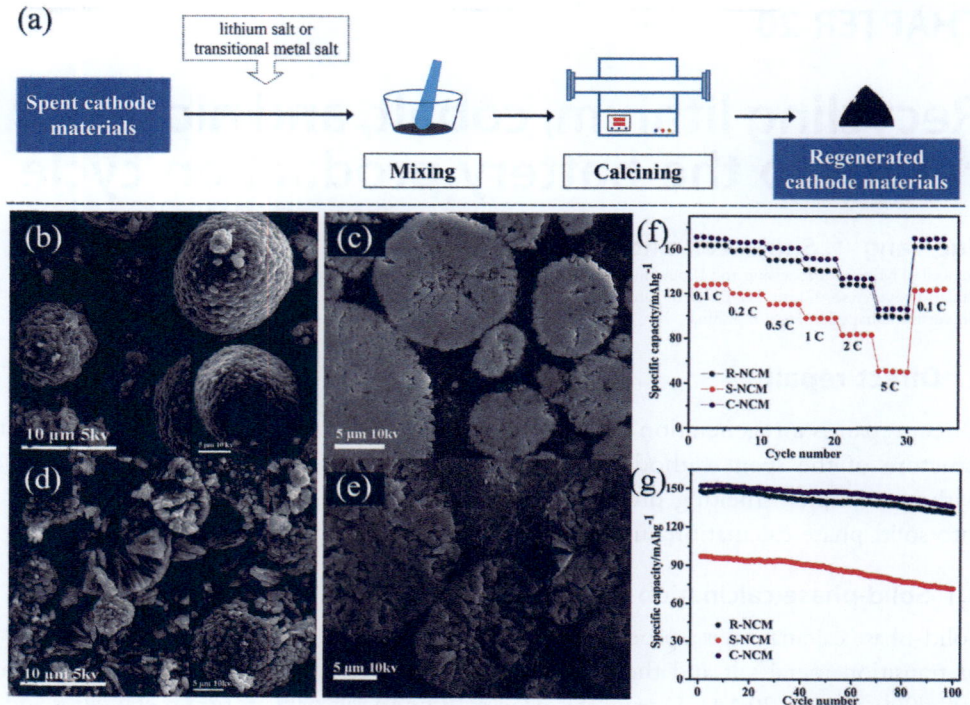

Figure 20.1 (A) Flowchart of regenerating cathode material by solid phase calcination. SEM images of (B, C) regenerated Li(Ni$_{1/3}$Mn$_{1/3}$Co$_{1/3}$)O$_2$ and (D, E) spent Li(Ni$_{1/3}$Mn$_{1/3}$Co$_{1/3}$)O$_2$ material, (F) cycling performance and (G) rate performance of regenerated Li(Ni$_{1/3}$Mn$_{1/3}$Co$_{1/3}$)O$_2$ [2].

enters into the vacancy of lithium in the spent cathode materials (Fig. 20.2B) [4], repairing the material structure. As Shi et al. [4] studied, spent NCM 523 and NCM 111 material is first lithiated at 220°C for 4 h to restore the lithium content, and then the material structure is reconstructed at 850°C for 4 h in O$_2$ by a thermal annealing step. The discharge capacity of NCM 523 and NCM 111 is 128.3 and 122.6 mAh/g after 100 cycles at 1 C.

Direct repair of spent cathode materials by solid-phase calcination and the hydrothermal method has the common advantages of short processing time, simple operation, environmental friendliness, and economy. Compared with solid-phase calcination, the hydrothermal method supplements Li$^+$ to the crystal structure more uniformly, avoiding side reactions [5]. However, the process of direct repair still has shortcomings. On one hand, impurities, such as Al and Cu, have a tremendous impact on the performance of regenerated material. On the other hand, the failure degree of spent cathode material from different manufacturers is uniform, such as the degree of structural damage and the volume of impurities. Using a single repair system cannot guarantee consistency of the electrochemical performance of the product, and it is difficult to achieve large-scale production.

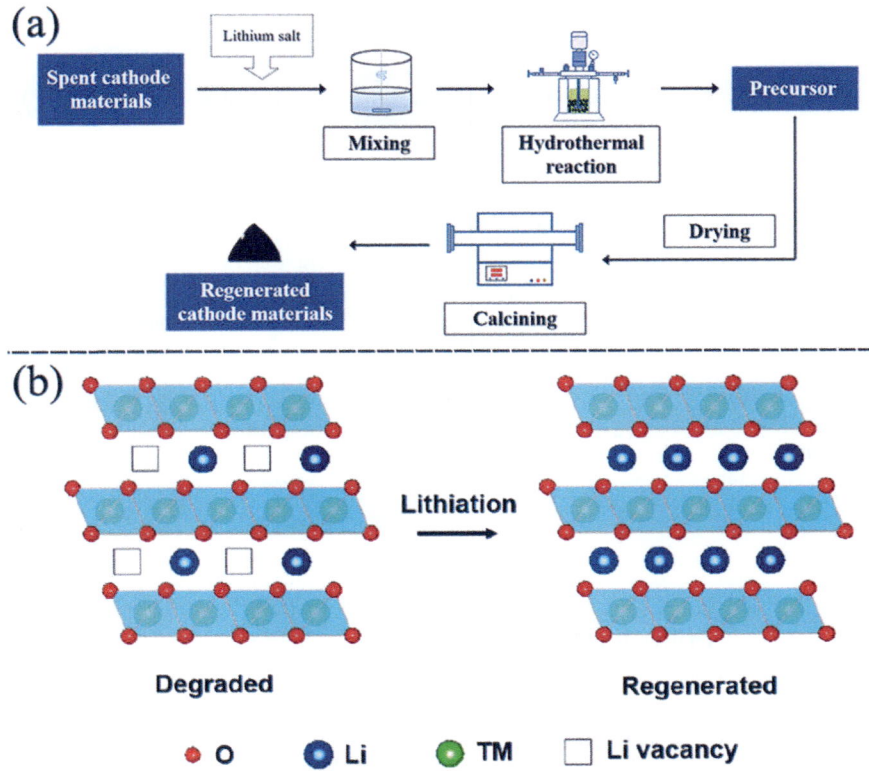

Figure 20.2 The process of (A) regenerating cathode materials and (B) mechanism of lithiation by hydrothermal method [4].

2. Extraction of metal followed by material regeneration

Compared with direct repair, extraction of metal followed by material regeneration is more adaptable to raw materials. It includes two processes—extraction of metal with hydrometallurgy method and material regeneration (Fig. 20.3).

2.1 Valuable metal leaching by hydrometallurgy

Before leaching valuable metals from spent cathode materials, pretreatment is needed (Fig. 20.4). The pretreatment process includes discharging, crushing, separating, and finally enriching to obtain the spent cathode materials. The leaching methods for recovering valuable metals from spent cathode materials are acid leaching, alkaline leaching, and bioleaching.

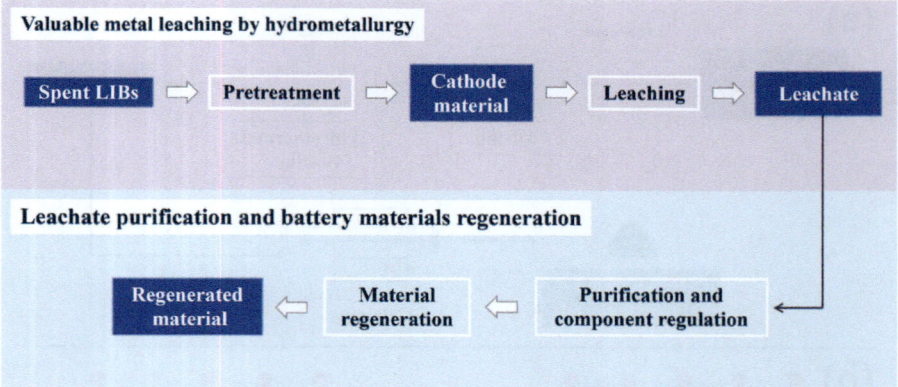

Figure 20.3 Flowchart of regenerating cathode material of spent lithium ion battery by hydrometallurgy.

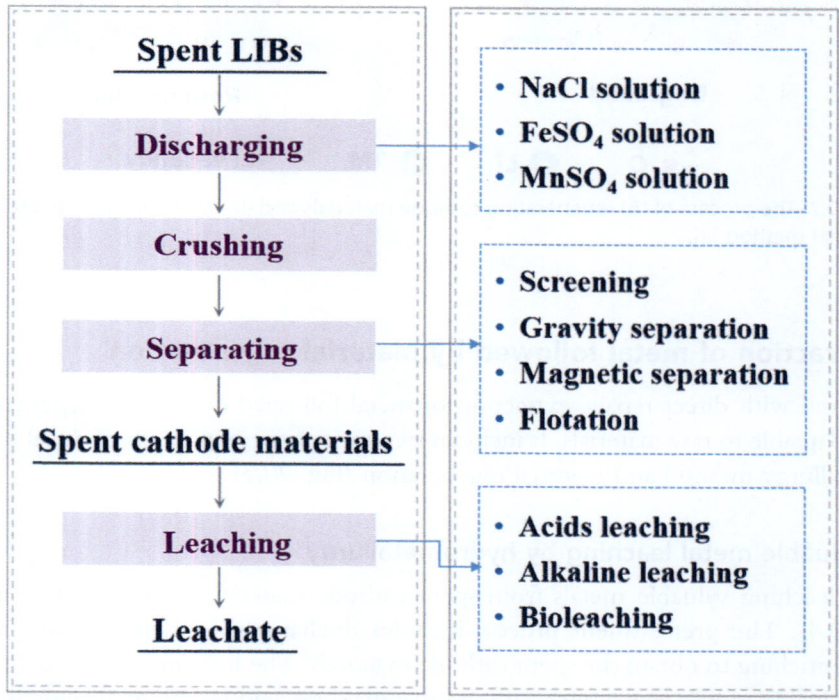

Figure 20.4 Pretreatment and leaching process of recovering valuable metals from spent LIBs.

2.1.1 Leaching kinetics

The two main models for studying leaching kinetics are shrinking-core and diffusion. The shrinking-core model of leaching spent cathode materials with citric acid is shown in Fig. 20.5 [6]. This leaching process can be divided into five steps [6]:

(1) The extractant molecules diffuse from the bulk solution to the solid–liquid interface (external diffusion).
(2) The extractant molecules further diffuse to the surface of the unreacted core (internal diffusion).
(3) The extractant molecules react with the core, and metals exist in the form of aqueous ions (chemical reaction process).
(4) The surface of the unreacted core becomes covered with a layer of insoluble substances, and the metal ions diffuse to the solid–liquid surface (internal diffusion).
(5) Metal ions further diffuse to the bulk solution (external diffusion).

Obviously, the reaction rate of the shrinking-core model is controlled by chemical reaction (step 3) and diffusion (steps 1, 2, 4, and 5). When the reaction is a first-order reaction, the rate equation controlled by the chemical reaction can be expressed as follows [7]:

$$1 - (1 - X)^{1/3} = kt \tag{20.1}$$

where X is the leaching efficiency, k is the rate constant, and t is the reaction time.

The rate equation controlled by internal diffusion (Eq. 20.2) and external diffusion (Eq. 20.3) can be expressed as follows [7]:

$$1 - (1 - X)^{2/3} - \frac{2}{3}X = kt \tag{20.2}$$

$$X = kct \tag{20.3}$$

where t is the leaching agent concentration.

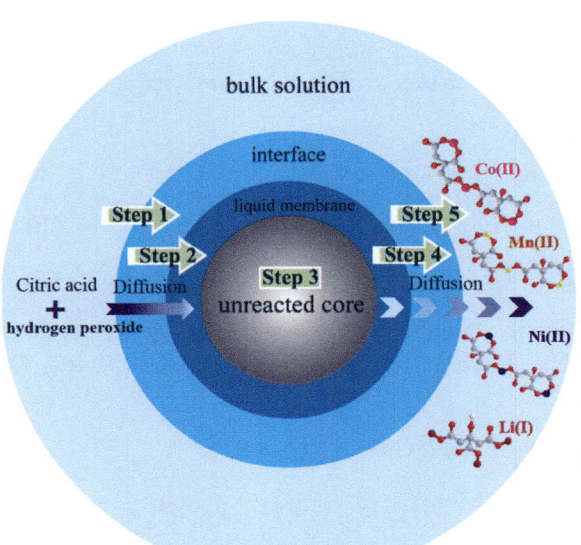

Figure 20.5 Shrinking-core model for leaching metal ions with citric acid [6].

In addition, the apparent activation energy (activation energy) of the leaching reaction can be used to roughly judge the control steps of the leaching reaction process. The activation energy of the leaching reaction can be expressed by rate constant and temperature, that is, the Arrhenius equation [7]:

$$k = Ae^{-Ea/RT} \tag{20.4}$$

where A is the frequency factor, Ea is the activation energy (J/mol), $R = 8.3145$ J/mol/K, and T is the temperature in K.

2.1.2 Acid leaching

Acid leaching is one of the most commonly used leaching methods for spent cathode materials. According to differences in the leaching agent, acid leaching is divided into inorganic and organic acid leaching.

For inorganic acid leaching, sulfuric acid (H_2SO_4), hydrochloric acid (HCl), and nitric acid (HNO_3) are frequently used as leaching agents; meanwhile, a certain amount of a reducing agent, such as H_2O_2 or Na_2SO_3, is required. The leaching reaction of H_2SO_4 and H_2O_2 leaching spent $LiNi_{0.5}Co_{0.2}Mn_{0.3}O_2$ materials is as follows [8]:

$$2LiNi_{0.5}Co_{0.2}Mn_{0.3}O_2 + 6H^+ + H_2O_2 \xrightarrow{\Delta}$$

$$2Li^+ + Ni^{2+} + 0.4Co^{2+} + 0.6Mn^{2+} + 4H_2O + O_2\uparrow \tag{20.5}$$

$$2H_2O_2 \rightarrow 2H_2O + O_2\uparrow \tag{20.6}$$

Leaching spent cathode materials by inorganic acid is easy to control, has high leaching efficiency, and is economically feasible. It is widely used in industrial applications to leach spent cathode materials.

Compared with inorganic acid leaching, organic acid leaching is generally more environmentally friendly. The commonly used organic acid and its leaching mechanism are shown in Fig. 20.6 [5,6,9,10].

Citric acid is a tribasic carboxylic acid with strong acidity. The three-stage ionization occurs in an aqueous solution [11]:

$$H_3Cit = H^+ + H_2Cit^- \quad pk_1 = 3.13 \tag{20.7}$$

$$H_2Cit^- = H^+ + HCit_2 \quad pk_2 = 4.76 \tag{20.8}$$

$$HCit_2^- = H^+ + Cit_3^- \quad pk3 = 6.40 \tag{20.9}$$

When using citric acid as a leaching agent, a certain amount of reducing agent is required (H_2O_2 or D-glucose ($C_6H_{12}O_6$)). The leaching reaction of $LiNi_{1/3}Co_{1/3}Mn_{1/3}O_2$ cathode materials is as follows [11]:

$$18LiNi_{1/3}Co_{1/3}Mn_{1/3}O_2 + 18H_3Cit + C_6H_{12}O_6 =$$

$$6Li_3Cit + 2Ni_3(Cit)_2 + 2Co_3(Cit)_2 + 2Mn_3(Cit)_2 + 33H_2O + 6CO_2 \tag{20.10}$$

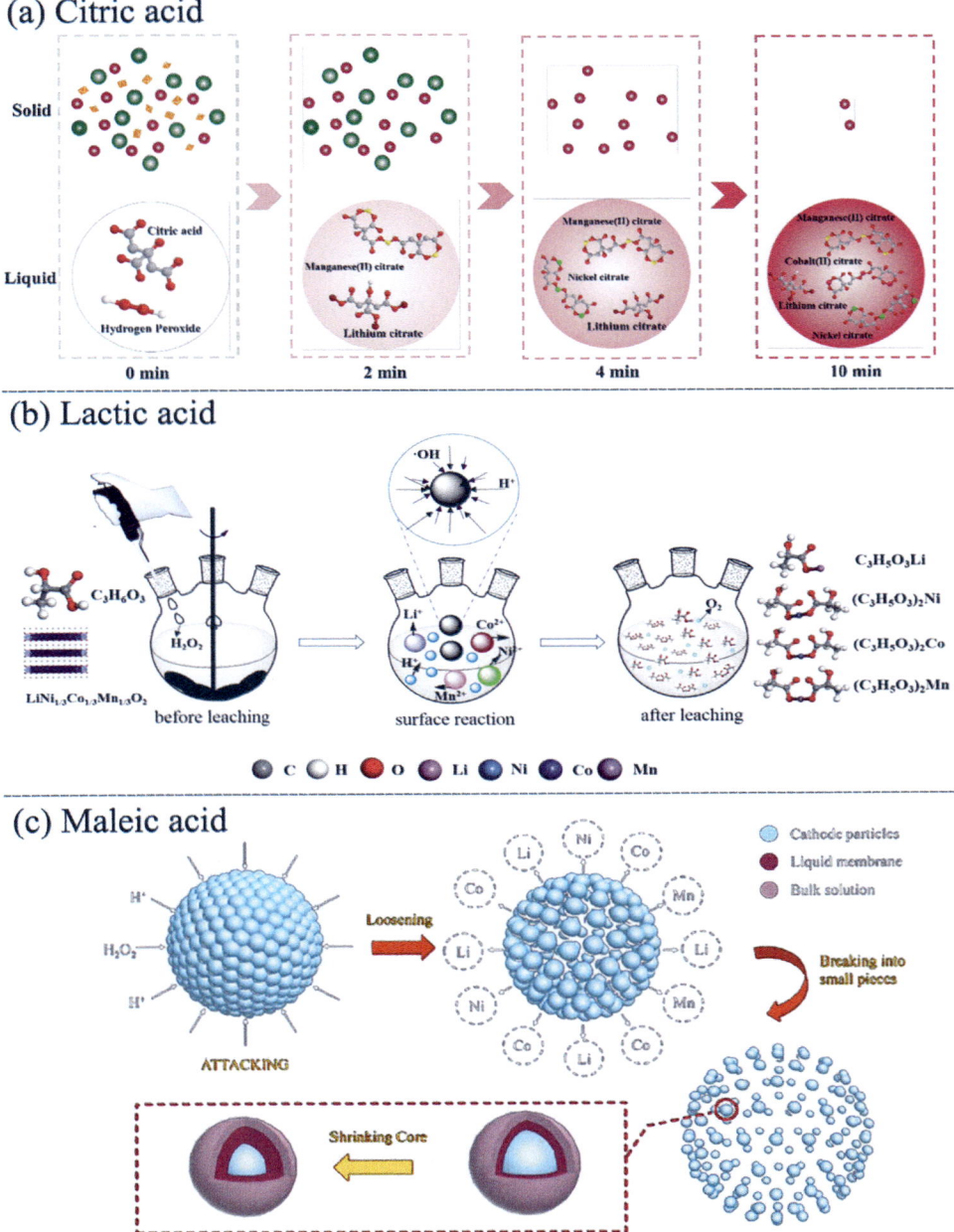

Figure 20.6 Mechanism diagram of leaching cathode material by (A) citric acid, (B) lactic acid, and (C) maleic acid [5,6,9,10].

During the leaching reaction, D-glucose is gradually oxidized as a reducing agent and converted to CO_2 and H_2O. The leaching process is green and environmentally safe and thus does not produce substances that are harmful to the environment.

Oxalic acid is a dicarboxylic acid. The two-stage ionization occurs in an aqueous solution [12]:

$$H_2C_2O_4 \leftrightarrow H^+ + HC_2O_4^- \quad pk_1 = 5.9 \times 10^{-2} \tag{20.11}$$

$$HC_2O_4^- \leftrightarrow H^+ + C_2O_4^{2-} \quad pk_2 = 6.4 \times 10^{-5} \tag{20.12}$$

From the ionization constant, the acidity of oxalic acid is strong and can provide H^+ for the reaction, which is conducive to the leaching reaction. In addition, oxalic acid has the property of precipitation. Nickel, cobalt, and manganese can be coprecipitated to avoid forming single-metal oxalate [12]. After leaching, the oxide precipitate of nickel–cobalt–manganese is used to regenerate the precursor:

$$4H_2C_2O_4 + 2LiNi_{1/3}Co_{1/3}Mn_{1/3}O_2 =$$
$$Li_2C_2O_4 + 2\left(Ni_{1/3}Co_{1/3}Mn_{1/3}\right)C_2O_4 + 4H_2O + 2CO_2 \tag{20.13}$$

2.1.3 Alkaline leaching

The leaching reaction of alkaline leaching is described in Eq. (20.14) [13]. In the alkaline leaching process, ammonia or ammonium salt is always used as a leaching agent, and ammonium sulfate ((NH_4)$_2SO_3$) or sodium sulfite (Na_2SO_3) is used as a reducing agent. The transition metals nickel, cobalt, and manganese can react with NH_3 to form complexes—Eqs. (20.15)–(20.17). Due to the different complexing abilities with ammonia, nickel and cobalt are more easily leached and separated from manganese [13]:

$$2LiNi_xCo_yMn_{1-x-y}O_2 + SO_3^{2-} + (z1+z2)NH_3 + 3H_2O =$$
$$2Li^+ + 2x[Ni(NH_3)_{z1}]^{2+} + 2y[Co(NH_3)_{z2}]^{2+} + 2(1-x-y)Mn^{2+} + SO_4^{2-}$$
$$+ 6OH^- \tag{20.14}$$

$$lgKf^\theta[Ni(NH_3)_x]^{2+} = 40.87 \tag{20.15}$$

$$lgKf^\theta[Co(NH_3)x]^{2+} = 26.97 \tag{20.16}$$

$$lgKf^\theta[Mn(NH_3)_x]^{2+} = 5.54 \tag{20.17}$$

The pH value is an important factor affecting leaching because the dissolving ranges of $[Co(NH_3)_x]^{2+}$ and $[Ni(NH_3)_x]^{2+}$ are 9–11 and 8.5–10.5, respectively [14]. To maintain a stable pH value, ammonium sulfate $((NH_4)_2SO_4)$ or ammonium carbonate $((NH_4)_2CO_3)$ is used as a pH buffer.

Manganese remains in leaching residue in two forms. One is $(NH_4)_2Mn(SO_3)_2 \cdot H_2O$ (Eq.20.18) [13], which depends on the presence of SO_3^{2-} in the reactants, and the other is $Mn(OH)_2$ and Mn_3O_4 (Eqs. (20.19) and (20.20)) [13]:

$$Mn^{2+} + 2NH_4^+ + 2SO_3^{2-} + H_2O = (NH_4)_2Mn(SO_3)_2 \cdot H_2O \tag{20.18}$$

$$Mn^{2+} + 2OH^- = Mn(OH)_2 \tag{20.19}$$

$$6Mn(OH)_2 + O_2 = 2Mn_3O_4 + 6H_2O \tag{20.20}$$

2.1.4 Bioleaching

Bioleaching oxidizes or reduces valuable metals in spent cathode materials through the oxidation and reduction characteristics of microorganisms. Compared with chemical leaching methods, bioleaching consumes less energy and is more environmentally friendly.

Aspergillus niger produces organic acids in the metabolic process, such as oxalic acid, malic acid, citric acid, and gluconic acid. These organic acid are used as leaching agents in the leaching of cathode materials. The reaction is shown in Fig. 20.7 [15].

For *Acidithiobacillus thiooxidans*, the dissolution process of metal can be expressed as follows [16]:

$$S^0 + H_2O + 1.5O_2 \rightarrow H_2SO_4 \tag{20.21}$$

$$H_2SO_4 + M \rightarrow MSO_4 + 2H^+ \quad (M = Co) \tag{20.22}$$

For *Acidithiobacillus ferrooxidans*, Fe^{2+} is oxidized to Fe^{3+}, and the final reaction formula for leaching $LiCoO_2$ can be expressed as follows [17]:

$$4Fe^{2+} + O_2 + 4H^+ \rightarrow 4Fe^{3+} + 4H_2O \tag{20.23}$$

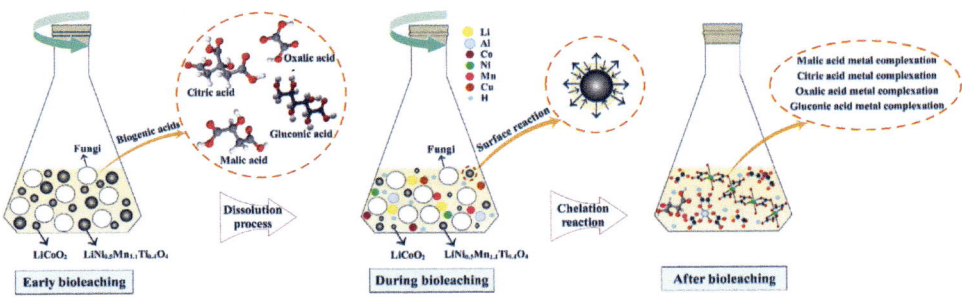

Figure 20.7 The possible reactions process of *Aspergillus niger* and cathode materials [15].

$$2FeSO_4 + 2LiCoO_2 + 4H_2SO_4 \rightarrow Fe_2(SO_4)_3 + 2CoSO_4 + Li_2SO_4 + 4H_2O \quad (20.24)$$

2.2 Leachate purification and battery material regeneration

After leaching, the leachate is first purified, and then the molar ratio of metals in the solution is adjusted by adding pure salt. Finally, the leachate of the adjusted components is used as the raw material solution to regenerate the battery materials.

2.2.1 Leachate purification

The most common impurities in the leachate of spent cathode materials are copper and aluminum. Currently, the impurity removal approach mainly includes precipitation and solvent extraction. Precipitation can be used to remove aluminum and copper ions in leachate. Aluminum ions are generally removed by hydroxide precipitation at a pH of around 6.5. Copper ions can be removed by hydroxide precipitation at a pH around 6.5 or sulfide precipitation at a pH > 1 (Fig. 20.8) [19].

Another method commonly used to remove copper from leachate is solvent extraction. Mextral 5640, Cynanex301, bis-2,4,4-trimethylpentyl phosphinic acid (Cynanex272), LIX 63, and LIX 84 are frequently used as extracting agents, Cu^{2+} reacts with the agent to form a stable chelate. Taking Mextral 5640 as an example, the reaction in copper extraction is as follows [18]:

$$Cu^{2+}_{(aq)} + nHR_{(org)} = CuR_2(HR)_{(n-2)(org)} + 2H^+_{(aq)} \quad (20.25)$$

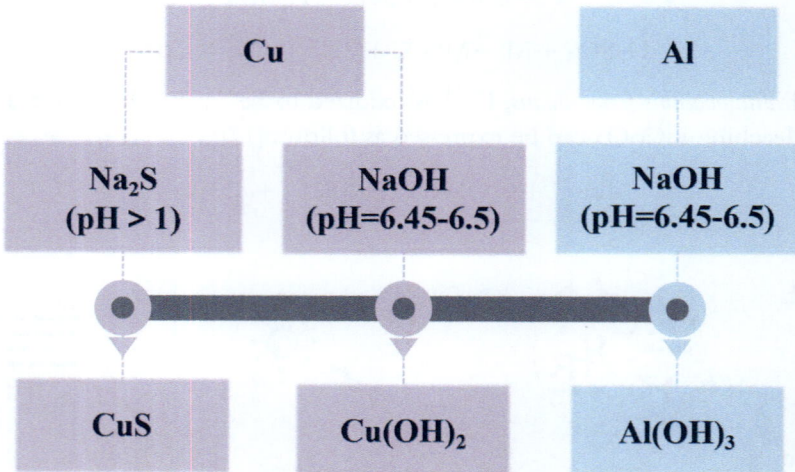

Figure 20.8 Removal of copper and aluminum by hydroxide and sulfide precipitation methods [19].

2.2.2 Cathode material regeneration

After purification and element adjustment, the cathode materials are generally regenerated by precipitation followed by a solid-state reaction.

Hydroxide precipitation is performed using OH^- as the precipitation agent. The purified leachate, ammonia, and OH^- are mixed to generate the hydroxide precipitation, and the pH range is 10–11. The reaction is as follows [20]:

$$M^{2+} + nNH_3 \cdot H_2O = [M(NH_3)_n]^{2+} + nH_2O \quad (M = Ni, Co \text{ and } Mn) \quad (20.26)$$

$$[M(NH_3)_n]^{2+} + 2OH^- = M(OH)_{2(s)} + nNH_3 \quad (20.27)$$

Then, mixing $LiOH \cdot H_2O$ and precursor in a molar ratio for Li:(Ni/Co/Mn) of 1.05, the mixture is calcined to regenerate cathode materials. The reaction is as follows [20]:

$$4M(OH)_2 + 4LiOH \cdot H_2O + O_{2(g)} = 4LiMO_2 + 10H_2O\uparrow \quad (20.28)$$

Carbonate precipitation is performed using CO_3^{2-} as the precipitation agent and ammonia as a complexing agent. The reaction is as follows [21]:

$$M^{2+} + nNH_3 \cdot H_2O = [M(NH_3)_n]^{2+} + nH_2O \quad (M = Ni, Co \text{ and } Mn) \quad (20.29)$$

$$[M(NH_3)_n]^{2+} + CO_3^{2-} + nH_2O = MCO_{3(s)} + nNH_3 \cdot H_2O \quad (20.30)$$

Taking Li_2CO_3 as a lithium source, the molar ratio of Li:(Ni/Co/Mn) can be maintained at 1.06. The cathode material is regenerated in two steps, and the reaction is as follows [21]:

$$6MCO_{3(s)} + O_{2(g)} \rightarrow 2M_3O_{4(s)} + 6CO_{2(g)} \quad (M = Ni, Co \text{ and } Mn) \quad (20.31)$$

$$4M_3O_{4(s)} + 6Li_2CO_{3(s)} + O_{2(g)} \rightarrow 12LiMO_{2(s)} + 6CO_{2(g)} \quad (20.32)$$

Oxalate precipitation uses oxalic acid as a leaching agent and precipitant to react directly with the spent cathode materials, and the oxalate is formed on particle surfaces. With reaction progress, the particles of spent cathode materials are gradually changed into oxalate. Then, the cathode materials are regenerated by calcination. The reaction is as follows [12]:

$$4H_2C_2O_4 + 2LiNi_xMn_yCo_{1-x-y}O_2 =$$
$$2(Ni_xMn_yCo_{1-x-y})C_2O_{4(s)} + 4H_2O + 2CO_2 + Li_2C_2O_4 \quad (20.33)$$

Similar with carbonate precipitation, the regenerated cathode material is obtained by a lithium source and precursor in two steps, and the reaction is as follows:

$$3(Ni_xMn_yCo_{1-x-y})C_2O_{4(s)} + 2O_{2(g)} \rightarrow (Ni_xMn_yCo_{1-x-y})_3O_{4(s)} + 6CO_{2(g)} \quad (20.34)$$

$$4(Ni_xMn_yCo_{1-x-y})_3O_{4(s)} + 6Li_2CO_{3(s)} + O_{2(g)} \rightarrow 12Li(Ni_xMn_yCo_{1-x-y})O_{2(s)}$$
$$+ 6CO_{2(g)}$$

(20.35)

2.2.3 Impact of impurities

Impurities are a key problem that cannot be ignored during the recycling process, especially the influence of trace impurities on the microstructure of regenerated materials, which profoundly determines the quality of regenerated materials. Before regenerating cathode materials, the impurities must be removed from the leachate by precipitation, solvent extraction, or other methods. However, some impurities mixed into the regenerated materials, such as aluminum, copper, and iron, affect the performance and even the crystal structure of regenerated materials.

2.2.3.1 Copper

The impurity of Cu comes from the Cu foil collector. Cu can reduce the degree of cation mixing due to the ion radius of Cu^{2+} (0.73 A°) being similar to that of Li^+ (0.76 A°) [22]. In addition, research shows that Cu does not affect the crystal structure of the regenerated material but decreases the particle size gradually as the content of Cu increases (Fig. 20.9) [23]. This is due to the formation of insoluble $Cu(OH)_2$ during precursor synthesis, which hinders the formation of nickel–cobalt–manganese hydroxide. When the Cu content increases to a certain value, the performance of regenerated cathode materials decreases but is still higher than for regenerated materials without Cu. The capacity retention of regenerated cathode materials with 2.5% Cu is slightly better than that with 5% Cu. Similarly, Wu et al. [24] compared the electrochemical performance of regenerated Li[NiCo$_{1-x}$Cu$_x$Mn]$_{1/3}$O$_2$ materials with various levels of Cu. The test results show that the cycling stability and rate performance of Li[NiCo$_{1-x}$Cu$_x$Mn]$_{1/3}$O$_2$ is greater than for Li[NiCoMn]$_{1/3}$O$_2$. The capacity retention of Li[NiCo$_{0.04}$Cu$_{0.06}$Mn]$_{1/3}$O$_2$ (x = 0.06) is 95.87% after 50 cycles at 1C, but it is 90.99% for Li[NiCoMn]$_{1/3}$O$_2$ (x = 0) under the same conditions.

Figure 20.9 SEM images of regenerated LiNi$_{1/3}$Co$_{1/3}$Mn$_{1/3}$O$_2$ with (A) 0% Cu, (B) 2.5% Cu and (C) 5% Cu [23].

2.2.3.2 Aluminum

The impurity of Al comes from the collector of Al foil. The experimental results show that a small amount of Al will enhance the stability of crystal structure and thermal stability of regenerated cathode materials [25]. The improvement of cycle stability is due to the strong Al—O binding will retain the original structure during the Li^+ intercalation and deintercalation when synthesizing ternary cathode materials [6]. Ren et al. [26] studied the effect of different content of Al on the performance of regenerated $Li(Ni_{1/3}Co_{1/3}Mn_{1/3})_{1-x}Al_xO_2$ material (0.01 $\leq$ x $\leq$ 0.05). The discharge capacity decreases with the amount of Al increasing. When x $\leq$ 0.03, the discharge capacity reaches 140 mAh/g at 0.1C but is only 100 mAh/g when x = 0.05. Zhang et al. [27] found that the regenerated $LiNi_{1/3}Co_{1/3}Mn_{1/3-x}Al_xO_2$ (x = 0, 0.02, 0.04) cathode materials can form a typical layered structure, the presence of Al can reduce the cations mixing. The discharge capacity of $LiNi_{1/3-0.02}Co_{1/3}Mn_{1/3}Al_{0.02}O_2$ reaches 154.9 mAh/g after 100 cycles at 0.2 C, which is 14.3 times higher than that without Al (Fig. 20.10).

2.2.3.3 Iron

The impurity of Fe comes from the shell of the battery. Many investigations have demonstrated that Fe is conducive to promoting the electrochemical performance of regenerated cathode materials. Park et al. [28] adopted hydroxide coprecipitation and calcination to regenerate $Li[Ni_{1/3}Co_{1/3}Mn_{1/3}]Fe_xO_2$ (x = 0, x = 0.05, x = 0.25, x = 1) cathode materials. When the Fe content is low, the discharge capacity, cycle performance, and capacity retention of regenerated materials are higher than for regenerated materials without Fe. However, when the content of Fe is equal to 1%, the migration of Li^+ is hindered in the intercalation and deintercalation process due to cation mixing. At this time, the capacity retention of regenerated materials is highest, reaching 97.3% (Fig. 20.11). In addition, the suitable amount of Fe ($Li[Ni_{1/3}Co_{1/3}Mn_{1/3}]Fe_xO_2$,

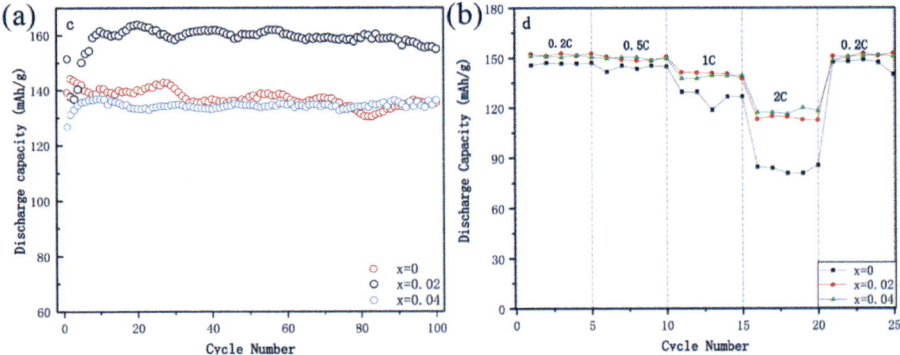

Figure 20.10 (A) cycling performance and (B) rate performance of $LiNi_{1/3-x}Co_{1/3}Mn_{1/3}Al_xO_2$ (x = 0, 0.02, 0.04) [27].

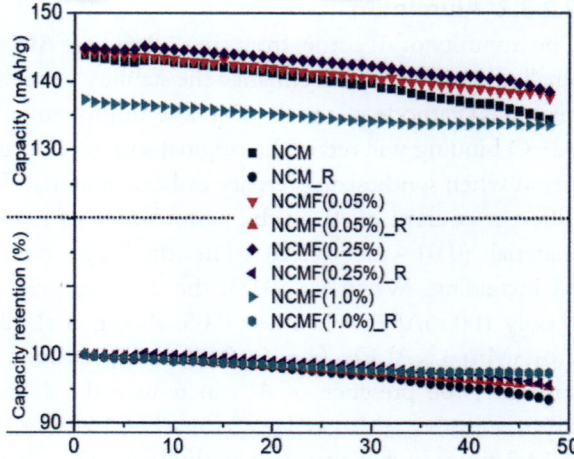

Figure 20.11 The discharge capacity and capacity retention of regenerated Li[Ni$_{1/3}$Co$_{1/3}$Mn$_{1/3}$]Fe$_x$O$_2$ cathode materials at 1 C (x = 0, 0.05, 0.25, 1) [28].

x = 0.05) will reduce the particle diameter and make the surface of the synthetic material uniform because Fe enters the crystal structure of cathode materials in the form of incorporation.

2.2.3.4 Magnesium

The impurity of Mg comes from battery scraps. During the leaching process, Mg effortlessly enters into the leachate and is difficult to remove, affecting the performance of regenerated cathode materials. Many researchers have paid attention to studying the effect of Mg on regenerated materials and have reached the consistent conclusion that Mg improves the cycle stability and rate performance of regenerated cathode materials [29]. Taking regenerated Li[(Ni$_{1/3}$Co$_{1/3}$Mn$_{1/3}$)$_{1-x}$Mg$_x$]O$_2$ material as an example, the first charge—discharge capacity slightly decreases, but the capacity retention of the material increases in the presence of Mg (Fig. 20.12A,C) [29]. The cycling property is enhanced when the value of x is less than 0.01 but reduced at a high level (x > 0.025) (Fig. 20.12 B,D) because the uniform distribution of Mg^{2+} in the material is difficult.

Current research shows that trace impurities improve the electrochemical performance of the regenerated cathode materials when regenerating from the leachate. The suitable content of various impurities and the corresponding electrochemical performance of regenerated materials are summarized in Table 20.1.

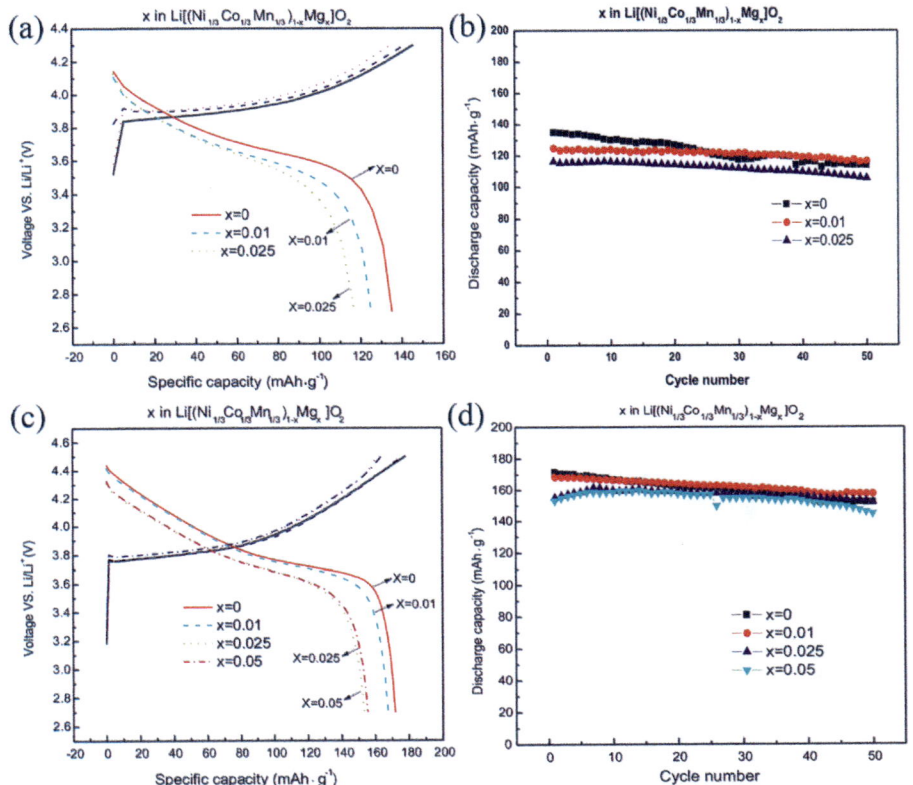

Figure 20.12 The initial charge-discharge curves at (A) 0.2 C and (C) 0.5 C, cycling performance at (B) 0.2 C and (D) 0.5 C of Li[(Ni$_{1/3}$Co$_{1/3}$Mn$_{1/3}$)$_{1-x}$Mg$_x$]O$_2$ (x = 0, 0.01, 0.025, 0.05) [29].

By exploring the influence of impurities removal on the performance of regenerated material, a collaborative feedback mechanism between the design of regenerated products and the depth of impurities removal is formed. By controlling the content of impurities in the leachate, the range of impurities in the regenerated material is determined. The problems of difficult impurities removal and the loss of the primary metals in the traditional process of impurities removal are solved, and self-doping regenerated material with excellent performance is obtained directly.

Table 20.1 Impact of impurities on performance of regenerated cathode materials.

Impurity	Regenerated cathode materials	Optimal conditions	Initial discharge capacities/(mAh/g)	Capacity retention	Ref.
Cu	$Li[Ni_{1/3}Co_{1/3}Mn_{1/3}]O_2$	0.015% (≤0.088 mol/L)	174.77 (0.1C)	93.07% (50 cycles, 1C)	[22]
	$LiNi_{1/3}Co_{1/3}Mn_{1/3}O_2$	0.25% (≤0.156 mol/L)	140.00 (0.1C)	84.00% (100 cycles, 2C)	[23]
	$Li[NiCo_{1-x}Cu_xMn]_{1/3}O_2$	X = 0.06 (≤0.06 mol/L)	161.00 (0.1C)	95.87% (50 cycles, 1C)	[24]
	$LiNi_{1/3-x}Co_{1/3}Mn_{1/3}Cu_x]O_2$	X = 0.02	180.40 (0.2C)	90.30% (50 cycles, 0.2C)	[30]
	$Li_{1.25}Mn_{0.50}Ni_{0.125}Co_{0.125-x}Cu_xO_2$	X = 0.05	225.20 (0.1C)	92.09% (50 cycles, 0.1C)	[31]
Fe	$Li[Ni_{1/3}Co_{1/3}Mn_{1/3}]Fe_xO_2$	X = 0.25% (≤0.005 mol/L)	162.00 (0.1C)	97.00% (50 cycles, 1C)	[28]
	$Li[Ni_{(1/3-x)}Fe_xCo_{1/3}Mn_{1/3}]O_2$	X = 0.05 (≤0.005 mol/L)	146.20 (0.5C)	87.30% (50 cycles, 0.5C)	[32]
	$LiNi_{1/3}Fe_xCo_{1/6}Mn_{1/3}O_2$	X = 1/6	150.00 (0.1C)	85.30% (30 cycles, 0.1C)	[33]
Mg	$Li[(Ni_{1/3}Co_{1/3}Mn_{1/3})_{1-x}Mg_x]O_2$	X = 0.01 (≤360 mg/L)	167.00 (0.5C)	95.80% (50 cycles, 0.5C)	[29]
	$Li_{1.17}Ni_{0.25-x}Mn_{0.58}Mg_xO_2$	X = 0.02	156.80 (1C)	95.1% (100 cycles, 2C)	[34]
Al	$Li[Ni_{0.78}Co_{0.1}Mn_{0.1}Al_{0.02}]O_2$	X = 0.02 (≤ 2 mol%)	173.00 (1C)	96.15% (100 cycles, 1C)	[35]
	$LiNi_{1/3}Co_{1/3}Mn_{1/3-x}Al_xO_2$	X = 0.06	186.59 (0.1C)	96.1% (30 cycles, 0.1C)	[25]
	$LiNi_{1/3}Mn_{1/3}Co_{1/3-x}Al_xO_2$	X = 0.03	197.6	96.4%	[36]

Examples of different impurity elements are distinguished by grey shade.

References

[1] X.X. Zhang, Q. Xue, L. Li, E. Fan, F. Wu, R.J. Chen, Sustainable recycling and regeneration of cathode scraps from industrial production of lithium-ion batteries, ACS Sustain. Chem. Eng. 4 (12) (2016) 7041−7049.

[2] H.M. Zhou, X.X. Zhao, C.J. Yin, J. Li, Regeneration of $LiNi_{0.5}Co_{0.2}Mn_{0.3}O_2$ cathode material from spent lithium-ion batteries, Electrochim. Acta 291 (2018) 142−150.

[3] Y. Pranolo, W. Zhang, C.Y. Cheng, Recovery of metals from spent lithium-ion battery leach solutions with a mixed solvent extractant system, Hydrometallurgy 102 (1−4) (2010) 37−42.

[4] Y. Shi, G. Chen, F. Liu, X.J. Yue, Z. Chen, Resolving the compositional and structural defects of degraded $LiNi_xCo_yMn_zO_2$ particles to directly regenerate high-performance lithium-ion battery cathodes, ACS Energy Lett. 3 (7) (2018) 1683−1692.

[5] Y.L. Zhao, X.Z. Yuan, L.B. Jiang, J. Wen, H. Wang, R.P. Guan, J.J. Zhang, G.M. Zeng, Regeneration and reutilization of cathode materials from spent lithium-ion batteries, Chem. Eng. J. 383 (2020) 123089.

[6] L. Li, Y.F. Bian, X.X. Zhang, Y.B. Guan, E. Fan, F. Wu, R.J. Chen, Process for recycling mixed-cathode materials from spent lithium-ion batteries and kinetics of leaching, Waste Manag. 71 (2018) 362−371.

[7] Z.Y. Fang, Leaching, Metallurgical industry press, Beijing, 2007, pp. 8−9.

[8] Y. Yang, S.Y. Lei, S.L. Song, W. Sun, L.S. Wang, Stepwise recycling of valuable metals from Ni-rich cathode material of spent lithium-ion batteries, Waste Manag. 102 (2020) 131−138.

[9] L. Li, Y.F. Bian, X.X. Zhang, Q. Xue, E. Fan, F. Wu, R.J. Chen, Economical recycling process for spent lithium-ion batteries and macro- and micro-scale mechanistic study, J. Power Sources 377 (2018) 70−79.

[10] L. Li, E. Fan, Y.B. Guan, X.B. Zhang, Q. Xue, L. Wei, F. Wu, R.B. Chen, Sustainable recovery of cathode materials from spent lithium-ion batteries using lactic acid leaching system, ACS Sustain. Chem. Eng. 5 (6) (2017) 5224−5233.

[11] X.P. Chen, B.L. Fan, L.P. Xu, T. Zhou, J.R. Kong, An atom-economic process for the recovery of high value-added metals from spent lithium-ion batteries, J. Clean. Prod. 112 (2016) 3562−3570.

[12] X.X. Zhang, Y.F. Bian, S.Y. Xu, E. Fan, Q. Xue, Y.B. Guan, F. Wu, L. Li, R.J. Chen, Innovative application of acid leaching to regenerate $Li(Ni_{1/3}Co_{1/3}Mn_{1/3})O_2$ cathodes from spent lithium-ion batteries, ACS Sustain. Chem. Eng. 6 (5) (2018) 5959−5968.

[13] K. Meng, Y. Cao, B. Zhang, X. Ou, D.M. Li, J.F. Zhang, X.B. Ji, Comparison of the ammoniacal leaching behavior of layered $LiNi_xCo_yMn_{1-x-y}O_2$ (x=1/3, 0.5, 0.8) cathode materials, ACS Sustain. Chem. Eng. 7 (8) (2019) 7750−7759.

[14] Y.Q. Wang, N. An, L. Wen, L. Wang, X.T. Jiang, F. Hou, Y.X. Yin, J. Liang, Recent progress on the recycling technology of Li-ion batteries, J. Energy Chem. 55 (2021) 391−419.

[15] N. Bahaloo-Horeh, S.M. Mousavi, M. Baniasadi, Use of adapted metal tolerant *Aspergillus niger* to enhance bioleaching efficiency of valuable metals from spent lithium-ion mobile phone batteries, J. Clean. Prod. 197 (2018) 1546−1557.

[16] B.K. Biswal, U.U. Jadhav, M. Madhaiyan, L.H. Ji, E.H. Yang, B. Cao, Biological leaching and chemical precipitation methods for recovery of Co and Li from spent lithium-ion batteries, ACS Sustain. Chem. Eng. 6 (9) (2018) 12343−12352.

[17] J.J. Roy, S. Madhavi, B. Cao, Metal extraction from spent lithium-ion batteries (LIBs) at high pulp density by environmentally friendly bioleaching process, J. Clean. Prod. 280 (2021) 124242.

[18] J.M. Nan, D.M. Han, X.X. Zuo, Recovery of metal values from spent lithium-ion batteries with chemical deposition and solvent extraction, J. Power Sources 152 (1) (2005) 278−284.

[19] Y.L. Yao, M.Y. Zhu, Z. Zhao, B.H. Tong, Y.Q. Fan, Z.S. Hua, Hydrometallurgical processes for recycling spent lithium-ion batteries: a critical review, ACS Sustain. Chem. Eng. 6 (11) (2018) 13611−13627.

[20] Y. Yang, G.Y. Huang, M. Xie, S.M. Xu, Y.H. He, Synthesis and performance of spherical $LiNi_xCo_yMn_{1-x-y}O_2$ regenerated from nickel and cobalt scraps, Hydrometallurgy 165 (2016) 358−369.

[21] L.P. He, S.Y. Sun, J.G. Yu, Performance of $LiNi_{1/3}Co_{1/3}Mn_{1/3}O_2$ prepared from spent lithium-ion batteries by a carbonate co-precipitation method, Ceram. Int. 44 (1) (2018) 351–357.
[22] M. Jo, S. Park, J. Song, K. Kwon, Incorporation of Cu into $Li[Ni_{1/3}Co_{1/3}Mn_{1/3}]O_2$ cathode: elucidating its electrochemical properties and stability, J. Alloys Compd. 764 (2018) 112–121.
[23] Q. Sa, J.A. Heelan, Y. Lu, D. Apelian, Y. Wang, Copper impurity effects on $LiNi_{(1/3)}Mn_{(1/3)}Co_{(1/3)}O_2$ cathode material, ACS Appl. Mater. Interfaces 7 (37) (2015) 20585–20590.
[24] Z.J. Wu, D. Wang, Z.F. Gao, H.F. Yue, W.M. Liu, Effect of Cu substitution on structures and electrochemical properties of $Li[NiCo_{1-x}Cu_xMn]_{1/3}O_2$ as cathode materials for lithium ion batteries, Dalton Trans. 44 (42) (2015) 18624–18631.
[25] Y.H. Ding, P. Zhang, Z.L. Long, Y. Jiang, F. Xu, Morphology and electrochemical properties of Al doped $LiNi_{1/3}Co_{1/3}Mn_{1/3}O_2$ nanofibers prepared by electrospinning, J. Alloys Compd. 487 (1–2) (2009) 507–510.
[26] J. Ren, R.H. Li, Y.L. Liu, Y.R. Cheng, D.Y. Mu, R.J. Zheng, J.C. Liu, C.S. Dai, The impact of aluminum impurity on the regenerated lithium nickel cobalt manganese oxide cathode materials from spent LIBs, New J. Chem. 41 (19) (2017) 10959–10965.
[27] Z.H. Zhang, J.H. Qiu, M. Yu, C.Z. Jin, B. Yang, G.H. Guo, Performance of Al-doped $LiNi_{1/3}Co_{1/3}Mn_{1/3}O_2$ synthesized from spent lithium ion batteries by sol-gel method, Vacuum 172 (2020) 109105.
[28] S. Park, D. Kim, H. Ku, M. Jo, S. Kim, J. Song, J. Yu, K. Kwon, The effect of Fe as an impurity element for sustainable resynthesis of $Li[Ni_{1/3}Co_{1/3}Mn_{1/3}]O_2$ cathode material from spent lithium-ion batteries, Electrochim. Acta 296 (2019) 814–822.
[29] Y.Q. Weng, S.M. Xu, G.Y. Huang, C.Y. Jiang, Synthesis and performance of $Li[(Ni_{1/3}Co_{1/3}Mn_{1/3})_{(1-x)}Mg_x]O_2$ prepared from spent lithium ion batteries, J. Hazard Mater. 246 (2013) 163–172.
[30] L. Yang, F.Z. Ren, Q.G. Feng, G.R. Xu, X.B. Li, Y.C. Li, E.Q. Zhao, J.J. Ma, S.M. Fan, Effect of Cu doping on the structural and electrochemical performance of $LiNi_{1/3}Co_{1/3}Mn_{1/3}O_2$ cathode materials, J. Electron. Mater. 47 (7) (2018) 3996–4002.
[31] Y.G. Sorboni, H. Arabi, A. Kompany, Effect of Cu doping on the structural and electrochemical properties of lithium-rich $Li_{1.25}Mn_{0.50}Ni_{0.125}Co_{0.125}O_2$ nanopowders as a cathode material, Ceram. Int. 45 (2) (2019) 2139–2145.
[32] H.J. Li, G. Chen, B. Zhang, J. Xu, Advanced electrochemical performance of $Li[Ni_{(1/3-x)}Fe_xCo_{1/3}Mn_{1/3}]O_2$ as cathode materials for lithium-ion battery, Solid State Commun. 146 (3–4) (2008) 115–120.
[33] Y.S. Meng, Y.W. Wu, B.J. Hwang, Y. Li, G. Ceder, Combining ab initio computation with experiments for designing new electrode materials for advanced lithium batteries: $LiNi_{1/3}Fe_{1/6}Co_{1/6}Mn_{1/3}O_2$, ACS Appl. Mater. Interfaces 151 (8) (2004) A1134–A1140.
[34] T.F. Yi, Y.M. Li, S.Y. Yang, Y.R. Zhu, Y. Xie, Improved cycling stability and fast charge-discharge performance of cobalt-free lithium-rich oxides by magnesium-doping, ACS Appl. Mater. Interfaces 8 (47) (2016) 32349–32359.
[35] S.J. Do, P. Santhoshkumar, S.H. Kang, K. Prasanna, Y.N. Jo, C.W. Lee, Al-doped $Li[Ni_{0.78}Co_{0.1}Mn_{0.1}Al_{0.02}]O_2$ for high performance of lithium ion batteries, Ceram. Int. 45 (6) (2019) 6972–6977.
[36] Z.Y. Li, H.L. Zhang, The improvement for the electrochemical performances of $LiNi_{1/3}Co_{1/3}Mn_{1/3}O_2$ cathode materials for lithium-ion batteries by both the Al-doping and an advanced synthetic method, Int. J. Electrochem. Sci. 14 (4) (2019) 3524–3534.

CHAPTER 21

Effects of imperfect separation of recycled cathode active materials on remanufactured lithium-ion battery performance

Hammad Al-Shammari[1] and Siamak Farhad[2]

[1]Department of Mechanical Engineering, Jouf University, Sakaka, Saudi Arabia; [2]Advanced Energy & Manufacturing Laboratory, Mechanical Engineering Department, University of Akron, Akron, OH, United States

1. Introduction

The lithium-ion battery (LIB) is an electric energy-storage system for cellular phones, laptops, electric vehicles [1–3], electric aircraft ([4–9] (two articles)), renewable storage [10], and so forth. Their high energy density, high power density, and relativity longer life spans have allowed LIBs to become the most used electrical energy-storage system in the world [11]. These advantages of LIBs have led to rapidly increasing demand for this technology among consumers [12]. Therefore, the rapid growth in the market for LIBs will soon result in considerable growth in the number of retired LIBs [13]. Spent LIB scrap containing toxic electrolytes and heavy metals cannot be disposed of in landfills because of its harmful ingredients [14]. In addition, the spent batteries contain valuable metals such as Ni, Mn, Li, and Co. These materials will likely be wasted if no appropriate action is taken [15]. Therefore, the recycling of spent LIBs is considered an optimistic approach to improved economics and environmental damage prevention.

In general, an LIB consists of five major components: an anode or a negative electrode, a cathode or positive electrode, an electrolyte, a separator, and current collectors [16,17]. Among these major components, the cathode accounts for more than 40% of the total cost of the battery [18] due to the cost of the cathode active materials. Improvements in the energy density of cathode active materials have been pursued by the scientific community for a long time [19]. Many researchers have studied the combination of two or three cathode active materials to either improve the performance of the electrode or reduce the cost of the materials. In one research work [20], the authors have investigated a cathode mixture of $LiMn_2O_4$ (LMO) and $LiNi_{0.8}Co_{0.15}Al_{0.05}O_2$ (NCA). NCA shows a high capacity profile and a reasonable lifetime but poor thermal stability at elevated temperatures. On the other hand, LMO shows good thermal stability, high nominal voltage, and low cost but has relatively low capacity. Therefore, the authors believed that each cathode's unique property would be improved when blended. Another research study has been conducted to investigate the blended electrode of two cathode active materials, LMO and LiNi$_{1/3}$Co$_{1/3}$Mn$_{1/3}$O$_2$ (NMC) [21]. The results have shown that the low capacity of LMO was enhanced when blended with NMC. The mixing of three cathode active materials has been studied in research conducted on ternary-blended cathodes of NMC, LMO, and LiMn$_x$Fe$_{1-x}$PO$_4$ (LMFP) [22]. Of the many blended scenarios, the best performance was found in a blended electrode of 75% NMC, 12.5% LMFP, and 12.5% LMO, resulting in the characteristics of pure NMC, which has the best performance, and the highest cost compared with the other types. Besides the studies cited above, many studies have been published on blended cathode systems with the same purpose [23–26].

However, most previous blending attempts have been conducted to decrease the cost or increase the electrochemical performance of the cathode. In this research, our focus is to study the electrochemical performance of the blending/mixture of the five most used cathode types: LiFePO$_4$ (LFP), LMO, NMC, NCA, and LiCoO$_2$ (LCO). This study is a continuation of studies by the authors for recycling LIBs [27–29] based on a paper we recently published [30]. The authors' previous studies have indicated that separation between the five cathode active materials is possible. However, a concern remains that the separation performance cannot reach 100% in practice, as a trace of several other active materials can still be seen in the separated active material. Therefore, this chapter presents the performance of LIBs made from recycled cathode materials based on the fact that the separation process is imperfect. In this case, a single regenerated cathode active material containing a few other types of cathode active materials is used to make new LIBs. This chapter presents the results of the modeling and computer simulation. A pseudo-two-dimensional model based on the porous theory proposed by Newman [31,32] has been adopted for the modeling. The model has been completely explained in the previous works of the authors [30,33–35] and [36–39]. For this reason, the chapter does not emphasize the details of the model. The simulation has been performed in COMSOL Multiphysics.

2. Brief summary of the mathematical model

The half-cell LIB contains lithium foil as the negative electrode or anode, a positive electrode or cathode, a separator, a cathode current collector, and liquid electrolyte, as shown in Fig. 21.1. The electrochemical reaction in the cathode during charging and discharging is explained in Eq. (21.1) [40]. During discharging, the lithium ions move from the anode into the cathode active materials, whereas during charging, the lithium ions move back from the cathode into the lithium foil. In Eq. (21.1), $\mathcal{M}$ represents the cathode active materials:

$$x\text{Li}^+ + xe^- + \underset{\text{discharge}}{\overset{\text{charge}}{\rightleftarrows}} \text{Li}_x\mathcal{M} \quad (21.1)$$

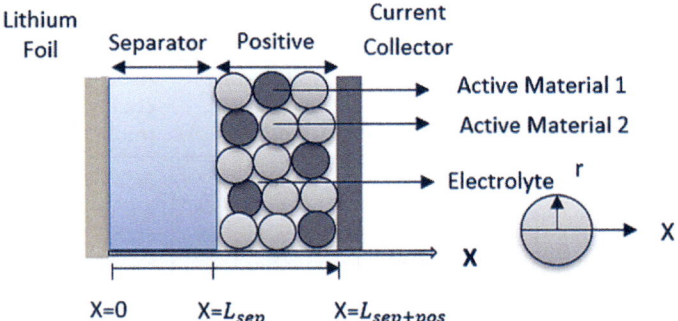

Figure 21.1 Schematic of the half-cell battery with multiple active materials [30].

A model considering multiple or a mixture of active materials is used to simulate the electrochemical performance of the blended active materials, as illustrated in Fig. 21.1. The blended cathode consists of spherical particles from different active materials mixed homogeneously and then coated onto an aluminum current collector to work as a positive electrode for the half-cell battery.

3. Experiments
3.1 Half-cell fabrication

For the experiment, half-cells were assembled in an argon-filled glove box. Lithium foil was used as a counterelectrode. The $LiPF_6$ electrolyte used in the cells was purchased from Sigma Aldrich. The cathode active materials, acetylene black and PVDF, were purchased from MTI Corporation. The cathode materials were mixed homogenously with PVDF and acetylene black in a ratio of 92:4:4, respectively. The mixture was added to NMP solvent purchased from Sigma Aldrich at a solid-to-liquid ratio of 1:2 before coating on an aluminum current collector. The cathodes were made of a mixture of fresh cathode active materials (or we assumed that the material regeneration process was ideal) with one dominant cathode active material to simulate the condition that the separation was not ideal. In fact, the cathodes were made from a dominant cathode active material and a minor percentage consisting of an equal mixture of the four other cathode active materials.

4. Results and discussion

In this section, the results of the computer simulation are presented. The results are generated based on the important assumption that the efficiency of the regeneration process to recover the recycled cathode active materials is 100% (ideal regeneration process). This means that the capacity of the recycled active material after the regeneration reaches the capacity of the fresh material. We studied the five most common cathode active materials for LIBs, namely, LCO, LMO, NCA, NMC, and LFP. The modeling parameters for these cathode active materials are listed in Table 21.1.

Table 21.1 Model parameters of the most commonly used cathode active materials in lithium-ion batteries [41–45].

Active material	Density [kg/m^3]	D$_{50}$ [μm]	Max state of charge [mol/m^3]	Initial state-of-charge [mol/m^3]	Reaction rate coefficient κ_i [mol/s.m^2]	Theoretical capacity [mAh/g]
LMO	4280	25	23,670.6	6000	5e^{-10}	148.2
LCO	5050	12	51,555	23,750	1e^{-7}	273.8
NMC	4770	10.5	51,385	21,500	1e^{-11}	279.5
NCA	4450	13.6	46,319	8067.9	1e^{-10}	278.9
LFP	3600	3.5	22,806	1000	3.63e^{-11}	169.9

To obtain the electrochemical performance of the half-cells made from these regenerated cathode active materials, pure LCO, LMO, NMC, NCA, and LFP powders were mixed homogeneously at different mass ratios. Two different mass ratios of 90% and 80% were considered for the dominant active materials, and 2.5% and 5% of each of the other four active materials were mixed with the dominant material, respectively. For example, for the dominant LCO cathode, we mixed 90 wt% LCO with 2.5 wt% NMC, 2.5 wt% LMO, 2.5 wt% NCA, and 2.5 wt% LFP for sample cathode 1, and we mixed 80 wt% LCO with 5 wt% NMC, 5 wt% LMO, 5 wt% NCA, and 5 wt% LFP for sample cathode 2. We did the same for dominant LMO, NMC, NCA, and LFP cathode samples 1 and 2. Obviously, in each half-cell, the dominant cathode active material has a higher percentage than the other cathode active materials. After preparing all the samples, the half-cells were cycled between voltage limits of 4.2 and 2.9 V with a 1C rate at 25°C.

Fig. 21.2 shows the performance of the half-cells for the five cathode active materials. As shown, when NCA is the dominant active material, the battery capacity decreases with a decrease in the NCA percentage from 90% to 80%. The reduction in the capacity is because NCA has a higher capacity than the other four cathode active materials. When LCO and NMC are the dominant materials, there is a slight reduction in the capacity when the percentage of the dominant active material is reduced. This insignificant reduction in capacity is due to these two cathode materials having a capacity in the average range among the cathode active materials.

When the dominant material is LMO, the capacity of half-cells increases with decreases in the percentage of LMO. This increase is due to the difference between the capacity of LMO and the other four cathode active materials. In fact, LMO has the lowest capacity among the materials, while it has the highest nominal voltage of 3.9 V.

The worst-case scenario is found when the dominant material is LFP. The capacity of the half-cells decreases significantly. This reduction in capacity is due to LFP having the lowest nominal voltage (3.2 V) among the five cathode materials, as shown in Fig. 21.2.

Several experiments were performed to validate the modeling results. Readers are referred to the authors' other publication [30] for the validation details.

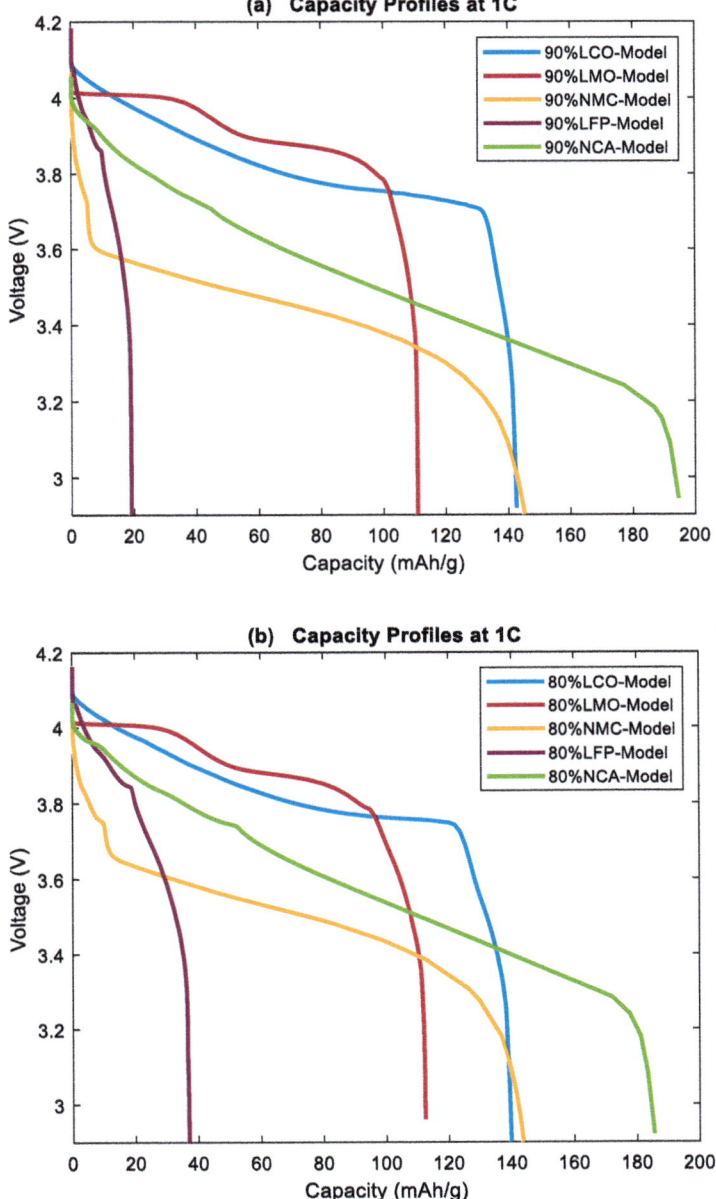

Figure 21.2 Discharge profile of half-cells made from a dominant cathode active material mixed with other types of cathode active materials [30].

5. Conclusions

This chapter presented the performance of LIBs made from recycled cathode active materials based on an imperfect separation between different active materials in recycling facilities. A single regenerated cathode active material containing small portions of other cathode active materials was used to make new batteries. The results were generated through modeling and computer simulation, and experiments were conducted for validation purposes. The cathode active materials studied in this chapter were the five most common cathode active materials on the market: LMO, LCO, NMC, NCA, and LFP. The results indicated that new LIBs could be fabricated with a mixture of regenerated cathode active materials even when cathode active materials had not been ideally separated from each other. However, the separation process should be performed to minimize the amount of LFP in the mixture. More studies are required to prove that making new batteries with a mixture of regenerated cathode active materials is technically and economically feasible. This is an ongoing study by the authors.

References

[1] I. Kay, R. Esmaeeli, S.R. Hashemi, A. Mahajan, S. Farhad, Recycling Li-ion batteries: robot disassembly of electric vehicle battery systems, in: Proceeding of the ASME 2019, Internal Mechanical Engineering Congress and Exposition (IMECE), Salt Lake City, UT, USA, November 2019, https://doi.org/10.1115/IMECE2019-11949.

[2] A.H. Mohammed, M. Alhadri, W. Zakri, H. Aliniagerdroudbari, R. Esmaeeli, S.R. Hashemi, G. Nadkarni, S. Farhad, Design and Comparison of Cooling Plates for a Prismatic Lithium-Ion Battery for Electrified Vehicles, SAE, 2018, https://doi.org/10.4271/2018-01-1188. Technical Paper, 2018-01-1188.

[3] E. Samadani, S. Farhad, S. Panchal, R. Fraser, M. Fowler, Modeling and Evaluation of Li-Ion Battery Performance Based on the Electric Vehicle Field Tests, SAE, 2014, https://doi.org/10.4271/2014-01-1848. Technical Paper.

[4] M. Alhadri, R. Esmaeeli, A.H. Mohammed, W. Zakri, S.R. Hashemi, H. Aliniagerdroudbari, H. Barua, S. Farhad, Studying the degradation of lithium-ion batteries using an empirical model for aircraft applications, in: ASME Power & Energy Conference, Lake Buena, Florida, United States, June 25–27, 2018, https://doi.org/10.1115/POWER2018-7428.

[5] S.R. Hashemi, A. Bahadoran Baghbadorani, R. Esmaeeli, A. Mahajan, S. Farhad, Machine learning-based model for lithium-ion batteries in BMS of electric/hybrid electric aircraft, Int. J. Energy Res. (2021), https://doi.org/10.1002/er.6197. Accepted Manuscript.

[6] S.R. Hashemi, A.M. Mahajan, S. Farhad, Online estimation of battery model parameters and state of health in electric and hybrid aircraft application, Energy 229 (15) (2021) 120699, https://doi.org/10.1016/j.energy.2021.120699.

[7] S.R. Hashemi, R. Esmaeeli, H. Aliniagerdroudbari, M. Alhadri, H. Al-Shammari, A. Mahajan, S. Farhad, New intelligent battery management system for drones, in: Proceeding of the ASME 2019, Internal Mechanical Engineering Congress and Exposition (IMECE), Salt Lake City, UT, USA, November 2019, https://doi.org/10.1115/IMECE2019-10479.

[8] S.R. Hashemi, R. Esmaeeli, A. Nazari, H. Aliniagerdroudbari, M. Alhadri, W. Zakri, A.H. Mohammed, A. Mahajan, S. Farhad, A fast diagnosis methodology for typical faults of a lithium-ion battery in electric and hybrid electric aircraft, ASME J. Electrochem. Energy Convers. Storage 17 (1) (2020), https://doi.org/10.1115/1.4044956.

[9] S.R. Hashemi, A. Nazari, R. Esmaeeli, H. Aliniagerdroudbari, M. Alhadri, W. Zakri, A.H. Mohammed, A. Mahajan, S. Farhad, Fast fault diagnosis of a lithium-ion battery for hybrid electric aircraft, in: ASME Power & Energy Conference, Lake Buena, Florida, United States, June 25—27, 2018, https://doi.org/10.1115/POWER2018-7476.

[10] M. Alhadri, W. Zakri, R. Esmaeeli, A.H. Mohammed, S.R. Hashemi, H. Barua, H. Aliniagerdroudbari, S. Farhad, Analysis of second-life of a lithium-ion battery in an energy storage system connected to a wind turbine, in: 2019 IEEE Power and Energy Conference at Illinois (PECI), 2019, pp. 1—8, https://doi.org/10.1109/PECI.2019.8698782.

[11] Q. Meng, Y. Zhang, P. Dong, A combined process for cobalt recovering and cathode material regeneration from spent $LiCoO_2$ batteries: process optimization and kinetics aspects, Waste Manag. 71 (2018) 372—380, https://doi.org/10.1016/j.wasman.2017.10.030.

[12] Y. Tang, H. Xie, B. Zhang, X. Chen, Z. Zhao, J. Qu, P. Xing, H. Yin, Recovery and regeneration of $LiCoO_2$-based spent lithium-ion batteries by a carbothermic reduction vacuum pyrolysis approach: controlling the recovery of CoO or Co, Waste Manag. 97 (2019) 140—148, https://doi.org/10.1016/j.wasman.2019.08.004.

[13] F. Pagnanelli, E. Moscardini, P. Altimari, T. Abo Atia, L. Toro, Leaching of electrodic powders from lithium ion batteries: optimization of operating conditions and effect of physical pretreatment for waste fraction retrieval, Waste Manag. 60 (2017) 706—715, https://doi.org/10.1016/j.wasman.2016.11.037.

[14] X. Chen, H. Ma, C. Luo, T. Zhou, Recovery of valuable metals from waste cathode materials of spent lithium-ion batteries using mild phosphoric acid, J. Hazard Mater. 326 (2017) 77—86, https://doi.org/10.1016/j.jhazmat.2016.12.021.

[15] B. Huang, Z. Pan, X. Su, L. An, Recycling of lithium-ion batteries: recent advances and perspectives, J. Power Sources (2018), https://doi.org/10.1016/j.jpowsour.2018.07.116.

[16] S. Farhad, A. Nazari, Introducing the energy efficiency map of lithium-ion batteries, Int. J. Energy Res. 43 (2019) 931—944, https://doi.org/10.1002/er.4332.

[17] E. Foreman, W. Zakri, M.H. Sanatimoghaddam, A. Modjtahedi, S. Pathak, N.G. Garafolo, S. Farhad, A review of inactive materials and components of flexible lithium-ion batteries, Adv. Sustain. Syst. 1 (11) (2017), https://doi.org/10.1002/adsu.201700061.

[18] A. Ritchie, W. Howard, Recent developments and likely advances in lithium-ion batteries, J. Power Sources 162 (2006) 809—812, https://doi.org/10.1016/j.jpowsour.2005.07.014.

[19] W.A. Appiah, J. Park, L. van Khue, Y. Lee, J. Choi, M.H. Ryou, Y.M. Lee, Comparative study on experiments and simulation of blended cathode active materials for lithium ion batteries, Electrochim. Acta 187 (2016) 422—432, https://doi.org/10.1016/j.electacta.2015.11.029.

[20] P. Albertus, J. Christensen, J. Newman, Experiments on and modeling of positive electrodes with multiple active materials for lithium-ion batteries, J. Electrochem. Soc. 156 (2009), https://doi.org/10.1149/1.3129656.

[21] S.K. Jeong, J.S. Shin, K.S. Nahm, T. Prem Kumar, A.M. Stephan, Electrochemical studies on cathode blends of $LiMn_2O_4$ and Li $[Li_1/15Ni_1/5Co_2/5Mn_1/3O_2$, Mater. Chem. Phys. 111 (2008) 213—217, https://doi.org/10.1016/j.matchemphys.2008.03.032.

[22] N.M. Jobst, A. Hoffmann, A. Klein, S. Zink, M. Wohlfahrt-Mehrens, Ternary cathode blend electrodes for environmentally friendly lithium-ion batteries, ChemSusChem 13 (2020) 3928—3936, https://doi.org/10.1002/cssc.202000251.

[23] S.B. Chikkannanavar, D.M. Bernardi, L. Liu, A review of blended cathode materials for use in Li-ion batteries, J. Power Sources (2014), https://doi.org/10.1016/j.jpowsour.2013.09.052.

[24] H.S. Kim, S. il Kim, W.S. Kim, A study on electrochemical characteristics of $LiCoO_2/LiNi_1/3Mn_1/3Co_1/3O_2$ mixed cathode for Li secondary battery, Electrochim. Acta 52 (2006) 1457—1461, https://doi.org/10.1016/j.electacta.2006.02.045.

[25] T. Numata, C. Amemiya, T. Kumeuchi, M. Shirakata, M. Yonezawa, Advantages of Blending $LiNi_{0.8}Co_{0.2}O_2$ into $Li_{1\pm x} Mn_{2\pm x}O_4$ Cathodes, 2001, https://doi.org/10.1016/S0378-7753(01)00753-4.

[26] J.F. Whitacre, K. Zaghib, W.C. West, B.v. Ratnakumar, Dual active material composite cathode structures for Li-ion batteries, J. Power Sources 177 (2008) 528—536, https://doi.org/10.1016/j.jpowsour.2007.11.076.

[27] H. Al-Shammari, R. Esmaeeli, H. Aliniagerdroudbari, M. Alhadri, S.R. Hashemi, H. Zarrin, S. Farhad, Recycling lithium-ion battery: mechanical separation of mixed cathode active materials, in: ASME International Mechanical Engineering Congress and Exposition, Proceedings (IMECE), American Society of Mechanical Engineers (ASME), 2019, https://doi.org/10.1115/IMECE2019-10755.

[28] H. Al-Shammari, S. Farhad, Regeneration of Cathode Mixture Active Materials Obtained from Recycled Lithium Ion Batteries, SAE, 2020, https://doi.org/10.4271/2020-01-0864. Technical Paper 2020-01-0864.

[29] H. Al-Shammari, S. Farhad, Heavy liquids for rapid separation of cathode and anode active materials from recycled lithium-ion batteries, Resour. Conserv. Recycl. 174 (2021) 105749, https://doi.org/10.1016/j.resconrec.2021.105749. ISSN 0921-3449.

[30] H. Al-Shammari, S. Farhad, Performance of cathodes fabricated from mixture of active materials obtained from recycled lithium-ion batteries, Energies 15 (2) (2022) 410, https://doi.org/10.3390/en15020410.

[31] M. Doyle, T.F. Fuller, J. Newman, Modeling of galvanostatic charge and discharge of the lithium/polymer/insertion cell, J. Electrochem. Soc. 140 (1993), https://doi.org/10.1149/1.2221597.

[32] M. Doyle, J. Newman, The use of mathematical modeling in the design of lithium/polymer battery systems, Electrochim. Acta 40 (1995), https://doi.org/10.1016/0013-4686(95)00162-8.

[33] S. Farhad, Mathematical modeling for charging/discharging processes of batteries/nanobatteries, in: Book Chapter in Nanobatteries and Nanogenerators: Materials, Technologies and Applications, 2021, https://doi.org/10.1016/B978-0-12-821548-7.00003-8.

[34] A.G. Kashkooli, G. Lui, S. Farhad, D.U. Lee, K. Feng, A. Yu, Z. Chen, Nano-particle size effect on the performance of $Li_4Ti_5O_{12}$ spinel, Electrochim. Acta 196 (2016) 33–40, https://doi.org/10.1016/j.electacta.2016.02.153.

[35] M. Mastali, E. Foreman, A. Modjtahedi, E. Samadani, A. Amirfazli, S. Farhad, R. Fraser, M. Fowler, Electrochemical-thermal modeling and experimental validation of commercial graphite/$LiFePO_4$ pouch lithium-ion batteries, Int. J. Therm. Sci. 129 (2018) 218–230, https://doi.org/10.1016/j.ijthermalsci.2018.03.004.

[36] A.G. Kashkooli, A. Amirfazli, S. Farhad, D.U. Lee, S. Felicelli, H.W. Park, K. Feng, V.D. Andrade, Z. Chen, Representative volume element model of lithium-ion battery electrodes based on X-ray nano-tomography, J. Appl. Electrochem. 47 (3) (2017) 281–293, https://doi.org/10.1007/s10800-016-1037-y.

[37] A.G. Kashkooli, E. Foreman, S. Farhad, D. Un Lee, W. Ahn, K. Feng, V. De Andrade, Z. Chen, Morphological and electrochemical characterization of a nanostructure $Li_4Ti_5O_{12}$ electrode using multiple imaging mode synchrotron X-ray computed tomography, J. Electrochem. Soc. 164 (12) (2017) A2861–A2871, https://doi.org/10.1149/2.0101713jes.

[38] A.G. Kashkooli, E. Foreman, S. Farhad, D. Un Lee, W. Ahn, K. Feng, V. De Andrade, Z. Chen, Synchrotron X-ray nano computed tomography based simulation of stress evolution in $LiMn_2O_4$ electrodes, Electrochim. Acta 247 (2017c) 1103–1116, https://doi.org/10.1016/j.electacta.2017.07.089.

[39] H. Zarrin, S. Farhad, F. Hamdullahpur, V. Chabot, A. Yu, M. Fowler, Z. Chen, Effects of diffusive charge transfer and salt concentration gradient in electrolyte on Li-ion battery energy and power densities, Electrochim. Acta 125 (2014) 117–123, https://doi.org/10.1016/j.electacta.2014.01.022.

[40] V. Chabot, S. Farhad, Z. Chen, A.S. Fung, A. Yu, F. Hamdullahpur, Effect of electrode physical and chemical properties on lithium-ion battery performance, Int. J. Energy Res. 37 (2013) 1723–1736, https://doi.org/10.1002/er.3114.

[41] T. Dong, P. Peng, F. Jiang, Numerical modeling and analysis of the thermal behavior of NCM lithium-ion batteries subjected to very high C-rate discharge/charge operations, Int. J. Heat Mass Tran. 117 (2018) 261–272, https://doi.org/10.1016/j.ijheatmasstransfer.2017.10.024.

[42] C. Liu, H. Li, X. Kong, J. Zhao, Modeling analysis of the effect of battery design on internal short circuit hazard in $LiNi_{0.8}Co_{0.1}Mn_{0.1}O_2$/$SiO_x$-graphite lithium ion batteries, Int. J. Heat Mass Tran. 153 (2020), https://doi.org/10.1016/j.ijheatmasstransfer.2020.119590.

[43] D. Miranda, A. Gören, C.M. Costa, M.M. Silva, A.M. Almeida, S. Lanceros-Méndez, Theoretical simulation of the optimal relation between active material, binder and conductive additive for lithium-ion battery cathodes, Energy 172 (2019) 68–78, https://doi.org/10.1016/j.energy.2019.01.122.

[44] N. Paul, J. Keil, F.M. Kindermann, S. Schebesta, O. Dolotko, M.J. Mühlbauer, L. Kraft, S.v. Erhard, A. Jossen, R. Gilles, Aging in 18650-type Li-ion cells examined with neutron diffraction, electrochemical analysis and physico-chemical modeling, J. Energy Storage 17 (2018) 383–394, https://doi.org/10.1016/j.est.2018.03.016.

[45] P. Ramadass, B. Haran, R. White, B.N. Popov, Mathematical modeling of the capacity fade of Li-ion cells, J. Power Sources 123 (2003) 230–240, https://doi.org/10.1016/S0378-7753(03)00531-7.

CHAPTER 22

Mechanical and physical processes of battery recycling

Denis Manuel Werner[1], Thomas Mütze[2], Alexandra Kaas[3] and Urs A. Peuker[3]

[1]LIBREC AG, Oensingen, Switzerland; [2]Helmholtz Institute Freiberg for Resource Technology, Helmholtz-Zentrum Dresden-Rossendorf, Freiberg, Germany; [3]Institute of Mechanical Process Engineering and Mineral Processing, TU Bergakademie Freiberg, Freiberg, Germany

1. Introduction

Lithium-ion batteries (LIBs) play a significant role in the transition toward a greenhouse gas emission-free transport sector. In addition, LIBs are used in the area of stationary energy-storage systems, and for several portable applications in the electronics sector. Hence, LIB recycling is becoming increasingly important in terms of environmental, economic, geostrategic, and health aspects. The rising number of LIBs, which are currently manufactured and introduced into the market, require energy-efficient and sustainable waste management at EOL. Ideally, battery materials circulate in a closed loop within their life cycles [1–3].

Special aspects of LIB technology, such as its future and complex waste properties and the classification of remaining hazardous wastes, are challenging for all processes within the recycling chain for reaching high recycling efficiencies (REs). Nevertheless, several recycling technologies are already commercially used worldwide [4], but only a few are capable of achieving the material-specific REs that have been set by upcoming legal frameworks. Adopting mechanical processes, in particular, generates material fractions either for further metallurgical refinement to secondary raw materials or for the production of new batteries.

This chapter aims to present a general survey of physical methods used for LIB recycling. First, the general design and composition of LIBs are characterized and their waste management scheme along the recycling chain is explained. After that, the principles and areas of application of approved and promising physical methods are described in more detail. Finally, four process groups are clustered for the processing of specific material fractions, and applied methods are presented for various processing strategies. The entire consideration is based on materials properties provided after preparation and pretreatment, as well as the requirements from subsequent metallurgical refinement or direct recycling.

2. Lithium-ion batteries

2.1 Design and composition

LIBs consist of anode active materials (AMs) like graphite or amorphous carbon compounds and cathode AMs based on layered transition metal oxide. Common layered oxides are lithium in combination with nickel (Ni), cobalt (Co), manganese (Mn), or aluminum (Al) alone or with a different stoichiometry on one hand or iron phosphate on the other [5]. Together with binders and additives, AMs are coated on both sides of thin Al foils on the cathode side and foils of copper (Cu) on the anode side. The separator, a porous plastic foil, is located between the electrodes to prevent direct contact between the two electrodes. The pores of electrode-coating materials and separator foil are filled with an ion-conducting electrolyte. This is a multicomponent system consisting of mixed organic solvents, conductive salt, and other additives.

A hermetic housing encloses the components of the functional unit (electrodes, separator, and electrolyte) forming the battery cell. Commonly metal or rarely plastic-based materials are for the housing differing in thickness and design properties. The cells can be connected in series or parallel to each other to form either a single block or a module as a subunit of a larger battery system [4,6]. Additional peripheral components like a battery management system, cooling, packaging, electronic and electric parts on cell, module, or system level create an LIB. Besides the broad variety in materials [7], also a variety of product generations, joining techniques, and composites are known. Hence, EOL LIB systems vary in applied battery cell types, the number of battery cells, connections, as well as dimensions and mass, and finally in their material composition [3,7–16].

2.2 Waste management

EOL LIBs accumulate as production residues or mainly as consumer residues either during or after the end of their use phase due to moral or physical wear [17]. They are considered as future waste with technologically high complexity [18,19]. Potential hazards arise from the voltage and state of charge as well as hazardous and reactive components [4,20–22]. Especially, the components of the electrolyte require special care. The organic solvents can cause fire and explosion [9,23] and their hygroscopicity facilitates corrosion [11]. The conductive salt has limited chemical and thermal stability. Thus, further flammable and toxic gases may be generated as reactions or decomposition products during battery lifetimes, under abuse conditions, or in worst-case scenarios. As a result, fire and various chemical reactions can occur before, during, and after opening the battery cell [24,25].

Therefore, LIB disposal is challenging and must deal with the great heterogeneity and complexity in LIB structure and composition as well as the problematic ingredients of LIBs [21,26–28]. Methods to reduce, reuse, recycle, and recover materials or components must be applied to reduce the amount for disposal [29,30]. Therein, the main goal of recycling is to keep the materials in closed loops by economically generating secondary (raw) materials [31–33]. Simultaneous to the enrichment of elements in respective material fractions, the scrap volume reduction and transformation of hazardous substances in an ecological and economically sustainable way should keep up with prescribed REs and safety management [3,34,35].

In principle, the transformation from EOL LIBs to secondary (raw) materials follows the recycling chain with its four process stages: preparation, pretreatment, processing, and refining [4,36]. The temperature applied for thermal depollution allows clustering of the recycling technologies into three main processing routes: low, medium, and high. As a consequence, the effort for preparation, pretreatment, and processing, and thus the RE, varies for each route [4]. Therein, the enrichment of valuable metals (Co, Cu, Ni, lithium (Li)) and critical materials (Co, graphite, Li) in the black mass fraction is one of the primary incentives due to energy-intensive raw materials production, high market values, and resource shortages [22,27].

3. Physical processing of end-of-life lithium-ion batteries

Physical processing covers two fundamental unit operations, liberation and separation. Both are mostly mechanical methods sometimes supported by thermal or chemical methods [4,37]. In general, mechanical methods are more energy-efficient and economically affordable than thermal, chemical, and metallurgical methods [17]. Mechanical processing has become urgent because of the increasing number of waste batteries and their increasing size and weight, such as those in modern vehicles. Especially in that case, dismantling is not a long-term option due to difficulties in large-scale industrialization. Moreover, individual safety measures must be installed for each operator to avoid exposition to toxic volatile organic solvents or electric shocks [38]. Hence, mechanical liberation and physical separation steps are alternatives due to their simplicity, efficiency, flexibility, scalability, and high throughput [39,40].

3.1 Mechanical liberation

Mechanical liberation transfers a feed material to a storable, conveyable, and flowable bulk with an enlarged surface area [41]. Furthermore, it controls the particle properties which influence the efficiency of subsequent separation technologies [42,43]. In principle, decreasing particle sizes is often accompanied by a higher degree of liberation at the cost of less efficient (dry) separation processes [44].

Mechanical liberation of LIBs can be distinguished into crushing the battery modules or cells, milling electrode mixtures, and grinding black mass. The main purpose of crushing (Section 3.1.1) in terms of automated cell opening is the liberation of the complex compound structure by breaking up the bonds between the individual components or materials. This initial liberation enables an efficient subsequent enrichment of valuable materials to defined concentrates [27,37,44−46]. In contrast, milling (Section 3.1.2) is applied in most cases for the liberation of coating materials from the current collector foils (decoating) or changing the collector particle shape toward spheres (deformation). Milling can also aim at the refinement of the coating concentrate (black mass fraction) liberating the AMs from the binder.

3.1.1 Crushing

Crushing describes the mechanical battery opening to disintegrate the solid components like housing, separator, anode, and cathode. Disintegration can be also achieved by dismantling (manual, hybrid or automated cell opening). In opposition to dismantling, crushing reduces also the components in terms of particle size. A usual and desired side effect of crushing is the simultaneous liberation of coating materials [9,44,47] producing a so-called black mass. The liberation depends mainly on the temperature used during thermal depollution, the electrode type, the grid size of the crushing device, and stressing mode [47−49].

Crushing whole LIB systems or modules is usually performed in two steps [50,51] and occasionally in one [16,23,52]. In contrast, cells are crushed mostly in one stage due to different compound properties [1]. The initial crushing (precrushing) of the feed generates small-sized battery fragments so they fit into the openings of subsequent devices. The second, and commonly the final, crushing step aims to fully liberate the individual components [48]. Here, an outlet grid size controls the degree of liberation and upper particle size of the crushing product [44,46]. The throughput of crushing decreases with decreasing size of the grid openings. Common opening sizes range from 80 to 40 mm for precrushing [53,54] to 30 to 6 mm for final crushing. The optimal opening size depends on the characteristics of the feed material and therefore on the battery itself as well as pretreatment [9,15,47,54,55]. With respect to the machine used for final crushing, precrushing can also be carried out without any outlet grid.

The applied stressing modes are shearing for all kinds of feeds since batteries show a ductile material behavior [39,47,56]. As an alternative, cutting is used for pouch cells since their housing is much thinner. Tearing is sometimes used in combination with shearing for precrushing to condition batteries.

Slow rotating rotary shears apply shear and/or tear stress depending on the design of their tools. Fast rotating rotary shears (granulators [57]) usually shear and/or cut the materials [9,46,58]. Rotary shears were intensively investigated by Steinbild [53], Diekmann et al. [58], and Wuschke [9] for their feasibility and the required outlet grid size for liberation. The rotary speed of the shears is mostly below 5.0 m/s tip speed for precrushing. In contrast, final crushing is performed at low as well as high rotary speeds depending on the safety and crushing strategy, as well as feed material and required throughput (see Fig. 22.1).

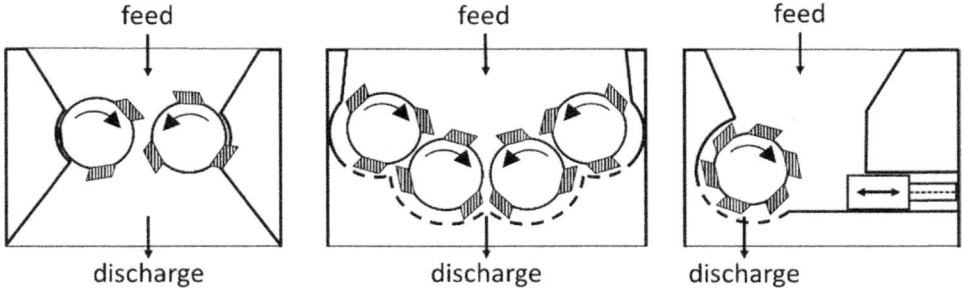

Figure 22.1 Schematic design and principles of precrushing (*left*) and final crushing (*middle and right*) according to [59].

The outlet grid size influences the residence time of the materials in the crushing chamber, and thus, the specific mechanical energy input [46,59]. If the openings in the outlet grid are too small or the stressing mode is not adjusted to the feed material, the crushed or liberated materials cannot be released from the crushing chamber. Especially, separator foils and sometimes pouch cell housings can block the outlet grid within the low-temperature route [4]. Those materials are often barely stressed by shearing due to their thin material thickness and elastic and ductile material properties. Therefore, those pieces remain bigger than the openings of the outlet grid size increasing the filling degree of the crushing chamber.

In final crushing, stress mechanisms based on slow or fast compression in combination with bending and tearing are also reported [55,60–64]. These stress combinations occur in hammer crushers or hammer mills. However, these machines are more likely to compact the particles, and consequently, trap other components and materials inside of them. In addition, this stress mode tends to create new compounds [47,50,54]. Moreover, the high rotation speeds together with the hammer tools generate a higher share of fine materials than rotary shears [65].

In addition to this conventional crushing, Zimmermann et al. [66] presented electrohydraulic fragmentation as an alternative. Electrohydraulic fragmentation requires dismantling to cell level to avoid extremely high amounts of energy for precrushing [67]. After that, a high voltage pulsed power creates in water shockwaves which stress the LIB cells [67]. This approach offers the advantages of selective material liberation along phase boundaries, which prevents uncontrolled size reduction and effort for further material processing. Disadvantages of this concept are the nonexistent availability of industrial-scale apparatus which lead to comparatively low throughputs and high energy costs.

In general, the complexity of crushing increases with decreasing dismantling depth (or depth of disassembly [28]) and depends strongly on the depollution strategy and the adequate choice of the process medium. Crushing under air or in an inert atmosphere are currently the favored options in the medium or low-temperature route, respectively. With that, a sophisticated crushing design can prevent an unwanted rise in temperature, fire, chemical reactions, or explosions [68,69]. However, the process medium influences also the process design and process control (i.e., continuous or discontinuous operation) and the overall operating and technological costs.

3.1.2 Milling

As the primary objective, milling targets the liberation of the coating materials from the metallic current collector foils. Therefore in literature, milling is also referred to as decoating, delamination, or pulverization [70] using impact stress, shearing, and/or bending. These mechanical stress modes in particular lead to selective comminution of the brittle coating materials on the one hand and deformation of the ductile metal foils on the other. The external forces induce tensions in the coating and between foils and coating to overcome cohesion and adhesion, respectively. After the liberation of the coating material, the selective deformation and compaction of the current collector foils toward spherical and compact particles is a secondary objective of milling. This deformation is used to enhance subsequent separation processes based on flow sorting in dry or wet media (Fig. 22.2).

The energy input for milling is significantly higher than for crushing because of size reduction to the low micrometer range and the deformation of the metal foils. An undesirable but often inevitable side effect of foils deformation is a simultaneous decrease in their particle size and a change in their breakage behavior from ductile to brittle. Milling is achieved rarely by vertical [72] and mostly by horizontal rotary impact mills like hammer mills [9,70], fine impact mills, and cutting mills [49,53]. In addition, electrohydraulic defragmentation (Section 3.1.1) is applied for decoating [66].

The active grinding tools in rotary impact mills like hammers, impact pins, or plates vary in size and design. These tools are combined with different milling chamber designs such as profiled wear surfaces, trapezoidal riffle sieves, or sieve rings. In combination with

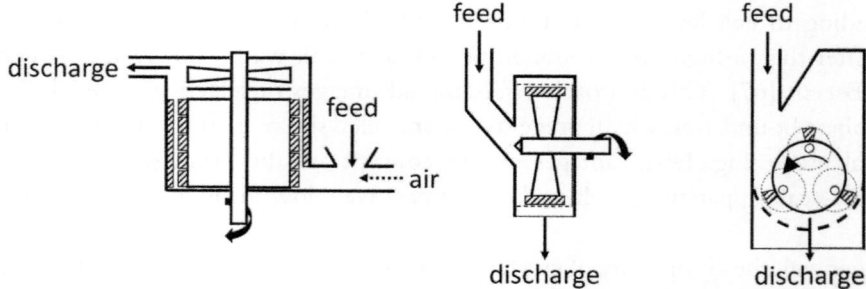

Figure 22.2 Schematic design and principle of a vertical (*left*) and horizontal (*middle*) fine impact mill with impact plates according to Ref. [42] and hammer mill according to [71].

high rotational speeds from 50 up to 140 m/s, complex impact stress conditions with superimposed shearing and bending lead to a complex stress situation and milling results regarding size, shape, and degree of deformation and liberation of the product particles.

The third objective of milling is the refinement of the coating concentrate (black mass fraction), aiming at size reduction, liberating the AMs from remaining binder mechanical activation of AMs for higher separation efficiency in subsequent physical separation or hydrometallurgical refining processes. Here, high-stress intensities by compression-shear and impact are applied to generate particle sizes in the range of 5–20 µm [73]. In laboratory scale, planetary ball mills are used for size reduction [74,75], liberation [76] and activation [77], wet high shear mixers [78,79] for the liberation for subsequent flotation in particular [80]. Besides that, nearly every ball mill or fine impact mill should be also capable of black mass refinement [42,75] even if they have not been included within the scope of research until now.

3.2 Physical separation

The liberated LIB components and materials are characterized by several differences in their physical properties. This enables physical separation to enrich valuable materials or deplete impurities in specific material fractions with higher purity [43,81]. Independently from the separation method and material characteristics, appropriate measures for dedusting and electrolyte separation must be installed [1,82].

Physical separation characteristics are size, shape, and density as well as magnetic, electrostatic, optic, and wetting properties of the liberated particles (cf. Table 22.1) [85–87]. Commonly used in LIB recycling are sieving, gravity separation [1,22,45], magnetic [16] and eddy current separation [88]. Moreover, electrostatic separation, sensor-based sorting, and flotation are currently investigated for their technical and economic feasibility [43,57,72,81,84]. In general, the technical effort is lower for dry processing whereas wet processing achieves higher separation efficiencies, especially for fine particle sizes [42,89]. However, wet processing results in higher environmental and economic loads than dry processing due to potential HF formation at elevated temperatures as well as subsequent dewatering and sludge processing or disposal costs [90].

The procedure, choice, and combination of separation processes for crushed EOL LIBs depend on their preparation, pretreatment, and degree of liberation, and thus, the composition of the input material [57]. After crushing, especially the particle shape of the liberated materials challenges further physical separation [67]. However, sieving is typically applied before sorting since all sorting processes are influenced by particle size.

3.2.1 Sieving

Sieving is the classification of particles into a fine and a coarse product by repeatedly comparing them with screen openings (Fig. 22.3). The separation is based on the materials' particle size, in particular the second main dimension of each specimen. In LIB

Table 22.1 Physical properties of the materials in LIB cells.

Component	Function	Material	Size x[a] (mm)	Shape[a]	Density ϱ (kg/m³)	γ/ϱ (Ω^{-1}m²kg^{-1})	Susceptibility χ (10^{-9} m³/kg)	Optic	Wettability
Cathode	Metal foil	Al	>0.5	Platy	2700[b]	13.0[c]	7.8[b]	Silver	—
	Coating	LFP...NMC	<0.5	Spheric-cuboid	3570...4999[d]	—	24...419[d]	Black	Hydrophil[e]
Anode	Metal foil	Cu	>0.5	Platy	8930[b]	6.3[c]	−0.8[b]	Auburn	—
	Coating	Graphite	<0.5	Spheric-cuboid	2260[b]	—	−0.5[d]	Gray	Hydrophob[e]
Housing	NSD contact	Cu	>1.0	Cuboid	8930[b]	6.3[c]	−0.8[b]	Auburn	—
	Electrical contact	Cu	>1.0	Cuboid	8930[b]	6.3[c]	−0.8[b]	Auburn	—
	Electrical contact	Al	>1.0	Cuboid	2700[b]	13.0[c]	7.8[b]	Silver	—
	Prismatic	Al	>1.0	Cuboid	2700[b]	13.0[c]	7.8[b]	Silver	—
	Round	Stainless steel	>1.0	Cuboid	7850[b]	0.1[c]	>3500[c]	Silver	—
	Pouch	PA/PP/Al	>1.0	Platy	1300	n. d.	n. d.	Matte silver	—
	Retainer	PP	>1.0	Cuboid	900...910[b]	—	—	Transparent	—
	Sleeve	PET	>1.0	Cuboid	920...960[b]	—	—	White	—
	NSD foil	PP	>1.0	Platy	900...910[b]	—	—	Turquoise	—
	Foils	PP	>1.0	Platy	900...910[b]	—	—	Turquoise	—
Separator	Foil	PP/PE	>1.0	Platy	920...930[b]	—	—	White	—

γ, electrical conductivity; *PA*, polyamide; *PP*, polypropylene; *PE*, polyethylene; *n.d.*, no data.
[a]Common values and forms after final crushing with shearing.
[b][83].
[c][42].
[d][76].
[e][84].

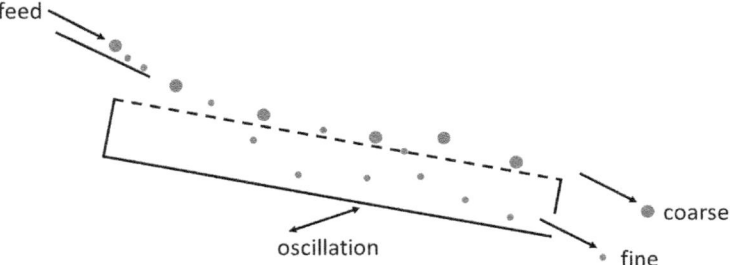

Figure 22.3 Schematic design and principle of a sieving device.

recycling, sieving is commonly used to separate the electrodes' coating materials (black mass) from the other components [44,57]. Thus, coarse particles like ones of the housing material, separator, and plastics as well as partly or fully decoated current collector foils are retained above the screen, while fine particles, in particular the coating materials, pass through [42,81].

Cut sizes for black mass separation are varying between 0.2 mm [81], 0.25 mm [5], 0.3 mm [91], 0.5 mm [15,44,92,93], and 1.0 mm [46,94]. The specific setting depends on the previous processing step like crushing, the arrangement of the sieving step within the process chain, and the subsequent refining strategy. Sieving requires dry material regarding the remaining organic solvents of the LIB electrolyte. The cut size influences especially the recovery of coating materials and the volume of impurities like copper and aluminum from the current collector foils [57,95]. Moreover, additional coarser screen decks are used to enrich other components [15,62] and to increase the separation efficiency and durability of the finer meshes. In addition, narrow size fractions also enhance subsequent mechanical separation processes. Typical sieves in LIB recycling are tumbling or vibrating screens [96], sometimes supported by air-jet configurations [49]. In addition, plan gyratory screens with circular vibration are applied for sieving [97].

3.2.2 Gravity separation

The particle density is the separation characteristic in gravity separation. Density differences between individual particles lead to different trajectories in static, stationary flowing, and pulsating fluids. A so-called light fraction is generated containing particles of comparatively low density and a heavy fraction with particles of high density [42]. Particle size and shape have a considerable impact on separation efficiency [98]. Hence, sieving is often necessary before gravity separation [17,83]. Separation by density is predominantly carried out in gravity fields. However, centrifugal forces have steadily gained importance in the fine particle range.

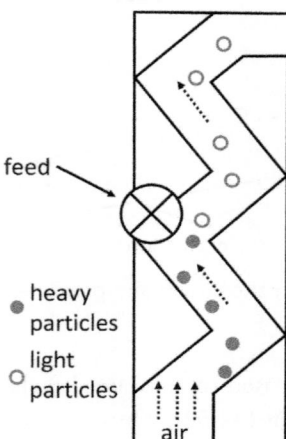

Figure 22.4 Schematic design and principle of a zigzag air classifier according to [99].

3.2.2.1 Airflow sorting in zigzag classifiers

Zigzag classifiers consist of an upright rectangular sorting tube designed in zigzag form with air flowing through from bottom to top (Fig. 22.4). Commonly, it is used in secondary raw material processing in the particle size range from 5 to 50 mm. In principle, the separation characteristic results from the velocity difference of the quasi-stationary airflow and the settling velocities of the particles. Therefore, the separation result depends on the flow regime as well as the density, size, and shape of the particles.

Pieces with high density and low thickness (platy or needle-shaped) might show the same trajectories as pieces with low density and high wall thickness (cubes or spheres) [98]. Consequently, the geometry of the pieces and the degree of liberation of LIB fragments have a bigger effect on the separation efficiency than the density differences alone. In LIB recycling, zigzag classifiers are commonly used to enrich the separator and other thin-walled plastic foils in one separation step regarding the pretreatment strategy. Moreover, electrodes and housing are separated in the respective light electrode concentrate and a heavy housing fraction [27,48].

3.2.2.2 Pneumatic tables

Pneumatic tables utilize a fluidized layer of fine-grained particles. The material is fed centrally on an inclined oscillating screen of narrow mesh openings or a porous surface (Fig. 22.5). The openings have to be smaller than the particles but can have a variety of geometries. The surfaces can consist of stainless steel or bronze. The material transport is enforced by the oscillation of the table as well as airflow from beneath. The complex combination of opening geometry, surface properties, and inclination, as well as oscillation rate, amplitude, and airflow, creates a layering of the material bed.

Therein, compact and heavy pieces accumulate in the lower layer and are carried upward the screen toward their removal. Flatty and lighter pieces are stronger fluidized and move downward to be dropped. Ideally, for maximum separation efficiency, the input material consists of particle sizes between 0.1 and 10 mm. If density differences are relatively low, it is recommended that narrower-sized classes are used. In addition, the throughput must be properly adjusted ensuring sufficient filling degree of the table for material transport and separation. Pneumatic tables have recently found application in industry for the separation of specially conditioned electrode fractions.

3.2.2.3 Centrifugal concentrators

Gravity is insufficient to separate efficiently the high-density cathode coating from the low-density anode coating in the black mass mixture due to the small particle sizes of these materials. Therefore, the centrifugal force field is used in wet or dry applications. The fluid determines the process design as well as the applied separation principle. Centrifugal separation in air or water is commonly used in processing primary resources, but can also be applied in the separation of black mass.

In dry separation, a spiral airflow, where fine or light particles follow the flow toward the center of the spiral and are discharged through a central opening (Fig. 22.6, left). The

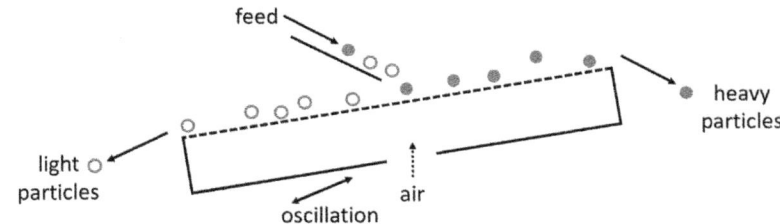

Figure 22.5 Schematic design and principle of a pneumatic table according to [99].

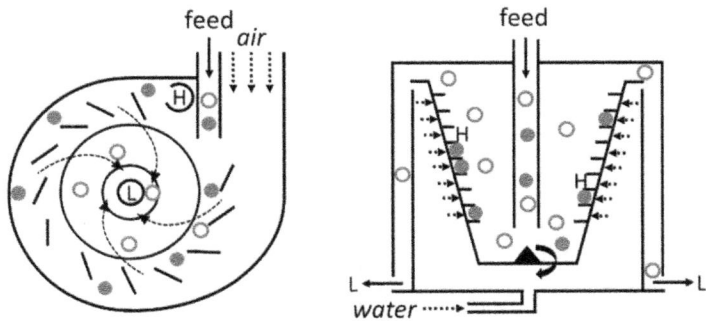

Figure 22.6 Schematic design and the principle of dry (*left, top view*) and wet (*right*) centrifugal concentrators according to Refs. [42,89], respectively; *H*, heavy particles; *L*, light particles.

coarse or heavy material remains at the outer circumference of the separation zone and is removed either continuously or discontinuously. A major parameter affecting the separation efficiency is the solid loading of the air, which can be influenced by the separation principle, material properties like composition and particle size, as well as the desired cut size. A rotor equipped with blades in the center of the separation zone improves the separation efficiency. However, the inflow has to be adjusted to the circumferential speed of the rotor to avoid undesirable vortex between the blades [42].

Wet separators consist of a cylindrical-conical vessel, which rotates at high speed around a horizontal or vertical axis (Fig. 22.6 right). The high-density particles are deposited on the outer wall of the vessel or in a series of riffles located along the vessel wall. Discharge of material is carried out discontinuously or by special measures while the machine is running. The low-density particles are flushed out with the water. It is generally used for material mixtures in which the heavy components make up less than 1% by weight [89].

3.2.2.4 Fluidized bed separators

Fluidized or teetered beds are used for hindered bed settling. Originally applied for the classification of coal in the size range 0.1–3 mm [42,100], it is currently investigated for density separation of black mass [65]. In general, a slurry is fed into a separation vessel and the particles settle against a uniform upward flow. The flow velocity is set to correspond with the settling velocity of the finest particles of the highest density, which creates an autogenous dense medium zone. Thereby, small interstices within these particles result in high interstitial velocities of the liquid carrying specific lighter particles upwards. In contrast, the specific heavy particles settle through the dense medium zone. They accumulate at the bottom of the vessel and are discharged via automatically controlled discharge devices [42] (see Fig. 22.7).

Fluidized bed technology is applied after the separation in hydro cyclones and before froth flotation. The level of the dense medium zone has to be permanently monitored to maintain the particle separation [89]. However, adequate conditioning of the particle size is required [100] and several approaches to improve the separation have been found [42].

3.2.3 Magnetic separation

Magnetic separation sorts particles according to their magnetic susceptibility into magnetic or nonmagnetic material. The separation equipment in this field can be categorized regarding the intensity of the magnetic field (low or high gradient), types of particle displacement (retention, lift-out, deflection sorting), and types of magnets (permanent and electromagnets). Besides particle susceptibility, particle volume also influences the resulting magnetic forces affecting the separator design and technology [42]. Overhead separators and drum separators are used for continuous dry applications with low- or high-intensity fields [72]. In contrast, the separation of fine coating particles (smaller

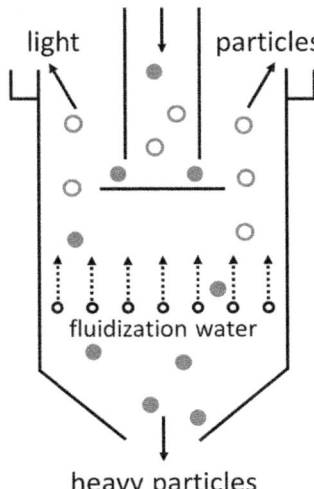

Figure 22.7 Schematic design and principle of a fluidized bed separator according to [42,65].

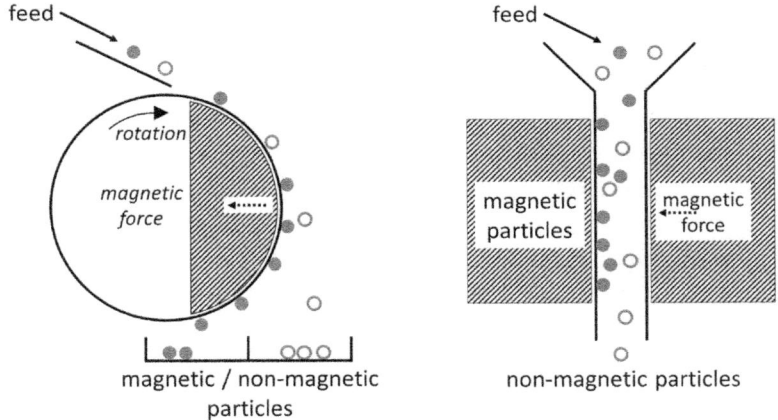

Figure 22.8 Schematic design and principle of magnetic separators with drum and high gradient design according to [42,99].

than 1 mm) is carried out with dry disc separators [76] or matrix separators in dry as well as wet media [101]. Here, high fields intensities are used with the retention principle [42] (see Fig. 22.8).

In LIB recycling, magnetic separation recovers coarse steel components (housing fragments, screws, peripheral materials) additionally protecting the subsequent process equipment against wear. The separation of the current collector materials is also feasible, but the sorting results are not satisfying [72]. Moreover, liberated coating particles are separated in a magnetic cathode concentrate and a nonmagnetic anode concentrate.

3.2.4 Eddy current separation

Eddy current separators sort particles due to their electrical conductivity [102,103]. Time-varying magnetic fields are induced within electrically conductive particle eddy currents, forming an opposing magnetic field and resulting in a repulsive force. This force depends on the one side of physical properties of the particles, such as shape, size, electrical resistance, and material density, and on the other side of the properties of the induced magnetic field like strength, range, and frequency. Here, the quotient of conductivity and density is directly proportional to the repulsive force [42,83,104] (see Fig. 22.9).

Eddy current separators are commonly used to separate nonmetallic materials like plastics and nonferrous metals such as copper and aluminum in the particle size range between 5 and 150 mm [83,104]. In principle, aluminum can be selectively separated as well [105], but unfavorable particle shape and wall thickness, in particular, prevent this application in industry. In battery processing, eddy current separation is generally used for further sorting of the nonmagnetic housing components.

3.2.5 Electrostatic separation

Electrostatic separation uses the differences of the surface's electrical conductivity or dielectric constant in an electrical field [98]. In principle, particles are charged by an electrical corona in the effective range between a corona electrode and the electrostatic counterelectrode. The magnitude of the maximum charge transferred is determined by the strength of the electric field, the particle size and shape, and the dielectric constant of the material. If the particles move into the range of the electrostatic counterelectrode, a differentiated discharge and thus trajectories occur. Parameters that influence the separation are the electrical conductivity, specific electrical resistance, size, shape, and surface moisture of particles [83].

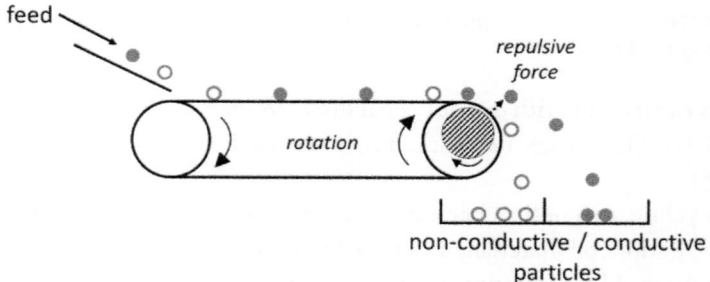

Figure 22.9 Schematic design and principle of an eddy current separator with eccentric pole drum according to [42,99].

In battery processing, corona-induced roll separators sort into a metal and nonmetal fraction. The first contains the electrode foils and coating materials, in the latter separator foils and plastics are enriched [64,85,106] (see Fig. 22.10).

3.2.6 Sensor-based sorting

Sensor-based or sensor-supported sorting determines the separation characteristic of individual particles via sensors [42]. In principle, all sensor-based sorting equipment contains four main processes, with each influencing overall separation efficiency [107]. Conveyor belts or chutes singularize the particles (1) and transport them past sensors (2). According to their detected characteristics and evaluation (3), they are subsequently ejected (4) pneumatically or mechanically. The evaluation of the measured particle properties and data is carried out via special computer software to initiate sorting [107,108].

Separation characteristics of the particles are optical properties, such as color, shape, surface texture, size and brightness, and reactions to transmitting magnetic or electromagnetic fields. Most often, the feed has to be preconditioned in terms of removal of dust and dirt by washing as well as removing spherical particles by ballistic separation, narrowing the particle sizes by sieving and drying of moist materials [42,109,110]. The coupling of several sensors simultaneously increases the detection efficiency [104] but also the overall costs for this separation method.

In battery recycling, sensor-based sorting is investigated to separate materials, such as current collector foils from each other with a combination of metal sensors and color cameras [53]. Moreover, there is a potential for the separation of nonmagnetic and either

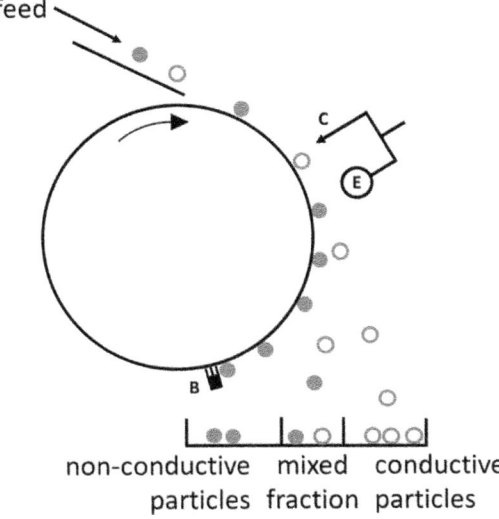

Figure 22.10 Schematic design and principle of a corona-induced roll separator according to Refs. [42,99]; C, corona electrode; E, electrostatic electrode; B, brush.

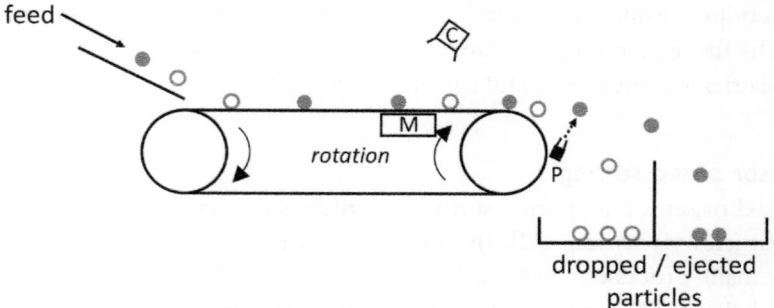

Figure 22.11 Schematic design and principle of a conveyor belt sorter according to Ref. [99]; *M*, metal sensor; *C*, color camera; *P*, pneumatic ejection unit.

nonconductive materials via visual spectroscopy or conductive housing materials via near-infrared sensors [72] (see Fig. 22.11).

3.2.7 Flotation

Flotation separates fine particles regarding differences in wettability [111–116]. This physicochemical separation uses water and rising air bubbles or oil droplets. Hydrophilic particles attach at the bubbles or droplets forming hetero-agglomerates, which rise to the top of the flotation cell forming a froth layer there. In contrast, hydrophilic particles settle down to the bottom of the cell. Flotation is used mainly in the size range between 50 and 500 μm. The separation efficiency is also influenced by the state of dispersion and conditioning in terms of the reagent regime. Reagents serve to control the particle hydrophobicity and are commonly collectors, frothers, depressants, and activators depending on the requirements.

In the context of LIB recycling, only froth flotation is the focus of interest to prepare black mass for metallurgical refining. Therein, a rotor and a stator cause turbulent flows in the cells, which results in particle suspension, air bubble generation, and particle-air collision [114]. In principle, graphite is naturally hydrophobic and can be recovered in the froth product, whereas the cathode coating is hydrophilic and remains as flotation residue [81,117]. However, the overall surface properties of the coating particles are strongly influenced by the degree of liberating regarding metal foils as well as the chemistry and structural properties of the cathode AMs [118]. Depending on the preprocessing, the wettability of the particles and the separation efficiency are also influenced by remaining electrode additives like polyvinylidene fluoride binder, carbon black, or organic solvents from the electrolyte [80]. Currently, worldwide research focuses on suitable depollution strategies in terms of binder and organic solvents removal [78] and reagents [57,117] to enhance separation efficiency as well as concentrate qualities (see Fig. 22.12).

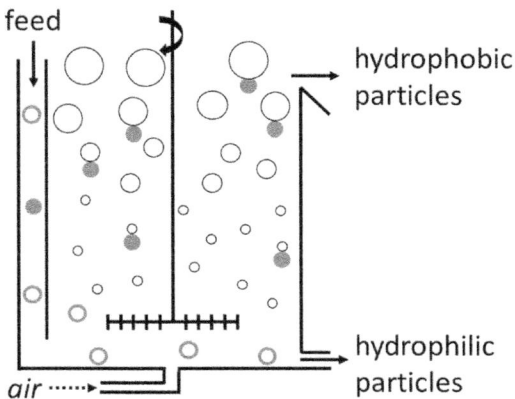

Figure 22.12 Schematic design and principle of froth flotation according to [42].

4. Process groups

A process group describes the connection of unit operations like liberation and separation in series, parallel, and as a circuit. The efficiency of a process group is determined by each unit operation included. At the beginning of mechanical processing, fragments are generated differing in physical properties according to their material composition and degree of liberation. This particle mixture is subsequently separated according to individual separation characteristics and into at least two material fractions [119]. Besides the separation characteristics, especially the degree of liberation, particle shape and size influence the efficiency of mechanical separation processes (Section 3.2) [120]. The material fractions can be a concentrate, which can be sold, or an intermediate product, which has to be further processed. For LIB processing, the batteries are crushed in a first process group to liberate at least the cell housing and the functional unit's components (Section 4.1, Fig. 22.13) [3]. Additionally, the disintegrated components are usually dried in accord with the previous depollution and processing strategy. Afterward, the material is enriched in first material fractions such as housing materials, a mixture of electrodes, and a mixture of AMs by combinations of physical separation processes [9,27,44,121,122]. In the low-temperature route [4], plastics (mainly the separator foils) are also separated.

The second process group focuses on further treatment of the housing (Section 4.2.1) and electrode fraction (Section 4.2.2). The latter usually still contains coatings that must be subsequently liberated and separated from the current collector foils [44,123]. A third process group separates or conditions the different electrode coating materials in the black mass for subsequent refining (Section 4.3) [27,57].

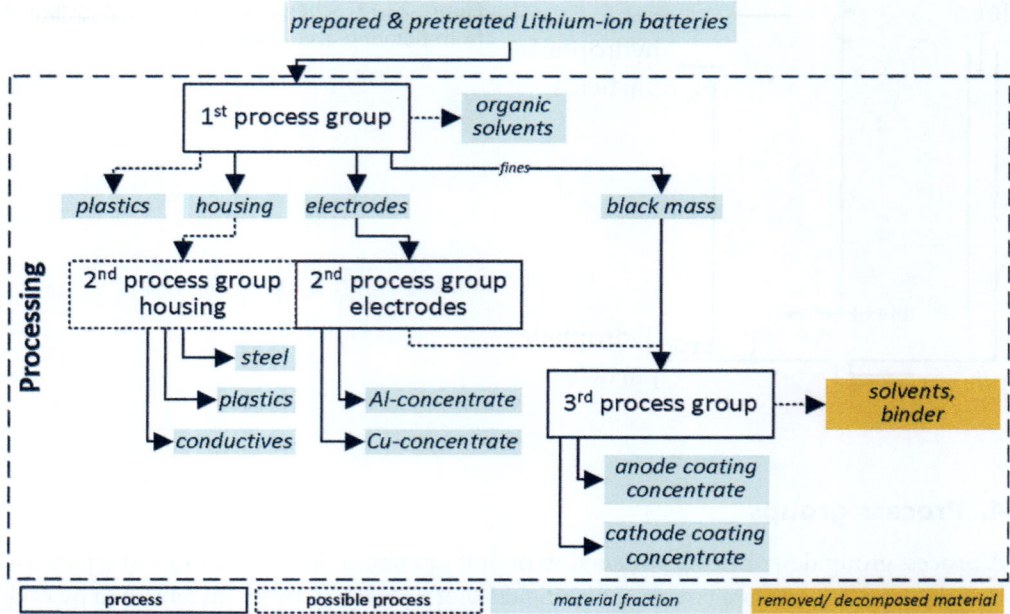

Figure 22.13 schematic material flows and fractions for the process group in lithium-ion battery recycling.

4.1 Initial liberation and enrichment

The goal of the first process group is the liberation of the cell's functional components to enrich them into first material fractions for further processing (Fig. 22.14). In general, mechanical processing of EOL LIBs becomes possible after special and complex pretreatment. The batteries are commonly manually disassembled to the module and sometimes to cell level, where assemblies, components, and materials from functional and structural parts are recovered for further use (second life) or special material-dedicated recycling processes [11,21]. Then, on the one hand, the batteries are only electrically depolluted (usually deep-discharged) before processing to remove the remaining energy (low-temperature route). On the other hand, the batteries are depolluted thermally above 500°C with [72,124,125] or without discharging (medium temperature route) [4]. Thereby, organic solvents are decomposed as well as the binder of the electrode coatings and plastics of the separator and housing components. Thus, short circuits occur when the electrodes get in contact leading to indirect energy removal.

The crushing step can be distinguished regarding the previous depollution strategy. Thermally depolluted batteries can be crushed safely in standard equipment with common dedusting. The latter is important due to heavy dust creation caused by the delamination of the fine electrode coatings. In contrast, crushing of only discharged batteries is

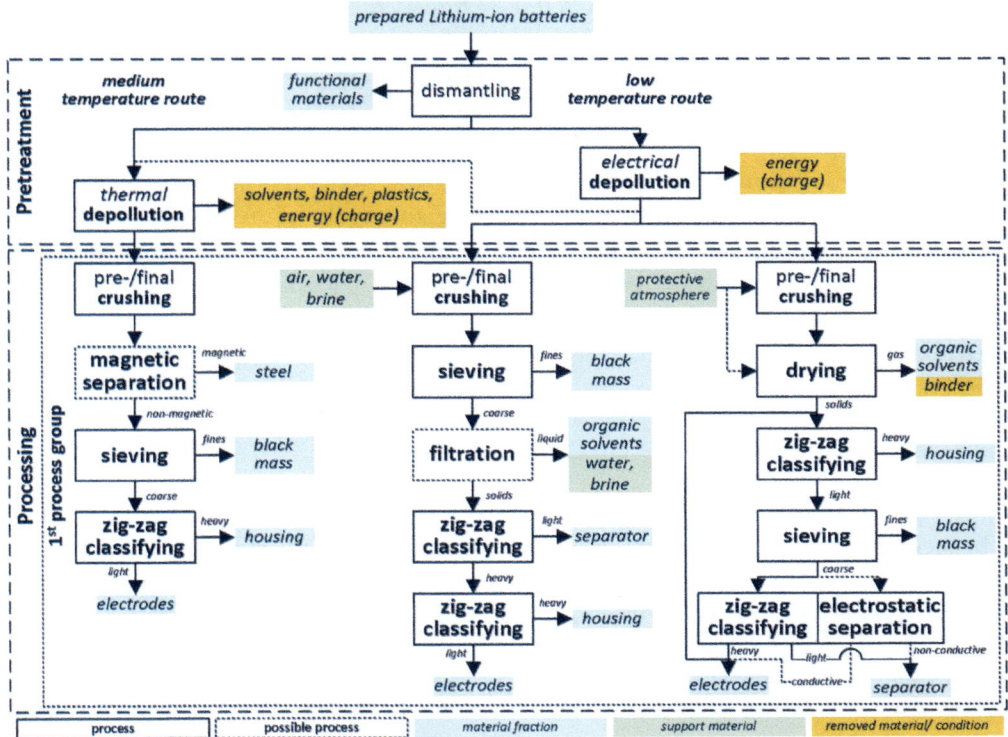

Figure 22.14 Schematic material flows and fractions in the first process group for medium- and low-temperature routes.

carried out either in the air [9] or under a protective atmosphere [96,126]. The first one requires high flow rates to dilute the evaporating solvents. Both strategies prevent fires or explosions during or after stressing and require an appropriate cleaning of the carrier gases from the solvents. In addition to dry liberation, also wet media such as water or brine solutions are used to open the batteries [4].

In the low-temperature route, the crushed product is dried [11] or fed directly to the subsequent separation process [9]. For both options, the process has to be hermetically sealed to prevent solvent release in the environment. Therein, at least the low volatile organic solvents can be recovered [11]. Physical separation in the medium-temperature route is performed by sieving and zigzag air classifiers to recover the black mass, the housing, and the electrode mixture [96,127]. If magnetic steel fragments are liberated during crushing, magnetic separation can be used to protect the fine meshes of the screen and to reduce the mass for subsequent separation processes [49,55,72].

The low-temperature route shows different possibilities to combine sieving and air classification. One reason is the selective decoating of anode coating during battery crushing as its adhesive forces on the Cu foil are lower compared with cathode coating on the Al foil [5,128]. Hence, the black mass created during the initial crushing consists of a higher graphite content, which can be easily recovered by sieving.

Consecutive two-step separation in zigzag air classifiers generates a separator concentrate containing the plastics, an electrode mixture, and a housing fraction [9,96]. The increasing use of pouch technology results in the recovery of the plastic-coated aluminum casing material mainly in the electrode and rarely in the housing fraction.

In contrast, drying the crushed material in a paddle dryer enhances decoating [53]. Here, the housing materials are removed first via zigzag separators [92]. In a second zigzag classifier step, the separator foils are recovered as light fractions from the electrode mixture (heavy fraction). Alternatively to second zigzag classifying, electrostatic separation is used to enrich separator foils in the nonconductive fraction and the electrode mixture in the conductive fraction, respectively [64,106,129].

4.2 Processing coarse fractions

The second process group covers the subsequent treatment of the coarse fractions like the housing concentrate and the electrode mixture. Even the separator fraction contains coarse fragments. Nevertheless, those are usually not processed due to their low intrinsic value, low share, and potential contribution to RE. One technically possibility to recycle the separator plastics PP and PE is selective dissolution via solvolysis lacking economy so far [36,83].

4.2.1 Processing the housing fraction

The mechanical processing of the housing fraction has not been a special focus of research worldwide so far, since this fraction is currently fed into established scrap processing. Except for pouch-type batteries, the housing fraction consists usually of the cell housing materials such as nickel-plated steel, aluminum, thick-walled plastics, and small quantities of copper from the electrical contacts. Depending on the dismantling depth, structural and functional metals and thick-walled plastics are also included originating from the system and module periphery [72]. In general, magnetic separation is applied to recover magnetic steels followed by eddy current separation to enrich metals and nonmetals (Fig. 22.15). Sensor-supported sorting can be used for further separation of metal and nonmetal fractions. Here, near-infrared sensors and hyperspectral cameras are suitable for the separation of different types of plastic. Color cameras, visual spectroscopy, and metal sensors are used as single devices or in combinations for sorting the metal fraction.

Different types of sensors, respective parameters, and sorting procedures have been evaluated [72]. However, parameters like throughput, product qualities, and revenues determine the choice of and investment in such an expensive separation technology. An alternative way of metal separation is compaction toward spheres by mechanical stress

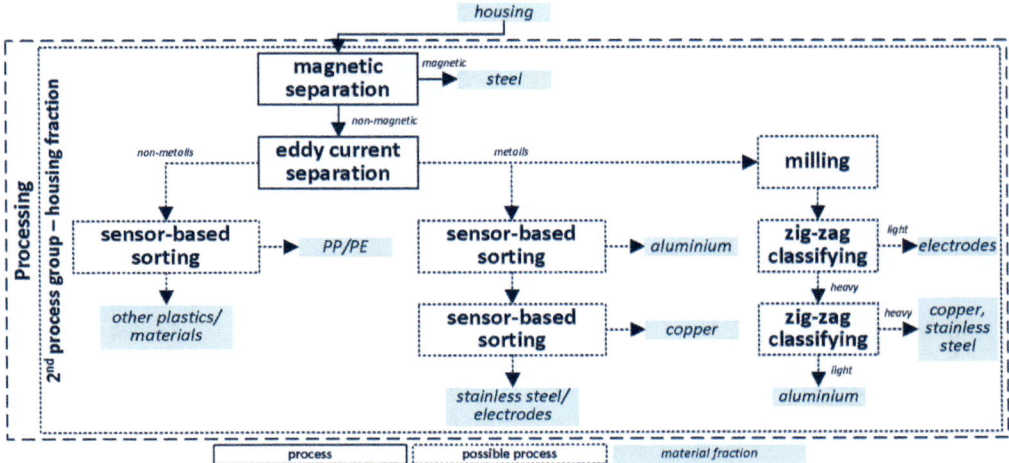

Figure 22.15 Schematic material flow and fractions in the second process group for further mechanical treatment of the housing fraction.

with a hammer mill followed by two-stage separation in a zigzag air classifier [50]. The economic feasibility of this variant remains questionable, too.

4.2.2 Processing the electrode fraction

In EOL-LIB recycling, processing of the electrode fraction is in the focus of intensive research activities for several years [9,47,53,130]. Besides partly decoated electrodes, this fraction contains also plastic and metal fragments originating from the housing and separators. The amount of remaining coating material on the current collector foils as well as of the plastics depends on the process route, processing strategy, and efficiency of the previous separation processes. Thus, the necessity of additional treatment varies and tends to increase with the increasing complexity of the fraction's composition. Hence, there exist manifold concepts for the process chain, especially in the low-temperature route. However, the absolute amount of the electrode fraction is relatively low with less than 30% of the cell mass, but in the future for many of the materials contained in this fraction will be required recovery rates by law.

Especially in the low-temperature route, one central task of electrode mixture treatment is the complete decoating of the current collector foils. This decoating is the basis for the subsequent separation of the coating from the current collector foils and the separation of anode and cathode current collectors. Therefore, several multistep processes are developed focusing on different input materials and separation targets (Fig. 22.16). Major common features are sieving for black mass separation and sorting of the metal foils due to differences in density, susceptibility, and optical particle properties [72].

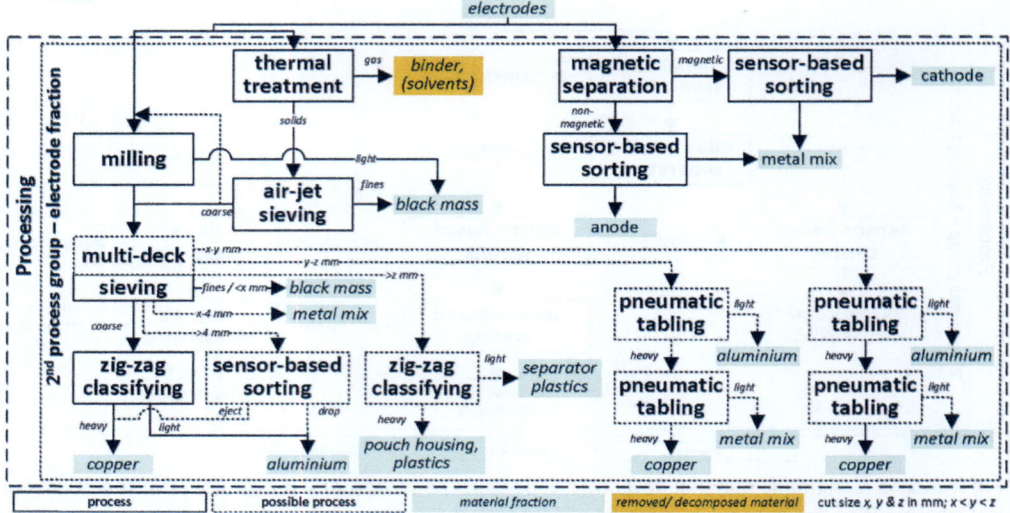

Figure 22.16 Schematic material flow and fractions in the second process group for further mechanical treatment of the electrodes fraction.

4.2.2.1 Decoating

One approach for the separation of current collector foils is density separation in airflows or hydroflows [9]. Here, one challenge is the similar settling velocity of the usually platy metal foils. Therefore, compaction toward spheres increases the differences in settling velocities and enhances the efficiency of gravity separation. This approach has been successfully applied in cable scrap and wire recycling [131] and can be also used for metal foils. Compaction and spherization are achieved by stressing in different kinds of impact mills.

Simultaneous to compaction, the coating materials are liberated from the current collector foils due to the selective breaking behaviors of the components (Section 3.1.2). However, the stressing in mills also generates some amount of fine Al and Cu particles depending on the stress mode and intensity, the equipment itself, and the composition of the electrode fraction [92]. These metal particles are considered on the one side as impurities for either direct recycling or further hydrometallurgical refining of the black mass, and on the other side as loss of valuable materials to the black mass [49]. Wuschke et al. [50] as well as Steinbild [53] and Diekmann et al. [92] propose compaction of the metal foils toward spheres in a discontinuous hammer mill or a continuously cutting mill. In general, any other type of continuous (fine) impact mill might suit the task [130].

In recent years, selective decoating is targeted by electrohydraulic fragmentation or high shear blending [79,132]. The latter is used for individual processing of anodes and cathodes generated by cell dismantling and manual components separation.

Hanisch et al. [49] introduce a thermomechanical method to reduce metal impurities in the black mass and increase decoating. Thereby, the binder is thermally decomposed at 500°C and the remaining solids are sieved with air jets. After thermal decomposition, the gentle impacts during sieving are sufficient for decoating and black mass separation.

4.2.2.2 Separation of decoated electrodes

After decoating and spherization, sieving is used to recover the coating material in a black mass fraction as fines. The coarse fraction of sieving contains the delaminated current collector foils which can be sorted in zigzag air classifiers [50,92] or hydro upstream sorters [9] into copper and aluminum concentrate. The latter achieves better separation results requiring high CAPEX and OPEX. Nevertheless, even the particles are deformed toward spheres, low compaction leads to overlapping settling properties. Consequently, one-step zigzag air classifiers achieve high product quality at a sufficient yield in either Al or Cu concentrate [92].

Alternative processing utilizes multideck sieving creating narrow size classes for density sorting of the medium-sized classes on pneumatic tables and the coarse size classes (>3…5 mm) in zigzag air classifiers. The selectively deformed copper and aluminum foils are enriched in the medium size factions. In contrast, the size of plastics, separator foils, and pouch housings is not influenced by impact stress and recovered in the coarse fraction. The classifier separates pouch housing together with heavy plastic pieces from the separator and light plastic pieces. The multiple narrow size classes enhance on the one hand the separation efficiency of the pneumatic tables but require on the other hand many sorting machines for relatively small throughputs, which increase the processing costs.

A third alternative is the separation of decoated, but nondeformed current collector foils by sensor-supported sorting. This technology requires a coarse feed, which can be generated by sieving at 4 mm enhancing the separation efficiency of the sensors [53]. However, the separation is negatively influenced by unfavorable flight characteristics during pneumatic ejection of the very thin metal foils [72]. Furthermore, the fine fraction below 4 mm would need special treatment, which is not addressed yet [72].

4.3 Processing black mass

The main task for the mechanical treatment of black mass fractions is to condition the material for subsequent metallurgical refining or direct recycling [3,133]. Hydrometallurgical treatment in particular is the subject of manifold international research activities [134], whereas upstream processing of the black mass is only gradually considered. Different valuables and impurities of the fed material as well as specifications for the products exist in this field depending on the refining strategy [72].

The black mass is usually produced as a fine fraction by sieving at cut sizes between 0.15 and 1.0 mm within the first and second process group. In principle, the coating materials of the electrodes are separated into graphite and an active material concentrate. The black mass is a compound of AMs, binder, and additives in the micrometer size range. They can contain remaining electrolyte components like conductive salt and organic solvents with respect to the applied pretreatment strategy in the first and second process group. Several technological proposals can be found in the literature (Fig. 22.17). Some of the physical separation processes require conditioning steps in terms of material liberation (attrition; grinding), binder and organic solvent removal (pyrolysis), size reduction (grinding), or chemical reactions (roasting).

Gellner [76] investigated dry processing of black mass via sieving and subsequent magnetic separation with a belt ring magnetic separator in the low-temperature route. She found that sieving already enriches graphite in the fines, whereas the cathode coating is recovered in the coarse magnetic fraction. The recovery of cathode coatings as heavy fractions via fluidized bed separation is sparsely described by Spangenberger and Gillard [65]. However, they mention that previous sieving at 150 μm and separating of only the coarse fraction increases the separation efficiency. Another dry processing method is separation in centrifugal fields via deflector wheel air classifier [53]. Therein, the cathode coating (LFP) is recovered at a cut size of 10 μm in the fine fraction with sufficient quality (88.9% by weight) and high yield (93.6% by weight).

A quite popular separation method in the research area is froth flotation utilizing differences in the wettability of particles. One crucial challenge is the negative influence of remaining organic solvents and binders on the wettability of the materials [57]. Hence, binder removal is performed via mechanical stressing and/or pyrolysis to recover graphite as froth fraction and cathode coating as tailings. The mechanical stressing aims at liberation via high shear or impact stress (attrition and grinding) [80], pyrolysis aims at the decomposition of the binder at elevated temperatures and low pressure.

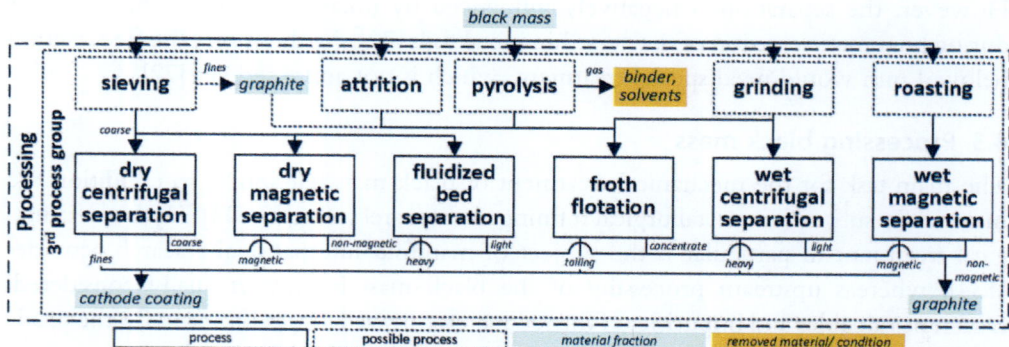

Figure 22.17 Schematic variants of the material flow and fractions in the third process group for further mechanical treatment of black mass.

Grinding was also applied before wet separation in centrifugal hydro classifiers after pyrolyzing the black mass. However, Gellner [76] reported an insufficient separation efficiency for the black mass of the low-temperature route. Ellis [101] published a patent for direct magnetic separation of black mass to separate the cathode coating as a magnetic product leaving from the graphite. He proposed a multistage magnetic separation with different field strengths so that the cathode materials can be selectively separated from each other according to nickel and cobalt content. In contrast, Li et al. [135] added oxygen-free roasting before wet magnetic separation to transfer LCO into pure cobalt, lithium carbonate, and graphite in a first step. Afterward, Lithium carbonate is dissolved, whereas cobalt is recovered as a magnetic fraction from the remaining graphite. However, this approach is only feasible for LCO cathodes so far.

5. Conclusion

Mechanical processing of EOL LIBs generates valuable battery materials in a safe, economic and ecological way. Most often, a combination of several liberation and separation steps is required for material enrichment due to the heterogeneity of the variety of materials, their physical properties, and size ranges. The processing can be subdivided into four process groups.

Processes and procedures within the first process group are already performed on industrial level generating housing, electrode, black mass concentrates. In the low-temperature route, also separator foil and electrolyte components are recovered. Further processing of the housing concentrate is not carried out due to low throughput and relatively high investment for the required sorting equipment. Several approaches exist for the processing of the electrode concentrate. A popular approach is currently milling with multistep physical separation in terms of sieving and gravity sorting in airflows.

Especially the separation of the black mass concentrate is in the focus of worldwide research activities. The midterm objective is to relieve subsequent hydrometallurgical refining for recycling on the (raw) material level to reduce expenses and increase RE. In the long term, efforts are being made to dispense with refining and to make both electrode AMs available for reuse in battery production in terms of recycling on the (functional) material level.

The importance of mechanical liberation and physical separation methods in the processing stage is steadily increasing. It enables compliance with the upcoming battery directive of the European Union. Direct recycling will be implemented in the medium term for production wastes. In contrast, reuse of all materials from consumption wastes is a long-term goal due to the high quality requirements for (secondary) raw materials in battery manufacturing and the continuous development of battery (active) materials.

References

[1] G. Zhao, Resource utilization and harmless treatment of power batteries, in: G. Zhao (Ed.), Reuse and Recycling of Lithium-Ion Power Batteries, John Wiley & Sons, Singapore, 2017, pp. 335–378.

[2] B. Friedrich, T. Träger, R. Weyhe, Recyclingtechnologien am Beispiel Batterien, Proceedings of 25, Aachener Kolloquium Abfallwirtschaft, Forum M, Aachen, 2012.

[3] E. Mossali, N. Picone, L. Gentilini, O. Rodrìguez, J.M. Perez, M. Colledani, Lithium-ion batteries towards circular economy: a literature review of opportunities and issues of recycling treatments, J. Environ. Manag. (2020) 264, https://doi.org/10.1016/j.jenvman.2020.110500.

[4] D. Werner, U.A. Peuker, T. Mütze, Recycling chain for spent lithium-ion batteries, Metals Open Acc. Metall. J. 10 (3) (2020), https://doi.org/10.3390/met10030316.

[5] S. Zhao, W. He, G. Li, Recycling technology and principle of spent lithium-ion battery, in: L. An (Ed.), Recycling of Spent Lithium-Ion Batteries — Processing Methods and Environmental Impacts, Springer Nature Switzerland AG, Cham, 2019, https://doi.org/10.1007/978-3-030-31834-5, pp. 1-26.

[6] R. Korthauer, in: R. Korthauer (Ed.), Handbuch Lithium-Ionen-Batterien, Springer Vieweg, Berlin Heidelberg, 2013.

[7] A. Kwade, W. Haselrieder, R. Leithoff, A. Modlinger, F. Dietrich, K. Droeder, Current status and challenges for automotive battery production technologies, Nat. Energy 3 (2018) 290–300.

[8] M. Chen, X. Ma, B. Chen, R. Arsenault, P. Karlson, N. Simon, Y. Wang, Recycling end-of-life electric vehicle lithium-ion batteries, Joule 3 (2019) 2622–2646, https://doi.org/10.1016/j.joule.2019.09.014.

[9] L. Wuschke, Mechanische Aufbereitung von Lithium-Ionen-Batteriezellen, Ph.D. Thesis, TU Bergakademie Freiberg, Freiberg, 2018.

[10] R. Weyhe, Verbundprojekt "Rückgewinnung der Rohstoffe aus Li-Ion Akkumulatoren", Accurec GmbH, Mühlheim, 2008.

[11] A. Kwade, J. Diekmann, in: A. Kwade, J. Diekmann (Eds.), Recycling of Lithium-Ion Batteries — The LithoRec Way, Springer, New York, NY, 2018, https://doi.org/10.1007/978-3-319-70572-9_12, pp. 312.

[12] A. Arnberger, K.-H. Gresslehner, R. Pomberger, A. Curtis, Recycling von lithium ionen batterien aus EVs & HEVs, in: Proceedings of DepoTech 2012, November 6, 2012. Leoben, AT.

[13] L. Gaines, J. Sullivan, A. Burnham, Life-cycle analysis for lithium-ion battery production and recycling. In: Proceedings of 90th Annual Meeting of the Transportation Research Board, (Washington, D.C), 2011.

[14] T. Georgi-Maschler, B. Friedrich, R. Weyhe, H. Heegn, M. Rutzc, Development of a recycling process for Li-ion batteries, J. Power Sources 207 (2012) 173–182.

[15] X. Wang, G. Gaustad, C.W. Babbitt, Targeting high value metals in lithium-ion battery recycling via shredding and size-based separation, Waste Manag. 51 (2016) 204–213.

[16] A. Arnberger, E. Coskun, B. Rutrecht, Recycling von lithium-ionen-batterien, in: K.J. Thomé-Kozmiensky, D. Goldmann (Eds.), Recycling und Rohstoffe, vol. 11, TK Verlag Karl Thomé-Kozmiensky: Neuruppin, 2018, pp. 583–599.

[17] G. Schubert, Aufbereitung Metallischer Sekundärrohstoffe — Aufkommen Charakterisierung Zerkleinerung, Springer Verlag, Wien, 2012.

[18] P. Pomberger, A. Ragossnig, Future waste — waste future, Waste Manag. Res. 32 (2014) 89–90.

[19] A. Rudolph, in: A. Rudolph (Ed.), Altproduktentsorgung aus betriebswirtschaftlicher Sicht, vol. 1, Physica-Verlag, Heidelberg, 1999.

[20] E. Rahimzei, Begleit- und Wirkungsforschung Schaufenster Elektromobilität (BuW) Ergebnispapier Nr. 37 — Sicherheit von Elektrofahrzeugen, Deutsches Dialog Institut GmbH, Frankfurt am Main, 2017.

[21] T. Elwert, F. Römer, K. Schneider, Q. Hua, M. Buchert, Recycling of batteries from electric vehicles, in: G. Pistoia, B. Liaw (Eds.), Behaviour of Lithium-Ion Batteries in Electric Vehicles — Battery Health, Performance, Safety, and Cost, Springer International Publishing AG, Cham, 2018.

[22] M. Gama, European Li-Ion Battery Advanced Manufacturing for Electric Vehicles - ELIBAMA. Electrodes and Cells Manufacturing White Paper, European Commission, 2014 online.
[23] C. Hanisch, J. Diekmann, A. Stieger, W. Haselrieder, A. Kwade, Recycling of lithium-ion batteries, in: J. Yan, L.F. Cabeza, R. Sioshansi (Eds.), Handbook of Clean Energy Systems, vol. 5, John Wiley & Sons, Ltd, Hoboken, NJ, 2015, pp. 2865–2888.
[24] R. Korthauer, in: R. Korthauer (Ed.), Lithium-Ion Batteries: Basics and Applications, Springer-Verlag GmbH, Berlin, 2019, https://doi.org/10.1007/978-3-662-53071-9.
[25] W. van Pels, Herausforderung hinsichtlich der Brandschutztechnologien in Recyclinganlagen insbesondere bei der Verarbeitung von Li-Ionen Traktionsbatterien. In: Proceedings of Recycling Und Sekundärrohstoffe, Thomé-Kozmiensky, Verlag GmbH, Berlin, Germany, 2020, pp. 496–505.
[26] K. Wegener, S. Andrew, A. Raatz, D. Klaus, C. Herrmann, Disassembly of electric vehicle batteries using the example of the Audi Q5 hybrid system, in: Proceedings of Conference on Assembly Technologies and Systems, Dresden, Germany, 2014, pp. 15–16, 05.
[27] C. Ekberg, M. Petranikova, Lithium batteries recycling, in: A. Chagnes, J. Swiatowska (Eds.), Lithium Process Chemistry – Resources, Extraction, Batteries and Recycling, Elsevier, Amsterdam, 2015, pp. 233–269, https://doi.org/10.1016/B978-0-12-801417-2.00007-4pp.
[28] T.E. Schwarz, W. Rübenbauer, B. Rutrecht, R. Pomberger, Forecasting real disassembly time of industrial batteries based on virtual MTM-UAS data. In: Proceedings of 25th CIRP Life Cycle Engineering (LCE) Conference, Copenhagen, Denmark, 2018, pp. 927–931.
[29] Gesetz zur Förderung der Kreislaufwirtschaft und Sicherung der umweltverträglichen Bewirtschaftung von Abfällen (Kreislaufwirtschaftsgesetz – KrWG), Deutscher Bundestag, Bonn, 2017.
[30] VDI-Fachbereich Umwelttechnik, VDI 4431 – life-cycle management in the manufacturing industry, in: VDI-Handbuch Ressourcenmanagement in der Umwelttechnik, VDI-Gesellschaft Energie und Umwelt, vol. 2001–07, Beuth, 2001.
[31] M. Ritthoff, Recycling und Ressourcenschonung stoffliche Aspekte der Kreislaufwirtschaft, in: NABU-DSD-Dialogforum Kreislaufwirtschaft, Wuppertal Institut, 2016.
[32] VDI-Fachbereich Produktentwicklung und Mechatronik, VDI 2243 – recycling-oriented product development, in: VDI-Handbuch Ressourcenmanagement in der Umwelttechnik, VDI-Gesellschaft Produkt- und Prozessgestaltung, vol. 2002–07, Beuth, 2002.
[33] T.R. Sojka, Sichere Aufbereitung von lithium-basierten batterien durch thermische Konditionierung. In: Proceedings of Recycling Und Sekundärrohstoffe, Thomé-Kozmiensky, Verlag GmbH, Berlin, Germany, 2020, pp. 506–523.
[34] Commission Regulation (EU) No 493/2012, European Commission, Brussels, 2012.
[35] J. Tytgat, The recycling efficiency of Li-ion EV batteries according to the European commission regulation, and the relation with the end-of-life vehicles directive recycling rate, in: Proceedings of International Battery, Hybrid and Fuel Cell Electric Vehicle Symposium, Barcelona, Spain, 2013, pp. 17–20, 11.
[36] H. Martens, G. Goldmann, in: H. Martens, G. Goldmann (Eds.), Recyclingtechnik, vol. 2, Springer Vieweg, Wiesbaden, 2016.
[37] A. Vezzini, Manufacturers, materials and recycling technologies, in: G. Pistoia (Ed.), Lithium-Ion Batteries Advances and Applications, Elsevier, Amsterdam, 2014, pp. 529–551.
[38] J. Xiao, J. Li, Z. Xu, Recycling metals from lithium ion battery by mechanical separation and vacuum metallurgy, J. Hazard Mater. 338 (2017) 124–131.
[39] T. Zhang, Y. He, F. Wang, H. Li, C. Duan, C. Wud, Surface analysis of cobalt-enriched crushed products of spent lithium-ion batteries by X-ray photoelectron spectroscopy, Separ. Purif. Technol. 138 (2014) 21–27.
[40] K. Huang, J. Li, Z. Xu, Enhancement of the recycling of waste Ni–Cd and Ni–MH batteries by mechanical treatment, Waste Manag. 31 (2011) 1292–1299, https://doi.org/10.1016/j.wasman.2011.01.006.
[41] K. Loehr, M. Melchiorre, Liberation of composite waste from manufactured products, Int. J. Miner. Process. 44 (1996) 143–153.
[42] H. Schubert, in: H. Schubert (Ed.), Handbuch der Mechanischen Verfahrenstechnik, WILEY-VCH Verlag GmbH & Co. KGaA, Weinheim, 2003.

[43] L. Li, X. Zhang, M. Li, R. Chen, F. Wu, K. Amine, J. Lu, The recycling of spent lithium-ion batteries: a review of current processes and technologies, Electrochem. Energy Rev. 1 (2018) 461–482.

[44] H. Pinegar, Y.R. Smith, Recycling of end-of-life lithium-ion batteries, part II: laboratory-scale research developments in mechanical, thermal, and leaching treatments, J. Sustain. Metall. 6 (2020), https://doi.org/10.1007/s40831-020-00265-8.

[45] S. Al-Thyabat, T. Nakamura, E. Shibata, A. Iizuka, Adaptation of minerals processing operations for lithium-ion (LiBs)and nickel metal hydride (NiMH) batteries recycling: critical review, Miner. Eng. 45 (2013) 4–17.

[46] L. Wuschke, H.-G. Jäckel, L. Leißner, U.A. Peuker, Crushing of large Li-ion battery cells, Waste Manag. 85 (2019) 317–326.

[47] M. Bertau, M. Fuhrland, C. Pätzold, N. Schmidt, B. Merkel, K. Bachmann, J. Gutzmer, H.-G. Jäckel, T. Leißner, U.A. Peuker, et al., Hybride Lithiumgewinnung, TU Berkakademie Freiberg, 2011.

[48] L. Wuschke, H.-G. Jäckel, M. Gellner, U.A. Peuker, Comminution of Li-ion traction batteries, in: Proceedings of European Symposium on Comminution and Classification, Gothenborg, Sweden, 2015, p. 6, 7-11.09.

[49] C. Hanisch, T. Loellhoeffel, J. Diekmann, K.J. Markley, W. Haselrieder, A. Kwade, Recycling of lithium-ion batteries: a novel method to separate coating and foil of electrodes, J. Clean. Prod. 108 (2015) 301–311.

[50] L. Wuschke, H.-G. Jäckel, U.A. Peuker, M. Gellner, Recycling of Li-ion batteries – a challenge, Recovery 4 (2015) 48–59.

[51] R. Weyhe, Recycling von lithium-ion-batterien, in: K.J. Thomé-Kozmiensky, D. Goldmann (Eds.), Recycling und Rohstoffe, vol. 6, TK Verlag Karl Thomé-Kozmiensky, Neuruppin, 2013, pp. 505–525.

[52] H. Pinegar, Y.R. Smith, End-of-Life lithium-ion battery component mechanical liberation and separation, J. Miner. Met. Mater. Soc. 71 (2019) 4447–4456.

[53] M. Steinbild, Recycling von Lithium-Ionen-Batterien – LithoRec II: Abschlussbericht der beteiligten Verbundpartner, Bundesministeriums für Umwelt, Naturschutz, 2017. Bau und Reaktorsicherheit.

[54] R. Weyhe, B. Friedrich, Demonstrationsanlage für ein kostenneutrales, ressourceneffizientes Processing ausgedienter Li-Ionen Batterien aus der Elektromobilität – EcoBatRec – Abschlussbericht zum Verbundvorhaben, IME Metallurgische Prozesstechnik und Metallrecycling, Aachen, 2016.

[55] E. Gratz, Q. Sa, D. Apelian, Y. Wang, A closed loop process for recycling spent lithium ion batteries, J. Power Sources 262 (2014) 255–262.

[56] G. Schubert, Comminution equipment for non-brittle waste and scrap, Aufbereitungstechnik 43 (2002) 6–23.

[57] R. Sommerville, J. Shaw-Stewart, V. Goodship, N. Rowson, E. Kendrick, A review of physical processes used in the safe recycling of lithium ion batteries, Sustain. Mater. Technol. 25 (2020) e00197.

[58] J. Diekmann, S. Sander, G. Sellin, M. Petermann, A. Kwade, Crushing of battery modules and cells, in: A. Kwade, J. Diekmann (Eds.), Recycling of Lithium-Ion Batteries – The LithoRec Way, Springer, New York, NY, 2018, https://doi.org/10.1007/978-3-319-70572-9_12, pp. 127-138.

[59] D. Woldt, G. Schubert, H.-G. Jäckel, Size reduction by means of low-speed rotary shears, Int. J. Miner. Process. 74 (2004) 405–415, https://doi.org/10.1016/j.minpro.2004.07.008.

[60] W.N. Smith, S. Swoffer, Process for Recovering and Regenerating Lithium Cathode Materal From Lithium-Ion Batteries, 2014, 11.11.2014.

[61] W.N. Smith, S. Swoffer, Recovery of Lithium Ion Batteries, 2013, 31.12.2013.

[62] A.J. da Costa, J. Fidel Matos, A. Moura Bernardes, I. Lourdes Müller, Beneficiation of cobalt, copper and aluminum from wasted lithium-ion batteries by mechanical processing, Int. J. Miner. Process. 145 (2015) 77–82.

[63] D.A. Bertuol, C. Toniasso, B.M. Jimenez, L. Meili, G.L. Dotto, E.H. Tanabe, M.L. Aguiar, Application of spouted bed elutriation in the recycling of lithium ion batteries, J. Power Sources 275 (2015) 627e632.

[64] A.V.M. Silveira, M.P. Santana, E.H. Tanabe, D.A. Bertuol, Recovery of valuable materials from spent lithium ion batteries using electrostatic separation, Int. J. Miner. Process. 169 (2017) 91–98.
[65] J. Spangenberger, S. Gillard, ReCell Advanced Battery Recycling Center, ReCell Advanced Battery Recycling Center, 2021.
[66] J. Zimmermann, D. Horn, R. Stauber, o. Gutfleisch, Intelligent Fragmentation of Li-Ionen Batteries by Shockwaves, ICM AG, Cham, Switzerland, 2016.
[67] T. Leißner, D. Hamann, L. Wuschke, H.-G. Jäckel, U.A. Peuker, High voltage fragmentation of composites from secondary raw materials – potential and limitations, Waste Manag. 74 (2018) 123–134.
[68] F. Stehmann, C. Bradtmöller, S. Scholl, Separation of the electrolyte—thermal drying, in: A. Kwade, J. Diekmann (Eds.), Recycling of Lithium-Ion Batteries – The LithoRec Way, Springer, New York, NY, 2018, pp. 139–153.
[69] D. Werner, T. Mütze, U.A. Peuker, Influence of cell opening methods on organic solvent removal during pretreatment in lithium-ion battery recycling, in: Waste Management & Research, Sage, 2021, pp. 1–12. Vol. October 2021.
[70] L. Gaines, J. Sullivan, A. Burnham, The future of automotive lithium-ion battery recycling: charting a sustainable course, Sustain. Mater. Technol. 1–2 (2014) 2–7.
[71] S. Sander, G. Schubert, Size reduction of metals by means of swing-hammer shredders, Chem. Eng. Technol. 26 (2003) 406–415, https://doi.org/10.1002/ceat.200390061.
[72] A. Arnberger, Entwicklung eines ganzheitlichen Recyclingkonzeptes für Traktionsbatterien basierend auf Lithium-Ionen-Batterien, Dissertation, Montanuniversität Leoben, Leoben, 2016.
[73] X. Zhang, Y. Xie, H. Cao, F. Nawaz, Y. Zhang, A novel process for recycling and resynthesizing $LiNi_{1/3}Co_{1/3}Mn_{1/3}O_2$ from the cathode scraps intended for lithium-ion batteries, Waste Manag. 34 (2014) 1715–1724.
[74] L. Li, J. Ge, R. Chen, F. Wu, S. Chen, X. Zhang, Environmental friendly leaching reagent for cobalt and lithium recovery from spent lithium-ion batteries, Waste Manag. 30 (2010) 2615–2621.
[75] S. Pavón, D. Kaiser, R. Mende, M. Bertau, The COOL-process—a selective approach for recycling lithium batteries, Metals Open Acc. Metall. J. 11 (2021) 259.
[76] M. Gellner, Mechanische Aufbereitung der Feinfraktion zerkleinerter Lithium-Ionen-Batterien, Dissertation, TU Bergakademie Freiberg, Freiberg, 2018.
[77] Y. Guo, Y. Li, X. Lou, J. Guan, Y. Li, X. Mai, H. Liu, C.X. Zhao, N. Wang, C. Yan, et al., Improved extraction of cobalt and lithium by reductive acid from spent lithium-ion batteries via mechanical activation process, J. Mater. Sci. 53 (2018) 13790–13800.
[78] A. Vanderbruggen, Recovery of graphite from spent lithium ion batteries. In: Proceedings of International Battery Recycling Conference, Salzburg, Austria, 2020.
[79] R. Zhan, T. Payne, T. Leftwich, K. Perrine, L. Pan, De-agglomeration of cathode composites for direct recycling of Li-ion batteries, Waste Manag. 105 (2020) 39–48.
[80] J. Yu, Y. He, Z. Ge, H. Li, W. Xie, S. Wang, A promising physical method for recovery of $LiCoO_2$ and graphite from spent lithium-ion batteries: grinding flotation, Separ. Purif. Technol. 190 (2018) 45–52.
[81] R. Zhan, Z. Oldenburg, L. Pan, Recovery of active cathode materials from lithium-ion batteries using froth flotation, Sustain. Mater. Technol. 1 (2018) e00062.
[82] L 334/17, Directive 2010/75/EU, European Union, Brussels, 2010.
[83] W. Nickel, in: W. Nickel (Ed.), Recyclinghandbuch: Strategien – Technologien – Produkte, VDI-Verlag, Düsseldorf, 1996.
[84] A. Vanderbruggen, M. Rudloph, Flotation of spherodized graphite from spent lithium ion batteries, in: Proceedings of MEI Flotation'19, Cape Town, South Africa.
[85] T. Zhang, Y. He, T. Wang, L. Ge, X. Zhu, H. Li, Chemical and process mineralogical characterizations of spent lithium-ion batteries: an approach by multi-analytical techniques, Waste Management 34 (2014) 1051–1058.
[86] H. Bi, H. Zhu, L. Zu, S. He, Y. Gao, S. Gao, Pneumatic separation and recycling of anode and cathode materials from spent lithium iron phosphate batteries, Waste Manag. Res. 37 (2019) 374–385.

[87] H. Schubert, Zu den Grundlagen des Sortierens: trennmerkmale — Wirkprinzipien — Makroprozesse — Mikroprozesse, Aufbereitungstechnik 45 (2004) 7—32.
[88] G. Granata, F. Pagnanelli, E. Moscardini, Z. Takacova, T. Havlik, L. Toro, Simultaneous recycling of nickel metal hydride, lithium ion and primary lithium batteries: accomplishment of European guidelines by optimizing mechanical pre-treatment and solvent extraction operations, J. Power Sources 212 (2012) 205—211, https://doi.org/10.1016/j.jpowsour.2012.04.016.
[89] B.A. Wills, J.A. Finch, Wills' Mineral Processing Technology, vol. 8, Elsevier, 2016.
[90] M. Habib, N.J. Miles, P. Hall, Recovering metallic fractions from waste electrical and electronic equipment by a novel vibration system, Waste Manag. 33 (2013) 722—729.
[91] J. Lin, C. Fan, I. Chang, J. Shiu, Clean Process of Recovering Metals from Waste Lithium Ion Batteries, Industrial Technology Research Institute, Taiwan, 2003. USOO6514311B1.
[92] J. Diekmann, S. Sander, G. Sellin, A. Kwade, Material separation, in: A. Kwade, J. Diekmann (Eds.), Recycling of Lithium-Ion Batteries — The LithoRec Way, Springer, New York, NY, 2018, pp. 207—217, https://doi.org/10.1007/978-3-319-70572-9_12.
[93] F. Tedjar, J.-C. Foudraz, Method for the Mixed Recycling of Lithium-Based Anode Batteries and Cells, Recupyl, France, 2010. U.S. Patent 7,820,317.
[94] L. Wuschke, H.-G. Jäckel, D. Borsdorff, D. Werner, U.A. Peuker, M. Gellner, Zur mechanischen Aufbereitung von Li-Ionen-Batterien, Berg- Huttenmannische Monatsh. 161 (2016) 267—276.
[95] S. Krüger, C. Hanisch, A. Kwade, M. Winter, S. Nowak, Effect of impurities caused by a recycling process on the electrochemical performance of Li[Ni0.33Co0.33Mn0.33]O$_2$, J. Electroanal. Chem. 726 (2014) 91—96, https://doi.org/10.1016/j.jelechem.2014.05.017.
[96] J. Diekmann, C. Hanisch, L. Froböse, G. Schälicke, T. Loellhoeffel, A.-S. Fölster, A. Kwade, Ecological recycling of lithium-ion batteries from electric vehicles with focus on mechanical processes, J. Electrochem. Soc. 164 (2017).
[97] R. Zheng, W. Wang, Y. Dai, Q. Ma, Y. Liu, D. Mu, R. Li, J. Ren, C. Dai, A closed-loop process for recycling LiNixCoyMn(1xy)O$_2$ from mixed cathode materials of lithium-ion batteries, Green Energy Environ. 2 (2017) 42—50.
[98] A.C. Kasper, N.C. de Freitas Juchneski, H.M. Veit, Mechanical processing, in: H.M. Veit, A.M. Bernardes (Eds.), Electronic Waste: Recycling Techniques, Springer International Publishing, Cham, 2015, https://doi.org/10.1007/978-3-319-15714-6.
[99] VDI-Fachbereich Umwelttechnik, VDI 2343 Blatt 4 — recycling elektrischer und elektronischer Geräte — Aufbereitung, in: VDI manual Resource Management for Environmental Technologies, VDI-Gesellschaft Energie und Umwelt, vol. 2012-01, Beuth, 2012, p. 61.
[100] G. Ozbayoglu, Energy production from coal, in: I. Dincer (Ed.), Comprehensive Energy Systems, vol. 3, Elsevier, 2018.
[101] T.W. Ellis, J.A. Montenegro, Magnetic Separation of Electrochemical Cell Materials, RSR Technologies, Inc., United States, 2019. US 2019 / 0039075 A1.
[102] H. Schubert, Wirbelstromsortierung — Grundlagen, Scheider, Anwendungen, Aufbereitungstechnik 35 (1994) 553—562.
[103] H. Schubert, Zum Trennmerkmal bei der Wirbelstromsortierung, Aufbereitungstechnik 42 (2001).
[104] M. Kranert, Einführung in die Kreislaufwirtschaft: Planung — Recht — Verfahren, 5 ed., Springer Vieweg, Wiesbaden, 2017 https://doi.org/10.1007/978-3-8348-2257-4.
[105] J.-R. Lin, C. Fan, I.L. Chang, J.-Y. Shiu, Clean Process of Recovering Metals From Waste Lithium Ion Batteries, Industrial Technology Research Institute, Taiwan, 2003. USOO6514311B1.
[106] R. Sommerville, Application of magnetic processing in Li-ion battery recycling. In: Proceedings of International Battery Recycling Conference, Salzburg, Austria, 2020.
[107] T. Pretz, J. Julius, Stand der Technik und Entwicklung bei der berührungslosen Sortierung von Abfällen, in: Österreichische Wasser- und Abfallwirtschaft, vol. 60, 2008, pp. 105—112.
[108] H. Wotruba, Stand der Technik der sensorgestützten Sortierung, BHM Berg- Hüttenmännische Monatsh. 153 (2008) 221—224.
[109] B. Küppers, J.C. Hernández Parrodi, C. Garcia Lopez, R. Pomberger, D. Vollprecht, Potential of sensor-based sorting in enhanced landfill Mining, Detritus 8 (2019) 24—30.

[110] H. Wotruba, H. Harbeck, Sensor-based sorting, in: Ullmann's Encyclopedia of Industrial Chemistry, vol. 32, Wiley-VCH, 2010, pp. 395–404.
[111] H. Schubert, Hydrophobie, hydrophobe Wechselwirkungen, hydrophober Effekt und Heterokoagulation sowie ihre Rolle bei Aufbereitungsprozessen, Aufbereitungstechnik 46 (2005).
[112] H. Schubert, Zur optimalen Hydrodynamik der Fein- und Feinstkornflotation, Aufbereitungstechnik 48 (2007) 30–47.
[113] H. Schubert, C. Bischofberger, Zu den Mikroprozessen Blasenzerteilen und Korn-Blase-Haftung in mechanischen Flotationsapparaten sowie Konsequenzen für die Maßstabsübertragung der Makroprozesse, Aufbereitungstechnik 37 (1996) 193–201.
[114] H. Schubert, Nanobubbles, hydrophobic effect, heterocoagulation and hydrodynamics in flotation, Int. J. Miner. Process. 78 (2005) 11–21, https://doi.org/10.1016/j.minpro.2005.07.002.
[115] H. Schubert, C. Bischofberger, On the microprocesses air dispersion and particle-bubble attachment in flotation machines as well as consequences for the scale-up of macroprocesses, Int. J. Miner. Process. 52 (1998) 245–259.
[116] H. Schubert, On the optimization of hydrodynamics in fine particle flotation, Miner. Eng. 21 (2008) 930–936, https://doi.org/10.1016/j.mineng.2008.02.012.
[117] F. Larouche, F. Tedjar, K. Amouzegar, G. Houlachi, P. Bouchard, G.P. Demopoulos, K. Zaghib, Progress and status of hydrometallurgical and direct recycling of Li-ion batteries and beyond, Materials Vol. 13 (2020).
[118] A. Vanderbruggen, E. Gugala, R. Blannin, K. Bachmann, R. Serna-Guerrero, M. Rudolph, Automated mineralogy as a novel approach for the compositional and textural characterization of spent lithium-ion batteries, Miner. Eng. Vol. 169 (2021).
[119] T. Leißner, Beitrag zur Kennzeichnung von Aufschluss und Trennerfolg am Beispiel der Magnetscheidung, Dissertation, TU Bergakademie Freiberg, Freiberg, 2016.
[120] J. Cui, E. Forssberg, Mechanical recycling of waste electric and electronic equipment: a review, J. Hazard Mater. B99 (2003) 243–263.
[121] C. Hanisch, W. Haselrieder, A. Kwade, Recycling von lithium-ionen-batterien — das Projekt LithoRec, in: K.J. Thomé-Kozmienski, D. Goldmann (Eds.), Recycling und Rohstoffe, vol. 5, TK Verlag, Neuruppin, 2012, pp. 691–698.
[122] G. Harper, R. Sommerville, E. Kendrick, L. Driscoll, P. Slater, R. Stolkin, A. Walton, P. Christensen, O. Heidrich, S. Lambert, et al., Recycling lithium-ion batteries from electric vehicles, Nature 575 (2019) 75–86, https://doi.org/10.1038/s41586-019-1682-5.
[123] F. Pagnanelli, E. Moscardini, T. Altimari, T. Abo Atia, L. Toro, Leaching of electrodic powders from lithium ion batteries: optimization of operating conditions and effect of physical pretreatment for waste fraction retrieval, Waste Manag. 60 (2017) 706–715.
[124] G. Lombardo, B. Ebin, B.-M. Steenari, M. Alemrajabi, I. Karlsson, M. Petranikova, Comparison of the effects of incineration, vacuum pyrolysis and dynamic pyrolysis on the composition of NMC-lithium battery cathode-material production scraps and separation of the current collector, Resour. Conserv. Recycl. Vol. 164 (2021).
[125] B. Friedrich, L. Schwich, E-mobility batteries recycling — how to reach the demanded recycling efficiency? In: Proceedings of VDMA Webinar, 2021.
[126] J. Valio, Critical Review on Lithium Ion Battery Recycling Technologies, Master Thesis, Aalto University, Helsinki, 2017.
[127] T. Träger, B. Friedrich, R. Weyhe, Recovery concept of value metals from automotive lithium-ion batteries, Chem. Ing. Tech. 87 (2015) 1550–1557.
[128] F. Jeschull, D. Brandell, M. Wohlfahrt-Mehrens, M. Memm, Water-soluble binders for lithium-ion battery graphite electrodes: slurry rheology, coating adhesion, and electrochemical performance, Energy Technol. Vol. 5 (2017) 2108–2118. Wiley-VCH: Weinheim, Germany.
[129] R. Sommerville, P. Zhu, M.A. Rajaeifar, O. Heidrich, V. Goodship, E. Kendrick, A qualitative assessment of lithium ion battery recycling processes, Resour. Conserv. Recycl. Vol. 165 (2021).
[130] D. Werner, S. Schrader, T. Mütze, U.A. Peuker, Untersuchungen zur mechanischen Entschichtung von Elektroden aus Lithium-Ionen-Altbatterien. In: Proceedings of Recy & DepoTech-Konferenz, Leoben, Austria, 2020. pp. 107–112.

[131] L. Li, G. Liu, D. Pan, W. Wang, Y. Wu, T. Zuo, Overview of the recycling technology for copper-containing cables, Resour. Conserv. Recycl. 126 (2017) 132–140.

[132] H. Shin, R. Zhan, K.S. Dhindsa, L. Pan, T. Han, Electrochemical performance of recycled cathode active materials using froth flotation-based separation process, J. Electrochem. Soc. 167 (2020).

[133] A. Chagnes, Fundamentals in electrochemistry and hydrometallurgy, in: A. Chagnes, J. Swiatowska (Eds.), Lithium Process Chemistry – Resources, Extraction, Batteries and Recycling, Elsevier, Amsterdam, 2015, https://doi.org/10.1016/B978-0-12-801417-2.00007-4, p. 41.80.

[134] J. Guan, Y. Li, Y. Guo, R. Su, G. Gao, H. Song, H. Yuan, B. Liang, Z. Guo, Mechanochemical process enhanced cobalt and lithium recycling from wasted lithium-ion batteries, Am. Chem. Soc. ACS Sustain. Chem. Eng. 5 (2017) 1026–1032.

[135] J. Li, G. Wang, Z. Xu, Environmentally-friendly oxygen-free roasting/wet magneticseparation technology for in situ recycling cobalt, lithium carbonateand graphite from spent $LiCoO_2$/graphite lithium batteries, J. Hazard Mater. 302 (2016) 97–104.

Index

'*Note:* Page numbers followed by "f" indicate figures and "t" indicate tables.'

A

Acetylene black (AB), 292–293
Acid and alkali leaching (AAL), 298–299
Acid leaching process (AL process), 192–197, 293–294, 298–299, 333, 377, 432–434, 433f. *See also* Redox leaching process
  cobalt extraction by, 393–400
  inorganic acid, 194–197
  organic acid, 197
Acid reflux
  SWCNTs, 377
  treatment, 377
Acid-leached graphite (AG), 295
Acid-leaching process, 295
"Acid/basic treatment", 58
Acidithiobacillus species, 223
  A. ferrooxidans, 95, 220–221, 311, 333, 435–436
Acidolysis, 223
Active electrode materials, 382
Active materials (AM), 129, 382, 456
Additives, 178
  additive-assisted roasting method, 132
Air battery, 9–11
  aluminum, 10–11
  lithium, 10
  zinc, 9–10
Air classification process, 362–365
Air-jet separator, 142
Airflow sorting in zigzag classifiers, 464
Alkali leaching method, 256
Alkali soaking method, 255–256
  research results of pretreatment processes, 256t
Alkaline battery, 4–7, 79, 217, 323, 330–331
Alkaline button battery, 5
  series product lineup, 5, 6t
  structure, 5, 6f
Alkaline dry batteries, 5. *See also* Alkaline battery
Alkaline electrolyte, 322–323, 331
Alkaline leaching process, 116, 197–203, 306, 434–435. *See also* Redox leaching process
  copper mechanism, 201–202
  hydrometallurgy, 312
  iron and aluminum mechanism, 203
  lithium mechanism, 202
  manganese mechanism, 202–203
  nickel and cobalt mechanism, 200–201
Alkaline-manganese batteries, 3
Alkaline secondary battery, 12–16
Alkaline zinc battery, 322–323
Alkaline zinc-manganese batteries, 3
All-carbon dual-ion battery (ACDIB), 298
Alumina, 249, 306–307
Aluminum (Al), 218, 349, 367–368, 456, 468, 474
  air battery, 10–11
  composite foils, 350
  foil, 253–255
  impurities impact on regeneration method, 439
  mechanism, 203
  melts, 355–356
  removal, 213
American Battery Metals Corporation, 342
3-aminopropyl triethoxysilane (APTES), 161–162
Ammonia (NH_4OH), 312, 386
  leaching method, 197–199, 312
Ammonium carbonate ((NH_4)$_2CO_3$), 311
Ammonium chloride (NH_4Cl), 311, 321–322
Ammonium sulfate ((NH_4)$_2SO_4$), 311
Amorphous carbon compounds, 456
Angled base fiber arrays, 148–149
Annealing treatment method, 144–145
Anode, 249, 258, 303, 445–446
  materials
    hydrometallurgical technology, 117
    in lithium-ion batteries, 289
    pyrometallurgical technology, 112
    recycling, 289–291
    recycling methods for, 292–299
    separation of, 274–276
    specific data for characteristics of, 290t
  recycling, 336–337
  validation of anode active materials mixture separation, 277–280, 278f
Aqueous sulfuric acid, 324–325

Artificially assembled nanostructured materials systems, 374
Ascorbic acids, 310
Aspergillus niger, 95, 207, 224, 311, 435, 435f
Atmosphere-assisted roasting method, 129–130
Atomic energy battery, 27–28
Autotrophic bacteria, 222–223
Autotrophic bioleaching, 95
Autotrophs, 229

B

Batteries, 53, 124–125, 217, 250–252, 321, 349, 351, 472
  battery-grade graphite, 292
  biohydrometallurgy, 94–96
  bioleaching, 230
    processes of spent batteries using various microorganisms, 231t–232t
  chemical battery, 3–26
  deep eutectic solvents, 93–94
  hydrometallurgy processes of NiMH for extraction of rare earth elements, 81f
  ionic liquids, 82–85
  nanohydrometallurgy, 87–89
  nanostructures, 54–55
  organic acids, 89–93
  physical battery, 26–28
  Pourbaix diagram of Co-S-H_2O system, 81f
  recovery, 53
  supercritical fluids, 85–87
  systems, 351
  technologies, 28, 330
Battery casing materials, recycling of
  generating housing fraction by mechanical processing, 352–357
  housing fraction and case disassembly, 351–352
  housing fraction processing, 357–365
  process flowchart, 357–358, 366f
    content of valuable material, 366t
  structure of cell housings, 350–351
  target, 349–350
Battery production cycle
  direct repair, 427–428
    hydrothermal method, 427–428
    solid-phase calcination, 427
  extraction of metal, 429–441
Battery recycling, 127–135
  approaches for recycling battery cathode materials, 304–316
  battery market, 125–127
  challenges and perspectives, 135–136
  evolution of portable electronic devices, 124f
  general recycling strategies, 128
  LIBs, 456–457
  physical processing of end-of-life lithium-ion batteries, 457–470
    mechanical liberation, 457–461
    physical separation, 461–470
  process groups, 471–479
    initial liberation and enrichment, 472–474
    processing black mass, 477–479
    processing coarse fractions, 474–477
  recycling of electrode materials, 129–133
  recycling of electrolyte and binders, 133–135
  remanufacturing, and reuse battery recycling, 64
Binders, 253–254
  recycling of, 133–135
Bio-inspired directional adhesives, 146–149
  directional polyurethane microfiber array, 150f
  experimental climbing machine for testing anisotropic adhesive structures, 148f
  microscopic image of arrays of 35-mm diameter angled polyurethane microfibers, 149f
Bio-inspired nanotechnology
  bio-inspired directional adhesives, 146–149
  combating interfacial delamination, 143–146
    microscopic image of copper hydroxide nanofiber, 144f
    microscopic images of nickel foam and nickel foam surface, 146f
    patterned vanadium oxide films with island heights, 145f
  experimental setup for combined thermal and mechanical process, 142f
  interface for easy-to-recycle batteries, 149–153
  underlying mechanisms, 153–154
Bioaccumulation, 220
Biochemical method, 70, 71t
Biogenic ferric sulfate mechanism, 226–227
Biogenic organic acids, 227–229
  complexation reactions of organic acids with Ka values, 228t
Biogenic sulfuric acid mechanism, 226–227
Biohydrometallurgy process, 46–47, 82, 94–96, 117–118, 219
  approaches for recycling spent lithium-ion batteries, 219
  bioleaching mechanism, 224–230

environmental aspects of inappropriate spent LIBs management, 218–219
microbes in bioleaching, 222–224
reuse of valuable metals in battery production, 233–239
   regeneration process, 237–239
   separation process, 233–237
spent battery bioleaching, 230
spent LIBs as precious secondary resource, 218
types, 219–222
   bioaccumulation, 220
   bioleaching, 220–222
   biosorption, 219–220
Bioleaching, 220–222, 333, 435–436, 435f
employs microorganisms, 311
mechanism, 224–230
   direct mechanism, 224–225
   indirect mechanism, 225–229
microbes in, 222–224
Biomass, 96
Biosorption, 96, 219–220
Biphasic Pickering emulsion, 162
Bis-(2, 4, 4-trimethyl-pentyl) phosphonic acid, 236
Bisphenol A (BPA), 298–299
Black mass processing, 477–479
Blended cathode, 447
Bottom-up approach, 70
Brine solutions, 472–473
Buoyancy force, 268–269

C

Calcination, solid-phase, 427
Calcium chloride, 177
Carbon
anode materials, 376
atoms, 291
carbonaceous nanomaterials recycling, 376–380
   soluble graphene nanosheet recycling, 379–380
   SWCNTs recycling, 376–379
nanomaterials, 68f, 376
nanotubes, 166, 376
thermal reduction process, 176
Carbon-based anode materials, 249, 259
Carbonate
precipitation, 235
Carbothermal reduction, 175

Carboxymethylcellulose sodium (CMC), 292–293
Cathode, 247–249, 258, 303, 327, 445–447
approaches for recycling battery cathode materials, 304–316
cathode preparation process, 304f
challenges and perspectives, 316
coating, 478
electrode, 314
material regeneration, 437–438
recycling, 335
separation of, 274–276, 305–307
   alkaline leaching, 306
   organic solvent dissolution, 306
   thermal treatment, 306–307
Cathode active materials, 172–173, 181, 266
mixture separation, validation of, 278f, 280–281
Cathode materials, 387
hydrometallurgical technology, 114–117
   alkaline leaching, 116
   chemical precipitation, 116–117
   electrodeposition, 117
   inorganic leaching, 115
   organic leaching, 115–116
   solvent extraction, 116
pyrometallurgical technology, 112
Cell housing
materials, 474
structure of, 350–351
   light microscope image, 351f
   lithium-ion battery cells, 350f
   materials list for battery poles, 351t
Cell shell, 303
Cell types, 356–357
Cellphones, 123–124
Cellulose, 166
Centrifugal concentrators, 465–466
Centrifugal force field, 463, 465
Centrifugal separation methods, 159, 465
Centrifugation, 162
Centrifuge speed, 277
Cesium, 176–177
Charge accumulators. *See* Secondary batteries
Chemical approach for recycling batteries, 111–118
biometallurgical technology, 117–118
hydrometallurgical technology, 113–117
pyrometallurgical technology, 112

Chemical battery
  fuel cell, 23—26
  primary battery, 3—12
  secondary battery, 12—23
Chemical precipitation process, 116—117, 233—235, 234t, 313
  separation, 331
Chemical reduction method, 70, 71t
Chemical solution precipitation. *See* Sol-gel process
Chemical vapor deposition method, 70, 71t
Chemolithoautotrophic bacteria, 222
Chemolithotrophic autotrophic bacteria, 222—223
Chemolithotrophic prokaryotes, 95
Chitosan, 161
Chlorination roasting method, 132, 176—177, 181
Chlorine gas (Cl_2), 299
Chromobacterium violaceum, 224
Circular economy, 53
  of battery metals, 392—393
Citric acid, 311, 382, 432, 433f
Clerici solution, 265, 276—277
  heavy liquid, 272, 279—281
  viscosity, 279—280
Closed-loop industrial hydrometallurgical process, 392—393
Coarse fractions processing, 474—477
  processing electrode fraction, 475—477
  processing housing fraction, 474—475
Cobalt (Co), 178, 180, 218, 349, 383, 391, 456
  countercurrent batch simulation studies for cobalt and lithium separation, 412
  extraction by acid leaching, 393—400
    H_2O_2 effect on cobalt leaching from $LiCoO_2$, 395—397
    H_2SO_4 concentration effect on cobalt leaching from $LiCoO_2$, 398—399
    leaching time effect on cobalt leaching from $LiCoO_2$, 397
    pulp density effect on cobalt leaching from $LiCoO_2$, 399—400
    temperature effect on cobalt leaching from $LiCoO_2$, 398
  pH-extraction isotherm for cobalt from $LiCoO_2$ leach liquor, 402—404
  purification from loaded Cyanex 272, 413—414
  recovery, 392
  recycling of, 427—444
  selective separation of cobalt by solvent extraction from $LiCoO_2$ leach liquor, 401—416
  separation and extraction of, 214
  stripping and cobalt sulfate recovery, 414—415
Cobalt ferrite nanoparticle recycling, 381—382
Cobalt mechanism, 200—201
Cobalt oxalate, 392—393
  synthesis of cobalt oxalate and cobalt oxide powder by precipitation stripping, 416—418
Cobalt oxide, 174, 392—393
  synthesis of cobalt oxalate and cobalt oxide powder by precipitation stripping, 416—418
Cobalt sulfate, 383, 392—393
Cobaltous hydroxide precipitation, 382
Collection methods, 376
Color cameras, 474
Combating interfacial delamination, 143—146
Compaction, 476
Complexing reaction, 185
Complexolysis, 223—224
Compound annual growth rate (CAGR), 322, 391
Computer simulation, 447
Concentrated hydrochloric acid, 377
Conductive materials (CFs), 260
Constructive method. *See* Bottom-up approach
Contact mechanism. *See* Direct mechanism
Conventional crushing process, 459
Conventional in situ reduction roasting method, 176
Conventional organic/aqueous biphasic system, 162—163
Conventional recycling techniques, 330—339, 376. *See also* Green recycling techniques
  alkaline and zinc—carbon battery, 330—331
    hydrometallurgical process, 331
    pyrometallurgical process, 330—331
  lead—acid battery, 331—333
  Li-ion battery, 337—339
  Ni—Cd battery, 333
  Ni—MH battery, 334—337
Conveyor belts, 469
Copper (Cu), 218, 456, 468
  foil, 253—254
  impurities impact on regeneration method, 438
  mechanism, 201—202
  removal, 209—211
    by extraction reaction, 210—211
    by replacement precipitation, 209—210
    by sulfide precipitation, 210

Coprecipitation method, 315
Coprecipitators, 237–238
Corona-induced roll separators, 469
Countercurrent batch simulation studies for cobalt and lithium separation, 412
Cradle-to-grave approach, 391
Crushed electrode material, 385
Crushing
   classification process, 361–362
   strategy, 458–460
Current collector, 445–446
   foil, 141–142, 469–470
   metallic, 142
Cyanex 272, 392–393, 401–402
   effect of Cyanex 272 concentration and volume of $LiCoO_2$ leach liquor, 404, 405t
   lithium scrubbing and cobalt purification from loaded, 413–414
   solvent extraction, 416
Cyanide, 224
Cyanogenic bacteria, 224

D

Deactivation methods, 252
Decoated electrodes, separation of, 477
Decoating, 475–477
Deep eutectic solvents (DESs), 93–94
Defects in industry recycling, 299–300
Delamination. See Milling
Demulsification process, 162–163
Depollution strategy, 472–473
Di-(2-ethylhexyl) phosphoric acid (D2EHPA), 236
Diaphragm materials, 249
Dimethyl carbonate (DMC), 249–250, 252
Dimethyl sulfoxide (DMSO), 87, 249–250, 252
Direct acid leaching, 206
Direct bacterial leaching, 206
Direct mechanical crushing technique, 43
Direct mechanism, 224–225, 225f. See also Indirect mechanism
Direct recycling process, Li-ion battery, 338–339
Direct repair, 427–428
   hydrothermal method, 427–428, 429f
   solid-phase calcination, 427, 428f
Direct roasting method, 129–130
Directional adhesion, 147–148
Discharging regime, 43
Disintegration process, 458
Dismantling process, 180
Dissolved O_2, 222
DMAC. See N-dimethylacetamide (DMAC)
DMF. See N-dimethylformamide (DMF)
Donnan dialysis-based technology (DD-based technology), 340
Drag force, 268–269
Drug delivery, 159–160
Dry batteries, 3
"Dry cell", 321
Dry crushing process, 43
Dry disc separators, 466–467
Dry metallurgy. See Pyrometallurgy
Dry separation, 465–466
"Dualfoil" model, 35

E

E-waste management, 392–393
(E)-4-((2-mercaptophenyl)diazenyl)-2-nitrosonaphthalen-1-ol (MPDN), 340–341
(E)-5-((1, 3, 4-thiadiazol-2-yl) diazenyl) benzene-1, 3-diol (TDDB), 340–341
Eddy current
   separation, 44, 468
   separators, 468
Effective leaching, 183
Eh–pH diagram of the hydrometallurgy process, 205–206
Electric vehicles (EVs), 33, 125–126, 217, 291–292, 391
Electrical energy-storage system, 445
Electro-Fenton process, 298–299
Electrochemical cathode reduction method, 341
Electrochemical impedance spectroscopy (EIS), 35
Electrochemical leaching process, 205–206
Electrochemical model-based techniques, 35
Electrochemical performance of cathode, 446
Electrochemically active materials, 142–143
Electrode
   active substances, 253–254
   containing cobalt, 381
   films, 142–143
   fraction processing, 475–477
      decoating, 476–477
      separation of decoated electrodes, 477
   materials, 373
      recycling of, 129–133
   tuning of electrode topography, 146

Electrode nanomaterials, 247
  carbonaceous nanomaterials recycling, 376—380
  closed-loop materials cycle, 375f
  metal nanomaterials recycling, 380—388
  structure and principle of LIBs, 374f
Electrodeposition process, 117, 313—314
  cathode electrode regeneration and reutilization, 314f
Electrodes, 377
Electrohydraulic fragmentation, 459, 476—477
Electrolyte, 249—250, 253—254, 303, 323, 445—446
  recovery, 86
  recycling of, 133—135
Electronic devices, 105
Electronic storage battery, 20—23
  primary battery, 21—22, 21f
  sodium-sulfur battery, 20—21, 20f
  zinc-bromine flow battery, 22—23, 22f
Electrostatic separation method, 44, 259—260, 461, 468—469
  roll-type electrostatic separator, 260f
Emulsification process, 162—163
Emulsified oil, 160
End-of-life batteries (EoL batteries), 141, 351, 377, 391
  manufacturing, 141
End-of-life lithium-ion batteries (EOL LIBs), 33—34, 35f, 43, 79—80, 456—470
  recycling, 475
Energy storage, 125—126
Environmental aspects of inappropriate spent LIBs management, 218—219
Environmental pollution, 218—219
Environmental Protection Agency, 217
Ethanol, 385
Ethyl carbonate, 249—250, 252
2-ethylhexyl phosphonic acid mono-2-ethylhexyl ester (PC88A), 236
European Union (EU), 217
Existing discharge methods, 316
Extracellular polymeric substances (EPSs), 224
Extraction
  copper removal by Extraction reaction, 210—211
  of materials, 45—46
    biohydrometallurgical processes, 46
    hydrometallurgical processes, 45
    pyrometallurgical processes, 45—46
  of metal, 429—441
  of recycled batteries
    alkaline leaching, 434—435
    bioleaching, 435—436
    leaching by hydrometallurgy, 429—436
  method, 293—294
  technologies, 207—215

F

Fast rotating rotary shears, 458
$FePO_4$ preparation, 213—214
Filter cake materials, 264
Filtration method, 162
Flocculation technique, 160
Flotation, 461, 470
  separation methods, 258—259, 259f
Fluidized bed separators, 466
Fluidized bed technology, 466
Fluorine removal, 213
Foils deformation, 460
Froth floatation technique, 44
Fuel cell
  molten carbonate fuel cell (MCFC), 24—25
  phosphoric acid fuel cell (PAFC), 23—24
  polymer electrolyte fuel cell, 26
  solid oxide fuel cell (SOFC), 25—26, 26f
Fungi, heterotrophic bacteria and, 223—224

G

Gecko foot-hairs, 147—148
Gluconic acid, 311
Gluconobacter oxydans, 224
Goethite method, iron removal by, 212
Gold NP (Au NP), 166
"Grafting through" technique, 165—166
Graphene, 55, 67—69, 379
  graphene-based nanomaterials, 55
  layers, 296—297
  nanosheet recycling, 379—380
    formation of, 380f
Graphene oxide (GO), 67—69, 107, 166, 296—297, 379
  with hydroxide, 380
  sheets, 380
Graphite, 289, 291, 456, 470
  anode active material, 277—278
  in lithium-ion batteries, 291—292
    graphite, 291
    recycle graphite, 291—292
    theoretical capacity of, 291

Gravitational force, 268–269
Gravity, 465
  separation, 44, 463–466
    airflow sorting in zigzag classifiers, 464
    centrifugal concentrators, 465–466
    fluidized bed separators, 466
    pneumatic tables, 464–465
Green chemistry principles, 181
Green recycling techniques, 340–342. *See also* Conventional recycling techniques
  donnan dialysis, 340
  electrochemical cathode reduction method, 341
  hierarchical mesosponge γ-Al2O3 monolith extraction, 340–341
  prerecycling processes, 341–342
Grinding process, 479

H

Half-cell fabrication process, 447
Heat-activated reserve battery, 11
Heat treatment process, 173
Heavy liquids, 273
  selection, 270–272
    centrifugal device with fixed angle rotor, 271f
Hetero-agglomerates, 470
Heterotrophic bacteria and fungi, 223–224
Heterotrophic microorganisms, 227
Heterotrophs, 229, 311
Hierarchical mesosponge γ-Al2O3 monolith extraction technology, 340–341
High shear blending, 476–477
High-capacity electrode materials, 376–377
High-efficiency ball milling process, 385
High-temperature heat treatment process, 181
High-temperature melting process, 174–175
High-temperature metallurgical process, 173
High-temperature smelting process, 181
Housing fraction
  and case disassembly, 351–352
  generation by mechanical processing, 352–357
    composition of housing products, 356t
    housing fraction components, 356f
    LithoRec process, 353f
    relationship between energy and piece size, 354f
    required comminution energy, 354f
    settling velocity distribution of different cells, 355f
  process, 357–365, 474–475

    air classification, 362–365
    eddy current separation, 358–361
    liberation comminution, 357–358
    particle size distribution of components, 362f
    selective crushing and sieve classification, 361–362
Hybrid electric vehicles, 374–375
Hydrocyanic acid (HCN), 224, 379
  treatment, 377
Hydrogen peroxide (H_2O_2), 298–299
  effect on cobalt leaching from $LiCoO_2$, 395–397
Hydrometallurgical process, 45, 47, 53, 58–59, 80, 113–117, 219, 264, 307–314, 330, 429–436, 430f. *See also* Pyrometallurgical process,
  acid leaching, 432–434, 433f
  alkaline and zinc–carbon battery, 331
  alkaline leaching, 312, 434–435
  anode, 117
  bioleaching, 311, 435–436, 435f
  cathode materials, 114–117
  hydrometallurgical recovery, 183–190
    advantages, 190
    leaching, 183–185
    precipitation, 185–190
    process flow of, 184f
  inorganic acid leaching, 308–309
  leaching, 192–207, 431–432, 431f
  lead–acid battery, 332–333
  Li-ion battery, 337–338
  Ni–Cd battery, 333
  Ni–MH battery, 335–337
  of NiMH for extraction of rare earth elements, 81f
  organic acid leaching, 309–310
  separation and extraction technologies, 207–215
    impurity removal, 209–213
    precursor preparation, 213–215
  separation and purification, 312–314
    chemical precipitation, 313
    electrodeposition, 313–314
    solvent extraction, 313
  status of technological research on hydrometallurgical recovery, 190–191
  waste lithium-ion battery hydrometallurgical recycling, 191t
Hydrophilic particles, 470
Hydrothermal method, 70, 427–428, 429f

Hydroxide precipitation, 233
Hyperspectral cameras, 474

I

Ihleite, iron removal using, 211–212
Impurities, 185
  Al, 439
  Cu, 438
  Fe, 439–440
  Mg, 440–441
  removal, 209–213, 292–294
    aluminum removal, 213
    copper removal, 209–211
    fluorine removal, 213
    iron removal, 211–212
In situ reduction roasting, 181
Incineration. *See* Pyrometallurgy
Indirect bacterial leaching, 206
Indirect mechanism, 224–229, 225f
  biogenic ferric sulfate and biogenic sulfuric acid mechanism, 226–227
  biogenic organic acids, 227–229
  mixed culture, 229–230
Industrial waste, 70, 160
Industry recycling, defects in, 299–300
Industry waste valorization
  cobalt extraction by acid leaching, 393–400
  process development, 422
  selective separation of cobalt by solvent extraction from $LiCoO_2$ leach liquor, 401–416
    cobalt stripping and cobalt sulfate recovery, 414–415
    countercurrent batch simulation studies for cobalt and lithium separation, 412
    effect of Cyanex 272 concentration and volume of $LiCoO_2$ leach liquor, 404
    lithium scrubbing and cobalt purification from loaded Cyanex 272, 413–414
    McCabe–Thiele cobalt extraction isotherm from $LiCoO_2$ leach liquor, 410–411
    metal extractability behavior and modeling, 404–410
    pH-extraction isotherm for cobalt from $LiCoO_2$ leach liquor, 402–404
    solvent extraction, 416
  valorization through oxide synthesis, 416–422
  waste characterization, 393
Industry-scale separation process, 272–273, 273f
Initial liberation, 472–474

Inorganic acid, 194–197
  leaching, 308–309, 309t
Inorganic leaching method, 115
Intensification, 181
Interface for easy-to-recycle batteries, 149–153
  adhesion between the patterned current collectors and composite films, 152f
  patterned surfaces, 151f
  peel energy, 152f
Interfacial delamination, combating, 143–146
Internal resistance (IR), 33–34
Ionic liquids (ILs), 82–85
  application in LIBs, 84t
  with different anions and the same cation, with distinct hydrophobicity properties, 83f
  main cations and anions present in the composition of, 83f
Iron (Fe), 218
  iron-oxidizing bacteria, 224–225
  mechanism, 203
  removal, 211–212
    by goethite method, 212
    using ihleite, 211–212
    using iron hydroxide, 211

J

Jarosite, 229

K

Kalman filter-based algorithms, 35
KIBs. *See* Potassium-ion batteries (KIBs)

L

Lab-scale process, 272
Lactic acid, 311, 432, 433f
Langmuir model, 62–63
Laser ablation, 70, 71t
LCO. *See* Lithium cobaltate ($LiCoO_2$)
LCST. *See* Lower critical solution temperature (LCST)
Le Clancy batteries. *See* Zinc–manganese battery
Leachate purification, 436, 436f
Leaching, 183–185, 192–207
  acid leaching process, 192–197, 432–434, 433f
  agent, 307
  alkaline leaching, 197–203, 312, 434–435
  approaches for Li-ion batteries, 337–338
  bioleaching, 311
  electrochemical leaching process, 205–206

by hydrometallurgy, 429–436
    acid leaching, 432–434
    leaching kinetics, 431–432
  inorganic acid leaching, 308–309
  leaching-solvent extraction process, 392
  microbial leaching process, 206–207
  organic acid leaching, 309–310
  redox leaching process, 203–205
Lead-acid batteries, 16, 17t
Lead dioxide (PbO_2), 324–325
Lead sulfate ($PbSO_4$), 324–325
Lead–acid batteries, 79, 105, 324–325, 331–333, 332f
  hydrometallurgical process, 332–333
  pyrometallurgical process, 331–332
Least-squares-based algorithms, 35
Le Clancy batteries, 3
Liberation
  comminution process, 357–358
  regime, 43
Light fraction, 463
$LiMnO_2$. See Lithium manganese dioxide ($Li-MnO_2$)
$LiMn_xFe_{1-x}PO_4/C$ cathode materials, regeneration of, 384–386
$LiNi_{0.6}Co_{0.2}Mn_{0.2}O_2$ from spent lithium-ion batteries, regeneration of, 386–388
$LiPF_6$ electrolyte, 447
Lithiated graphite (LiC_6), 106
Lithium (Li), 178, 218, 324, 349
  air battery, 10
  battery, 324
  extraction, 176–177
  foil, 446–447
  mechanism, 202
  metal, 175
  prioritized separation of, 215
  recovery, 292
  recycling of, 427–444
  salt extraction, 215
  scrubbing from loaded Cyanex 272, 413–414
Lithium carbon fluoride ($LieCF_x$), 324
Lithium carbonate (LI_2CO_3), 177, 342
Lithium cobalt oxide. See Lithium cobaltate ($LiCoO_2$)
Lithium cobaltate ($LiCoO_2$), 106, 247–249, 303, 446–447
  concentration effect on cobalt leaching from, 398–399

  effect of Cyanex 272 concentration and volume of $LiCoO_2$ leach liquor, 404
  H_2O_2 effect on cobalt leaching from, 395–397
  leaching time effect on cobalt leaching from, 397
  McCabe–Thiele cobalt extraction isotherm from $LiCoO_2$ leach liquor, 410–411
  pH-extraction isotherm for cobalt from $LiCoO_2$ leach liquor, 402–404
  pulp density effect on cobalt leaching from, 399–400
  selective separation of cobalt by solvent extraction from $LiCoO_2$ leach liquor, 401–416
  separating experiment of, 279–280
  separation of, 277, 281
  temperature effect on cobalt leaching from, 398
  waste, 392
Lithium ions (Li^+), 446
Lithium iron disulfide ($LieFeS_2$), 324
Lithium iron phosphate (LFP), 108–109, 129, 247–249, 276, 280, 384
  separating experiment of, 279–280
Lithium-iron phosphate secondary battery, 19–20
Lithium manganese dioxide ($Li-MnO_2$), 247–249, 324
Lithium-manganese dioxide cell, 7–8
Lithium manganese oxide (LMO), 108–109, 129, 276
  separation of, 281
Lithium nickel cobalt aluminum (NCA), 125–126, 276–277
  separation of, 281
Lithium nickel cobalt manganese (NCM), 125–126
Lithium nickel cobalt manganese oxide separation, 277
Lithium nickel cobalt manganese oxide-111 separation, 281
Lithium nickel oxide ($LiNiO_2$)), 108–109, 129
Lithium polymer battery, 18–19
Lithium secondary battery, 18–23
Lithium sulfide batteries, 123–124
Lithium thionyl chloride ($LieSOCl_2$), 324
Lithium thionyl chloride battery, 8–9, 8f
Lithium titanate ($Li_4Ti_5O_{12}$), 327
Lithium-cobalt oxide ($LiCoO_2$), 327
Lithium-ion batteries (LIBs), 33, 79, 106, 123, 125–126, 141, 171, 217, 247, 263, 274–276, 289, 291–292, 303, 326–327, 337–340, 349, 373, 391, 445, 455–457

Lithium-ion batteries (LIBs) (*Continued*)
  anode materials in, 289
  approaches for recycling spent LIBs, 219
  cathode, 172−173
  challenges and future trends, 46−48
  components and constituent elements at issue, 86f
  design and composition, 456
  direct recycling processs, 338−339
  disposal, 456−457
  environmental aspects of inappropriate spent LIBs management, 218−219
  extraction of materials, 45−46
  graphite materials in
  housing of LIB cells, 350
  hydrometallurgical process, 337−338
    leaching, 337−338
    metal separation, 338
  imperfect separation of recycled cathode active material
    experiments, 447
    mathematical model, 446−447
    model parameters of most commonly used cathode active materials in, 448t
    results, 447−448
  ionic liquids application in, 84t
  metallic materials in batteries, 328t−329t
  nanomaterials separation from lithium-ion batteries
    anode, 249
    cathode, 247−249
    diaphragm, 249
    electrolyte, 249−250
    failure mechanisms, 250−252
    other structural components, 250
    pretreatment process, 252−254
    separation of nanomaterials, 258−260
    separation of nanomaterials from collector fluid, 254−258
  physical properties of the materials in, 462t
  processing end-of-life, 35f
  pyrometallurgical process, 337
  recycling, 40−45
  remanufacturing, 36−39
    process flowchart, 37f
  repurposing, 39−40
  spent LIBs as precious secondary resource, 218
  state-of-health estimation, 35−36
  technology, 455
  waste management, 456−457

Lithium-iron-phosphate (LiFePO$_4$), 327
Lithium-manganese oxide (LiMn$_2$O$_4$), 327
Lithium-metal batteries. *See* Lithium battery (Li battery)
Lithium-nickel oxide (LiNiO$_2$), 327
Lithium-nickel-cobalt-aluminium oxide (LiNiCoAlO$_2$), 327
Lithium-sulfur dioxide (LieSO$_2$), 324
Lithiumnickel-cobalt-manganese oxide (LiNiCoMnO$_2$), 327
Low-temperature route, 473
Lower critical solution temperature (LCST), 165

M
Machine learning (ML), 36
Magnesium (Mg), 209
  Mg impurities impact on regeneration method, 440−441
Magnetic field, 159, 162
Magnetic nanohydrometallurgy. *See* Nanohydrometallurgy
Magnetic nanoparticles (MNP), 161
Magnetic separation, 466−467, 474
Malic acid, 227−229, 311, 432, 433f
Manganese (Mn), 178, 180, 218, 456
  mechanism, 202−203
  separation and extraction of, 214
Manganese dioxide battery (MnO$_2$ battery), 321−323
Manual dismantling process, 305
Materials separation
  validation of anode and cathode active materials mixture separation, 277−280
  validation of separation of mixture of cathode active materials, 280−281
MATLAB 6. 5 Interactive Curve Fitting procedure, 404−407
Matrix separators, 466−467
McCabe−Thiele cobalt extraction isotherm from LiCoO$_2$ leach liquor, 410−411
MCFC. *See* Molten carbonate fuel cell (MCFC)
Mechanical liberation, 457−461
  crushing, 458−460
  milling, 460−461
Mechanical methods, 457
Mechanical milling process, 70, 71t
Mechanical processing, generating housing fraction by, 352−357

Mesophase-carbon microbeads (MCMBs), 289–291
Metal extraction, material regeneration, 430f
  cathode material regeneration, 437–438
  hydrometallurgy, 429–436, 430f
    acid leaching, 432–434, 433f
    alkaline leaching, 434–435
    bioleaching, 435–436, 435f
    leaching kinetics, 431–432, 431f
  impurities, 438–441
    aluminum, 439, 439f
    copper, 438, 438f
    iron, 439–440, 440f
    magnesium, 440–441, 441f, 442t
  leachate purification, 436, 436f
Metal nanomaterials recycling, 380–388
  cobalt ferrite nanoparticle recycling, 381–382
  fabrication of nano-Co_3S_4, 383–384
  regeneration of $Co1_{+x}Al_zZn_xZr_xFe_{2-3x}O_4$ from spent lithium-ion batteries, 386
  regeneration of $LiMn_xFe_{1-x}PO_4/C$ cathode materials, 384–386
  regeneration of $LiNi_{0.6}Co_{0.2}Mn_{0.2}O_2$ from spent lithium-ion batteries, 386–388
Metal nitride lithium manganese nitride ($Li7MnN_4$), 249
Metal oxide
  cathode, 259
  NPs, 54
Metal recycling, technologies for, 330–342
  conventional recycling techniques, 330–339
  recent green recycling techniques, 340–342
Metal sulfates, 178
Metal sulfides, 55
Metal(s), 445, 457
  components, 54
  extractability behavior and modeling, 404–410
  ions, 178
  nanotechnology for metal recovery from batteries, 64–69
  sensors, 474
  separation for Li-ion batteries, 338
Metallic materials, commercial batteries and recyclability of, 321–327
  primary batteries, 321–324
  secondary batteries, 324–327
Metallic nanoparticles (metallic NPs), 69–70
  NP synthesis techniques, 71t
  synthesis of metallic nanoparticles and other materials, 69–70
Microbes in bioleaching, 222–224
  chemolithotrophic autotrophic bacteria, 222–223
  heterotrophic bacteria and fungi, 223–224
Microbial bioleaching, 224
Microbial leaching process, 206–207
Microbial metabolites, 221
Microwave-assisted technique, 134
Milling process, 457–458, 460–461
Mining process, 160
Mixed culture of microorganisms, 229–230
Mixed electrode material, 385
ML. See Machine learning (ML)
MNP. See Magnetic nanoparticles (MNP)
Model-based techniques, 35
Modeling parameters, 447
Molten carbonate fuel cell (MCFC), 24–25, 25f
Molten salt battery, 11–12
Multistage magnetic separation, 479

N

N-dimethylacetamide (DMAC), 256, 306
N-dimethylformamide (DMF), 256, 306
N-methyl-pyrrolidone (NMP), 256, 266–268, 303
Nano-Co_3S_4, fabrication of, 383–384
Nano/microelectrode active materials, separating battery
  materials, 265
  modeling, 266–273
  results, 274–281
Nanoassembly systems, 374
Nanohydrometallurgy process, 53, 61–64, 61f, 87–89
  battery recycling, remanufacturing, and reuse, 64
  core-shell support used in, 62f
  effect of loading superparamagnetic nanostructures, 63
  noble metal concentration, 62–63
  effect of solution pH and contact time, 63
  effect of temperature, 63–64
Nanolithography, 70, 71t
Nanomaterials, 53, 55, 373–375
  failure mechanisms, 250–252
    existing degradation mechanisms for rechargeable batteries, 251f
  pretreatment process, 252–254

Nanomaterials (*Continued*)
    deactivation treatment, 252–253
    mechanical dismantling and shredding, 253–254
    processes for recovering and synthesizing, 55–64
        hydrometallurgy process, 58–59
        nanohydrometallurgy, 61–64
        nanomaterials in battery components, 56t
        pyrometallurgy, 59
        recycled battery nanomaterials and methodologies, 57t
        steps in hydrometallurgy and pyrometallurgy processes for recovering and synthesizing, 57f
    recycling falls, 376
    separation from lithium-ion batteries
        anode, 249
        cathode, 247–249
        diaphragm, 249
        electrolyte, 249–250
        other structural components, 250
    separation of nanomaterials, 258–260
        from collector fluid, 254–258
    used in nanohydrometallurgy, 61–62
Nanoparticles (NPs), 161, 373–374
Nanorods, 373–374
Nanoscale surface morphologies, 144–145
Nanosheets, 373–374
Nanosolids, 380–381
Nanostructure electrodes, 374–375
Nanostructured colloidal solvents, 159
Nanostructured materials, 89
Nanotechnology, 54, 87. *See also* Bio-inspired nanotechnology
    battery nanostructures, 54–55
    for metal recovery from batteries, 64–69
        graphene and graphene oxide, 67–69
        nanofibers, 66
    processes for recovering and synthesizing nanomaterials, 55–64
        hydrometallurgy process, 58–59
        nanohydrometallurgy, 61–64
        pyrometallurgy, 59
    synthesis of metallic nanoparticles and other materials, 69–70
Nanotubes, 373–374
Nanowires, 373–374
NCA. *See* Lithium nickel cobalt aluminum (NCA)

NCM. *See* Lithium nickel cobalt manganese (NCM)
Near-infrared sensors, 474
Negative electrode, 445–446
$Ni_{0.8}Co_{0.15}Al_{0.05}O_2$ (NCA), 445–447
Nickel (Ni), 178, 180, 218, 349, 456
    ions, 383
    mechanism, 200–201
    nickel cobalt lithium manganate electrode material, 386–387
    nickel-plated steel, 474
    recycling of, 427–444
    separation and extraction of, 214
Nickel-cadmium battery, 12–13
Nickel hydroxide ($Ni[OH]_2$), 174, 325–326
    cathode, 326
Nickel manganese cobalt oxide (NMC), 47
Nickel-metal hydride battery, 13–16
    advantages, 15–16
    electron transfer, 14f
    positive and negative chemical reaction, 15t
Nickel–cadmium battery (Ni–Cd battery), 79, 217, 325–326, 333
    hydrometallurgical process, 333
    pyrometallurgical process, 333
Nickel–cobalt–aluminum ternary material ($LiNi_xCo_yAl_{1-x-y}O_2$), 247–249
Nickel–cobalt–manganese ternary cathode ($LiNi_xCo_yMn_{1-x-y}O_2$), 247–249
Nickel–metal hydride battery (Ni–MH battery), 79, 105–106, 217, 326, 334–337
    hydrometallurgical process, 335–337
        anode recycling, 336–337
        cathode recycling, 335
    pyrometallurgical process, 334–335
Nitration roasting method, 132–133
4-nitrophenol (4-NP), 166
$Ni_xCo_yMn_{1-x-y}O_2$ preparation, 214–215
    prioritized separation of lithium and preparation of, 215
    separation and extraction of nickel, cobalt, and manganese, 214
Noble metal concentration, 62–63
Non-destructive recovery technique, 134–135
Non-rechargeables battery, 217
Nonaqueous electrolyte, 327
Nonconductive material (NCF), 260
Noncontact mechanism. *See* Indirect mechanism
Nongraphite materials, 289

Nonprecious elements, 54
Nonrenewable sources, 125–126
Nutrient, 222

O
Oil–water
    emulsion, 160
    pollution, 160
    separation, 160, 165–166
One-step method, 221
OnTo recycling process, 82
Optical extractors, 340–341
Organic acids, 82, 89–93, 197, 223, 310–311
    leaching, 309–310, 311t
        conditions, efficiency rates, and batteries treated by, 90t–91t
        efficiencies of spent lithium-ion materials in organic acid leaching system, 198t–199t
Organic electrolyte battery, 7–9
Organic leaching, 115–116
Organic solvents, 252, 472
    cathode materials, 257f
    dissolution, 306
    soaking method, 256–257
Organic/biphasic catalysis, 162–163
Oxalate
    coprecipitation, 437
    precipitation, 235
Oxalic acids, 310
Oxidization leaching, 204
Oxygen plasma treatment, 144–145

P
Particle
    density, 463
    size of solid waste, 222
Patterned vanadium oxide films, 144–145
Penicillium, 221
    P. chrysogenum, 224, 311
    P. notatum, 207
    P. simplicissimum, 224, 311
Petrochemicals, 160
pH value in bioleaching, 221
pH-based stimulus-responsive materials, 160
pH-extraction isotherm for cobalt from $LiCoO_2$ leach liquor, 402–404
pH-responsive nanomaterials, 160. *See also* Thermo-responsive nanomaterials
    future trends, 167

    recycling and applications, 160–164
pH-responsive oil–water Pickering system, 163–164
pH-responsive Pickering emulsion, 162
pH/thermal stimuli-responsive nanomaterials, 159
Phase separation, 159
Phase separation technique, 163–164
Phosphoric acid fuel cell (PAFC), 23–24, 24f
Photocell, 26
Physical approach for recycling batteries, 109–110
Physical battery
    atomic energy, 27–28
    solar, 26–27
    thermoelectric, 27
Physical discharge method, 43
Physical separation method, 43, 461–470
    eddy current separation, 468
    electrostatic separation, 468–469
    flotation, 470
    gravity separation, 463–466
    magnetic separation, 466–467
    sensor-based sorting, 469–470
    sieving, 461–463
Phytolacca Americana, 96
Pickering emulsion/organic biphasic system, 162–163
Plante battery, 16–18
Plasmonic NPs, 69–70
Plastic separator, 266–268
Pneumatic tables, 464–465
Poly (2-dimethylaminoethyl methacrylate) (PDMAEMA), 160
Poly (acrylic acid), 160
Poly(2-dimethylaminoethyl methacrylate), 161
Poly(butyl acrylate), 161
Poly(glycidyl methacrylate), 166
Poly(N-isopropyl acrylamide) (PNIPAM), 161, 165–166
Polyelectrolytes, 160
Polymer, 159, 161, 249, 379
    polymer-based nanosorbents, 88
    polymer-derived ZnO nanoadsorbents, 88
    recycling, 87
Polymer electrolyte fuel cell, 26
Polymer lithium battery. *See* Lithium–polymer battery
Polystyrene, 161
Polytetrafluoroethylene, 172–173

Polyvinylidene fluoride (PVDF), 79, 108–109, 172–173, 247–249, 256, 306
  dissolution in supercritical CO_2 process, 87f
Portable devices, 123
Positive electrode, 445–446
Potassium, 176–177
Potassium hydroxide (KOH), 322–323
Potassium-ion batteries (KIBs), 297
Pouch cells, 38–39, 349
Pourbaix diagram, 80–81, 81f
Precipitation, 96, 183, 185–190, 215
  extraction, 333
  metal separation, 338
  solubility of common metallic compounds in water, 186t–189t
  stripping
    chemistry of precipitation stripping and cobalt oxide powder synthesis, 420–422
    synthesis of cobalt oxalate and cobalt oxide powder by, 416–418
Precursor preparation, 213–215
  $FePO_4$ preparation, 213–214
  lithium salt extraction, 215
  $Ni_xCo_yMn_{1-x-y}O_2$ preparation, 214–215
Preexisting crushing techniques, 43
Prerecycling processes, 341–342
Pretreatment process, 85–86, 304–307, 316
  discharging, 304–305
  dismantling, 305
  pyrometallurgy application in, 172–173
  separation of cathode materials, 305–307
Primary batteries, 3–12, 105, 321–324. *See also* Secondary batteries
  air, 9–11
  alkaline, 4–7
  alkaline zinc/manganese dioxide battery, 322–323
  lithium battery, 324
  molten salt, 11–12
  organic electrolyte, 7–9
  storage, 11
  zinc-manganese, 3–4
  zinc/silver-oxide battery, 323
  zinc–carbon battery, 321–322
Prismatic LIB cells, 38, 349, 352
Process development, 422
  recovery of high-purity cobalt from sulfate leach liquor of waste $LiCoO_2$, 423f
Process groups, 471–479

Propylene carbonate, 249–250, 252
Proton exchange membrane fuel cell (PEMFC), 26
Pseudomonas strains, 224
Pulp density effect on cobalt leaching from $LiCoO_2$, 399–400
Pulverization. *See* Milling
Purification process, 237–238, 312–314
Pyrometallurgical process, 45–46, 53, 59, 80, 112, 172, 219, 263, 307, 330, 392. *See also* Hydrometallurgical process
  alkaline and zinc–carbon battery, 330–331
  anode materials, 112
  application in pretreatment process, 172–173
  cathode materials, 112
  lead–acid battery, 331–332
  Li-ion battery, 337
  Ni–Cd battery, 333
  Ni–MH battery, 334–335
  technology, 173–180
    chlorination roasting, 176–177
    high-temperature melting, 174–175
    reduction roasting, 175–176
    sulfate roasting, 178–180
Pyrometallurgical recovery process, 173–174
Pyrometallurgical recycling processes, 349

Q

Quasi-stationary settling velocity, 363–364

R

Radioisotope battery. *See* Atomic energy battery
Radioisotope thermoelectric generator. *See* Atomic energy battery
Rare metals, 183
Reaction equation, 175
Reaction process, 172–173
Rechargeable batteries, 217
Recovery technology, 181, 250–252
Recursive least-squares-based models, 35
Recycle graphite, 291–292
Recycle spent LIBs, 263
Recycled cathode active material separation from LIB
  experiments, 447
    half-cell fabrication, 447
    mathematical model, 446–447
Recycled from spent LIBs, 379
Recycled SWCNTs, 377–379

Recycling, 141–142, 172
    for anode materials, 292–299
    application after recycling, 297–299
    battery, 64
    chain, 457
    of electrode materials, 129–133
    of electrolyte and binders, 133–135
    impurities removal, 292–294
        scanning electron microscopy images, 294f
    LIBs, 40–45
        discharging regime, 43
        liberation regime, 43
        methodologies for, 60f
        separation regime, 43–45
    metallic materials, 327
    method, 377–379
        for spent LIB materials, 264
    need for recycling used batteries, 106–108
    pH-responsive nanomaterials, 160–164
    pretreatment separation techniques for, 111f
    process, 174, 289
    spent lithium-ion batteries, approaches for, 219
    strategies for recycling battery components, 108–118
        chemical approach, 111–118
        physical approach, 109–110
    structural changes, 294–297
        peak reflection, 295f
        scanning electron microscopy images, 296f
    thermo-responsive nanomaterials, 165–166
Recycling battery
    anode materials
        defects in industry recycling, 299–300
        graphite materials in lithium-ion batteries, 291–292
        imperative of anode materials recycling, 289–291
        in lithium-ion batteries, 289
        prospects and challenges, 300–301
        recycling methods for anode materials, 292–299
    cathode materials, 304–316
        hydrometallurgy, 307–314
        pretreatment, 304–307
        pyrometallurgy, 307
        regeneration, 314–316
    interface for easy-to-recycle batteries, 149–153
    metallic materials
        commercial batteries and recyclability of metallic materials, 321–327
        technologies for metal recycling, 330–342
Recycling efficiencies (RE), 455
Redox leaching process, 203–205. *See* Alkaline leaching process; Acid leaching process
    oxidization leaching, 204
    reduction leaching, 204–205
Redox reaction, 184
Redoxolysis, 223
Reduced graphene oxide (rGO), 128, 298
Reduction leaching process, 204–205
Reduction roasting technology, 175–176
Regeneration method, 237–239, 314–316, 447
    $Co_{1+x}Al_zZn_xZr_xFe_{2-3x}O_4$ from spent lithium-ion batteries, 386
    coprecipitation method, 315
    of $LiMn_xFe_{1-x}PO_4/C$ cathode materials, 384–386
    $LiNi_{0.6}Co_{0.2}Mn_{0.2}O_2$ from spent lithium-ion batteries, 386–388
    sol-gel method, 316
    solid-state method, 315
Remanufacturing
    battery, 64
    LIBs, 36–39
        process flowchart, 37f
    technologies, 39
Replacement precipitation, copper removal by, 209–210
Repurposing LIBs, 39–40, 41t
Resource recovery process, 180
Resource regeneration process, 181
Reusability, 270–271
Reuse
    battery, 64
    of valuable metals in battery production, 233–239
Roasting process, 174
Round cells, 349
Rubidium, 176–177

S

Safety protocols for remanufacturing, 39
Scanning electron microscope (SEM), 265, 293–294
Scrubbing test, 401–402
Secondary batteries, 105, 324–327, 325f. *See also* Primary batteries

Secondary batteries (*Continued*)
　alkaline secondary battery, 12−16
　fuel cell, 23−26
　lead−acid battery, 324−325
　lithium-ion battery, 327
　lithium secondary battery, 18−23
　Ni−Cd battery, 325−326
　nickel−metal hydride battery, 326
　plante battery, 16−18
Self-propagating combustion method, 382
Semiconductor NPs, 69
Sensor-based sorting, 461, 469−470
Sensor-supported sorting, 474
Separation, 207−215, 457, 471
　battery nano/microelectrode active materials, 263−265
　of cathode materials, 305−307, 305f
　from collector fluid, 254−258
　　solvent soaking, 255−257
　　thermal treatment, 254−255
　efficiency, 470
　electrostatic separation, 259−260
　equation, 270
　flotation separation methods, 258−259
　of impurity ions, 237
　materials, 265
　　density and typical particle size of LIBs, 266t
　model, 274−277
　　of anode and cathode materials, 274−276, 275f
　　of LFP from cathode mixture, 276
　　of lithium cobalt oxide and lithium nickel cobalt manganese oxide, 277
　　of LMO from cathode mixture, 276
　　of NCA from cathode mixture, 276−277
　modeling, 266−273
　　heavy liquid selection, 270−272
　　Industry-scale separation process, 272−273
　　separation process, 266−268
　　Stokes' law, 268−270
　of nanomaterials, 258−260
　process, 233−237, 265−268, 312−314, 446
　　chemical precipitation, 233−235
　　solvent extraction, 236−237
　　for spent materials, 267f
　regime, 43−45
　results, 274−281
　　separation model, 274−277
　　validation of anode and cathode active materials mixture separation, 277−280
　　validation of separation of mixture of cathode active materials, 280−281
Separator, 303, 445−446
Sieve classification process, 361−362
Sieve rings, 460−461
Sieving process, 461−463
Silica NP, 162−163
Silicon dioxide, 249
Silver oxide battery, 6−7, 323
Silver−zinc battery. *See* Silver oxide battery
"Similarity solubility", 256
Similarity-intermiscibility theory, 306
Simple dissolution reaction, 184
Single cation extraction systems, 331
Single-walled carbon nanotubes recycling (SWCNTs recycling), 376−379, 378f
Slow rotating rotary shears, 458
Small sealed lead battery, 16−18
Smart polymers, 159
Smart water treatment, 159
Soda ash, 331
Sodium hydroxide (NaOH), 253, 323
Sodium polytungstate (SPT), 265, 271−272
Sodium rare earth (sodium RE), 336−337
Sodium sulfide, 253
Sodium-ion batteries (SIBs), 297
Sodium-sulfur battery, 20−21, 20f
Solar battery, 26−27
Solar chip, 26
Sol-gel method, 70, 71t, 238, 316
　metal separation, 338
Solid electrolyte interface (SEI), 291
Solid graphite oxide, 380
Solid oxide fuel cell (SOFC), 25−26, 26f
Solid-phase calcination, 427, 428f
Solid-state batteries, 123−124
Solid-state method, 315
Solvent based separation method, 142−143
Solvent extraction, 116, 233, 236−237, 313, 331, 333, 392
　metal separation, 338
　reagents for separation of specific metals, 236t
　selective separation of cobalt by solvent extraction from $LiCoO_2$ leach liquor, 401−416
Solvent separation methods, 159
Solvent soaking, 255−257
　alkali soaking method, 255−256
　mechanical treatment, 257−258
　organic solvent soaking method, 256−257

Solvothermal hydrolysis process, 70, 71t, 161
Spent battery powder (SBP), 128
Spent lithium-ion batteries (SLIBs), 171, 174, 266, 316, 340–341, 382, 384
 approaches for recycling, 219
 environmental aspects of inappropriate spent LIBs management, 218–219
 as precious secondary resource, 218
 regeneration of $Co_{1+x}Al_zZn_xZr_xFe_{2-3x}O_4$ from, 386
 regeneration of $LiNi_{0.6}Co_{0.2}Mn_{0.2}O_2$ from, 386–388
Spherization, 476
Sputtering method, 70, 71t
Stability constant, 199
 of various nickel, cobalt, and copper complexes, 199t
State of health (SoH), 34–35
 estimation, 35–36
Stimuli-responsive polymeric materials, 159
Stimuli-responsiveness, 161
Stokes' law, 268–270
 equation, 270–271
 fallen particle in liquid, 269f
Storage battery, 11
Storage cells. See Secondary batteries
Stress mechanisms, 459
Stripping test, 401–402
Structural changes of battery recycling, remanufacturing, and reusing, 294–297
Styrenebutadiene rubber (SBR), 292–293
Subsequent separation process, 473
Sulfate roasting method, 178–180
Sulfidation roasting, 181
Sulfide precipitation, copper removal by, 210
Sulfuric acid (H_2SO_4), 174, 379
 concentration effect on cobalt leaching from $LiCoO_2$, 398–399
 leaching, 384
Sulfurization roasting method, 178
Supercapacitors, 124–125
Supercritical carbon dioxide ($SC-CO_2$), 86
Supercritical fluids, 82, 85–87
Superoleophilic sponges, 160
Superparamagnetic nanostructures, 63
Support vector regression model, 36
Surface-initiated atom transfer radical polymerization (SI-ATRP), 161, 165–166
Sustainability of battery recycling, 126–127
Sustainable resource development, 292
Sustainable waste battery recycling, 392–393
Synthesis methods, 376, 379

T

Target elements, 185–190
Task-ILs, 85
Technologies for metal recycling, 330–342
 conventional recycling techniques, 330–339
 recent green recycling techniques, 340–342
Temperature
 effect on cobalt leaching from $LiCoO_2$, 398
 effect on nanohydrometallurgy process, 63–64
 impact on bioleaching process, 221–222
 variation, 165
Temporary loss of cell capacity, 325–326
Terminal velocity of solid particles, 269–270
Ternary LIBs, 171
Ternary LIBs, 172
Ternary lithium oxides ($LiNi_xCo_yMn_zO_2$), 129
Thermal battery. See Molten salt battery
Thermal oxidation of SWCNTs, 377
Thermal processes, 330
Thermal treatment approach, 254–255, 292–294
 cathode materials, 306–307
 TG-DSC curves, 255f
Thermal-responsive polymer poly(N-vinyl caprolactam), 165
Thermoelectric battery, 27
Thermo-responsive nanomaterials, 159, 165. See also pH-responsive nanomaterials
 future trends, 167
 recycling and applications, 165–166
Thermophilic bacteria, 230
Thick-walled plastics, 474
Time-varying magnetic fields, 468
Tin (Sn), 143
 Sn-based intermetallic compounds, 249
Titanium dioxide (TiO_2), 163–164
Top-down methods, 70
Toxco process, 80
Traditional recovery method, 314
Trapezoidal riffle sieves, 460–461
Tri-n-butyl-phosphoric acid (TBP), 401–402
Trust-region algorithms, 404–408
Two-step method, 221

U

Ultra-high-temperature method, 295
Ultrafiltration/membranes, 159
Umicore recycling process, 80, 174
Uninter-ruptible power sources (UPSs), 326
United States Bureau of Mines, 176–177
Upper critical solution temperature, 165
Used batteries, 106–107

V

Vacuum pyrolysis, 172–173
Validation
   of anode and cathode active materials mixture separation, 277–280
      experiment for separating anode and cathode mixture, 278–279
      experiment for separating of lithium iron phosphate and lithium cobalt oxide, 279–280
   of cathode active materials mixture separation, 280–281
      separation of lithium manganese oxide and lithium nickel cobalt aluminum oxide, 281
      separation of lithium nickel cobalt manganese oxide-111 and lithium cobalt oxide, 281
Valorization through oxide synthesis, 416–422
   chemistry of precipitation stripping and cobalt oxide powder synthesis, 420–422
   synthesis of cobalt oxalate and cobalt oxide powder by precipitation stripping, 416–418
Valuable metals reuse in battery production, 233–239
Van der Waals forces, 144
Visual spectroscopy, 474

W

Waste
   LIBs, 392–393
   $LiCoO_2$, 392
   lithium-ion battery, 192
      hydrometallurgical recycling, 191, 191t
   management, 456–457
Waste battery recycling methods, 107, 392–393
Waste electrical and electronic equipment (WEEE), 218
Water, 472–473
   treatment, 160
   water-soluble molecules, 379
Weak selective leaching, 309
WEEE. *See* Waste electrical and electronic equipment (WEEE)
"Wet cell", 321
Wet media, 472–473
Wet separators, 466

X

X-ray diffraction (XRD), 265, 295
X-ray fluorescence (XRF), 341–342

Z

Zigzag classifiers, airflow sorting in, 464
Zinc (Zn)
   air battery, 9–10
   anode, 321–322
   batteries, 217
Zinc-bromine flow battery, 22–23, 22f
Zinc chloride ($ZnCl_2$), 321–322
Zinc–carbon battery (Zn–C battery), 321–322, 322f, 330–331
Zinc-manganese battery, 3–4
Zinc–oxygen battery. *See* Zinc–air battery
Zinc–silver oxide battery. *See* Silver oxide battery

CPI Antony Rowe
Eastbourne, UK
August 31, 2023